STUDENT SOLUTIONS MANUAL F(

MW00379170

PRECALCULUS

FUNCTIONS & GRAPHS

NINTH EDITION

Jeffery A. Cole
Anoka-Ramsey Community College

BROOKS/COLE

TM

THOMSON LEARNING

Australia • Canada • Mexico • Singapore • Spain • United Kingdom • United States

BROOKS/COLE

THOMSON LEARNING

Assistant Editor: Julie Foster
Marketing Manager: Leah Thomson
Marketing Associate: Maria Salinas
Production Coordinator: Stephanie Andersen
Permissions Editor: Sue Ewing
Advertising: Samantha Cabaluna

Cover Design: Roger Knox
Cover Photo: Hiroshi Yagi/Photonica
Print Buyer: Micky Lawler
Typesetting: Lifland et al., Bookmakers
Printing and Binding: Webcom Ltd.

Printed in Canada

10   9   8   7   6   5   4   3   2   1

ISBN: 0-534-37897-8

PREFACE

This *Student's Solutions Manual* contains selected solutions and strategies for solving typical exercises in the text, *Precalculus: Functions and Graphs, Ninth Edition*, by Earl W. Swokowski and Jeffery A. Cole.

In each exercise set, all odd-numbered solutions are included, with an emphasis on the solutions of the applied "word" problems. In the review exercise section at the end of each chapter, all odd-numbered solutions and some even-numbered solutions are included. For the discussion exercises at the end of each chapter, all odd-numbered solutions are included. I have tried to illustrate enough solutions so that the student will be able to obtain an understanding of all types of problems in each section.

A significant number of today's students are involved in various outside activities, and find it difficult, if not impossible, to attend all class sessions. This manual should help meet the needs of these students. In addition, it is my hope that this manual's solutions will enhance the understanding of all readers of the material and provide insights to solving other exercises.

I would appreciate any feedback concerning errors, solution correctness, solution style, or manual style—comments from students using previous editions have greatly strengthened the text's supplements as well as the text itself. These and any other comments may be sent directly to me at the address below or in care of the publisher: Brooks/Cole Thomson Learning, 511 Forest Lodge Road, Pacific Grove, CA 93950.

I would like to thank: Joan Cole, my wife, for proofing various features of the manual; George Morris, of Scientific Illustrators, for creating the mathematically precise art package; and Sally Lifland and Gail Magin, of Lifland et al., Bookmakers, for assembling the final manuscript. I dedicate this book to my children, Becky and Brad.

Jeffery A. Cole

Anoka-Ramsey Community College

11200 Mississippi Blvd. NW

Coon Rapids, MN 55433

Table of Contents

Table of Contents

Table of Contents

To the Student

This manual is a text supplement and should be read along *with* the text. Read all exercise solutions in this manual since explanations of concepts are given and then appear in subsequent solutions. All concepts necessary to solve a particular problem are not reviewed for every exercise. If you are having difficulty with a previously covered concept, look back to the section where it was covered for more complete help. The writing style I have used in this manual reflects the way I explain concepts to my own students. It is not as mathematically precise as that of the text, including phrases such as "goes down" or "touches and turns around." My students have told me that these terms help them understand difficult concepts with ease.

Lengthier explanations and more steps are given for the more difficult problems. Additional information that my students have found helpful is included—see page 14. The guidelines given in the text are followed for some solutions—see page 119.

In the review sections, the solutions are somewhat abbreviated since more detailed solutions were given in previous sections. However, this is not true for the word problems in these sections since they are unique. In easier groups of exercises, representative solutions are shown. Occasionally, alternate solutions are also given.

All figures have been plotted using computer software, offering a high degree of precision. The calculator graphs are from various TI screens. When possible, each piece of art was made with the same scale to show a realistic and consistent graph.

This manual was done using EXP: *The Scientific Word Processor*. I have used a variety of display formats for the mathematical equations, including centering, vertical alignment, and flushing text to the right. I hope that these make reading and comprehending the material easier for you.

Notations

The following notations are used in the manual.

Note: { Notes to the student pertaining to hints on solutions, common mistakes, or

conventions to follow. }

{ }	{ comments to the reader are in braces }
LS	{ Left Side of an equation }
RS	{ Right Side of an equation }
$\approx$	{ approximately equal to }
$\Rightarrow$	{ implies, next equation, logically follows }
$\Leftrightarrow$	{ if and only if, is equivalent to }
$\bullet$	{ bullet, used to separate problem statement from solution or explanation }
$\star$	{ used to identify the answer to the problem }
§	{ *section* references }
$\forall$	{ For all, i.e., $\forall x$ means "for all x". }
$\mathbb{R} - \{a\}$	{ The set of all real numbers except a. }
$\therefore$	{ therefore }
QI–QIV	{ quadrants I, II, III, IV }

Chapter 1: Topics from Algebra

1 (a) Since x and y have opposite signs, the product xy is negative.

(b) Since $x^2 > 0$ and $y > 0$, $x^2 y > 0$.

(c) Since $x < 0$ { x is negative } and $y > 0$ { y is positive }, $\frac{x}{y}$ is negative.

Thus, $\frac{x}{y} + x$ is the sum of two negatives, which is *negative*.

(d) Since $y > 0$ and $x < 0$, $y - x > 0$.

3 (a) Since -7 is to the left of -4 on a coordinate line, $-7 \boxed{<} -4$.

(b) Using a calculator, we see that $\frac{\pi}{2} \approx 1.5708$. Hence, $\frac{\pi}{2} \boxed{>} 1.57$.

(c) $\sqrt{225} \boxed{=} 15$ *Note:* $\sqrt{225} \neq \pm 15$

5 (a) Since $\frac{1}{11} = 0.\overline{09}$, $\frac{1}{11} \boxed{>} 0.09$. (b) Since $\frac{2}{3} = 0.\overline{6}$, $\frac{2}{3} \boxed{>} 0.6666$.

(c) Since $\frac{22}{7} = 3.\overline{142857}$ and $\pi \approx 3.141593$, $\frac{22}{7} \boxed{>} \pi$.

7 (a) "x is negative" is equivalent to $x < 0$. We symbolize this by writing

$$x \text{ is negative } \Leftrightarrow x < 0.$$

(b) y is nonnegative $\Leftrightarrow y \geq 0$

(c) q is less than or equal to $\pi \Leftrightarrow q \leq \pi$

(d) d is between 4 and 2 $\Leftrightarrow 2 < d < 4$

(e) t is not less than 5 $\Leftrightarrow t \geq 5$

(f) The negative of z is not greater than 3 $\Leftrightarrow -z \leq 3$

(g) The quotient of p and q is at most 7 $\Leftrightarrow \frac{p}{q} \leq 7$

(h) The reciprocal of w is at least 9 $\Leftrightarrow \frac{1}{w} \geq 9$

(i) The absolute value of x is greater than 7 $\Leftrightarrow |x| > 7$

Note: An informal definition of absolute value that may be helpful is

$$|\, something\, | = \begin{cases} itself & \text{if } itself \text{ is positive or zero} \\ -(itself) & \text{if } itself \text{ is negative} \end{cases}$$

9 (a) $|-3 - 2| = |-5| = -(-5)$ {since $-5 < 0$} $= 5$

(b) $|-5| - |2| = -(-5) - 2 = 5 - 2 = 3$

(c) $|7| + |-4| = 7 + [-(-4)] = 7 + 4 = 11$

11 (a) $(-5)|3 - 6| = (-5)|-3| = (-5)[-(-3)] = (-5)(3) = -15$

(b) $|-6|/(-2) = -(-6)/(-2) = 6/(-2) = -3$

(c) $|-7| + |4| = -(-7) + 4 = 7 + 4 = 11$

13 (a) Since $(4 - \pi)$ is positive, $|4 - \pi| = 4 - \pi$.

(b) Since $(\pi - 4)$ is negative, $|\pi - 4| = -(\pi - 4) = 4 - \pi$.

(c) Since $(\sqrt{2} - 1.5)$ is negative, $|\sqrt{2} - 1.5| = -(\sqrt{2} - 1.5) = 1.5 - \sqrt{2}$.

15 (a) $d(A, B) = |7 - 3| = |4| = 4$ (b) $d(B, C) = |-5 - 7| = |-12| = 12$

     (c) $d(C, B) = d(B, C) = 12$ (d) $d(A, C) = |-5 - 3| = |-8| = 8$

17 (a) $d(A, B) = |1 - (-9)| = |10| = 10$ (b) $d(B, C) = |10 - 1| = |9| = 9$

     (c) $d(C, B) = d(B, C) = 9$ (d) $d(A, C) = |10 - (-9)| = |19| = 19$

Note: Exer. 19–24: Since $|a| = |-a|$, the answers could have a different form.

     For example, $|-3 - x| \geq 8$ is equivalent to $|x + 3| \geq 8$.

19 $A = x$ and $B = 7$, so $d(A, B) = |7 - x|$.

                 Thus, "$d(A, B)$ is less than 5" can be written as $|7 - x| < 5$.

21 $d(A, B) = |-3 - x| \Rightarrow |-3 - x| \geq 8$

23 $d(A, B) = |x - 4| \Rightarrow |x - 4| \leq 3$

25 Pick an arbitrary value for x that is less than -3, say -5.

     Since $3 + (-5) = -2$ is negative, we conclude that if $x < -3$, then $3 + x$ is negative.

                 Hence, $|3 + x| = -(3 + x) = -x - 3$.

27 If $x < 2$, then $2 - x > 0$, and $|2 - x| = 2 - x$.

29 If $a < b$, then $a - b < 0$, and $|a - b| = -(a - b) = b - a$.

31 Since $x^2 + 4 > 0$ for every x, $|x^2 + 4| = x^2 + 4$.

33 LS $= \dfrac{ab + ac}{a} = \dfrac{ab}{a} + \dfrac{ac}{a} = b + c$ $\boxed{\neq}$ RS $(b + ac)$.

35 LS $= \dfrac{b + c}{a} = \dfrac{b}{a} + \dfrac{c}{a}$ $\boxed{=}$ RS.

37 LS $= (a \div b) \div c = \dfrac{a}{b} \cdot \dfrac{1}{c} = \dfrac{a}{bc}$. RS $= a \div (b \div c) = a \div \dfrac{b}{c} = a \cdot \dfrac{c}{b} = \dfrac{ac}{b}$. LS $\boxed{\neq}$ RS

39 LS $= \dfrac{a - b}{b - a} = \dfrac{-(b - a)}{b - a} = -1$ $\boxed{=}$ RS.

41 (a) On the TI-83 Plus, the absolute value feature is choice 1 under MATH, NUM.

     Enter abs$(3.2^2 - \sqrt{}(3.15))$. On the TI-86, abs is F5 under MATH, NUM.

                 $\left|3.2^2 - \sqrt{3.15}\right| \approx 8.4652$

     (b) $\sqrt{(15.6 - 1.5)^2 + (4.3 - 5.4)^2} \approx 14.1428$

43 (a) $\dfrac{1.2 \times 10^3}{3.1 \times 10^2 + 1.52 \times 10^3} \approx 0.6557 = 6.557 \times 10^{-1}$

     (b) $(1.23 \times 10^{-4}) + \sqrt{4.5 \times 10^3} \approx 67.08 = 6.708 \times 10^1$

45 Construct a right triangle with sides of lengths $\sqrt{2}$ and 1. The hypotenuse will have length $\sqrt{(\sqrt{2})^2 + 1^2} = \sqrt{3}$. Next construct a right triangle with sides of lengths $\sqrt{3}$ and $\sqrt{2}$. The hypotenuse will have length $\sqrt{(\sqrt{3})^2 + (\sqrt{2})^2} = \sqrt{5}$.

47 The large rectangle has area $=$ width $\times$ length $= a(b + c)$. The sum of the areas of the two small rectangles is $ab + ac$. Since the areas are the same, we have $a(b + c) = ab + ac$.

49 (a) Since the decimal point is 5 places to the right of the first nonzero digit,

$$427{,}000 = 4.27 \times 10^5.$$

(b) Since the decimal point is 8 places to the left of the first nonzero digit,

$$0.000\ 000\ 098 = 9.8 \times 10^{-8}.$$

51 (a) Moving the decimal point 5 places to the right, we have $8.3 \times 10^5 = 830{,}000$.

(b) Moving the decimal point 12 places to the left, we have

$$2.9 \times 10^{-12} = 0.000\ 000\ 000\ 002\ 9.$$

53 Since the decimal point is 24 places to the left of the first nonzero digit,

$$0.000\ 000\ 000\ 000\ 000\ 000\ 000\ 001\ 7 = 1.7 \times 10^{-24}.$$

55 It is helpful to write the units of any fraction, and then "cancel" those units to determine the units of the final answer.

$$\frac{186{,}000 \text{ miles}}{\text{second}} \cdot \frac{60 \text{ seconds}}{1 \text{ minute}} \cdot \frac{60 \text{ minutes}}{1 \text{ hour}} \cdot \frac{24 \text{ hours}}{1 \text{ day}} \cdot \frac{365 \text{ days}}{1 \text{ year}} \cdot 1 \text{ year} \approx 5.87 \times 10^{12} \text{ mi}$$

57 $\dfrac{\dfrac{1.01 \text{ grams}}{\text{mole}}}{\dfrac{6.02 \times 10^{23} \text{ atoms}}{\text{mole}}} \cdot 1 \text{ atom} = \dfrac{1.01 \text{ grams}}{6.02 \times 10^{23}} \approx 0.1678 \times 10^{-23} \text{ g} = 1.678 \times 10^{-24} \text{ g}$

59 $\dfrac{24 \text{ frames}}{\text{second}} \cdot \dfrac{60 \text{ seconds}}{1 \text{ minute}} \cdot \dfrac{60 \text{ minutes}}{1 \text{ hour}} \cdot 48 \text{ hours} = 4.1472 \times 10^6 \text{ frames}$

61 (a) $\text{IQ} = \dfrac{\text{mental age (MA)}}{\text{chronological age (CA)}} \times 100 = \dfrac{15}{12} \times 100 = 125.$

(b) CA = 15 and IQ = 140 $\Rightarrow$

$$140 = \frac{\text{MA}}{15} \times 100 \quad \Rightarrow \quad \frac{140}{100} = \frac{\text{MA}}{15} \quad \Rightarrow \quad \text{MA} = \frac{15(140)}{100} = 21.$$

63 (a) 1 ft^2 = 144 in^2, so the force on one square foot of a wall is

$$144 \text{ in}^2 \times 1.4 \text{ lb/in}^2 = 201.6 \text{ lb}.$$

(b) The area of the wall is $40 \times 8 = 320 \text{ ft}^2$, or $320 \text{ ft}^2 \times 144 \text{ in}^2/\text{ft}^2 = 46{,}080 \text{ in}^2$.

The total force is $46{,}080 \text{ in}^2 \times 1.4 \text{ lb/in}^2 = 64{,}512 \text{ lb}$.

Converting to tons, we have $64{,}512 \text{ lb}/(2000 \text{ lb/ton}) = 32.256 \text{ tons}$.

1.2 Exercises

1 $\left(-\frac{2}{3}\right)^4 = \left(-\frac{2}{3}\right) \cdot \left(-\frac{2}{3}\right) \cdot \left(-\frac{2}{3}\right) \cdot \left(-\frac{2}{3}\right) = \frac{16}{81}$

Note: Do not confuse $(-x)^4$ and $-x^4$ since $(-x)^4 = x^4$ and $-x^4$ is the negative of x^4.

3 $\dfrac{2^{-3}}{3^{-2}} = \dfrac{3^2}{2^3} = \dfrac{9}{8}$ *Note:* Remember that negative exponents don't necessarily equate to negative results—that is, $2^{-3} = \dfrac{1}{2^3} = \dfrac{1}{8}$, not $-\dfrac{1}{8}$.

$\boxed{5}$ $-2^4 + 3^{-1} = -16 + \frac{1}{3} = -\frac{48}{3} + \frac{1}{3} = \frac{-47}{3}$

$\boxed{7}$ $16^{-3/4} = 1/16^{3/4} = 1/(\sqrt[4]{16})^3 = 1/2^3 = \frac{1}{8}$

$\boxed{9}$ $(-0.008)^{2/3} = (\sqrt[3]{-0.008})^2 = (-0.2)^2 = 0.04 = \frac{4}{100} = \frac{1}{25}$

$\boxed{11}$ $(\frac{1}{2}x^4)(16x^5) = (\frac{1}{2} \cdot 16)x^{4+5} = 8x^9$

$\boxed{13}$ A common mistake is to write $x^3 x^2 = x^6$, and another is to write $(x^2)^3 = x^5$.

     The following solution illustrates the proper use of the exponent rules.

$$\frac{(2x^3)(3x^2)}{(x^2)^3} = \frac{(2 \cdot 3)x^{3+2}}{x^{2 \cdot 3}} = \frac{6x^5}{x^6} = 6x^{5-6} = 6x^{-1} = \frac{6}{x}$$

$\boxed{15}$ $(\frac{1}{6}a^5)(-3a^2)(4a^7) = \frac{1}{6} \cdot (-3) \cdot 4 \cdot a^{5+2+7} = -2a^{14}$

$\boxed{17}$ $\dfrac{(6x^3)^2}{(2x^2)^3} \cdot (3x^2)^0 =$

     $\dfrac{6^2 x^{3 \cdot 2}}{2^3 x^{2 \cdot 3}} \cdot 1$ { an expression raised to the zero power is equal to 1 } $= \dfrac{36x^6}{8x^6} = \dfrac{36}{8} = \dfrac{9}{2}$

$\boxed{19}$ $(3u^7 v^3)(4u^4 v^{-5}) = 12u^{7+4}v^{3+(-5)} = 12u^{11}v^{-2} = \dfrac{12u^{11}}{v^2}$

$\boxed{21}$ $(8x^4 y^{-3})(\frac{1}{2}x^{-5}y^2) = 4x^{-1}y^{-1} = \dfrac{4}{xy}$

$\boxed{23}$ $(\frac{1}{3}x^4 y^{-3})^{-2} = (\frac{1}{3})^{-2}(x^4)^{-2}(y^{-3})^{-2} = (\frac{3}{1})^2 x^{-8}y^6 = 3^2 x^{-8}y^6 = \dfrac{9y^6}{x^8}$

$\boxed{25}$ $(3y^3)^4(4y^2)^{-3} = 81y^{12} \cdot 4^{-3}y^{-6} = 81y^6 \cdot \dfrac{1}{64} = \dfrac{81}{64}y^6$

$\boxed{27}$ $(-2r^4 s^{-3})^{-2} = (-2)^{-2}r^{-8}s^6 = \dfrac{s^6}{(-2)^2 r^8} = \dfrac{s^6}{4r^8}$

$\boxed{29}$ $(5x^2 y^{-3})(4x^{-5}y^4) = 20x^{-3}y = \dfrac{20y}{x^3}$

$\boxed{31}$ $\left(\dfrac{3x^5 y^4}{x^0 y^{-3}}\right)^2$ { remember that $x^0 = 1$ } $= \dfrac{9x^{10}y^8}{y^{-6}} = 9x^{10}y^{8+6} = 9x^{10}y^{14}$

$\boxed{33}$ $(4a^{3/2})(2a^{1/2}) = 4 \cdot 2a^{(3/2)+(1/2)} = 8a^{4/2} = 8a^2$

$\boxed{35}$ $(3x^{5/6})(8x^{2/3}) = 3 \cdot 8x^{(5/6)+(4/6)} = 24x^{9/6} = 24x^{3/2}$

$\boxed{37}$ $(27a^6)^{-2/3} = 27^{-2/3}a^{-12/3} = \dfrac{a^{-4}}{27^{2/3}} = \dfrac{1}{(\sqrt[3]{27})^2 a^4} = \dfrac{1}{3^2 a^4} = \dfrac{1}{9a^4}$

$\boxed{39}$ $(8x^{-2/3})x^{1/6} = 8x^{(-4/6)+(1/6)} = 8x^{-3/6} = \dfrac{8}{x^{1/2}}$

$\boxed{41}$ $\left(\dfrac{-8x^3}{y^{-6}}\right)^{2/3} = \dfrac{(-8)^{2/3}(x^3)^{2/3}}{(y^{-6})^{2/3}} = \dfrac{(\sqrt[3]{-8})^2 x^{(3)(2/3)}}{y^{(-6)(2/3)}} = \dfrac{(-2)^2 x^2}{y^{-4}} = \dfrac{4x^2}{y^{-4}} = 4x^2 y^4$

43 $\left(\dfrac{x^6}{9y^{-4}}\right)^{-1/2} = \dfrac{x^{-3}}{9^{-1/2}y^2} = \dfrac{9^{1/2}x^{-3}}{y^2} = \dfrac{3}{x^3y^2}$

45 $\dfrac{(x^6y^3)^{-1/3}}{(x^4y^2)^{-1/2}} = \dfrac{(x^6)^{-1/3}(y^3)^{-1/3}}{(x^4)^{-1/2}(y^2)^{-1/2}} = \dfrac{x^{-2}y^{-1}}{x^{-2}y^{-1}} = 1$

47 $\sqrt[4]{x^3} = (x^3)^{1/4} = x^{3\cdot(1/4)} = x^{3/4}$

49 $\sqrt[3]{(a+b)^2} = [(a+b)^2]^{1/3} = (a+b)^{2/3}$

51 $\sqrt{x^2+y^2} = (x^2+y^2)^{1/2}$ *Note:* $\sqrt{x^2+y^2} \neq x+y$

53 (a) $4x^{3/2} = 4x^1x^{1/2} = 4x\sqrt{x}$

(b) $(4x)^{3/2} = (4x)^1(4x)^{1/2} = (4x)^1\,4^{1/2}x^{1/2} = 4x\cdot2\cdot x^{1/2} = 8x\sqrt{x}$

55 (a) $8 - y^{1/3} = 8 - \sqrt[3]{y}$ (b) $(8-y)^{1/3} = \sqrt[3]{8-y}$

57 $\sqrt{81} = \sqrt{9^2} = 9$

59 $\sqrt[5]{-64} = \sqrt[5]{-32}\,\sqrt[5]{2} = \sqrt[5]{(-2)^5}\,\sqrt[5]{2} = -2\sqrt[5]{2}$

61 In the denominator, you would like to have $\sqrt[3]{2^3}$. How do you get it? Multiply by $\sqrt[3]{2^2}$, or, equivalently, $\sqrt[3]{4}$. Of course, we have to multiply the numerator by the same value so that we don't change the value of the given fraction.

$$\frac{1}{\sqrt[3]{2}} = \frac{1}{\sqrt[3]{2}}\cdot\frac{\sqrt[3]{4}}{\sqrt[3]{4}} = \frac{\sqrt[3]{4}}{\sqrt[3]{2\cdot4}} = \frac{\sqrt[3]{4}}{\sqrt[3]{8}} = \frac{\sqrt[3]{4}}{2} = \frac{1}{2}\sqrt[3]{4}$$

63 $\sqrt{9x^{-4}y^6} = (9x^{-4}y^6)^{1/2} = 9^{1/2}(x^{-4})^{1/2}(y^6)^{1/2} = 3x^{-2}y^3 = \dfrac{3y^3}{x^2}$

65 $\sqrt[3]{8a^6b^{-3}} = 2a^2b^{-1} = \dfrac{2a^2}{b}$

Note: For exercises similar to those in 67–74, pick a multiplier that will make all of the exponents of the terms in the denominator a multiple of the index.

67 The index is 2. Choose the multiplier to be $\sqrt{2y}$ so that the denominator contains only terms with even exponents. $\sqrt{\dfrac{3x}{2y^3}} = \sqrt{\dfrac{3x}{2y^3}}\cdot\dfrac{\sqrt{2y}}{\sqrt{2y}} = \sqrt{\dfrac{6xy}{4y^4}} = \dfrac{\sqrt{6xy}}{2y^2}$, or $\dfrac{1}{2y^2}\sqrt{6xy}$

69 The index is 3. Choose the multiplier to be $\sqrt[3]{3x^2}$ so that the denominator contains only terms with exponents that are multiples of 3.

$$\sqrt[3]{\dfrac{2x^4y^4}{9x}} = \sqrt[3]{\dfrac{2x^4y^4}{9x}}\cdot\dfrac{\sqrt[3]{3x^2}}{\sqrt[3]{3x^2}} = \dfrac{\sqrt[3]{6x^6y^4}}{\sqrt[3]{27x^3}} = \dfrac{\sqrt[3]{x^6y^3}\,\sqrt[3]{6y}}{3x} = \dfrac{x^2y\,\sqrt[3]{6y}}{3x} = \dfrac{xy}{3}\sqrt[3]{6y}$$

71 The index is 4. Choose the multiplier to be $\sqrt[4]{3x^2}$ so that the denominator contains only terms with exponents that are multiples of 4.

$$\sqrt[4]{\frac{5x^8y^3}{27x^2}} = \sqrt[4]{\frac{5x^8y^3}{27x^2}} \cdot \frac{\sqrt[4]{3x^2}}{\sqrt[4]{3x^2}} = \frac{\sqrt[4]{15x^{10}y^3}}{\sqrt[4]{81x^4}} = \frac{\sqrt[4]{x^8}\sqrt[4]{15x^2y^3}}{3x} = \frac{x^2\sqrt[4]{15x^2y^3}}{3x}$$
$$= \frac{x}{3}\sqrt[4]{15x^2y^3}$$

73 The index is 5. Choose the multiplier to be $\sqrt[5]{4x^2}$ so that the denominator contains only terms with exponents that are multiples of 5.

$$\sqrt[5]{\frac{5x^7y^2}{8x^3}} = \sqrt[5]{\frac{5x^7y^2}{8x^3}} \cdot \frac{\sqrt[5]{4x^2}}{\sqrt[5]{4x^2}} = \frac{\sqrt[5]{x^5}\sqrt[5]{20x^4y^2}}{2x} = \frac{x\sqrt[5]{20x^4y^2}}{2x} = \frac{1}{2}\sqrt[5]{20x^4y^2}$$

75 $\sqrt[4]{(3x^5y^{-2})^4} = 3x^5y^{-2} = \dfrac{3x^5}{y^2}$

77 $\sqrt[5]{\dfrac{8x^3}{y^4}}\sqrt[5]{\dfrac{4x^4}{y^2}} = \sqrt[5]{\dfrac{8x^3}{y^4}}\sqrt[5]{\dfrac{4x^4}{y^2}} \cdot \dfrac{\sqrt[5]{y^4}}{\sqrt[5]{y^4}} = \dfrac{\sqrt[5]{32x^5}\sqrt[5]{x^2y^4}}{\sqrt[5]{y^{10}}} = \dfrac{\sqrt[5]{32x^5}\sqrt[5]{x^2y^4}}{y^2} = \dfrac{2x}{y^2}\sqrt[5]{x^2y^4}$

79 $\sqrt[3]{3t^4v^2}\sqrt[3]{-9t^{-1}v^4} = \sqrt[3]{-27t^3v^6} = -3tv^2$

81 $\sqrt{x^6y^4} = \sqrt{(x^3)^2(y^2)^2} = \sqrt{(x^3)^2}\sqrt{(y^2)^2} = |x^3||y^2| = |x^3|y^2$ since y^2 is always

nonnegative. *Note:* $|x^3|$ could be written as $x^2|x|$.

83 $\sqrt[4]{x^8(y-1)^{12}} = \sqrt[4]{(x^2)^4((y-1)^3)^4} = |x^2||(y-1)^3| = x^2|(y-1)^3|$,

or $x^2(y-1)^2|(y-1)|$

85 $(a^r)^2 = a^{2r} \boxed{\neq} a^{(r^2)}$ since $2r \neq r^2$ for every r.

87 $(ab)^{xy} = a^{xy}b^{xy} \boxed{\neq} a^xb^y$ for every x and y.

89 $\sqrt[n]{\dfrac{1}{c}} = \left(\dfrac{1}{c}\right)^{1/n} = \dfrac{1^{1/n}}{c^{1/n}} \boxed{=} \dfrac{1}{\sqrt[n]{c}}$

91 (a) $(-3)^{2/5} = [(-3)^2]^{1/5} = 9^{1/5} \approx 1.5518$

   (b) $(-5)^{4/3} = [(-5)^4]^{1/3} = 625^{1/3} \approx 8.5499$

93 (a) $\sqrt{\pi+1} \approx 2.0351$ (b) $\sqrt[3]{15.1} + 5^{1/4} \approx 3.9670$

95 $\$200(1.04)^{180} \approx \$232,825.78$

97 $W = 230$ kg $\Rightarrow L = 0.46\sqrt[3]{W} = 0.46\sqrt[3]{230} \approx 2.82$ m

99 $b = 75$ and $w = 180 \Rightarrow W = \dfrac{w}{\sqrt[3]{b-35}} = \dfrac{180}{\sqrt[3]{75-35}} \approx 52.6.$

   $b = 120$ and $w = 250 \Rightarrow W = \dfrac{w}{\sqrt[3]{b-35}} = \dfrac{250}{\sqrt[3]{120-35}} \approx 56.9.$

It is interesting to note that the 75-kg lifter can lift 2.4 times his/her body weight and the 120-kg lifter can lift approximately 2.08 times his/her body weight, but

the formula ranks the 120-kg lifter as the superior lifter.

101	$W = 0.1166h^{1.7}$	Height	64	65	66	67	68	69	70	71
		Weight	137	141	145	148	152	156	160	164

Height	72	73	74	75	76	77	78	79
Weight	168	172	176	180	184	188	192	196

1.3 Exercises

1 $(2u + 3)(u - 4) + 4u(u - 2) = (2u^2 - 5u - 12) + (4u^2 - 8u) = 6u^2 - 13u - 12$

3 Divide the denominator into each term of the numerator.

$$\frac{8x^2y^3 - 10x^3y}{2x^2y} = \frac{8x^2y^3}{2x^2y} - \frac{10x^3y}{2x^2y} = 4y^2 - 5x$$

5 We recognize this product as the difference of two squares.

$$(2x + 3y)(2x - 3y) = (2x)^2 - (3y)^2 = 4x^2 - 9y^2$$

7 When squaring binomials, try to become very efficient at determining the three terms—mentally think of the phrase "square of the first term, twice the product of the two terms, square of the second term," until it becomes second nature.

$$(3x + 2y)^2 = (3x)^2 + 2(3x)(2y) + (2y)^2 = 9x^2 + 12xy + 4y^2$$

9 $(\sqrt{x} + \sqrt{y})(\sqrt{x} - \sqrt{y}) = (\sqrt{x})^2 - (\sqrt{y})^2 = x - y$

11 Use Product Formula (3) on text page 35.

$$\begin{aligned}(x - 2y)^3 &= (x)^3 + 3(x)^2(-2y) + 3(x)(-2y)^2 + (-2y)^3 \\ &= x^3 + 3(x^2)(-2y) + 3(x)(4y^2) - 8y^3 \\ &= x^3 - 6x^2y + 12xy^2 - 8y^3\end{aligned}$$

13 We recognize this as a trinomial that may be able to be factored into the product of two binomials. $8x^2 - 53x - 21 = (8x + 3)(x - 7)$

15 The factors for $x^2 + 3x + 4$ would have to be of the form $(x + \_)$ and $(x + \_)$.

The factors of 4 are 1 & 4 and 2 & 2, but their sums are 5 and 4, respectively.

Thus, $x^2 + 3x + 4$ is *irreducible*.

17 $4x^2 - 20x + 25 = (2x - 5)(2x - 5) = (2x - 5)^2$

19 $x^4 - 4x^2 = x^2(x^2 - 4) = x^2(x^2 - 2^2) = x^2(x + 2)(x - 2)$

21 We recognize this expression as the difference of two cubes.

$$\begin{aligned}64x^3 - y^6 &= (4x)^3 - (y^2)^3 \\ &= (4x - y^2)\left[(4x)^2 + (4x)(y^2) + (y^2)^2\right] \\ &= (4x - y^2)(16x^2 + 4xy^2 + y^4)\end{aligned}$$

[23] We recognize this expression as the sum of two cubes.

$$64x^3 + 27 = (4x)^3 + (3)^3$$
$$= (4x + 3)\left[(4x)^2 - (4x)(3) + (3)^2\right]$$
$$= (4x + 3)(16x^2 - 12x + 9)$$

[25] First we will factor out the greatest common factor, 3.

$$3x^3 + 3x^2 - 27x - 27 = 3(x^3 + x^2 - 9x - 9)$$

Since there are more than 3 terms, we will next try to factor by grouping.

$$3(x^3 + x^2 - 9x - 9) = 3[x^2(x + 1) - 9(x + 1)]$$
$$= 3(x^2 - 9)(x + 1)$$
$$= 3(x + 3)(x - 3)(x + 1)$$

[27] We could treat this expression as the difference of two squares or the difference of two cubes. Factoring as the difference of two squares and then as the sum and difference of two cubes leads to the following:

$$a^6 - b^6 = (a^3)^2 - (b^3)^2$$
$$= (a^3 + b^3)(a^3 - b^3)$$
$$= (a + b)(a - b)(a^2 - ab + b^2)(a^2 + ab + b^2)$$

[29] We might first try to factor this expression by grouping since it has more than 3 terms, but this would prove to be unsuccessful. Instead, we will group the terms containing x and the constant term together, and then proceed as in Example 2 part (c).

$$x^2 + 4x + 4 - 9y^2 = (x^2 + 4x + 4) - 9y^2$$
$$= (x + 2)^2 - (3y)^2$$
$$= (x + 2 + 3y)(x + 2 - 3y)$$

[31] $\dfrac{y^2 - 25}{y^3 - 125} = \dfrac{(y + 5)(y - 5)}{(y - 5)(y^2 + 5y + 25)} = \dfrac{y + 5}{y^2 + 5y + 25}$

[33] $\dfrac{9x^2 - 4}{3x^2 - 5x + 2} \cdot \dfrac{9x^4 - 6x^3 + 4x^2}{27x^4 + 8x} = \dfrac{(3x + 2)(3x - 2)}{(3x - 2)(x - 1)} \cdot \dfrac{x^2(9x^2 - 6x + 4)}{x(27x^3 + 8)}$

$\qquad\qquad = \dfrac{(3x + 2)(3x - 2)}{(3x - 2)(x - 1)} \cdot \dfrac{x^2(9x^2 - 6x + 4)}{x(3x + 2)(9x^2 - 6x + 4)} = \dfrac{x}{x - 1}$

[35] $\dfrac{2}{3s + 1} - \dfrac{9}{(3s + 1)^2} = \dfrac{2(3s + 1)}{(3s + 1)^2} - \dfrac{9}{(3s + 1)^2} = \dfrac{6s + 2 - 9}{(3s + 1)^2} = \dfrac{6s - 7}{(3s + 1)^2}$

[37] $\dfrac{2}{x} + \dfrac{3x + 1}{x^2} - \dfrac{x - 2}{x^3} = \dfrac{2x^2}{x^3} + \dfrac{(3x + 1)x}{x^3} - \dfrac{(x - 2)}{x^3} = \dfrac{2x^2 + 3x^2 + x - x + 2}{x^3} = \dfrac{5x^2 + 2}{x^3}$

$\boxed{39}$ $\dfrac{3t}{t+2} + \dfrac{5t}{t-2} - \dfrac{40}{t^2-4} = \dfrac{3t}{t+2} + \dfrac{5t}{t-2} - \dfrac{40}{(t+2)(t-2)}$

$$= \frac{3t(t-2)}{(t+2)(t-2)} + \frac{5t(t+2)}{(t+2)(t-2)} - \frac{40}{(t+2)(t-2)}$$

$$= \frac{3t^2 - 6t + 5t^2 + 10t - 40}{(t+2)(t-2)}$$

$$= \frac{8t^2 + 4t - 40}{(t+2)(t-2)}$$

$$= \frac{4(2t+5)(t-2)}{(t+2)(t-2)} = \frac{4(2t+5)}{t+2}$$

$\boxed{41}$ $\dfrac{4x}{3x-4} + \dfrac{8}{3x^2-4x} + \dfrac{2}{x} = \dfrac{4x(x) + 8 + 2(3x-4)}{x(3x-4)} = \dfrac{4x^2 + 6x}{x(3x-4)} = \dfrac{2x(2x+3)}{x(3x-4)} = \dfrac{2(2x+3)}{3x-4}$

$\boxed{43}$ $\dfrac{2x}{x+2} - \dfrac{8}{x^2+2x} + \dfrac{3}{x} = \dfrac{2x(x) - 8 + 3(x+2)}{x(x+2)} = \dfrac{2x^2 + 3x - 2}{x(x+2)} = \dfrac{(2x-1)(x+2)}{x(x+2)} = \dfrac{2x-1}{x}$

$\boxed{45}$ $3 + \dfrac{5}{u} + \dfrac{2u}{3u+1} = \dfrac{3u(3u+1) + 5(3u+1) + 2u(u)}{u(3u+1)}$

$$= \frac{9u^2 + 3u + 15u + 5 + 2u^2}{u(3u+1)} = \frac{11u^2 + 18u + 5}{u(3u+1)}$$

$\boxed{47}$ $\dfrac{2x+1}{x^2+4x+4} - \dfrac{6x}{x^2-4} + \dfrac{3}{x-2} = \dfrac{(2x+1)(x-2) - 6x(x+2) + 3(x^2+4x+4)}{(x+2)^2(x-2)} =$

$$\frac{-x^2 - 3x + 10}{(x+2)^2(x-2)} = -\frac{x^2 + 3x - 10}{(x+2)^2(x-2)} = -\frac{(x+5)(x-2)}{(x+2)^2(x-2)} = -\frac{x+5}{(x+2)^2}$$

$\boxed{49}$ The lcd of the entire expression is ab.

Thus, we will multiply both the numerator and denominator by ab.

$$\frac{\frac{b}{a} - \frac{a}{b}}{\frac{1}{a} - \frac{1}{b}} = \frac{\left(\frac{b}{a} - \frac{a}{b}\right) \cdot ab}{\left(\frac{1}{a} - \frac{1}{b}\right) \cdot ab} = \frac{b^2 - a^2}{b - a} = \frac{(b+a)(b-a)}{b-a} = a + b$$

$\boxed{51}$ $\dfrac{y^{-1} + x^{-1}}{(xy)^{-1}} = \dfrac{\frac{1}{y} + \frac{1}{x}}{\frac{1}{xy}} = \dfrac{\left(\frac{1}{y} + \frac{1}{x}\right) \cdot xy}{\left(\frac{1}{xy}\right) \cdot xy} \{\,\text{lcd is } xy\,\} = \dfrac{\frac{1}{y} \cdot xy + \frac{1}{x} \cdot xy}{\frac{1}{xy} \cdot xy} = \dfrac{x+y}{1} = x + y$

$\boxed{53}$ $\dfrac{\frac{r}{s} + \frac{s}{r}}{\frac{r^2}{s^2} - \frac{s^2}{r^2}} = \dfrac{\left(\frac{r}{s} + \frac{s}{r}\right) \cdot r^2 s^2}{\left(\frac{r^2}{s^2} - \frac{s^2}{r^2}\right) \cdot r^2 s^2} = \dfrac{r^3 s + r s^3}{r^4 - s^4} = \dfrac{rs(r^2 + s^2)}{(r^2 + s^2)(r^2 - s^2)} = \dfrac{rs}{r^2 - s^2}$

$\boxed{55}$ $\dfrac{(x+h)^2 - 3(x+h) - (x^2 - 3x)}{h} = \dfrac{x^2 + 2xh + h^2 - 3x - 3h - x^2 + 3x}{h} = \dfrac{2xh + h^2 - 3h}{h} =$

$$\frac{h(2x + h - 3)}{h} = 2x + h - 3$$

57 $\dfrac{\dfrac{3}{x-1}-\dfrac{3}{a-1}}{x-a}=\dfrac{\dfrac{3(a-1)-3(x-1)}{(x-1)(a-1)}}{x-a}=\dfrac{\dfrac{3a-3-3x+3}{(x-1)(a-1)}}{x-a}=\dfrac{3a-3x}{(x-1)(a-1)(x-a)}=$

$$\dfrac{3(a-x)}{(x-1)(a-1)(x-a)}=-\dfrac{3}{(x-1)(a-1)}\left\{\text{Note that, in general, } \tfrac{a-x}{x-a}=-1.\right\}$$

59 $\dfrac{\dfrac{1}{(x+h)^3}-\dfrac{1}{x^3}}{h}=\dfrac{\dfrac{x^3-(x+h)^3}{(x+h)^3x^3}}{h}=\dfrac{x^3-(x+h)^3}{hx^3(x+h)^3}=$

$$\dfrac{[x-(x+h)][x^2+x(x+h)+(x+h)^2]}{hx^3(x+h)^3}=\dfrac{-h(3x^2+3xh+h^2)}{hx^3(x+h)^3}=-\dfrac{3x^2+3xh+h^2}{x^3(x+h)^3}$$

61 The conjugate of $\sqrt{t}-5$ is $\sqrt{t}+5$. Multiply the numerator and the denominator by the conjugate of the denominator. This will eliminate the radical in the denominator.

$$\dfrac{\sqrt{t}+5}{\sqrt{t}-5}=\dfrac{\sqrt{t}+5}{\sqrt{t}-5}\cdot\dfrac{\sqrt{t}+5}{\sqrt{t}+5}=\dfrac{(\sqrt{t})^2+2\cdot5\sqrt{t}+5^2}{(\sqrt{t})^2-5^2}=\dfrac{t+10\sqrt{t}+25}{t-25}$$

63 We must recognize $\sqrt[3]{a}-\sqrt[3]{b}$ as the first factor of the product formula for the difference of two cubes, $x^3-y^3=(x-y)(x^2+xy+y^2)$. The second factor is then

$$\left(\sqrt[3]{a}\right)^2+\left(\sqrt[3]{a}\right)\left(\sqrt[3]{b}\right)+\left(\sqrt[3]{b}\right)^2=\sqrt[3]{a^2}+\sqrt[3]{ab}+\sqrt[3]{b^2}.$$

$$\dfrac{1}{\sqrt[3]{a}-\sqrt[3]{b}}=\dfrac{1}{\sqrt[3]{a}-\sqrt[3]{b}}\cdot\dfrac{\sqrt[3]{a^2}+\sqrt[3]{ab}+\sqrt[3]{b^2}}{\sqrt[3]{a^2}+\sqrt[3]{ab}+\sqrt[3]{b^2}}=\dfrac{\sqrt[3]{a^2}+\sqrt[3]{ab}+\sqrt[3]{b^2}}{a-b}$$

65 $\dfrac{\sqrt{a}-\sqrt{b}}{a^2-b^2}=\dfrac{\sqrt{a}-\sqrt{b}}{a^2-b^2}\cdot\dfrac{\sqrt{a}+\sqrt{b}}{\sqrt{a}+\sqrt{b}}=\dfrac{a-b}{(a+b)(a-b)(\sqrt{a}+\sqrt{b})}=\dfrac{1}{(a+b)(\sqrt{a}+\sqrt{b})}$

67 $\dfrac{\sqrt{2(x+h)+1}-\sqrt{2x+1}}{h}=\dfrac{\sqrt{2(x+h)+1}-\sqrt{2x+1}}{h}\cdot\dfrac{\sqrt{2(x+h)+1}+\sqrt{2x+1}}{\sqrt{2(x+h)+1}+\sqrt{2x+1}}$

$$=\dfrac{(2x+2h+1)-(2x+1)}{h(\sqrt{2(x+h)+1}+\sqrt{2x+1})}=\dfrac{2}{\sqrt{2(x+h)+1}+\sqrt{2x+1}}$$

69 $\dfrac{4x^2-x+5}{x^{2/3}}=\dfrac{4x^2}{x^{2/3}}-\dfrac{x}{x^{2/3}}+\dfrac{5}{x^{2/3}}=4x^{4/3}-x^{1/3}+5x^{-2/3}$

71 $\dfrac{(x^2+2)^2}{x^5}=\dfrac{x^4+4x^2+4}{x^5}=\dfrac{x^4}{x^5}+\dfrac{4x^2}{x^5}+\dfrac{4}{x^5}=x^{-1}+4x^{-3}+4x^{-5}$

Note: Exercises 73–90 are worked using the factoring concept given as the third method of simplification in Example 7.

73 The *smallest* exponent that appears on the factor x is -3.

$$x^{-3}+x^2\ \{\text{factor out } x^{-3}\}\ =x^{-3}(1+x^{2-(-3)})=x^{-3}(1+x^5)=\dfrac{1+x^5}{x^3}$$

75 $x^{-1/2} - x^{3/2} = x^{-1/2}(1 - x^2) = \dfrac{1 - x^2}{x^{1/2}}$

77 $(2x^2 - 3x + 1)(4)(3x + 2)^3(3) + (3x + 2)^4(4x - 3)$

$= (3x + 2)^3[12(2x^2 - 3x + 1) + (3x + 2)(4x - 3)]$

$= (3x + 2)^3[(24x^2 - 36x + 12) + (12x^2 - x - 6)] = (3x + 2)^3(36x^2 - 37x + 6)$

79 The smallest exponent that appears on the factor $(x^2 - 4)$ is $-\frac{1}{2}$ and the smallest

exponent that appears on the factor $(2x + 1)$ is 2. Thus, we will factor out

$(x^2 - 4)^{-1/2}(2x + 1)^2$.

$(x^2 - 4)^{1/2}(3)(2x + 1)^2(2) + (2x + 1)^3(\frac{1}{2})(x^2 - 4)^{-1/2}(2x)$

$\qquad = (x^2 - 4)^{-1/2}(2x + 1)^2[6(x^2 - 4) + x(2x + 1)]$

If you are unsure of this factoring, it is easy to visually check at this stage by merely

multiplying the expression—that is, we mentally add the exponents on the factor

$(x^2 - 4)$, $-\frac{1}{2}$ and 1, and we get $\frac{1}{2}$, which is the exponent we started with.

Proceeding, we may simplify as follows:

$$= \frac{(2x + 1)^2(8x^2 + x - 24)}{(x^2 - 4)^{1/2}}$$

81 $(3x + 1)^6(\frac{1}{2})(2x - 5)^{-1/2}(2) + (2x - 5)^{1/2}(6)(3x + 1)^5(3)$

$= (3x + 1)^5(2x - 5)^{-1/2}[(3x + 1) + 18(2x - 5)] \qquad \{\text{factor out } (3x + 1)^5(2x - 5)^{-1/2}\}$

$= (3x + 1)^5(2x - 5)^{-1/2}[(3x + 1) + (36x - 90)]$

$= \dfrac{(3x + 1)^5(39x - 89)}{(2x - 5)^{1/2}}$

83 $\dfrac{(6x + 1)^3(27x^2 + 2) - (9x^3 + 2x)(3)(6x + 1)^2(6)}{(6x + 1)^6}$

$= \dfrac{(6x + 1)^2[(6x + 1)(27x^2 + 2) - 18(9x^3 + 2x)]}{(6x + 1)^6}$

$= \dfrac{[(162x^3 + 12x + 27x^2 + 2) - (162x^3 + 36x)]}{(6x + 1)^4} = \dfrac{27x^2 - 24x + 2}{(6x + 1)^4}$

85 $\dfrac{(x^2 + 2)^3(2x) - x^2(3)(x^2 + 2)^2(2x)}{[(x^2 + 2)^3]^2} = \dfrac{(x^2 + 2)^2(2x)[(x^2 + 2)^1 - x^2(3)]}{(x^2 + 2)^6} =$

$$\dfrac{2x(x^2 + 2 - 3x^2)}{(x^2 + 2)^4} = \dfrac{2x(2 - 2x^2)}{(x^2 + 2)^4} = \dfrac{4x(1 - x^2)}{(x^2 + 2)^4}$$

87 $\dfrac{(x^2 + 4)^{1/3}(3) - (3x)(\frac{1}{3})(x^2 + 4)^{-2/3}(2x)}{[(x^2 + 4)^{1/3}]^2} = \dfrac{(x^2 + 4)^{-2/3}[3(x^2 + 4) - 2x^2]}{(x^2 + 4)^{2/3}} = \dfrac{x^2 + 12}{(x^2 + 4)^{4/3}}$

$\boxed{89}$ $\dfrac{(4x^2+9)^{1/2}(2) - (2x+3)(\frac{1}{2})(4x^2+9)^{-1/2}(8x)}{[(4x^2+9)^{1/2}]^2} =$

$$\dfrac{(4x^2+9)^{-1/2}[2(4x^2+9) - 4x(2x+3)]}{(4x^2+9)} = \dfrac{18-12x}{(4x^2+9)^{3/2}} = \dfrac{6(3-2x)}{(4x^2+9)^{3/2}}$$

$\boxed{91}$ Table $Y_1 = \dfrac{113x^3 + 280x^2 - 150x}{22x^3 + 77x^2 - 100x - 350}$ and $Y_2 = \dfrac{3x}{2x+7} + \dfrac{4x^2}{1.1x^2 - 5}$.

x	Y_1	Y_2
1	-0.6923	-0.6923
2	-26.12	-26.12
3	8.0392	8.0392
4	5.8794	5.8794
5	5.3268	5.3268

The values for Y_1 and Y_2 agree. Therefore, the two expressions might be equal.

$\boxed{93}$ The dimensions of I are (x) and $(x-y)$. The area of I is $(x-y)x$, and the area of II is $(x-y)y$. The area $A = x^2 - y^2 = (x-y)x + (x-y)y = \underline{(x-y)(x+y)}$.

$\boxed{95}$ (a) For the 25-year-old female, use

$\quad\quad C_f = 66.5 + 13.8w + 5h - 6.8y$ with $w=59$, $h=163$, and $y=25$.

$\quad\quad C_f = 66.5 + 13.8(59) + 5(163) - 6.8(25) = 1525.7$ calories

For the 55-year-old male, use

$\quad\quad C_m = 655 + 9.6w + 1.9h - 4.7y$ with $w=75$, $h=178$, and $y=55$.

$\quad\quad C_m = 655 + 9.6(75) + 1.9(178) - 4.7(55) = 1454.7$ calories

(b) As people age they require fewer calories. The coefficients of w and h are positive because large people require more calories.

1.4 Exercises

$\boxed{1}$ $4x - 3 = -5x + 6 \;\Rightarrow\; 4x + 5x = 6 + 3 \;\Rightarrow\; 9x = 9 \;\Rightarrow\; x = 1$

$\boxed{3}$ $(3x-2)^2 = (x-5)(9x+4) \;\Rightarrow\; 9x^2 - 12x + 4 = 9x^2 - 41x - 20 \;\Rightarrow$

$$29x = -24 \;\Rightarrow\; x = -\tfrac{24}{29}$$

Note: You may have solved equations such as those in Exercise 5 using a process called *cross-multiplication* in the past. This method is sufficient for problems of the form $\frac{P}{Q} = \frac{R}{S}$, but the following guidelines for solving an equation containing rational expressions apply to rational equations of a more complex form.

> ### Guidelines for Solving an Equation Containing Rational Expressions

(1) Determine the lcd of the rational expressions.

(2) Find the values of the variable that make the lcd zero. These are *not* solutions, because they yield at least one zero denominator when substituted into the given equation.

(3) Multiply each term of the equation by the lcd and simplify, thereby eliminating all of the denominators.

(4) Solve the equation obtained in guideline 3.

(5) The solutions of the given equation are the solutions found in guideline 4, with the exclusion of the values found in guideline 2.

$\boxed{5}$ $\left[\dfrac{3x+1}{6x-2} = \dfrac{2x+5}{4x-13}\right] \cdot (6x-2)(4x-13) \ \Rightarrow \ 12x^2 - 35x - 13 = 12x^2 + 26x - 10 \ \Rightarrow$

$$-3 = 61x \ \Rightarrow \ x = -\tfrac{3}{61} \ \{\text{note that } x \neq \tfrac{1}{3}, \tfrac{13}{4}\}$$

$\boxed{7}$ $\left[\dfrac{4}{x+2} + \dfrac{1}{x-2} = \dfrac{5x-6}{x^2-4}\right] \cdot (x+2)(x-2) \ \Rightarrow \ 4(x-2) + x + 2 = 5x - 6 \ \Rightarrow \ 0 = 0.$

This is an identity, and the solutions consist of every number in the domains of the given expressions. Thus, the solutions are all real numbers except ± 2, which we denote by $\mathbb{R} - \{\pm 2\}$.

$\boxed{9}$ $\left[\dfrac{5}{2x+3} + \dfrac{4}{2x-3} = \dfrac{14x+3}{4x^2-9}\right] \cdot (2x+3)(2x-3) \ \Rightarrow$

$5(2x-3) + 4(2x+3) = 14x + 3 \ \Rightarrow \ 18x - 3 = 14x + 3 \ \Rightarrow \ 4x = 6 \ \Rightarrow \ x = \tfrac{3}{2},$

which is not in the domain of the given expressions. No solution

$\boxed{11}$ $75x^2 + 35x - 10 = 0$ { factor out the gcf, 5 } $\Rightarrow$

$5(15x^2 + 7x - 2) = 0$ { divide by 5 } $\Rightarrow$

$$15x^2 + 7x - 2 = 0 \ \Rightarrow \ (3x+2)(5x-1) = 0 \ \Rightarrow \ x = -\tfrac{2}{3}, \tfrac{1}{5}$$

$\boxed{13}$ $\left[\dfrac{2x}{x+3} + \dfrac{5}{x} - 4 = \dfrac{18}{x^2+3x}\right] \cdot x(x+3) \ \Rightarrow \ 2x(x) + 5(x+3) - 4(x^2+3x) = 18 \ \Rightarrow$

$$0 = 2x^2 + 7x + 3 \ \Rightarrow \ (2x+1)(x+3) = 0 \ \Rightarrow$$

$$x = -\tfrac{1}{2} \ \{-3 \text{ is not in the domain of the given expressions}\}$$

$\boxed{15}$ $25x^2 = 9 \ \Rightarrow \ x^2 = \tfrac{9}{25} \ \Rightarrow \ x = \pm\sqrt{\tfrac{9}{25}} = \pm\tfrac{3}{5}$

$\boxed{17}$ $(x-3)^2 = 17 \ \Rightarrow \ x - 3 = \pm\sqrt{17} \ \Rightarrow \ x = 3 \pm \sqrt{17}$

$\boxed{19}$ To solve the equation $x^2 + 4x + 2 = 0$, use the quadratic formula,

$$x = \frac{-b \pm \sqrt{b^2 - 4ac}}{2a}, \text{ with } a = 1, b = 4, \text{ and } c = 2.$$

$$x = \frac{-(4) \pm \sqrt{(4)^2 - 4(1)(2)}}{2(1)} = \frac{-4 \pm \sqrt{16 - 8}}{2} = \frac{-4 \pm \sqrt{8}}{2} = \frac{-4 \pm 2\sqrt{2}}{2} = -2 \pm \sqrt{2}$$

Note: A common mistake is to not divide 2 into both terms of the numerator.

$\boxed{21}$ $|3x - 2| + 3 = 7 \Rightarrow |3x - 2| = 4 \Rightarrow 3x - 2 = 4 \text{ or } 3x - 2 = -4 \Rightarrow$

$$3x = 6 \text{ or } 3x = -2 \Rightarrow x = 2 \text{ or } x = -\tfrac{2}{3}$$

$\boxed{23}$ $3|x + 1| - 2 = -11 \Rightarrow 3|x + 1| = -9 \Rightarrow |x + 1| = -3.$

Since the absolute value of an expression is nonnegative, $|x + 1| = -3$ has no solution.

$\boxed{25}$ $9x^3 - 18x^2 - 4x + 8 = 0 \Rightarrow 9x^2(x - 2) - 4(x - 2) = 0 \Rightarrow$

$$(9x^2 - 4)(x - 2) = 0 \Rightarrow x = \pm\tfrac{2}{3}, 2$$

$\boxed{27}$ $y^{3/2} = 5y \Rightarrow y^{3/2} - 5y = 0 \Rightarrow y(y^{1/2} - 5) = 0 \Rightarrow y = 0 \text{ or } y^{1/2} = 5.$

$$y^{1/2} = 5 \Rightarrow (y^{1/2})^2 = 5^2 \Rightarrow y = 25. \quad y = 0, 25$$

Note: The following guidelines may be helpful when solving radical equations.

Guidelines for Solving a Radical Equation

(1) Isolate the radical. If we cannot get the radical isolated on one side of the equals sign because there is more than one radical, then we will split up the radical terms as evenly as possible on each side of the equals sign. For example, if there are 2 radicals, we put one on each side; if there are 3 radicals, we put 2 on one side and 1 on the the other.

(2) Raise both sides to the same power as the root index. *Note:* Remember here that

$$\boxed{(a + b\sqrt{n})^2 = a^2 + 2ab\sqrt{n} + b^2 n} \text{ and that it is } not \quad a^2 + b^2 n.$$

(3) If your equation contains no radicals, proceed to part (4). If there are still radicals in the equation, go back to part (1).

(4) Solve the resulting equation.

(5) Check the answers found in part (4) in the original equation to determine the valid solutions. *Note:* You may check the solutions in any equivalent equation of the original equation, i.e., an equation which occurs prior to raising both sides to a power. Also, extraneous solutions are introduced when raising both sides to an even power. Hence, all solutions *must* be checked in this case. Checking solutions when raising each side to an odd power is up to the individual instructor.

$\boxed{29}$ $\sqrt{7-x} = x - 5$ {square both sides} $\Rightarrow$

$\quad 7 - x = x^2 - 10x + 25$ {set equal to zero} $\Rightarrow$

$$x^2 - 9x + 18 = 0 \text{ \{factor\}} \quad \Rightarrow \quad (x-3)(x-6) = 0 \quad \Rightarrow \quad x = 3, 6.$$

Check $x = 6$: LS $= \sqrt{7-6} = 1$; RS $= 6 - 5 = 1$.

$\qquad$ Since both sides have the same value, $x = 6$ is a valid solution.

Check $x = 3$: LS $= \sqrt{7-3} = 2$; RS $= 3 - 5 = -2$.

$\qquad$ Since both sides do not have the same value, $x = 3$ is an extraneous solution.

$\boxed{31}$ $x = 3 + \sqrt{5x-9} \Rightarrow x - 3 = \sqrt{5x-9} \Rightarrow x^2 - 6x + 9 = 5x - 9 \Rightarrow$

$$x^2 - 11x + 18 = 0 \quad \Rightarrow \quad (x-2)(x-9) = 0 \quad \Rightarrow \quad x = 2, 9.$$

Check $x = 2$: LS $= 2$, RS $= 3 + 1 = 4$. LS $\neq$ RS $\Rightarrow$ $x = 2$ is an extraneous solution.

Check $x = 9$: LS $= 9$, RS $= 3 + 6 = 9$. LS $=$ RS $\Rightarrow$ $x = 9$ is a valid solution.

$\boxed{33}$ This equation is quadratic in form, and we can use the quadratic formula to solve for y^2 (*not* y) as follows:

$$5y^4 - 7y^2 + 1 = 0 \Rightarrow y^2 = \frac{7 \pm \sqrt{29}}{10} \cdot \frac{10}{10} = \frac{70 \pm 10\sqrt{29}}{100} \Rightarrow y = \pm\frac{1}{10}\sqrt{70 \pm 10\sqrt{29}}$$

Alternatively, let $u = y^2$ and solve $5u^2 - 7u + 1 = 0$.

$\boxed{35}$ $36x^{-4} - 13x^{-2} + 1 = 0 \Rightarrow (4x^{-2} - 1)(9x^{-2} - 1) = 0 \Rightarrow x^{-2} = \frac{1}{4}, \frac{1}{9} \Rightarrow$

$$x^2 = 4, 9 \Rightarrow x = \pm 2, \pm 3$$

Alternatively, let $u = x^{-2}$ and solve $36u^2 - 13u + 1 = 0$.

$\boxed{37}$ $3x^{2/3} + 4x^{1/3} - 4 = 0 \Rightarrow (3x^{1/3} - 2)(x^{1/3} + 2) = 0 \Rightarrow \sqrt[3]{x} = \frac{2}{3}, -2 \Rightarrow$

$$x = \frac{8}{27}, -8$$

Alternatively, let $u = x^{1/3}$ and solve $3u^2 + 4u - 4 = 0$.

$\boxed{39}$ (a) $x^{5/3} = 32 \Rightarrow (x^{5/3})^{3/5} = (32)^{3/5} \Rightarrow x = (\sqrt[5]{32})^3 = 2^3 = 8$

$\quad$ (b) $x^{4/3} = 16 \Rightarrow (x^{4/3})^{3/4} = \pm(16)^{3/4} \Rightarrow x = \pm(\sqrt[4]{16})^3 = \pm 2^3 = \pm 8$

$\quad$ (c) $x^{2/3} = -36 \Rightarrow (x^{2/3})^{3/2} = \pm(-36)^{3/2} \Rightarrow x = \pm(\sqrt{-36})^3,$

$\qquad\qquad\qquad\qquad\qquad$ which are not real numbers. No real solutions

$\quad$ (d) $x^{3/4} = 125 \Rightarrow (x^{3/4})^{4/3} = (125)^{4/3} \Rightarrow x = (\sqrt[3]{125})^4 = 5^4 = 625$

$\quad$ (e) $x^{3/2} = -27 \Rightarrow (x^{3/2})^{2/3} = (-27)^{2/3} \Rightarrow x = (\sqrt[3]{-27})^2 = (-3)^2 = 9,$

$\qquad\qquad\qquad\qquad\qquad$ which is an extraneous solution. No real solutions

41 (a) For this exercise, we must recognize the equation as a quadratic in x, that is,

$$(A)x^2 + (B)x + (C) = 0,$$

where A is the coefficient of x^2, B is the coefficient of x, and C is the collection of all terms that do not contain x^2 or x.

$$4x^2 - 4xy + 1 - y^2 = 0 \;\Rightarrow\; (4)x^2 + (-4y)x + (1 - y^2) = 0 \;\Rightarrow$$

$$x = \frac{4y \pm \sqrt{16y^2 - 16(1 - y^2)}}{2(4)} = \frac{4y \pm \sqrt{16[y^2 - (1 - y^2)]}}{2(4)} =$$

$$\frac{4y \pm 4\sqrt{2y^2 - 1}}{2(4)} = \frac{y \pm \sqrt{2y^2 - 1}}{2}$$

(b) Similar to part (a), we must now recognize the equation as a quadratic equation in y. $4x^2 - 4xy + 1 - y^2 = 0 \;\Rightarrow\; (-1)y^2 + (-4x)y + (4x^2 + 1) = 0 \;\Rightarrow$

$$y = \frac{4x \pm \sqrt{16x^2 + 4(4x^2 + 1)}}{-2} = \frac{4x \pm 2\sqrt{8x^2 + 1}}{-2} = -2x \pm \sqrt{8x^2 + 1}$$

43 (a) $x = \dfrac{-4{,}500{,}000 \pm \sqrt{4{,}500{,}000^2 - 4(1)(-0.96)}}{2} \approx 0$ and $-4{,}500{,}000$

(b) $x = \dfrac{-b \pm \sqrt{b^2 - 4ac}}{2a} \cdot \dfrac{-b \mp \sqrt{b^2 - 4ac}}{-b \mp \sqrt{b^2 - 4ac}} = \dfrac{b^2 - (b^2 - 4ac)}{2a(-b \mp \sqrt{b^2 - 4ac})} =$

$$\frac{4ac}{2a(-b \mp \sqrt{b^2 - 4ac})} = \frac{2c}{-b \mp \sqrt{b^2 - 4ac}}. \text{ The root near zero was obtained in part}$$

(a) using the plus sign, In the second formula, it corresponds to the minus sign.

$$x = \frac{2(-0.96)}{-4{,}500{,}000 - \sqrt{4{,}500{,}000^2 - 4(1)(-0.96)}} \approx 2.13 \times 10^{-7}$$

45 $A = P + Prt \;\Rightarrow\; A - P = Prt \;\Rightarrow\; r = \dfrac{A - P}{Pt}$

47                    $S = \dfrac{p}{q + p(1 - q)}$ given

$Sq + Sp(1 - q) = p$ eliminate the fraction

$Sq + Sp - Spq = p$ multiply terms

$Sq - Spq = p - Sp$ isolate terms containing q

$Sq(1 - p) = p(1 - S)$ factor

$q = \dfrac{p(1 - S)}{S(1 - p)}$ solve for q

49 $\dfrac{1}{f} = \dfrac{1}{p} + \dfrac{1}{q}$ { multiply by the lcd fpq } $\;\Rightarrow$

$$pq = fq + fp \;\Rightarrow\; pq - fq = fp \;\Rightarrow\; q(p - f) = fp \;\Rightarrow\; q = \frac{fp}{p - f}$$

51 $K = \frac{1}{2}mv^2 \;\Rightarrow\; v^2 = \frac{2K}{m} \;\Rightarrow\; v = \pm\sqrt{\frac{2K}{m}} \;\Rightarrow\; v = \sqrt{\frac{2K}{m}}$ since $v > 0$.

53 $A = 2\pi r(r+h) \;\Rightarrow\; A = 2\pi r^2 + 2\pi rh \;\Rightarrow\; 2\pi r^2 + 2\pi rh - A = 0 \;\Rightarrow$

$(2\pi)r^2 + (2\pi h)r - A = 0$ { a quadratic in r with $a = 2\pi$, $b = 2\pi h$, and $c = -A$ } $\;\Rightarrow$

$$r = \frac{-(2\pi h) \pm \sqrt{(2\pi h)^2 - 4(2\pi)(-A)}}{2(2\pi)}$$

$$= \frac{-2\pi h \pm \sqrt{4\pi^2 h^2 + 8\pi A}}{2(2\pi)} = \frac{-2\pi h \pm 2\sqrt{\pi^2 h^2 + 2\pi A}}{2(2\pi)} = \frac{-\pi h \pm \sqrt{\pi^2 h^2 + 2\pi A}}{2\pi}.$$

Since $r > 0$, we must use the plus sign, and $r = \dfrac{-\pi h + \sqrt{\pi^2 h^2 + 2\pi A}}{2\pi}$.

55 $d = \frac{1}{2}\sqrt{4R^2 - C^2} \;\Rightarrow\; 2d = \sqrt{4R^2 - C^2} \;\Rightarrow\; (2d)^2 = (\sqrt{4R^2 - C^2})^2 \;\Rightarrow$

$4d^2 = 4R^2 - C^2 \;\Rightarrow\; C^2 = 4R^2 - 4d^2 \;\Rightarrow\; C^2 = 4(R^2 - d^2) \;\Rightarrow$

$$C = \pm 2\sqrt{R^2 - d^2} \;\Rightarrow\; C = 2\sqrt{R^2 - d^2} \text{ since } C > 0$$

57 Let x denote the number of months needed to recover the cost of the insulation. The savings in one month is 10% of $60 = $6. We want to know when the total savings (given by $6x$) will equal the cost of the insulation ($1080).

$$6x = 1080 \;\Rightarrow\; x = 180 \text{ months (or 15 yr)}.$$

59 (a) Let t denote the desired number of seconds. The distance walked by each child is $1.5t$ and $2t$, respectively. $1.5t + 2t = 224 \;\Rightarrow\; t = 64$ sec

(b) $64(1.5) = 96$ m and $64(2) = 128$ m, respectively

61 Let x denote the distance to the target. We know the total time involved and need a formula for time. Solving $d = rt$ for t gives us $t = d/r$.

$\text{Time}_{\text{to target}} + \text{Time}_{\text{from target}} = \text{Time}_{\text{total}} \;\Rightarrow$

$\dfrac{x}{3300} + \dfrac{x}{1100} = 1.5$ { multiply by the lcd, 3300 } $\;\Rightarrow$

$$x + 3x = 1.5(3300) \;\Rightarrow\; 4x = 4950 \;\Rightarrow\; x = 1237.5 \text{ ft}.$$

63 Use the formula for the area of a trapezoid (see the endpapers in the front of the text). $A = \frac{1}{2}(b_1 + b_2)h \;\Rightarrow\; 5 = \frac{1}{2}(3 + b_2)(1) \;\Rightarrow\; 10 = 3 + b_2 \;\Rightarrow\; b_2 = 7$ ft.

65 (a) $h = 5280$ and $T_0 = 70 \;\Rightarrow\; T = 70 - \left(\dfrac{5.5}{1000}\right)5280 = 40.96°\text{F}$.

(b) $T = 32 \;\Rightarrow\; 32 = 70 - \left(\dfrac{5.5}{1000}\right)h \;\Rightarrow\; h = (70 - 32)\left(\dfrac{1000}{5.5}\right) \approx 6909$ ft.

67 $B = 55$ and $h = 10{,}000 - 4000 = 6000 \;\Rightarrow\; T = 55 - \left(\dfrac{3}{1000}\right)(6000) = 37°\text{F}$.

69 (a) $v = 55 \;\Rightarrow\; d = v + (v^2/20) = 55 + (55^2/20) = 206.25$ ft

(b) $d = 120 \Rightarrow 120 = v + (v^2/20) \Rightarrow$

$\qquad 2400 = 20v + v^2 \{\text{multiply by the lcd, 20}\} \Rightarrow$

$\qquad\qquad\qquad v^2 + 20v - 2400 = 0 \Rightarrow (v + 60)(v - 40) = 0 \Rightarrow v = 40 \text{ mi/hr}$

71 (a) The distances of the northbound and eastbound planes are $100 + 200t$ and $400t$,

respectively. Using the Pythagorean theorem,

$$d = \sqrt{(100 + 200t)^2 + (400t)^2} = \sqrt{100^2(1 + 2t)^2 + 100^2(4t)^2} = 100\sqrt{20t^2 + 4t + 1}.$$

(b) $d = 500 \Rightarrow 500 = 100\sqrt{20t^2 + 4t + 1} \Rightarrow 5 = \sqrt{20t^2 + 4t + 1} \Rightarrow$

$\qquad 5^2 = 20t^2 + 4t + 1 \Rightarrow 5t^2 + t - 6 = 0 \Rightarrow$

$\qquad\qquad\qquad (5t + 6)(t - 1) = 0 \Rightarrow t = 1$ hour after 2:30 P.M., or 3:30 P.M.

73 Let x denote the number of \$10 reductions in price. The revenue is equal to the

number of units sold times the cost per unit. To obtain expressions for the number

of units sold and the cost per unit, it is helpful to write some specific values in a table

format.

x	0	1	2	$\cdots$
units sold	15	17	19	$\cdots$
unit price	300	290	280	$\cdots$

Since the number of units sold starts at 15 and increases by 2 when x increases by 1,

we may denote the number of units sold by $(15 + 2x)$. Similarly, $(300 - 10x)$ is the

price per unit.

Revenue $= (\# \text{ of units}) \times (\text{unit price}) \Rightarrow$

$7000 = (15 + 2x)(300 - 10x) \Rightarrow 700 = -2x^2 + 45x + 450 \Rightarrow$

$2x^2 - 45x + 250 = 0 \Rightarrow (2x - 25)(x - 10) = 0 \Rightarrow x = 10$ or 12.5.

$\qquad\qquad$ The selling price is $\$300 - \$10(10) = \$200$, or $\$300 - \$10(12.5) = \$175$.

75 (a) Area$_\text{capsule}$ = Area$_\text{sphere}$ {the two ends are hemispheres} + Area$_\text{cylinder}$ =

$$4\pi r^2 + 2\pi rh = 4\pi\left(\tfrac{1}{4}\right)^2 + 2\pi\left(\tfrac{1}{4}\right)\left(2 - \tfrac{1}{2}\right) = \tfrac{\pi}{4} + \tfrac{3\pi}{4} = \pi \text{ cm}^2.$$

Area$_\text{tablet}$ = Area$_\text{top and bottom}$ + Area$_\text{cylinder}$ = $2\pi r^2 + 2\pi r\left(\tfrac{1}{2}\right) = 2\pi r^2 + \pi r$.

Equating the two surface areas yields $2\pi r^2 + \pi r = \pi \Rightarrow$

$\qquad 2r^2 + r - 1 = 0 \Rightarrow (2r - 1)(r + 1) = 0 \Rightarrow r = \tfrac{1}{2}$, and the diameter is 1 cm.

(b) Volume$_\text{capsule}$ = Volume$_\text{sphere}$ + Volume$_\text{cylinder}$ =

$$\tfrac{4}{3}\pi r^3 + \pi r^2 h = \tfrac{4}{3}\pi\left(\tfrac{1}{4}\right)^3 + \pi\left(\tfrac{1}{4}\right)^2\tfrac{3}{2} = \tfrac{\pi}{48} + \tfrac{3\pi}{32} = \tfrac{11\pi}{96} \approx 0.360 \text{ cm}^3.$$

Volume$_\text{tablet}$ = Volume$_\text{cylinder}$ = $\pi r^2 h = \pi\left(\tfrac{1}{2}\right)^2\tfrac{1}{2} = \tfrac{\pi}{8} \approx 0.393 \text{ cm}^3.$

77 $k = 10^5$ and $c = \frac{1}{2} \Rightarrow Q = kP^{-c} = 10^5 P^{-1/2} \Rightarrow Q = 10^5/\sqrt{P} \Rightarrow$

$$\sqrt{P} = \frac{10^5}{Q} \Rightarrow P = \left(\frac{10^5}{Q}\right)^2 = \left(\frac{100,000}{5000}\right)^2 = (20)^2 = 400 \text{ cents, or, } \$4.00.$$

79 $\text{Cost}_{\text{underwater}} + \text{Cost}_{\text{overland}} = \text{Cost}_{\text{total}} \Rightarrow$

$7500 \cdot (\text{underwater miles}) + 6000 \cdot (\text{overland miles}) = 35,000 \Rightarrow$

$7500\sqrt{x^2 + 1} + 6000(5 - x) = 35,000 \Rightarrow 15\sqrt{x^2 + 1} = 12x + 10 \Rightarrow$

$225(x^2 + 1) = 144x^2 + 240x + 100 \Rightarrow 81x^2 - 240x + 125 = 0 \Rightarrow$

$x = \dfrac{240 \pm \sqrt{17,100}}{162} = \dfrac{40 \pm 5\sqrt{19}}{27} \approx 2.2887, \ 0.6743$ mi. There are two possible routes.

81 The y-values are increasing rapidly and can best be described by equation (4), $y = x^3 - x^2 + x - 10$.

83 (a) Let $Y_1 = D_1 = 6.096L + 685.7$ and

$Y_2 = D_2 = 0.00178L^3 - 0.072L^2 + 4.37L + 719.$

Table each equation and compare them to the actual values.

x (L)	Y_1	Y_2	Summer
0	686	719	720
10	747	757	755
20	808	792	792
30	869	833	836
40	930	893	892
50	991	980	978
60	1051	1106	1107

Comparing Y_1 (D_1) with Y_2 (D_2) we can see that the linear equation D_1 is not as accurate as the cubic equation D_2.

(b) $L = 35 \Rightarrow D_2 = 0.00178(35)^3 - 0.072(35)^2 + 4.37(35) + 719 \approx 860$ min.

85 The volume of the box is $V = hw^2 = 25$, where h is the height and w is the length of a side of the square base. The amount of cardboard will be minimized when the surface area of the box is a minimum. The surface area is given by $S = w^2 + 4wh$. Since $h = 25/w^2$, we have $S = w^2 + 100/w$. Form a table for w and S.

w	S	w	S
3.4	40.972	3.7	40.717
3.5	40.821	3.8	40.756
3.6	40.738	3.9	40.851

The minimum surface area is $S \approx 40.717$ when $w \approx 3.7$ and $h = 25/w^2 \approx 1.8$.

1.5 Exercises

1 $(5 - 2i) + (-3 + 6i) = [5 + (-3)] + (-2 + 6)i = 2 + 4i$

3 $(7 - 6i) - (-11 - 3i) = (7 + 11) + (-6 + 3)i = 18 - 3i$

5 $(3 + 5i)(2 - 7i) \;= (3 + 5i)2 + (3 + 5i)(-7i)$ { distributive property }

$\qquad\qquad\qquad = 6 + 10i - 21i - 35i^2$ { multiply terms }

$\qquad\qquad\qquad = (6 - 35i^2) + (10 - 21)i$ $\left\{ \begin{array}{l} \text{group the real parts} \\ \text{and the imaginary parts} \end{array} \right\}$

$\qquad\qquad\qquad = (6 + 35) - 11i$ $\{\, i^2 = -1 \,\}$

$\qquad\qquad\qquad = 41 - 11i$

7 $(1 - 3i)(2 + 5i) = (2 - 15i^2) + (5 - 6)i = (2 + 15) - i = 17 - i$

9   Use the special product formula for $(x - y)^2$ on the inside front cover of the text.

$\qquad (5 - 2i)^2 = 5^2 - 2(5)(2i) + (2i)^2 = 25 - 20i + 4i^2 = (25 - 4) - 20i = 21 - 20i$

11 $i(3 + 4i)^2 = i\big[3^2 + 2(3)(4i) + (4i)^2\big] = i\big[(9 - 16) + 24i\big] = i(-7 + 24i) = -24 - 7i$

13 $(3 + 4i)(3 - 4i)$ { note that this difference of squares ... } $=$

$\qquad 3^2 - (4i)^2 = 9 - (-16) = \{ \ldots \text{ becomes a "sum of squares" } \} \; 9 + 16 = 25$

15   Since $i^k = 1$ if k is a multiple of 4, we will write i^{43} as $i^{40}i^3$, knowing that i^{40} will

reduce to 1. $i^{43} = i^{40}i^3 = (i^4)^{10}(-i) = 1^{10}(-i) = -i$

17   Since $i^k = 1$ if k is a multiple of 4, we will write i^{73} as $i^{72}i^1$, knowing that i^{72} will

reduce to 1. $i^{73} = i^{72}i = (i^4)^{18}i = 1^{18}i = i$

19   Multiply both the numerator and the denominator by the conjugate of the denominator to eliminate all i's in the denominator. The new denominator is the sum of the squares of the coefficients—in this case, 2^2 and 4^2.

$$\frac{3}{2 + 4i} \cdot \frac{2 - 4i}{2 - 4i} = \frac{6 - 12i}{4 - (-16)} = \frac{6 - 12i}{20} = \frac{3}{10} - \frac{3}{5}i$$

21   Multiply both the numerator and the denominator by the conjugate of the denominator to eliminate all i's in the denominator. The new denominator is the sum of the squares of the coefficients—in this case, 6^2 and 2^2.

$$\frac{1 - 7i}{6 - 2i} \cdot \frac{6 + 2i}{6 + 2i} = \frac{(6 + 14) + (2 - 42)i}{36 - (-4)} = \frac{20 - 40i}{40} = \frac{1}{2} - i$$

23 $\dfrac{-4 + 6i}{2 + 7i} \cdot \dfrac{2 - 7i}{2 - 7i} = \dfrac{(-8 + 42) + (28 + 12)i}{4 - (-49)} = \dfrac{34 + 40i}{53} = \dfrac{34}{53} + \dfrac{40}{53}i$

25   Multiplying the denominator by i will eliminate the i's in the denominator.

$$\frac{4 - 2i}{-5i} = \frac{4 - 2i}{-5i} \cdot \frac{i}{i} = \frac{4i - 2i^2}{-5i^2} = \frac{2 + 4i}{5} = \frac{2}{5} + \frac{4}{5}i$$

27 Use the special product formula for $(x+y)^3$ on the inside front cover of the text.

$$(2+5i)^3 = (2)^3 + 3(2)^2(5i) + 3(2)(5i)^2 + (5i)^3$$

$$= (8+150i^2) + (60i + 125i^3) = (8-150) + (60-125)i = -142-65i$$

29 A common mistake is to multiply $\sqrt{-4}\,\sqrt{-16}$ and obtain $\sqrt{64}$, or 8.

The correct procedure is $\sqrt{-4}\,\sqrt{-16} = \sqrt{4}\,i \cdot \sqrt{16}\,i = (2i)(4i) = 8i^2 = -8$.

$$(2-\sqrt{-4})(3-\sqrt{-16}) = (2-2i)(3-4i) = (6-8) + (-6i - 8i) = -2 - 14i$$

31 $\dfrac{4+\sqrt{-81}}{7-\sqrt{-64}} = \dfrac{4+9i}{7-8i} \cdot \dfrac{7+8i}{7+8i} = \dfrac{(28-72)+(32+63)i}{49-(-64)} = \dfrac{-44+95i}{113} = -\dfrac{44}{113} + \dfrac{95}{113}i$

33 $\dfrac{\sqrt{-36}\,\sqrt{-49}}{\sqrt{-16}} = \dfrac{(6i)(7i)}{4i} \cdot \dfrac{-i}{-i} = \dfrac{(-42)(-i)}{-4i^2} = \dfrac{42i}{4} = \dfrac{21}{2}i$

35 We need to equate the real parts and the imaginary parts on each side of " $=$ ".

$$4+(x+2y)i = x + 2i \quad\Rightarrow\quad 4 = x \text{ and } x + 2y = 2 \quad\Rightarrow$$

$$x = 4 \text{ and } 4 + 2y = 2 \quad\Rightarrow\quad 2y = -2 \quad\Rightarrow\quad y = -1, \text{ so } x = 4 \text{ and } y = -1.$$

37 $(2x-y) - 16i = 10 + 4yi \quad\Rightarrow\quad 2x - y = 10 \text{ and } -16 = 4y \quad\Rightarrow$

$y = -4 \text{ and } 2x - (-4) = 10 \quad\Rightarrow\quad 2x + 4 = 10 \quad\Rightarrow\quad 2x = 6 \quad\Rightarrow\quad x = 3,$

$$\text{so } x = 3 \text{ and } y = -4.$$

39 $x^2 - 6x + 13 = 0 \quad\Rightarrow$

$$x = \frac{-(-6) \pm \sqrt{(-6)^2 - 4(1)(13)}}{2(1)} = \frac{6 \pm \sqrt{36-52}}{2} = \frac{6 \pm \sqrt{-16}}{2} = \frac{6 \pm 4i}{2} = 3 \pm 2i$$

41 $x^2 + 4x + 13 = 0 \quad\Rightarrow$

$$x = \frac{-4 \pm \sqrt{4^2 - 4(1)(13)}}{2(1)} = \frac{-4 \pm \sqrt{16-52}}{2} = \frac{-4 \pm \sqrt{-36}}{2} = \frac{-4 \pm 6i}{2} = -2 \pm 3i$$

43 $x^2 - 5x + 20 = 0 \quad\Rightarrow$

$$x = \frac{-(-5) \pm \sqrt{(-5)^2 - 4(1)(20)}}{2(1)} = \frac{5 \pm \sqrt{25-80}}{2} = \frac{5 \pm \sqrt{-55}}{2} = \frac{5}{2} \pm \frac{1}{2}\sqrt{55}\,i$$

45 $4x^2 + x + 3 = 0 \quad\Rightarrow$

$$x = \frac{-1 \pm \sqrt{1^2 - 4(4)(3)}}{2(4)} = \frac{-1 \pm \sqrt{1-48}}{8} = \frac{-1 \pm \sqrt{-47}}{8} = -\frac{1}{8} \pm \frac{1}{8}\sqrt{47}\,i$$

47 Solving $x^3 = -125$ would only give us the solution $x = -5$. So first we need to factor

$x^3 + 125$ as the sum of cubes. $x^3 + 125 = 0 \quad\Rightarrow\quad (x+5)(x^2 - 5x + 25) = 0 \quad\Rightarrow$

$$x = -5 \text{ or } x = \frac{5 \pm \sqrt{25-100}}{2} = \frac{5 \pm 5\sqrt{3}\,i}{2}. \text{ The three solutions are } -5, \frac{5}{2} \pm \frac{5}{2}\sqrt{3}\,i.$$

49 $x^4 = 256 \quad\Rightarrow\quad x^4 - 256 = 0 \quad\Rightarrow\quad (x^2 - 16)(x^2 + 16) = 0 \quad\Rightarrow$

$$x^2 = 16, -16 \quad\Rightarrow\quad x = \pm\sqrt{16}, \pm\sqrt{-16} \quad\Rightarrow\quad x = \pm 4, \pm 4i$$

51 $4x^4 + 25x^2 + 36 = 0 \Rightarrow (x^2 + 4)(4x^2 + 9) = 0 \Rightarrow$

$$x^2 = -4, -\tfrac{9}{4} \Rightarrow x = \pm\sqrt{-4}, \pm\sqrt{-\tfrac{9}{4}} \Rightarrow x = \pm 2i, \pm\tfrac{3}{2}i$$

53 $x^3 + 3x^2 + 4x = 0 \Rightarrow x(x^2 + 3x + 4) = 0 \Rightarrow$

$$x = 0 \text{ or } x = \frac{-3 \pm \sqrt{9 - 16}}{2} = \frac{-3 \pm \sqrt{7}\,i}{2}. \text{ The three solutions are } 0, -\tfrac{3}{2} \pm \tfrac{1}{2}\sqrt{7}\,i.$$

Note: Exer. 55–60: Let $z = a + bi$ and $w = c + di$.

55 If $z = a + bi$ and $w = c + di$, then

$$\begin{aligned}
\overline{z + w} &= \overline{(a + bi) + (c + di)} && \text{definition of } z \text{ and } w \\
&= \overline{(a + c) + (b + d)i} && \text{write in complex number form} \\
&= (a + c) - (b + d)i && \text{definition of conjugate} \\
&= (a - bi) + (c - di) && \text{rearrange terms} \\
&= \bar{z} + \bar{w}. && \text{definition of conjugates of } z \text{ and } w
\end{aligned}$$

In the step described by "rearrange terms,"

we are really looking ahead to the terms we want to obtain, $\bar{z}$ and $\bar{w}$.

57 $\overline{z \cdot w} = \overline{(a + bi) \cdot (c + di)} = \overline{(ac - bd) + (ad + bc)i} =$

$(ac - bd) - (ad + bc)i = ac - adi - bd - bci = a(c - di) - bi(c - di) =$

$$(a - bi) \cdot (c - di) = \bar{z} \cdot \bar{w}$$

59 If $\bar{z} = z$, then $a - bi = a + bi$ and hence $-bi = bi$, or $2bi = 0$.

Thus, $b = 0$ and $z = a$ is real. Conversely, if z is real, then $b = 0$ and hence

$$\bar{z} = \overline{a + 0i} = a - 0i = a + 0i = z.$$

1.6 Exercises

Note: The bracket symbols "[" and "]", are used with $\leq$ or $\geq$ to denote that the endpoint of the interval is part of the solution. Parentheses, "(" and ")" are used with $<$ or $>$ and denote that the endpoint is *not* part of the solution.

1 $x < -2 \Leftrightarrow (-\infty, -2)$

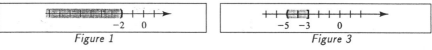

Figure 1   Figure 3

3 $-3 \geq x > -5 \Leftrightarrow -5 < x \leq -3 \Leftrightarrow (-5, -3]$

5 $(-5, 8] \Leftrightarrow -5 < x \leq 8$

7 $2x + 5 < 3x - 7 \Rightarrow -x < -12$ { Remember to change the direction of the inequality when multiplying or dividing by a negative value. } $\Rightarrow x > 12 \Leftrightarrow (12, \infty)$

⑨ $\left[3 \le \dfrac{2x-3}{5} < 7\right] \cdot 5$ { given inequality } $\Rightarrow$

$15 \le 2x - 3 < 35$ { multiply by the lcd, 5 } $\Rightarrow$

$18 \le 2x < 38$ { add 3 to all three parts } $\Rightarrow$

$9 \le x < 19$ { divide all three parts by 2 } $\Leftrightarrow$ $[9, 19)$ { equivalent interval notation }

⑪ By the law of signs, a quotient is positive if the sign of the numerator and the sign of the denominator are the same. Since the numerator is positive, $\dfrac{4}{3x+2} > 0$ $\Rightarrow$

$3x + 2 > 0$ $\Rightarrow$ $3x > -2$ $\Rightarrow$ $x > -\frac{2}{3}$ $\Leftrightarrow$ $\left(-\frac{2}{3}, \infty\right)$. The expression is never equal to 0 since the numerator is never 0. Thus, the solution of $\dfrac{4}{3x+2} \ge 0$ is $\left(-\frac{2}{3}, \infty\right)$.

⑬ $\dfrac{-2}{4-3x} > 0$ $\Rightarrow$ $4 - 3x < 0$ { denominator must also be negative } $\Rightarrow$ $4 < 3x$ $\Rightarrow$

$$x > \tfrac{4}{3} \Leftrightarrow \left(\tfrac{4}{3}, \infty\right)$$

⑮ $(1-x)^2 > 0$ $\forall x$ except 1. Thus, $\dfrac{2}{(1-x)^2} > 0$ has solution $\mathbb{R} - \{\,1\,\}$.

⑰ $|x+3| < 0.01$ $\Rightarrow$ $-0.01 < x + 3 < 0.01$ $\Rightarrow$ $-3.01 < x < -2.99$ $\Leftrightarrow$

$$(-3.01, -2.99)$$

⑲ $|3x-7| \ge 5$ $\Rightarrow$ $3x - 7 \ge 5$ or $3x - 7 \le -5$ $\Rightarrow$ $x \ge 4$ or $x \le \frac{2}{3}$ $\Leftrightarrow$

$$\left(-\infty, \tfrac{2}{3}\right] \cup [4, \infty)$$

㉑ Since $|7x+2| \ge 0$ $\forall x$, $|7x+2| > -2$ has solution $(-\infty, \infty)$.

㉓ $|3x-9| > 0$ $\forall x$ except when $3x - 9 = 0$, or $x = 3$. The solution is $(-\infty, 3) \cup (3, \infty)$.

㉕ $-2 < |x| < 4$ $\Rightarrow$ $|x| > -2$ *and* $|x| < 4$ $\Rightarrow$

$|x| < 4$ { since $|x|$ is *always* greater than -2 } $\Rightarrow$ $-4 < x < 4$ $\Leftrightarrow$ $(-4, 4)$

Note: Many solutions for exercises involving inequalities contain a sign diagram. You may want to read Example 7 again if you have trouble interpreting the sign diagrams.

㉗ $(3x+1)(5-10x) > 0$ has solutions in the interval $\left(-\frac{1}{3}, \frac{1}{2}\right)$. See *Diagram 27* for details concerning the signs of the individual factors and the resulting sign.

Resulting sign:	$\ominus$	$\oplus$	$\ominus$
Sign of $5 - 10x$:	$+$	$+$	$-$
Sign of $3x + 1$:	$-$	$+$	$+$
x values:	$-1/3$	$1/2$	

Diagram 27

㉙ $x^2 - x - 6 < 0$ $\Rightarrow$ $(x-3)(x+2) < 0$; $(-2, 3)$

Resulting sign:	$\oplus$	$\ominus$	$\oplus$
Sign of $x - 3$:	$-$	$-$	$+$
Sign of $x + 2$:	$-$	$+$	$+$
x values:	-2	3	

Diagram 29

$\boxed{31}$ $x(2x+3) \geq 5$ $\Rightarrow$ $2x^2 + 3x - 5 \geq 0$ $\Rightarrow$ $(2x+5)(x-1) \geq 0$; $(-\infty, -\frac{5}{2}] \cup [1, \infty)$

Resulting sign:	$\oplus$	$\ominus$	$\oplus$
Sign of $x - 1$:	$-$	$-$	$+$
Sign of $2x + 5$:	$-$	$+$	$+$
x values:	$-5/2$		1

$\boxed{\text{Diagram 31}}$

Note: Solving $x^2 <$ (or $>$) a^2 for $a > 0$ may be solved using factoring, that is,

$$x^2 - a^2 < 0 \Rightarrow (x+a)(x-a) < 0 \Rightarrow -a < x < a;$$ or by taking the square

root of each side, that is, $\sqrt{x^2} < \sqrt{a^2} \Rightarrow |x| < a \Rightarrow -a < x < a.$

$\boxed{33}$ *Note:* The most common mistake is not remembering that $\sqrt{x^2} = |x|$.

$$25x^2 - 9 < 0 \Rightarrow 25x^2 < 9 \Rightarrow x^2 < \frac{9}{25} \Rightarrow |x| < \frac{3}{5} \Rightarrow -\frac{3}{5} < x < \frac{3}{5} \Leftrightarrow \left(-\frac{3}{5}, \frac{3}{5}\right)$$

$\boxed{35}$ $\dfrac{x^2(x+2)}{(x+2)(x+1)} \leq 0$ $\Rightarrow$ $\dfrac{x^2}{x+1} \leq 0$ {we will exclude $x = -2$ since it makes the original

expression undefined} $\Rightarrow$ $\dfrac{1}{x+1} \leq 0$ {we can divide by x^2 since $x^2 \geq 0$ and we will

include $x = 0$ since it makes x^2 equal to zero and we want all solutions less than *or*

equal to zero} $\Rightarrow$ $x + 1 < 0$ {the fraction cannot equal zero and $x + 1$ must be

negative so that the fraction is negative} $\Rightarrow$ $x < -1$; $(-\infty, -2) \cup (-2, -1) \cup \{0\}$

$\boxed{37}$ $\dfrac{x^2 - x}{x^2 + 2x} \leq 0$ $\Rightarrow$ $\dfrac{x(x-1)}{x(x+2)} \leq 0$ $\Rightarrow$ $\dfrac{x-1}{x+2} \leq 0$ {we will exclude $x = 0$ from the

solution}; $(-2, 0) \cup (0, 1]$

Resulting sign:	$\oplus$	$\ominus$	$\oplus$
Sign of $x - 1$:	$-$	$-$	$+$
Sign of $x + 2$:	$-$	$+$	$+$
x values:	-2		1

$\boxed{\text{Diagram 37}}$

$\boxed{39}$ $\dfrac{x-2}{x^2 - 3x - 10} \geq 0$ $\Rightarrow$ $\dfrac{x-2}{(x-5)(x+2)} \geq 0$ {$x = 2$ is a solution since it makes the

fraction equal to zero, $x = 5$ and $x = -2$ are excluded since these values make the

fraction undefined}; $(-2, 2] \cup (5, \infty)$

Resulting sign:	$\ominus$	$\oplus$	$\ominus$	$\oplus$
Sign of $x - 5$:	$-$	$-$	$-$	$+$
Sign of $x - 2$:	$-$	$-$	$+$	$+$
Sign of $x + 2$:	$-$	$+$	$+$	$+$
x values:	-2		2	5

$\boxed{\text{Diagram 39}}$

41 $\dfrac{-3x}{x^2-9}>0 \Rightarrow \dfrac{x}{(x+3)(x-3)}<0$ { divide by -3 }; $(-\infty,-3)\cup(0,3)$

Resulting sign:	$\ominus$	$\oplus$	$\ominus$	$\oplus$
Sign of $x-3$:	$-$	$-$	$-$	$+$
Sign of x:	$-$	$-$	$+$	$+$
Sign of $x+3$:	$-$	$+$	$+$	$+$
x values:		-3	0	3

Diagram 41

43 $\dfrac{x+1}{2x-3}>2 \Rightarrow \dfrac{x+1-2(2x-3)}{2x-3}>0 \Rightarrow \dfrac{-3x+7}{2x-3}>0.$

From *Diagram 43*, the solution is $(\frac{3}{2},\frac{7}{3})$. Note that you should *not* multiply by the factor $2x-3$ as we did with rational *equations* because $2x-3$ may be positive or negative, and multiplying by it would require solving two inequalities. This method of solution tends to be more difficult than the sign diagram method.

Resulting sign:	$\ominus$	$\oplus$	$\ominus$
Sign of $-3x+7$:	$+$	$+$	$-$
Sign of $2x-3$:	$-$	$+$	$+$
x values:		$3/2$	$7/3$

Diagram 43

45 $\dfrac{1}{x-2}\geq\dfrac{3}{x+1} \Rightarrow \dfrac{1(x+1)-3(x-2)}{(x-2)(x+1)}\geq0 \Rightarrow \dfrac{-2x+7}{(x-2)(x+1)}\geq0;\ (-\infty,-1)\cup(2,\frac{7}{2}]$

Resulting sign:	$\oplus$	$\ominus$	$\oplus$	$\ominus$
Sign of $-2x+7$:	$+$	$+$	$+$	$-$
Sign of $x-2$:	$-$	$-$	$+$	$+$
Sign of $x+1$:	$-$	$+$	$+$	$+$
x values:		-1	2	$7/2$

Diagram 45

47 $\dfrac{x}{3x-5}\leq\dfrac{2}{x-1} \Rightarrow \dfrac{x(x-1)-2(3x-5)}{(3x-5)(x-1)}\leq0 \Rightarrow \dfrac{(x-2)(x-5)}{(3x-5)(x-1)}\leq0;\ (1,\frac{5}{3})\cup[2,5]$

Res. sign:	$\oplus$	$\ominus$	$\oplus$	$\ominus$	$\oplus$
$x-5$:	$-$	$-$	$-$	$-$	$+$
$x-2$:	$-$	$-$	$-$	$+$	$+$
$3x-5$:	$-$	$-$	$+$	$+$	$+$
$x-1$:	$-$	$+$	$+$	$+$	$+$
x values:		1	$5/3$	2	5

Diagram 47

49 $x^3>x \Rightarrow x^3-x>0 \Rightarrow x(x^2-1)>0 \Rightarrow x(x+1)(x-1)>0;\ (-1,0)\cup(1,\infty)$

Resulting sign:	$\ominus$	$\oplus$	$\ominus$	$\oplus$
Sign of $x-1$:	$-$	$-$	$-$	$+$
Sign of x:	$-$	$-$	$+$	$+$
Sign of $x+1$:	$-$	$+$	$+$	$+$
x values:		-1	0	1

Diagram 49

51 (a) $|x+5| = 3 \Rightarrow x+5 = 3$ or $x+5 = -3 \Rightarrow x = -2$ or $x = -8$.

(b) $|x+5| < 3$ has solutions between the values found in part (a), that is, $(-8, -2)$.

(c) The solutions of $|x+5| > 3$ are the portions of the real line that are not in

parts (a) and (b), that is, $(-\infty, -8) \cup (-2, \infty)$.

53 We could think of this statement as "the difference between w and 148 is at most 2." In symbols, we have $|w - 148| \le 2$. Intuitively, we know that this inequality must describe the weights from 146 to 150.

55 We want to know what condition will assure us that an object's image is at least 3 times as large as the object, or, equivalently, when $M \ge 3$.

$M \ge 3 \{f = 6\} \Rightarrow \dfrac{6}{6-p} \ge 3 \Rightarrow 6 \ge 18 - 3p \{\text{since } 6 - p > 0, \text{ we can multiply}$

by $6 - p$ and not change the direction of the inequality$\} \Rightarrow 3p \ge 12 \Rightarrow$

$$p \ge 4, \text{ but } p < 6 \text{ since } p < f. \text{ Thus, } 4 \le p < 6.$$

57 Let x denote the number of years before A becomes more economical than B. The costs are the initial costs plus the yearly costs times the number of years.

$\text{Cost}_A < \text{Cost}_B \Rightarrow 50,000 + 4000x < 40,000 + 5500x \Rightarrow 10,000 < 1500x \Rightarrow$

$$x > \tfrac{20}{3}, \text{ or } 6\tfrac{2}{3} \text{ yr.}$$

59 $s > 9 \Rightarrow -16t^2 + 24t + 1 > 9 \Rightarrow -16t^2 + 24t - 8 > 0 \Rightarrow$

$2t^2 - 3t + 1 < 0 \{\text{divide by } -8\} \Rightarrow (2t-1)(t-1) < 0 \Rightarrow \tfrac{1}{2} < t < 1$.

$$\text{The dog is more than 9 ft off the ground for } 1 - \tfrac{1}{2} = \tfrac{1}{2} \text{ sec.}$$

61 $d < 75 \Rightarrow v + \tfrac{1}{20}v^2 < 75 \{\text{multiply by } 20\} \Rightarrow 20v + v^2 < 1500 \Rightarrow$

$v^2 + 20v - 1500 < 0 \Rightarrow (v+50)(v-30) < 0 \{\text{use a sign diagram}\} \Rightarrow$

$$-50 < v < 30 \Rightarrow 0 \le v < 30 \{\text{since } v \ge 0\}$$

63 (a) 5 ft 9 in $= 69$ in. In a 40 year period, a person's height will decrease by $40 \times 0.024 = 0.96$ in ≈ 1 in. The person will be approximately one inch shorter, or 5 ft 8 in. at age 70.

(b) 5 ft 6 in $= 66$ in. In 20 years, a person's height ($h = 66$) will change by $0.024 \times 20 = 0.48$ in. Thus, $66 - 0.48 \le h \le 66 + 0.48 \Rightarrow 65.52 \le h \le 66.48$.

65 The numerator is equal to zero when $x = 2, 3$ and the denominator is equal to zero when $x = \pm 1$. From the table on the next page, the expression

$$Y_1 = \frac{(2-x)(3x-9)}{(1-x)(x+1)}$$

is positive when $x \in [-2, -1) \cup (1, 2) \cup (3, 3.5]$.

x	Y_1	x	Y_1
-2.0	20	1.0	ERROR
-1.5	37.8	1.5	1.8
-1.0	ERROR	2.0	0
-0.5	-35	2.5	-0.1429
0.0	-18	3.0	0
0.5	-15	3.5	0.2

1.7 Exercises

$\boxed{1}$ (a) $x = -2$ is the line parallel to the y-axis that intersects the x-axis at $(-2,\, 0)$.

(b) $y = 3$ is the line parallel to the x-axis that intersects the y-axis at $(0,\, 3)$.

(c) $x \geq 0$ { x is zero or positive }

is the set of all points to the right of and on the y-axis.

(d) $xy > 0$ { x and y have the same sign, that is, either both are positive or both are

negative } is the set of all points in quadrants I and III.

(e) $y < 0$ { y is negative } is the set of all points below the x-axis.

(f) $x = 0$ is the set of all points on the y-axis.

$\boxed{3}$ (a) $A(4,\, -3),\ B(6,\, 2) \ \Rightarrow\ d(A,\, B) = \sqrt{(6-4)^2 + [2-(-3)]^2} = \sqrt{4+25} = \sqrt{29}$

(b) $M_{AB} = \left(\dfrac{4+6}{2},\, \dfrac{-3+2}{2} \right) = \left(5,\, -\dfrac{1}{2} \right)$

$\boxed{5}$ We need to show that the sides satisfy the Pythagorean theorem. Finding the distances, we have $d(A,\, B) = \sqrt{98}$, $d(B,\, C) = \sqrt{32}$, and $d(A,\, C) = \sqrt{130}$. Since $d(A,\, C)$ is the largest of the three values, it must be the hypotenuse, hence, we need to check if $d(A,\, C)^2 = d(A,\, B)^2 + d(B,\, C)^2$. Since $(\sqrt{130})^2 = (\sqrt{98})^2 + (\sqrt{32})^2$, we know that $\triangle ABC$ is a right triangle. The area of a triangle is given by $A = \frac{1}{2}(\text{base})(\text{height})$. We can use $d(B,\, C)$ for the base and $d(A,\, B)$ for the height. Hence, area $= \frac{1}{2}bh = \frac{1}{2}(\sqrt{32})(\sqrt{98}) = \frac{1}{2}(4\sqrt{2})(7\sqrt{2}) = \frac{1}{2}(28)(2) = 28$.

$\boxed{7}$ We need to show that all 4 sides are the same length. Checking, we find that $d(A,\, B) = d(B,\, C) = d(C,\, D) = d(D,\, A) = \sqrt{29}$. This guarantees that we have a rhombus { a parallelogram with 4 equal sides }. Thus, we also need to show that adjacent sides meet at right angles. This can be done by showing that two adjacent sides and a diagonal form a right triangle. Using $\triangle ABC$, we see that $d(A,\, C) = \sqrt{58}$ and hence $d(A,\, C)^2 = d(A,\, B)^2 + d(B,\, C)^2$. We conclude that $ABCD$ is a square.

9 Let $B = (x, y)$. $A(-3, 8)$ $\Rightarrow$ $M_{AB} = \left(\dfrac{-3+x}{2}, \dfrac{8+y}{2}\right)$. $M_{AB} = C(5, -10)$ $\Rightarrow$

$-3 + x = 2(5)$ and $8 + y = 2(-10)$ $\Rightarrow$ $x = 13$ and $y = -28$. $B = (13, -28)$.

11 The perpendicular bisector of AB is the line that passes through the midpoint of segment AB and intersects segment AB at a right angle. The points on the perpendicular bisector are all equidistant from A and B. Thus, we need to show that $d(A, C) = d(B, C)$. Since each of these is $\sqrt{145}$, we conclude that C is on the perpendicular bisector of AB.

13 Let $Q(0, y)$ be an arbitrary point on the y-axis.

Applying the distance formula with Q and $P(5, 3)$, we have $6 = d(P, Q)$ $\Rightarrow$

$6 = \sqrt{(0-5)^2 + (y-3)^2}$ $\Rightarrow$ $36 = 25 + y^2 - 6y + 9$ $\Rightarrow$

$y^2 - 6y - 2 = 0$ $\Rightarrow$ $y = 3 \pm \sqrt{11}$. The points are $(0, 3 + \sqrt{11})$ and $(0, 3 - \sqrt{11})$.

15 With $P(a, 3)$ and $Q(5, 2a)$, $d(P, Q) > \sqrt{26}$ $\Rightarrow$ $\sqrt{(5-a)^2 + (2a-3)^2} > \sqrt{26}$ $\Rightarrow$

$25 - 10a + a^2 + 4a^2 - 12a + 9 > 26$ $\Rightarrow$ $5a^2 - 22a + 8 > 0$ $\Rightarrow$ $(5a - 2)(a - 4) > 0$.

Using *Diagram 15*, we see that $a < \frac{2}{5}$ or $a > 4$ will assure us that $d(P, Q) > \sqrt{26}$.

Resulting sign:	$\oplus$	$\ominus$	$\oplus$
Sign of $5a - 2$:	$-$	$+$	$+$
Sign of $a - 4$:	$-$	$-$	$+$
a values:		2/5	4

Diagram 15

17 As in Example 3, we expect the graph to be a line.

Creating a table of values similar to those in the text, we have:

x	-2	-1	0	1	2
y	-7	-5	-3	-1	1

By plotting these points and connecting them, we obtain *Figure 17*.

To find the x-intercept, let $y = 0$ in $y = 2x - 3$, and solve for x to get 1.5.

To find the y-intercept, let $x = 0$ in $y = 2x - 3$, and solve for y to get -3.

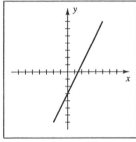

Figure 17

19 $y = -x + 1$ • x-intercept: $y = 0 \Rightarrow 0 = -x + 1 \Rightarrow x = 1$

y-intercept: $x = 0 \Rightarrow y = -0 + 1 = 1$

The graph is a line with x-intercept 1 and y-intercept 1.

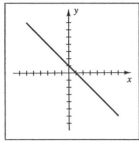

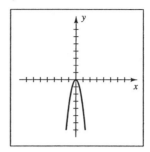

Figure 19   *Figure 21*

21 $y = -4x^2$ • Multiplying the y-values of $y = x^2$ by -4 gives us all negative y-values for $y = -4x^2$. The vertex of the parabola is at $(0, 0)$, so its x-intercept is 0 and its y-intercept is 0.

23 $y = 2x^2 - 1$ • x-intercepts: $y = 0 \Rightarrow 0 = 2x^2 - 1 \Rightarrow 1 = 2x^2 \Rightarrow$

$x^2 = \frac{1}{2} \Rightarrow x = \pm\sqrt{\frac{1}{2}}$, which can be written as $\pm\dfrac{1}{\sqrt{2}}$ or $\pm\dfrac{\sqrt{2}}{2}$ or $\pm\frac{1}{2}\sqrt{2}$

y-intercept: $x = 0 \Rightarrow y = -1$

Since we can substitute $-x$ for x in the equation and obtain an equivalent equation, we know the graph is symmetric with respect to the y-axis. We will make use of this fact when constructing our table. As in Example 4, we obtain a parabola.

x	± 2	$\pm\frac{3}{2}$	± 1	$\pm\frac{1}{2}$	0
y	7	$\frac{7}{2}$	1	$-\frac{1}{2}$	-1

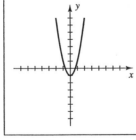

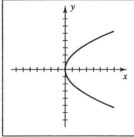

Figure 23   *Figure 25*

25 $x = \frac{1}{4}y^2$ • This graph is similar to the one in Example 7—multiplying by $\frac{1}{4}$ narrows the parabola $x = y^2$. The x-intercept is 0 and the y-intercept is 0.

$\boxed{27}$ $x = -y^2 + 3$ • x-intercept: $y = 0 \Rightarrow x = 3$;

y-intercepts: $x = 0 \Rightarrow 0 = -y^2 + 3 \Rightarrow y^2 = 3 \Rightarrow y = \pm\sqrt{3}$

Since we can substitute $-y$ for y in the equation and obtain an equivalent equation,
we know the graph is symmetric with respect to the x-axis. We will make use of this
fact when constructing our table. As in Example 7, we obtain a parabola.

x	-13	-6	-1	2	3
y	± 4	± 3	± 2	± 1	0

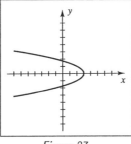

Figure 27   Figure 29

$\boxed{29}$ $y = -\frac{1}{2}x^3$ • The graph is similar to the graph of $y = \frac{1}{4}x^3$ in Example 8. The
negative has the effect of "flipping" the graph about the x-axis. The x-intercept is 0
and the y-intercept is 0.

$\boxed{31}$ $y = x^3 - 8$ • x-intercept: $y = 0 \Rightarrow 0 = x^3 - 8 \Rightarrow x^3 = 8 \Rightarrow x = \sqrt[3]{8} = 2$

y-intercept: $x = 0 \Rightarrow y = -8$

The effect of the -8 is to shift the graph of $y = x^3$ down 8 units.

x	-2	-1	0	1	2
y	-16	-9	-8	-7	0

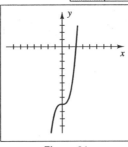

Figure 31   Figure 33

$\boxed{33}$ $y = \sqrt{x}$ • x-intercept 0; y-intercept 0

This is the top half of the parabola $x = y^2$.

An equation of the bottom half is $y = -\sqrt{x}$.

35 $y = \sqrt{x} - 4$ •

x-intercept: $y = 0 \Rightarrow 0 = \sqrt{x} - 4 \Rightarrow$

$\qquad\qquad 4 = \sqrt{x} \Rightarrow x = 16$

y-intercept: $x = 0 \Rightarrow y = -4$

x	0	1	4	9	16
y	-4	-3	-2	-1	0

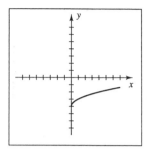

Figure 35

37 You may be able to do this exercise mentally. For example (using Exercise 17 with $y = 2x - 3$), we see that substituting $-x$ for x gives us $y = -2x - 3$; substituting $-y$ for y gives us $-y = 2x - 3$ or, equivalently, $y = -2x + 3$; and substituting $-x$ for x and $-y$ for y gives us $-y = -2x - 3$ or, equivalently, $y = 2x + 3$. None of the resulting equations are equivalent to the original equation, so there is no symmetry with respect to the y-axis, x-axis, or the origin.

(a) The graphs of the equations in Exercises 21 and 23 are symmetric with respect to the y-axis.

(b) The graphs of the equations in Exercises 25 and 27 are symmetric with respect to the x-axis.

(c) The graph of the equation in Exercise 29 is symmetric with respect to the origin.

39 $x^2 + y^2 = 11$ is a circle of radius $\sqrt{11}$ with center at the origin.

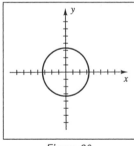   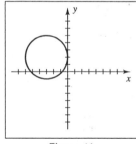   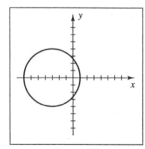

Figure 39   *Figure 41*   *Figure 43*

41 $(x + 3)^2 + (y - 2)^2 = 9$ is a circle of radius $r = \sqrt{9} = 3$ with center $C(-3, 2)$. A simple way to find the x- and y-coordinates of the center is to examine what values make the "squared" expressions equal to zero. For example, -3 makes the expression $(x + 3)$ equal to zero, so the x-coordinate of the center is -3.

43 $(x + 3)^2 + y^2 = 16$ is a circle of radius $r = \sqrt{16} = 4$ with center $C(-3, 0)$.

[45] $4x^2 + 4y^2 = 25$ $\Rightarrow$

$x^2 + y^2 = \frac{25}{4}$ is a circle of radius $r = \sqrt{\frac{25}{4}} = \frac{5}{2}$ with center $C(0, 0)$.

[47] As in Example 11, $y = -\sqrt{16 - x^2}$ is the lower half of the circle $x^2 + y^2 = 16$.

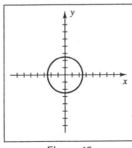   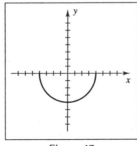   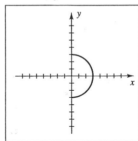

Figure 45   *Figure 47*   *Figure 49*

[49] $x = \sqrt{9 - y^2}$ is the right half of the circle $x^2 + y^2 = 9$.

[51] Center $C(2, -3)$, radius 5   •   An equation is $(x - 2)^2 + (y + 3)^2 = 5^2$,

or equivalently, $(x - 2)^2 + (y + 3)^2 = 25$.

[53] Center $C(\frac{1}{4}, 0)$, radius $\sqrt{5}$ • $(x - \frac{1}{4})^2 + y^2 = (\sqrt{5})^2 = 5$

[55] An equation of a circle with center $C(-4, 6)$ is $(x + 4)^2 + (y - 6)^2 = r^2$.

Since the circle passes through $P(1, 2)$, we know that $x = 1$ and $y = 2$ is one solution

of the general equation. Letting $x = 1$ and $y = 2$ yields $5^2 + (-4)^2 = r^2$ $\Rightarrow$ $r^2 = 41$.

An equation is $(x + 4)^2 + (y - 6)^2 = 41$.

[57] "Tangent to the y-axis" means that the circle will intersect the y-axis at exactly one

point. The distance from the center $C(-3, 6)$ to this point of tangency is 3 units—

this is the length of the radius of the circle. An equation is $(x + 3)^2 + (y - 6)^2 = 9$.

[59] Since the radius is 4 and $C(h, k)$ is in QII, $h = -4$ and $k = 4$.

An equation is $(x + 4)^2 + (y - 4)^2 = 16$.

[61] The center of the circle is the midpoint M of $A(4, -3)$ and $B(-2, 7)$. $M = (1, 2)$.

The radius of the circle is $\frac{1}{2} \cdot d(A, B) = \frac{1}{2}\sqrt{136} = \sqrt{34}$.

An equation is $(x - 1)^2 + (y - 2)^2 = 34$.

[63] $x^2 + y^2 - 4x + 6y = 36$ { complete the square on x and y } $\Rightarrow$

$x^2 - 4x + \underline{\;4\;} + y^2 + 6y + \underline{\;9\;} = 36 + \underline{\;4\;} + \underline{\;9\;}$ $\Rightarrow$

$(x - 2)^2 + (y + 3)^2 = 49$.   This is a circle with center $C(2, -3)$ and radius $r = 7$.

[65] $x^2 + y^2 + 4y - 117 = 0$ { complete the square on x and y } $\Rightarrow$

$x^2 + y^2 + 4y + \underline{\;4\;} = 117 + \underline{\;4\;}$ $\Rightarrow$ $x^2 + (y + 2)^2 = 121$.

This is a circle with center $C(0, -2)$ and radius $r = 11$.

$\boxed{67}$ $2x^2 + 2y^2 - 12x + 4y = 15$ { divide both sides by 2 } $\Rightarrow$

$x^2 + y^2 - 6x + 2y = \frac{15}{2}$ { complete the square on x and y } $\Rightarrow$

$x^2 - 6x + \underline{9} + y^2 + 2y + \underline{1} = \frac{15}{2} + \underline{9} + \underline{1}$ $\Rightarrow$

$(x-3)^2 + (y+1)^2 = \frac{35}{2}$. This is a circle with center $C(3, -1)$ and radius $r = \frac{1}{2}\sqrt{70}$.

$\boxed{69}$ $x^2 + y^2 + 4x - 2y + 5 = 0$ $\Rightarrow$ $x^2 + 4x + \underline{4} + y^2 - 2y + \underline{1} = -5 + \underline{4} + \underline{1}$ $\Rightarrow$

$(x+2)^2 + (y-1)^2 = 0$. $C(-2, 1)$; $r^2 = 0$ $\Rightarrow$ $r = 0$ (a point)

$\boxed{71}$ $x^2 + y^2 - 2x - 8y + 19 = 0$ $\Rightarrow$

$x^2 - 2x + \underline{1} + y^2 - 8y + \underline{16} = -19 + \underline{1} + \underline{16}$ $\Rightarrow$

$(x-1)^2 + (y-4)^2 = -2$. This is not a circle since r^2 cannot equal -2.

$\boxed{73}$ To obtain equations for the upper and lower halves, we solve the given equation for y

in terms of x. $x^2 + y^2 = 36$ $\Rightarrow$ $y^2 = 36 - x^2$ $\Rightarrow$ $y = \pm\sqrt{36 - x^2}$.

The upper half is $y = \sqrt{36 - x^2}$ and the lower half is $y = -\sqrt{36 - x^2}$.

To obtain equations for the right and left halves, we solve for x in terms of y.

$x^2 + y^2 = 36$ $\Rightarrow$ $x^2 = 36 - y^2$ $\Rightarrow$ $x = \pm\sqrt{36 - y^2}$.

The right half is $x = \sqrt{36 - y^2}$ and the left half is $x = -\sqrt{36 - y^2}$.

$\boxed{75}$ To obtain equations for the upper and lower halves, we solve the given equation for y

in terms of x. $(x-2)^2 + (y+1)^2 = 49$ $\Rightarrow$ $(y+1)^2 = 49 - (x-2)^2$ $\Rightarrow$

$y + 1 = \pm\sqrt{49 - (x-2)^2}$ $\Rightarrow$ $y = -1 \pm \sqrt{49 - (x-2)^2}$.

The upper half is $y = -1 + \sqrt{49 - (x-2)^2}$ and

the lower half is $y = -1 - \sqrt{49 - (x-2)^2}$.

To obtain equations for the right and left halves, we solve for x in terms of y.

$(x-2)^2 + (y+1)^2 = 49$ $\Rightarrow$ $(x-2)^2 = 49 - (y+1)^2$ $\Rightarrow$

$x - 2 = \pm\sqrt{49 - (y+1)^2}$ $\Rightarrow$ $x = 2 \pm \sqrt{49 - (y+1)^2}$. The right half is

$x = 2 + \sqrt{49 - (y+1)^2}$ and the left half is $x = 2 - \sqrt{49 - (y+1)^2}$.

$\boxed{77}$ We need to determine if the distance from P to C is *less than r*, *greater than r*, or

equal to r and hence, P will be *inside* the circle, *outside* the circle, or *on* the circle,

respectively.

(a) $P(2, 3)$, $C(4, 6)$ $\Rightarrow$ $d(P, C) = \sqrt{4+9} = \sqrt{13} < r$ { $r = 4$ } $\Rightarrow$ P is *inside* C.

(b) $P(4, 2)$, $C(1, -2)$ $\Rightarrow$ $d(P, C) = \sqrt{9 + 16} = 5 = r$ { $r = 5$ } $\Rightarrow$ P is *on* C.

(c) $P(-3, 5)$, $C(2, 1)$ $\Rightarrow$ $d(P, C) = \sqrt{25 + 16} = \sqrt{41} > r$ { $r = 6$ } $\Rightarrow$

P is *outside* C.

$\boxed{79}$ (a) To find the x-intercepts, let $y = 0$ and solve the resulting equation for x.

$$x^2 - 4x + 4 = 0 \;\Rightarrow\; (x - 2)^2 = 0 \;\Rightarrow\; x = 2.$$

(b) To find the y-intercepts, let $x = 0$ and solve the resulting equation for y.

$$y^2 - 6y + 4 = 0 \;\Rightarrow\; y = \frac{6 \pm \sqrt{36 - 16}}{2} = 3 \pm \sqrt{5}.$$

$\boxed{81}$ $x^2 + y^2 + 4x - 6y + 4 = 0 \;\Leftrightarrow\; (x + 2)^2 + (y - 3)^2 = 9$. This is a circle with

center $C(-2,\,3)$ and radius 3. The circle we want has the same center, $C(-2,\,3)$,

and radius that is equal to the distance from C to $P(2,\,6)$.

$$d(P,\,C) = \sqrt{16 + 9} = 5 \text{ and an equation is } (x + 2)^2 + (y - 3)^2 = 25.$$

$\boxed{83}$ The equation of circle C_2 is $(x - h)^2 + (y - 2)^2 = 2^2$. If we draw a line from the

origin to the center of C_2, we form a right triangle with hypotenuse $5 - 2$ $\{C_2$

radius $- C_1$ radius$\} = 3$ and sides of length 2 and h.   Thus,

$h^2 + 2^2 = 3^2 \;\Rightarrow\; h = \sqrt{5}.$

$\boxed{85}$ Assuming that the x- and y-values of each point of intersection are integers, and that

the ticks each represent one unit, we see that the intersection points are $(-3,\,-5)$ and

$(2,\,0)$. The viewing rectangle (VR) is $[-15,\,15]$ by $[-10,\,10]$.

$$Y_1 < Y_2 \text{ on } [-15,\,-3) \cup (2,\,15].$$

$\boxed{87}$ VR: $[-5,\,5]$ by $[-1.3,\,5.3]$. The intersection points are $(-1,\,1)$, $(0,\,0)$, and $(1,\,1)$.

$$Y_1 > Y_2 \text{ on } [-5,\,-1) \cup (1,\,5]. \quad Y_1 < Y_2 \text{ on } (-1,\,0) \cup (0,\,1).$$

$\boxed{89}$ (a) Plot the points $(1990,\ 54{,}871{,}330)$, $(1991,\ 55{,}786{,}390)$, $(1992,\ 57{,}211{,}600)$,

$(1993,\ 58{,}834{,}440)$, and $(1994,\ 59{,}332{,}200)$.

$$[1988,\ 1996] \text{ by } [54 \times 10^6,\ 61 \times 10^6,\ 10^6]$$

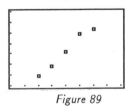

Figure 89

(b) The number of cable subscribers is increasing each year.

91 The viewing rectangles significantly affect the shape of the circle. Viewing rectangle (2) results in a graph that most looks like a circle.

$[-2, 2]$ by $[-2, 2]$

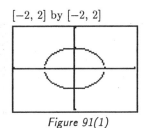

Figure 91(1)

$[-3, 3]$ by $[-2, 2]$

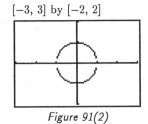

Figure 91(2)

$[-2, 2]$ by $[-5, 5]$

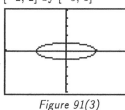

Figure 91(3)

$[-5, 5]$ by $[-2, 2]$

Figure 91(4)

93 Assign $x^3 - \frac{9}{10}x^2 - \frac{43}{25}x + \frac{24}{25}$ to Y_1. After trying a standard viewing rectangle, we see that the x-intercepts are near the origin and we choose the viewing rectangle $[-6, 6]$ by $[-4, 4]$. This is simply one choice, not necessarily the best choice. For most graphing calculator exercises, we have selected viewing rectangles that are in a $3:2$ proportion (horizontal:vertical) to maintain a true proportion. From the graph, there are three x-intercepts. Use a root feature to determine that they are -1.2, 0.5, and 1.6.

$[-6, 6]$ by $[-4, 4]$

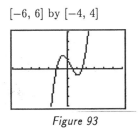

Figure 93

$[-3, 3]$ by $[-2, 2]$

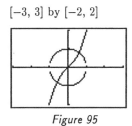

Figure 95

95 Make the assignments $Y_1 = x^3 + x$, $Y_2 = \sqrt{1 - x^2}$, and $Y_3 = -Y_2$.

From the graph, there are two points of intersection.

They are approximately $(0.6, 0.8)$ and $(-0.6, -0.8)$.

97 Depending on the type of graphing utility used, you may need to solve for y first.

$$x^2 + (y-1)^2 = 1 \Rightarrow y = 1 \pm \sqrt{1-x^2}; \quad (x-\tfrac{5}{4})^2 + y^2 = 1 \Rightarrow y = \pm\sqrt{1-(x-\tfrac{5}{4})^2}.$$

Make the assignments $Y_1 = \sqrt{1-x^2}$, $Y_2 = 1 + Y_1$, $Y_3 = 1 - Y_1$, $Y_4 = \sqrt{1-(x-\tfrac{5}{4})^2}$, and $Y_5 = -Y_4$. Be sure to "turn off" Y_1 before graphing. From the graph, there are two points of intersection. They are approximately $(0.999, 0.968)$ and $(0.251, 0.032)$.

$[-3, 3]$ by $[-2, 2]$

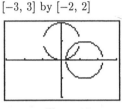

Figure 97

99 The cars are initially 4 miles apart. The distance between them decreases to 0 when they meet on the highway after 2 minutes. Then, the distance between them starts to increase until it is 4 miles after a total of 4 minutes.

$[0, 4]$ by $[0, 4]$ $[-50, 50, 10]$ by $[900, 1200, 100]$

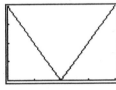   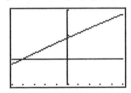

Figure 99   *Figure 101*

101   (a) $T = 20°C$ and $v = 1087\sqrt{\dfrac{T+273}{273}} \Rightarrow v = 1087\sqrt{\dfrac{20+273}{273}} \approx 1126$ ft/sec.

(b) Algebraically: $v = 1000 \Rightarrow 1000 = 1087\sqrt{\dfrac{T+273}{273}} \Rightarrow$

$$\sqrt{\dfrac{T+273}{273}} = \dfrac{1000}{1087} \Rightarrow \dfrac{T+273}{273} = \dfrac{1000^2}{1087^2} \Rightarrow T+273 = \dfrac{273 \cdot 1000^2}{1087^2} \Rightarrow$$

$$T = \dfrac{273 \cdot 1000^2}{1087^2} - 273 \approx -42°C.$$

Graphically: Graph $Y_1 = 1087\sqrt{(T+273)/273}$ and $Y_2 = 1000$.

At the point of their intersection, $T \approx -42°C.$

1.8 Exercises

$\boxed{1}$ $A(-3, 2)$, $B(5, -4)$ $\Rightarrow$ $m_{AB} = \dfrac{y_2 - y_1}{x_2 - x_1} = \dfrac{(-4) - 2}{5 - (-3)} = \dfrac{-6}{8} = -\dfrac{3}{4}$

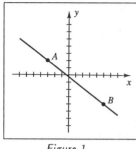

Figure 1

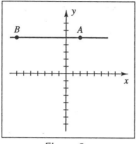

Figure 3

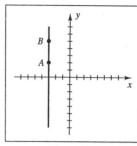

Figure 5

$\boxed{3}$ $A(2, 5)$, $B(-7, 5)$ $\Rightarrow$ $m_{AB} = \dfrac{y_2 - y_1}{x_2 - x_1} = \dfrac{5 - 5}{-7 - 2} = \dfrac{0}{-9} = 0$ { horizontal line }

$\boxed{5}$ $A(-3, 2)$, $B(-3, 5)$ $\Rightarrow$ $m_{AB} = \dfrac{y_2 - y_1}{x_2 - x_1} = \dfrac{5 - 2}{-3 - (-3)} = \dfrac{3}{0}$ $\Rightarrow$

m is undefined { vertical line }

$\boxed{7}$ To show that the polygon is a parallelogram, we must show that the slopes of opposite sides are equal. $A(-3, 1)$, $B(5, 3)$, $C(3, 0)$, $D(-5, -2)$ $\Rightarrow$

$$m_{AB} = \tfrac{1}{4} = m_{DC} \quad \text{and} \quad m_{DA} = \tfrac{3}{2} = m_{CB}.$$

$\boxed{9}$ To show that the polygon is a rectangle, we must show that the slopes of opposite sides are equal (parallel lines) and the slopes of two adjacent sides are negative reciprocals (perpendicular lines). $A(6, 15)$, $B(11, 12)$, $C(-1, -8)$, $D(-6, -5)$ $\Rightarrow$

$$m_{DA} = \tfrac{5}{3} = m_{CB} \quad \text{and} \quad m_{AB} = -\tfrac{3}{5} = m_{DC}.$$

$\boxed{11}$ $A(-1, -3)$ is 5 units to the left and 5 units down from $B(4, 2)$. The fourth vertex D will have the same relative position from $C(-7, 5)$, that is, 5 units to the left and 5 units down from C. Its coordinates are $(-7 - 5, 5 - 5) = (-12, 0)$.

$\boxed{13}$ $m = 3, -2, \tfrac{2}{3}, -\tfrac{1}{4}$ •

Lines with equation $y = mx$ pass through the origin.

Draw lines through the origin with slopes

3 { rise = 3, run = 1 }, -2 { rise = -2, run = 1 },

$\tfrac{2}{3}$ { rise = 2, run = 3 }, $-\tfrac{1}{4}$ { rise = -1, run = 4 }.

A negative "rise" can be thought of as a "drop."

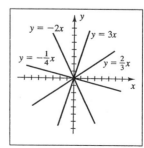

Figure 13

15 Draw lines through the point $P(3, 1)$ with slopes $\frac{1}{2}$ { rise $= 1$, run $= 2$ },

-1 { rise $= -1$, run $= 1$ }, and $-\frac{1}{5}$ { rise $= -1$, run $= 5$ }.

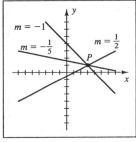   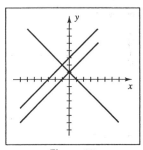

Figure 15   Figure 17

17 The line $y = 1x + 3$ has slope 1 and y-intercept 3.

The line $y = 1x + 1$ has slope 1 and y-intercept 1.

The line $y = -1x + 1$ has slope -1 and y-intercept 1.

19 (a) "Parallel to the y-axis" implies the equation is of the form $x = k$.

The x-value of $A(5, -2)$ is 5, hence $x = 5$ is the equation.

(b) "Perpendicular to the y-axis" implies the equation is of the form $y = k$.

The y-value of $A(5, -2)$ is -2, hence $y = -2$ is the equation.

21 Using the point-slope form, the equation of the line through $A(5, -3)$ with slope -4

is $y + 3 = -4(x - 5)$ $\Rightarrow$ $y + 3 = -4x + 20$ $\Rightarrow$ $4x + y = 17$.

23 $A(4, 0)$; slope -3 { use the point-slope form of a line } $\Rightarrow$

$y - 0 = -3(x - 4)$ $\Rightarrow$ $y = -3x + 12$ $\Rightarrow$ $3x + y = 12$.

25 $A(4, -5)$, $B(-3, 6)$ $\Rightarrow$ $m_{AB} = -\frac{11}{7}$. By the point-slope form, an equation of the

line is $y + 5 = -\frac{11}{7}(x - 4)$ $\Rightarrow$ $7(y + 5) = -11(x - 4)$ $\Rightarrow$

$7y + 35 = -11x + 44$ $\Rightarrow$ $11x + 7y = 9$.

27 $5x - 2y = 4$ $\Leftrightarrow$ $y = \frac{5}{2}x - 2$. Using the same slope, $\frac{5}{2}$, with $A(2, -4)$, gives us

$y + 4 = \frac{5}{2}(x - 2)$ $\Rightarrow$ $2(y + 4) = 5(x - 2)$ $\Rightarrow$ $2y + 8 = 5x - 10$ $\Rightarrow$ $5x - 2y = 18$.

29 $2x - 5y = 8$ $\Leftrightarrow$ $y = \frac{2}{5}x - \frac{8}{5}$. The slope of this line is $\frac{2}{5}$, so we'll use the negative

reciprocal, $-\frac{5}{2}$, for the slope of the new line. $y + 3 = -\frac{5}{2}(x - 7)$ $\Rightarrow$

$2(y + 3) = -5(x - 7)$ $\Rightarrow$ $2y + 6 = -5x + 35$ $\Rightarrow$ $5x + 2y = 29$.

31 $A(4, 0)$, $B(0, -3)$ $\Rightarrow$ $m = \frac{3}{4}$.

Using the slope-intercept form with $b = -3$ gives us $y = \frac{3}{4}x - 3$.

33 $A(5, 2)$, $B(-1, 4)$ $\Rightarrow$ $m = -\frac{1}{3}$.

$y - 2 = -\frac{1}{3}(x - 5)$ $\Rightarrow$ $y = -\frac{1}{3}x + \frac{5}{3} + 2$ $\Rightarrow$ $y = -\frac{1}{3}x + \frac{11}{3}$.

35 We need the line through the midpoint M of segment AB that is perpendicular to

segment AB. $A(3, -1)$, $B(-2, 6)$ $\Rightarrow$ $M_{AB} = (\frac{1}{2}, \frac{5}{2})$ and $m_{AB} = -\frac{7}{5}$.

$$y - \tfrac{5}{2} = \tfrac{5}{7}(x - \tfrac{1}{2}) \;\Rightarrow\; 7(y - \tfrac{5}{2}) = 5(x - \tfrac{1}{2}) \;\Rightarrow\; 7y - \tfrac{35}{2} = 5x - \tfrac{5}{2} \;\Rightarrow\; 5x - 7y = -15.$$

37 An equation of the line with slope -1 through the origin is

$$y - 0 = -1(x - 0), \text{ or } y = -x.$$

39 We can solve the given equation for y to obtain the slope-intercept form, $y = mx + b$.

$$2x = 15 - 3y \;\Rightarrow\; 3y = -2x + 15 \;\Rightarrow\; y = -\tfrac{2}{3}x + 5; \; m = -\tfrac{2}{3}, \, b = 5$$

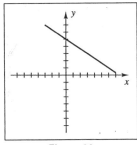

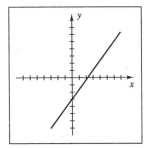

Figure 39   *Figure 41*

41 $4x - 3y = 9 \;\Rightarrow\; -3y = -4x + 9 \;\Rightarrow\; y = \tfrac{4}{3}x - 3; \; m = \tfrac{4}{3}, \, b = -3$

43 (a) An equation of the horizontal line with y-intercept 3 is $y = 3$.

   (b) An equation of the line through the origin with slope $-\tfrac{1}{2}$ is $y = -\tfrac{1}{2}x$.

   (c) An equation of the line with slope $-\tfrac{3}{2}$ and y-intercept 1 is $y = -\tfrac{3}{2}x + 1$.

   (d) An equation of the line through $(3, -2)$ with slope -1 is $y + 2 = -(x - 3)$.

             Alternatively, we have a slope of -1 and a y-intercept of 1, i.e., $y = -x + 1$.

45 Since we want to obtain a "1" on the right side of the equation, we will divide by 6.

$$[4x - 2y = 6] \cdot \tfrac{1}{6} \;\Rightarrow\; \frac{4x}{6} - \frac{2y}{6} = \frac{6}{6} \;\Rightarrow\; \frac{2x}{3} - \frac{y}{3} = 1 \;\Rightarrow\; \frac{x}{\frac{3}{2}} + \frac{y}{-3} = 1$$

                             The x-intercept is $\tfrac{3}{2}$ and the y-intercept is -3.

47 The radius of the circle is the vertical distance from the center of the circle to the line

          $y = 5$, that is, $r = 5 - (-2) = 7$. An equation is $(x - 3)^2 + (y + 2)^2 = 49$.

49 $L = 28 \;\Rightarrow\; 1.53t - 6.7 = 28 \;\Rightarrow\; t = \dfrac{28 + 6.7}{1.53} \approx 22.68$, or approximately 23 weeks.

51 (a) $L = 40 \;\Rightarrow\; W = 1.70(40) - 42.8 = 25.2$ tons

   (b) Error in $L = \pm 2 \;\Rightarrow\;$ Error in $W = 1.70(\pm 2) = \pm 3.4$ tons

53 (a) $y = mx = \dfrac{\text{change in } y \text{ from the beginning of the season}}{\text{change in } x \text{ from the beginning of the season}}(x) = \dfrac{5 - 0}{14 - 0}x = \dfrac{5}{14}x$.

   (b) $x = 162 \;\Rightarrow\; y = \tfrac{5}{14}(162) \approx 58$.

55 (a) Using the slope-intercept form,

$$W = mt + b = mt + 10.$$

$W = 30$ when $t = 3$ $\Rightarrow$ $30 = 3m + 10$ $\Rightarrow$

$m = \frac{20}{3}$ and $W = \frac{20}{3}t + 10$.

(b) $t = 6$ $\Rightarrow$ $W = \frac{20}{3}(6) + 10$ $\Rightarrow$ $W = 50$ lb

(c) $W = 70$ $\Rightarrow$ $70 = \frac{20}{3}t + 10$ $\Rightarrow$ $60 = \frac{20}{3}t$ $\Rightarrow$

$t = 9$ years old

(d) The graph has endpoints at $(0, 10)$ and $(12, 90)$.

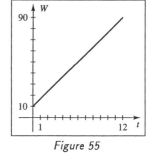

Figure 55

57 Using $(10, 2480)$ and $(25, 2440)$ { since an increase of 15°C lowers H by 40 }, we have

$$H - 2440 = \frac{2440 - 2480}{25 - 10}(T - 25) \;\Rightarrow\; H - 2440 = -\frac{8}{3}(T - 25) \;\Rightarrow$$

$$H - \frac{7320}{3} = -\frac{8}{3}T + \frac{200}{3} \;\Rightarrow\; H = -\frac{8}{3}T + \frac{7520}{3}.$$

59 (a) Using the slope-intercept form with $m = 0.032$ and $b = 13.5$,

we have $T = 0.032t + 13.5$.

(b) $t = 2000 - 1915 = 85$ $\Rightarrow$ $T = 0.032(85) + 13.5 = 16.22$°C.

61 (a) Expenses $(E) = (\$1000) + (5\% \text{ of } R) + (\$2600) + (50\% \text{ of } R)$ $\Rightarrow$

$$E = 1000 + 0.05R + 2600 + 0.50R \;\Rightarrow\; E = 0.55R + 3600.$$

(b) Profit $(P) = $ Revenue $(R) - $ Expenses (E) $\Rightarrow$ $P = R - (0.55R + 3600)$ $\Rightarrow$

$$P = R - 0.55R - 3600 \;\Rightarrow\; P = 0.45R - 3600.$$

(c) *Break even* means P would be 0. $P = 0$ $\Rightarrow$ $0 = 0.45R - 3600$ $\Rightarrow$

$$0.45R = 36''0 \;\Rightarrow\; \tfrac{45}{100}R = 3600 \;\Rightarrow\; R = 3600(\tfrac{100}{45}) = \$8000/\text{month}$$

63 The targets are on the x-ax   { which is the line $y = 0$ }.

To determine if a target is hit, set $y = 0$ and solve for x.

(a) $y - 2 = -1(x - 1)$ $\Rightarrow$ $x + y = 3$. $y = 0$ $\Rightarrow$ $x = 3$ and a creature is hit.

(b) $y - \frac{5}{3} = -\frac{4}{9}(x - \frac{3}{2})$ $\Rightarrow$ $4x + 9y = 21$. $y = 0$ $\Rightarrow$ $x = 5.25$ and no creature is hit.

65 $s = \dfrac{v_2 - v_1}{h_2 - h_1}$ $\Rightarrow$ $0.07 = \dfrac{v_2 - 22}{185 - 0}$ $\Rightarrow$ $v_2 = 22 + 0.07(185) = 34.95$ mi/hr.

67 The slope of AB is $\dfrac{-1.11905 - (-1.3598)}{-0.55 - (-1.3)} = 0.321$. Similarly, the slopes of BC and

CD are also 0.321. Therefore, the points all lie on the same line. Since the common

slope is 0.321, let $a = 0.321$. $y = 0.321x + b$ $\Rightarrow$ $-1.3598 = 0.321(-1.3) + b$ $\Rightarrow$

$b = -0.9425$. Thus, the points are linearly related by the equation

$$y = 0.321x - 0.9425.$$

69 $x - 3y = -58 \Leftrightarrow y = (x + 58)/3$ and $3x - y = -70 \Leftrightarrow y = 3x + 70$. Assign $(x + 58)/3$ to Y_1 and $3x + 70$ to Y_2. Using a standard viewing rectangle, we don't see the lines. Zooming out gives us an indication where the lines intersect and by using an intersect feature, we find that the lines intersect at $(-19, 13)$.

$[-30, 3, 2]$ by $[-2, 20, 2]$ $[-15, 15]$ by $[-10, 10]$

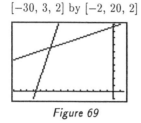   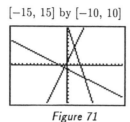

Figure 69   Figure 71

71 From the graph, we can see that the points of intersection are $A(-0.8, -0.6)$, $B(4.8, -3.4)$, and $C(2, 5)$. The lines intersecting at A are perpendicular since they have slopes of 2 and $-\frac{1}{2}$. Since $d(A, B) = \sqrt{39.2}$ and $d(A, C) = \sqrt{39.2}$, the triangle is isosceles. Thus, the polygon is a right isosceles triangle.

73 The data appear to be linear. Using the two arbitrary points $(-7, -25)$ and $(4.6, 12.2)$, the slope of the line is $\dfrac{12.2 - (-25)}{4.6 - (-7)} \approx 3.2$. An equation of the line is $y + 25 = 3.2(x + 7) \Rightarrow y = 3.2x - 2.6$. If we find the regression line on a calculator, we get the model $y \approx 3.20303x - 2.51553$.

$[-8, 5]$ by $[-27, 15, 5]$ $[1980, 1988]$ by $[300, 625, 100]$

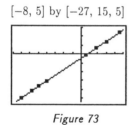   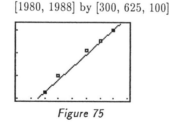

Figure 73   Figure 75

75 (a) Plot the points with the form (Year, Cost):

  $(1982, 325)$, $(1983, 400)$, $(1985, 510)$, $(1986, 550)$, and $(1987, 600)$

  (b) To find a first approximation for the line use the arbitrary points $(1982, 325)$ and $(1987, 600)$. The resulting line is $y = 55x - 108{,}685$. Adjustments may be made to this equation. If we find the regression line on a calculator, we get the model $y \approx 54.01163x - 106{,}714.477$.

  (c) Let $y = 55x - 108{,}685$. When $x = 1984$, $y = 435$ and when $x = 1995$, $y = 1040$.

1 If $x \leq -3$, then $x + 3 \leq 0$, and $|x + 3| = -(x + 3) = -x - 3$.

2 If $2 < x < 3$, then $x - 2 > 0$ { $x - 2$ is positive } and $x - 3 < 0$ { $x - 3$ is negative }.

Thus, $(x - 2)(x - 3) < 0$ { positive times negative is negative }, and since the absolute

value of an expression that is negative is the negative of the expression,

$$|(x - 2)(x - 3)| = -(x - 2)(x - 3), \text{ or, equivalently, } (2 - x)(x - 3).$$

3 $\left(\dfrac{a^{2/3}b^{3/2}}{a^2 b}\right)^6 = \dfrac{\left(a^{2/3}\right)^6 \left(b^{3/2}\right)^6}{\left(a^2 b\right)^6} = \dfrac{a^4 b^9}{a^{12} b^6} = \dfrac{b^3}{a^8}$

5 $\left(\dfrac{xy^{-1}}{\sqrt{z}}\right)^4 \div \left(\dfrac{x^{1/3}y^2}{z}\right)^3 = \dfrac{x^4 y^{-4}}{z^2} \cdot \dfrac{z^3}{xy^6} = \dfrac{x^3 z}{y^{10}}$

7 $\left[(a^{2/3}b^{-2})^3\right]^{-1} = (a^2 b^{-6})^{-1} = a^{-2} b^6 = \dfrac{b^6}{a^2}$

9 $\sqrt[3]{8x^5 y^3 z^4} = \sqrt[3]{(8x^3 y^3 z^3)(x^2 z)} = \sqrt[3]{8x^3 y^3 z^3}\,\sqrt[3]{x^2 z} = 2xyz\sqrt[3]{x^2 z}$

11 $\dfrac{1}{\sqrt{t}}\left(\dfrac{1}{\sqrt{t}} - 1\right) = \dfrac{1}{\sqrt{t}}\left(\dfrac{1}{\sqrt{t}} - \dfrac{\sqrt{t}}{\sqrt{t}}\right) = \dfrac{1}{\sqrt{t}}\left(\dfrac{1 - \sqrt{t}}{\sqrt{t}}\right) = \dfrac{1 - \sqrt{t}}{t}$

13 $\dfrac{\sqrt{12x^4 y}}{\sqrt{3x^2 y^5}} = \sqrt{\dfrac{12x^4 y}{3x^2 y^5}} = \sqrt{\dfrac{4x^2}{y^4}} = \dfrac{2x}{y^2}$

15 $(3x^3 - 4x^2 + x - 7) + (x^4 - 2x^3 + 3x^2 + 5) = x^4 + x^3 - x^2 + x - 2$

17 $(3a - 5b)(2a + 7b) = 6a^2 + 11ab - 35b^2$

19 $(13a^2 + 4b)(13a^2 - 4b) = (13a^2)^2 - (4b^2)^2 = 169a^4 - 16b^2$

21 $(3x + 2y)^2(3x - 2y)^2 = \left[(3x + 2y)(3x - 2y)\right]^2 = (9x^2 - 4y^2)^2 = 81x^4 - 72x^2 y^2 + 16y^4$

23 $60xw + 70w = 10w(6x + 7)$

25 $8x^3 + 64y^3 = 8(x^3 + 8y^3) = 8\left[(x)^3 + (2y)^3\right] = 8(x + 2y)(x^2 - 2xy + 4y^2)$

27 $p^8 - q^8 = (p^4)^2 - (q^4)^2$

$= (p^4 + q^4)(p^4 - q^4)$

$= (p^4 + q^4)(p^2 + q^2)(p^2 - q^2)$

$= (p^4 + q^4)(p^2 + q^2)(p + q)(p - q)$

29 $x^2 - 49y^2 - 14x + 49 = (x^2 - 14x + 49) - 49y^2$

$= (x - 7)^2 - (7y)^2$

$= (x - 7 + 7y)(x - 7 - 7y)$

31 $\dfrac{2}{4x - 5} - \dfrac{5}{10x + 1} = \dfrac{2(10x + 1) - 5(4x - 5)}{(4x - 5)(10x + 1)} = \dfrac{20x + 2 - 20x + 25}{(4x - 5)(10x + 1)} = \dfrac{27}{(4x - 5)(10x + 1)}$

$\boxed{33}$ $\dfrac{x + x^{-2}}{1 + x^{-2}} = \dfrac{x + \dfrac{1}{x^2}}{1 + \dfrac{1}{x^2}} = \dfrac{\left(x + \dfrac{1}{x^2}\right) \cdot x^2}{\left(1 + \dfrac{1}{x^2}\right) \cdot x^2} = \dfrac{x^3 + 1}{x^2 + 1}$. We could factor the numerator,

but since it doesn't lead to a reduction of the fraction, we leave it in this form.

$\boxed{35}$ $\dfrac{\dfrac{x}{x+2} - \dfrac{4}{x+2}}{x - 3 - \dfrac{6}{x+2}} = \dfrac{\dfrac{x-4}{x+2}}{\dfrac{(x-3)(x+2) - 6}{x+2}} = \dfrac{x-4}{x^2 - x - 12} = \dfrac{x-4}{(x+3)(x-4)} = \dfrac{1}{x+3}$

$\boxed{36}$ $\dfrac{(4 - x^2)(\frac{1}{3})(6x+1)^{-2/3}(6) - (6x+1)^{1/3}(-2x)}{(4 - x^2)^2} = \dfrac{2(6x+1)^{-2/3}[(4 - x^2) + x(6x+1)]}{(4 - x^2)^2}$

$$= \dfrac{2[4 - x^2 + 6x^2 + x]}{(6x+1)^{2/3}(4 - x^2)^2}$$

$$= \dfrac{2(5x^2 + x + 4)}{(6x+1)^{2/3}(4 - x^2)^2}$$

$\boxed{37}$ $\left[\dfrac{3x+1}{5x+7} = \dfrac{6x+11}{10x-3}\right] \cdot (5x+7)(10x-3) \;\Rightarrow\; (3x+1)(10x-3) =$

$(6x+11)(5x+7) \;\Rightarrow\; 30x^2 + x - 3 = 30x^2 + 97x + 77 \;\Rightarrow\; 96x = -80 \;\Rightarrow\; x = -\frac{5}{6}$

$\boxed{39}$ $x(3x+4) = 5 \;\Rightarrow\; 3x^2 + 4x - 5 = 0 \;\Rightarrow$

$$x = \dfrac{-4 \pm \sqrt{16 + 60}}{6} = \dfrac{-4 \pm 2\sqrt{19}}{6} = -\dfrac{2}{3} \pm \dfrac{1}{3}\sqrt{19}$$

$\boxed{41}$ $20x^3 + 8x^2 - 35x - 14 = 0 \;\Rightarrow\; 4x^2(5x+2) - 7(5x+2) = 0 \;\Rightarrow$

$$(4x^2 - 7)(5x + 2) = 0 \;\Rightarrow\; x = \pm\dfrac{1}{2}\sqrt{7},\, -\dfrac{2}{5}$$

$\boxed{43}$ $2|2x+1| + 1 = 19 \;\Rightarrow\; 2|2x+1| = 18 \;\Rightarrow\; |2x+1| = 9 \;\Rightarrow$

$2x + 1 = 9$ or $2x + 1 = -9 \;\Rightarrow\; 2x = 8$ or $2x = -10 \;\Rightarrow\; x = 4$ or $x = -5$

$\boxed{44}$ $\left[\dfrac{1}{x} + 6 = \dfrac{5}{\sqrt{x}}\right] \cdot x \;\Rightarrow\; 1 + 6x = 5\sqrt{x} \;\Rightarrow$

$6x - 5\sqrt{x} + 1 = 0$ {factoring or substituting would be appropriate} $\Rightarrow$

$$(2\sqrt{x} - 1)(3\sqrt{x} - 1) = 0 \;\Rightarrow\; \sqrt{x} = \tfrac{1}{2}, \tfrac{1}{3} \;\Rightarrow\; x = \tfrac{1}{4}, \tfrac{1}{9}$$

Check $x = \frac{1}{4}$: LS $= 4 + 6 = 10$; RS $= 5/\frac{1}{2} = 10 \;\Rightarrow\; x = \frac{1}{4}$ is a solution.

Check $x = \frac{1}{9}$: LS $= 9 + 6 = 15$; RS $= 5/\frac{1}{3} = 15 \;\Rightarrow\; x = \frac{1}{9}$ is a solution.

$\boxed{45}$ $\sqrt{7x+2} + x = 6 \;\Rightarrow\; (\sqrt{7x+2})^2 = (6 - x)^2 \;\Rightarrow\; 7x + 2 = 36 - 12x + x^2 \;\Rightarrow$

$x^2 - 19x + 34 = 0 \;\Rightarrow\; (x - 2)(x - 17) = 0 \;\Rightarrow$

$x = 2$ and 17 is an extraneous solution.

$\boxed{47}$ $10 - 7x < 4 + 2x \;\Rightarrow\; -9x < -6 \;\Rightarrow\; x > \frac{2}{3} \;\Leftrightarrow\; (\frac{2}{3}, \infty)$

$\boxed{49}$ $\dfrac{6}{10x+3} < 0 \;\Rightarrow\; 10x + 3 < 0$ {since $6 > 0$} $\;\Rightarrow\; x < -\frac{3}{10} \;\Leftrightarrow\; (-\infty, -\frac{3}{10})$

$\boxed{51}$ $2\,|\,3-x\,|\,+1>5 \;\Rightarrow\; 2\,|\,3-x\,|\,>4 \;\Rightarrow\; |\,3-x\,|\,>2 \;\Rightarrow$

$3-x>2$ or $3-x<-2 \;\Rightarrow\; 1>x$ or $5<x \;\Rightarrow\; x<1$ or $x>5 \;\Leftrightarrow$

$$(-\infty,\,1)\cup(5,\,\infty)$$

$\boxed{53}$ $10x^2+11x>6 \;\Rightarrow\; 10x^2+11x-6>0 \;\Rightarrow\; (2x+3)(5x-2)>0;\; (-\infty,\,-\tfrac{3}{2})\cup(\tfrac{2}{5},\,\infty)$

Resulting sign:	$\oplus$	$\ominus$	$\oplus$
Sign of $5x-2$:	$-$	$-$	$+$
Sign of $2x+3$:	$-$	$+$	$+$
x values:	$-3/2$		$2/5$

Diagram 53

$\boxed{55}$ $\dfrac{x^2(3-x)}{x+2}\le 0 \;\Rightarrow\; \dfrac{3-x}{x+2}\le 0$ { include 0 }; $(-\infty,\,-2)\cup\{\,0\,\}\cup[3,\,\infty)$

Resulting sign:	$\ominus$	$\oplus$	$\ominus$
Sign of $3-x$:	$+$	$+$	$-$
Sign of $x+2$:	$-$	$+$	$+$
x values:		-2	3

Diagram 55

$\boxed{57}$ $\dfrac{3}{2x+3}<\dfrac{1}{x-2} \;\Rightarrow\; \dfrac{3(x-2)-1(2x+3)}{(2x+3)(x-2)}<0 \;\Rightarrow\; \dfrac{x-9}{(2x+3)(x-2)}<0;$

$$(-\infty,\,-\tfrac{3}{2})\cup(2,\,9)$$

Resulting sign:	$\ominus$	$\oplus$	$\ominus$	$\oplus$
Sign of $x-9$:	$-$	$-$	$-$	$+$
Sign of $x-2$:	$-$	$-$	$+$	$+$
Sign of $2x+3$:	$-$	$+$	$+$	$+$
x values:		$-3/2$	2	9

Diagram 57

$\boxed{59}$ $x^3>x^2 \;\Rightarrow\; x^2(x-1)>0$ { $x^2\ge 0$ } $\;\Rightarrow\; x-1>0 \;\Rightarrow\; x>1 \;\Leftrightarrow\; (1,\,\infty)$

$\boxed{60}$ $(x^2-x)(x^2-5x+6)<0 \;\Rightarrow$

$x(x-1)(x-2)(x-3)<0;\;(0,\,1)\cup(2,\,3)$

Res. sign:	$\oplus$	$\ominus$	$\oplus$	$\ominus$	$\oplus$
$x-3$:	$-$	$-$	$-$	$-$	$+$
$x-2$:	$-$	$-$	$-$	$+$	$+$
$x-1$:	$-$	$-$	$+$	$+$	$+$
x:	$-$	$+$	$+$	$+$	$+$
x values:		0	1	2	3

Diagram 44

$\boxed{61}$ $F=\dfrac{\pi P R^4}{8VL} \;\Rightarrow\; R^4=\dfrac{8FVL}{\pi P} \;\Rightarrow\; R=\pm\sqrt[4]{\dfrac{8FVL}{\pi P}} \;\Rightarrow\; R=\sqrt[4]{\dfrac{8FVL}{\pi P}}$ since $R>0$

$\boxed{62}$ $V=\tfrac{1}{3}\pi h(r^2+R^2+rR) \;\Rightarrow\; r^2+Rr+R^2-\dfrac{3V}{\pi h}=0 \;\Rightarrow$

$(\pi h)r^2+(\pi hR)r+(\pi hR^2-3V)=0 \;\Rightarrow$

$$r=\frac{-\pi hR\pm\sqrt{\pi^2h^2R^2-4\pi h(\pi hR^2-3V)}}{2\pi h}=\frac{-\pi hR\pm\sqrt{12\pi hV-3\pi^2h^2R^2}}{2\pi h}.$$

Since $r>0$, we must use the plus sign, and $r=\dfrac{-\pi hR+\sqrt{12\pi hV-3\pi^2h^2R^2}}{2\pi h}$.

63 $(3+8i)^2 = 3^2 + 2(3)(8i) + (8i)^2 = (9-64) + 48i = -55 + 48i$

65 $\dfrac{6-3i}{2+7i} = \dfrac{6-3i}{2+7i} \cdot \dfrac{2-7i}{2-7i} = \dfrac{(12-21)+(-42-6)i}{53} = -\dfrac{9}{53} - \dfrac{48}{53}i$

67 Show that $d(A,B)^2 + d(A,C)^2 = d(B,C)^2$; that is, $(\sqrt{80})^2 + (\sqrt{5})^2 = (\sqrt{85})^2$.

$$\text{Area} = \tfrac{1}{2}bh = \tfrac{1}{2}(\sqrt{80})(\sqrt{5}) = \tfrac{1}{2}(4\sqrt{5})(\sqrt{5}) = 10.$$

69 Let $Q(0,y)$ be an arbitrary point on the y-axis. $13 = d(P,Q) \Rightarrow$

$13 = \sqrt{(0-12)^2 + (y-6)^2} \Rightarrow 169 = 144 + y^2 - 12y + 36 \Rightarrow$

$y^2 - 12y + 11 = 0 \Rightarrow (y-1)(y-11) = 0 \Rightarrow y = 1, 11.$

The points are $(0,1)$ and $(0,11)$.

71 The equation of a circle with center $C(7,-4)$ is $(x-7)^2 + (y+4)^2 = r^2$.

Letting $x = -3$ and $y = 3$ yields $(-10)^2 + 7^2 = r^2 \Rightarrow r^2 = 149$.

An equation is $(x-7)^2 + (y+4)^2 = 149$.

73 We need to solve the equation for x. $(x+2)^2 + y^2 = 9 \Rightarrow (x+2)^2 = 9 - y^2 \Rightarrow$

$x + 2 = \pm\sqrt{9-y^2} \Rightarrow x = -2 \pm \sqrt{9-y^2}.$

Choose the term with the minus sign for the left half.

75 (a) $6x + 2y + 5 = 0 \Leftrightarrow y = -3x - \tfrac{5}{2}$. Using the same slope, -3, with $A(\tfrac{1}{2}, -\tfrac{1}{3})$,

we have $y + \tfrac{1}{3} = -3(x - \tfrac{1}{2}) \Rightarrow 6y + 2 = -18x + 9 \Rightarrow 18x + 6y = 7.$

(b) Using the negative reciprocal of -3 for the slope,

$$y + \tfrac{1}{3} = \tfrac{1}{3}(x - \tfrac{1}{2}) \Rightarrow 6y + 2 = 2x - 1 \Rightarrow 2x - 6y = 3.$$

77 $x^2 + y^2 - 4x + 10y + 26 = 0 \Rightarrow$

$x^2 - 4x + \underline{\ 4\ } + y^2 + 10y + \underline{\ 25\ } = -26 + \underline{\ 4\ } + \underline{\ 25\ } \Rightarrow$

$(x-2)^2 + (y+5)^2 = 3 \Rightarrow C(2,-5)$. We want the equation of the line through

$(-3,0)$ and $(2,-5)$. $y - 0 = \frac{-5-0}{2+3}(x+3) \Rightarrow y = -1(x+3) \Rightarrow x + y = -3.$

79 $A(-1,2)$ and $B(3,-4) \Rightarrow M_{AB} = (1,-1)$ and $m_{AB} = -\tfrac{3}{2}$. We want the

equation of the line through $(1,-1)$ with slope $\tfrac{2}{3}$ { the negative reciprocal of $-\tfrac{3}{2}$ }.

$$y + 1 = \tfrac{2}{3}(x-1) \Rightarrow 3y + 3 = 2x - 2 \Rightarrow 2x - 3y = 5.$$

81 $x^2 + y^2 - 12y + 31 = 0 \Rightarrow x^2 + y^2 - 12y + \underline{\ 36\ } = -31 + \underline{\ 36\ } \Rightarrow x^2 + (y-6)^2 = 5.$

$$C(0,6); \ r = \sqrt{5}$$

$\boxed{83}$ $2y + 5x - 8 = 0 \Leftrightarrow y = -\frac{5}{2}x + 4$, a line with slope $-\frac{5}{2}$ and y-intercept 4.

x-intercept: $y = 0 \Rightarrow 2(0) + 5x - 8 = 0 \Rightarrow 5x = 8 \Rightarrow x = 1.6$

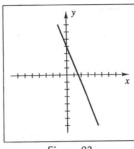
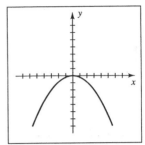

$\qquad\qquad$ *Figure 83* $\qquad\qquad\qquad\qquad\qquad$ *Figure 85*

$\boxed{85}$ $9y + 2x^2 = 0 \Leftrightarrow y = -\frac{2}{9}x^2$, a parabola opening down; x- and y-intercept 0

$\boxed{87}$ $y = \sqrt{1-x}$ $\quad\bullet\quad$ The radicand must be nonnegative for the radical to be defined.

$1 - x \geq 0 \Rightarrow 1 \geq x$, or equivalently, $x \leq 1$.

The domain is $(-\infty, 1]$ and the range is $[0, \infty)$.

x-intercept: $y = 0 \Rightarrow 0 = \sqrt{1-x} \Rightarrow 0 = 1 - x \Rightarrow x = 1$

y-intercept: $x = 0 \Rightarrow y = \sqrt{1-0} = 1$

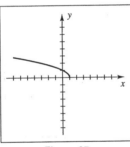
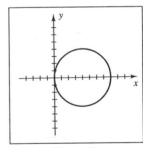

$\qquad\qquad$ *Figure 87* $\qquad\qquad\qquad\qquad\qquad$ *Figure 89*

$\boxed{89}$ $x^2 + y^2 - 8x = 0 \Leftrightarrow x^2 - 8x + \underline{16} + y^2 = \underline{16} \Leftrightarrow (x-4)^2 + y^2 = 16$;

$\quad$ a circle with center $C(4, 0)$ and radius $r = \sqrt{16} = 4$. x-intercepts:

$$y = 0 \Rightarrow x^2 - 8x = 0 \Rightarrow x(x - 8) = 0 \Rightarrow x = 0 \text{ and } 8, \text{ } y\text{-intercept: } 0$$

$\boxed{91}$ Using the intercept form of a line with x-intercept 5 and y-intercept -2, an equation

$\quad$ is $\frac{x}{5} + \frac{y}{-2} = 1$. Multiplying by 10 gives us $\left[\frac{x}{5} + \frac{y}{-2} = 1\right] \cdot 10$, or, equivalently,

$$2x - 5y = 10.$$

$\boxed{93}$ Let P denote the principal that will be invested, and r the yield rate of the stock

$\quad$ fund. Income$_{\text{stocks}}$ $-$ 28% federal tax $-$ 7% state tax $=$ Income$_{\text{bonds}}$ $\Rightarrow$

$\quad$ $(Pr) - 0.28(Pr) - 0.07(Pr) = 0.07186P$ { divide by P } $\Rightarrow$

$\quad\quad$ $1r - 0.28r - 0.07r = 0.07186 \Rightarrow 0.65r = 0.07186 \Rightarrow r \approx 0.11055$, or, 11.055%.

95 Let x denote the number of gallons of 20% solution, $120 - x$ the number of gallons of 50% solution. $20(x) + 50(120 - x) = 30(120)$ { all in % } $\Rightarrow$ $20 \cdot 120 = 30x$ $\Rightarrow$ $x = 80$. Use 80 gal of the 20% solution and 40 gal of the 50% solution.

97 (a) The eastbound car has distance $20t$ and the southbound car has distance

$$(-2 + 50t). \quad d^2 = (20t)^2 + (-2 + 50t)^2 \quad \Rightarrow \quad d = \sqrt{2900t^2 - 200t + 4}$$

(b) $104 = \sqrt{2900t^2 - 200t + 4}$ $\Rightarrow$ $2900t^2 - 200t - 10{,}812 = 0$ $\Rightarrow$

$$725t^2 - 50t - 2703 = 0 \quad \Rightarrow \quad t = \frac{50 \pm \sqrt{7{,}841{,}200}}{1450} \ \{\, t > 0 \,\} = \frac{5 + 2\sqrt{19{,}603}}{145} \approx$$

1.97, or approximately 11:58 A.M.

99 Let x denote the length of one side of an end.

(a) $V = lwh$ $\Rightarrow$ $48 = 6 \cdot x \cdot x$ $\Rightarrow$ $x^2 = 8$ $\Rightarrow$ $x = 2\sqrt{2}$ ft

(b) $S = lw + 2wh + 2lh$ $\Rightarrow$ $44 = 6x + 2(x^2) + 2(6x)$ $\Rightarrow$ $44 = 2x^2 + 18x$ $\Rightarrow$

$$x^2 + 9x - 22 = 0 \quad \Rightarrow \quad (x + 11)(x - 2) = 0 \quad \Rightarrow \quad x = 2 \text{ ft}$$

101 $v > 1100$ $\Rightarrow$ $1087\sqrt{\dfrac{T}{273}} > 1100$ $\Rightarrow$

$$\sqrt{\frac{T}{273}} > \frac{1100}{1087} \quad \Rightarrow \quad \frac{T}{273} > \frac{(1100)^2}{(1087)^2} \quad \Rightarrow \quad T > \frac{273(1100)^2}{(1087)^2} \quad \Rightarrow \quad T > 279.57 \text{ K}$$

103 The slope of the ramp should be between $\frac{1}{12}$ and $\frac{1}{20}$. If the rise of the ramp is 3 feet, then the run should be between $3 \times 12 = 36$ ft and $3 \times 20 = 60$ ft. The range of the ramp lengths should be from $L = \sqrt{3^2 + 36^2} \approx 36.1$ ft to $L = \sqrt{3^2 + 60^2} \approx 60.1$ ft.

105 The y-values are increasing slowly and can best be described by equation (3), $y = 3\sqrt{x - 0.5}$.

1 1 gallon ≈ 0.13368 ft^3 is a conversion factor that would help. The volume of the tank is 10,000 gallons ≈ 1336.8 ft^3. Use $V = \frac{4}{3}\pi r^3$ to determine the radius. $1336.8 = \frac{4}{3}\pi r^3 \Rightarrow r^3 = \frac{1002.6}{\pi} \Rightarrow r \approx 6.83375$ ft. Then use $S = 4\pi r^2$ to find the surface area. $S = 4\pi(6.83375)^2 \approx 586.85$ ft^2.

3 We first need to determine the term that needs to be added and subtracted. Since $25 = 5^2$, it makes sense to add and subtract $2 \cdot \underline{5}x = 10x$. Then we will obtain the square of a binomial—i.e., $(x^2 + 10x + 25) - 10x = (x+5)^2 - 10x$. We can now factor this expression as the difference of two squares,
$$(x+5)^2 - 10x = (x+5)^2 - (\sqrt{10x})^2 = (x+5+\sqrt{10x})(x+5-\sqrt{10x}).$$

5 Try $\dfrac{3x^2 - 4x + 7}{8x^2 + 9x - 100}$ with $x = 10^3$, 10^4, and 10^5. You get approximately 0.374, 0.3749, and 0.37499. The numbers seem to be getting closer to 0.375, which is the decimal representation for $\frac{3}{8}$, which is the ratio of the coefficients of the x^2 terms. In general, the quotients of this form get close to the ratio of leading coefficients as x gets larger.

7 Follow the algebraic simplification given.

   1 Write down his/her age   Denote the age with x.

   2 Multiply it by 2 $2x$

   3 Add 5 $2x + 5$

   4 Multiply this sum by 50 $50(2x + 5) = 100x + 250$

   5 Subtract 365 $(100x + 250) - 365 = 100x - 115$

   6 Add his/her height (in inches) $100x - 115 + y$, where y is the height

   7 Add 115 $100x - 115 + y + 115 = 100x + y$

As a specific example, suppose the age is 21 and the height is 68. The number obtained by following the steps is $100x + y = 2168$ and we can see that the first two digits of the result equal the age and the last two digits equal the height.

9 We need to solve the equation $x^2 - xy + y^2 = 0$ for x.

Use the quadratic formula with $a = 1$, $b = -y$, and $c = y^2$.
$$x = \frac{-(-y) \pm \sqrt{(-y)^2 - 4(1)(y^2)}}{2(1)} = \frac{y \pm \sqrt{y^2 - 4y^2}}{2} = \frac{y \pm \sqrt{-3y^2}}{2} = \frac{y \pm |y|\sqrt{3}\,i}{2}.$$

Since this equation has imaginary solutions, $x^2 - xy + y^2$ is not factorable over the reals. A similar argument holds for $x^2 + xy + y^2$.

11 (a) $\dfrac{1}{\frac{a+bi}{c+di}} = \dfrac{c+di}{a+bi} \cdot \dfrac{a-bi}{a-bi} = \dfrac{ac+bd+(ad-bc)i}{a^2+b^2} = \dfrac{ac+bd}{a^2+b^2} + \dfrac{ad-bc}{a^2+b^2}i = p+qi$

(b) Yes, try an example such as $3/4$. Let $a = 3$, $b = 0$, $c = 4$, and $d = 0$. Then, from part (a), $p + qi = \frac{12}{9} + \frac{0}{9}i = \frac{12}{9} = \frac{4}{3}$, which is the multiplicative inverse of $3/4$.

(c) a and b cannot both be 0 because then the denominator would be 0.

13 *Hint*: Try these examples to help you get to the general solution.

(1) $x^2 + 1 \geq 0$ {In this case, $a > 0$, $D = -4 < 0$, and by examining a sign chart with

$$x^2 + 1 \text{ as the only factor, we see that the solution is } x \in \mathbb{R}.\}$$

(2) $x^2 - 2x - 3 \geq 0$

(3) $-x^2 - 4 \geq 0$

(4) $-x^2 - 2x - 1 \geq 0$

(5) $-x^2 + 2x + 3 \geq 0$

General solutions categorized by a and D:

(1) $a > 0$, $D \leq 0$: solution is $x \in \mathbb{R}$

(2) $a > 0$, $D > 0$: let $x_1 = (-b - \sqrt{D})/(2a)$ and $x_2 = (-b + \sqrt{D})/(2a)$ $\Rightarrow$

$$\text{solution is } (-\infty, x_1] \cup [x_2, \infty)$$

(3) $a < 0$, $D < 0$: solution is $\{ \ \}$

(4) $a < 0$, $D = 0$: solution is $x = -b/(2a)$

(5) $a < 0$, $D > 0$: solution is $[x_1, x_2]$

15 To determine the x-coordinate of R,

we want to start at x_1 and go $\frac{m}{n}$ of the way to x_2. We could write this as

$$x_3 = x_1 + \tfrac{m}{n}\Delta x = x_1 + \tfrac{m}{n}(x_2 - x_1) = x_1 + \tfrac{m}{n}x_2 - \tfrac{m}{n}x_1 = \left(1 - \tfrac{m}{n}\right)x_1 + \tfrac{m}{n}x_2.$$

$$\text{Similarly, } y_3 = \left(1 - \tfrac{m}{n}\right)y_1 + \tfrac{m}{n}y_2.$$

Chapter 2: Functions

1. $f(x) = -x^2 - x - 4 \Rightarrow f(-2) = -4 + 2 - 4 = -6$, $f(0) = -4$, and $f(4) = -24$.

3. $f(x) = \sqrt{x-4} - 3x \Rightarrow f(4) = \sqrt{4-4} - 3(4) = \sqrt{0} - 12 = 0 - 12 = -12$.

   Similarly, $\qquad\qquad\qquad f(8) = \sqrt{8-4} - 3(8) = \sqrt{4} - 24 = 2 - 24 = -22$

   and $\qquad\qquad\qquad\qquad f(13) = \sqrt{13-4} - 3(13) = \sqrt{9} - 39 = 3 - 39 = -36$.

   Note that $f(a)$, with $a < 4$, would be undefined.

5. (a) $f(x) = 5x - 2 \Rightarrow f(a) = 5(a) - 2 = 5a - 2$

   (b) $f(-a) = 5(-a) - 2 = -5a - 2$

   (c) $-f(a) = -1 \cdot (5a - 2) = -5a + 2$

   (d) $f(a + h) = 5(a + h) - 2 = 5a + 5h - 2$

   (e) $f(a) + f(h) = (5a - 2) + (5h - 2) = 5a + 5h - 4$

   (f) $\dfrac{f(a+h) - f(a)}{h} = \dfrac{(5a + 5h - 2) - (5a - 2)}{h} = \dfrac{5h}{h} = 5$

7. (a) $f(x) = -x^2 + 4 \Rightarrow f(a) = -(a)^2 + 4 = -a^2 + 4$

   (b) $f(-a) = -(-a)^2 + 4 = -a^2 + 4$

   (c) $-f(a) = -1 \cdot (-a^2 + 4) = a^2 - 4$

   (d) $f(a + h) = -(a + h)^2 + 4 = -(a^2 + 2ah + h^2) + 4 = -a^2 - 2ah - h^2 + 4$

   (e) $f(a) + f(h) = (-a^2 + 4) + (-h^2 + 4) = -a^2 - h^2 + 8$

   (f) $\dfrac{f(a+h) - f(a)}{h} = \dfrac{(-a^2 - 2ah - h^2 + 4) - (-a^2 + 4)}{h} = \dfrac{-2ah - h^2}{h} = \dfrac{h(-2a - h)}{h}$

   $$= -2a - h$$

9. (a) $f(x) = x^2 - x + 3 \Rightarrow f(a) = (a)^2 - (a) + 3 = a^2 - a + 3$

   (b) $f(-a) = (-a)^2 - (-a) + 3 = a^2 + a + 3$

   (c) $-f(a) = -1 \cdot (a^2 - a + 3) = -a^2 + a - 3$

   (d) $f(a + h) = (a + h)^2 - (a + h) + 3 = a^2 + 2ah + h^2 - a - h + 3$

   (e) $f(a) + f(h) = (a^2 - a + 3) + (h^2 - h + 3) = a^2 + h^2 - a - h + 6$

   (f) $\dfrac{f(a+h) - f(a)}{h} = \dfrac{(a^2 + 2ah + h^2 - a - h + 3) - (a^2 - a + 3)}{h} = \dfrac{2ah + h^2 - h}{h} =$

   $$\dfrac{h(2a + h - 1)}{h} = 2a + h - 1$$

11. (a) $g(x) = 4x^2 \Rightarrow g\left(\dfrac{1}{a}\right) = 4\left(\dfrac{1}{a}\right)^2 = 4 \cdot \dfrac{1}{a^2} = \dfrac{4}{a^2}$

    (b) $g(a) = 4a^2 \Rightarrow \dfrac{1}{g(a)} = \dfrac{1}{4a^2}$

    (c) $g(\sqrt{a}) = 4(\sqrt{a})^2 = 4a$

    (d) $\sqrt{g(a)} = \sqrt{4a^2} = 2\,|\,a\,| = 2a$ since $a > 0$

13 (a) $g(x) = \dfrac{2x}{x^2+1} \Rightarrow g\left(\dfrac{1}{a}\right) = \dfrac{2(1/a)}{(1/a)^2+1} = \dfrac{2/a}{1/a^2+1} \cdot \dfrac{a^2}{a^2} = \dfrac{2a}{1+a^2} = \dfrac{2a}{a^2+1}$

(b) $\dfrac{1}{g(a)} = \dfrac{1}{\dfrac{2a}{a^2+1}} = \dfrac{a^2+1}{2a}$

(c) $g(\sqrt{a}) = \dfrac{2\sqrt{a}}{(\sqrt{a})^2+1} = \dfrac{2\sqrt{a}}{a+1}$

(d) $\sqrt{g(a)} = \sqrt{\dfrac{2a}{a^2+1}} \cdot \dfrac{\sqrt{a^2+1}}{\sqrt{a^2+1}} = \dfrac{\sqrt{2a(a^2+1)}}{a^2+1}$, or, equivalently, $\dfrac{\sqrt{2a^3+2a}}{a^2+1}$

15 (a) The domain of a function f is the set of all x-values for which the function is defined. In this case, the graph extends from $x = -3$ to $x = 4$. Hence, the domain is $[-3, 4]$.

(b) The range of a function f is the set of all y-values that the function takes on. In this case, the graph includes all values from $y = -2$ to $y = 2$. Hence, the range is $[-2, 2]$.

(c) $f(1)$ is the y-value of f corresponding to $x = 1$. In this case, $f(1) = 0$.

(d) If we were to draw the horizontal line $y = 1$ on the same coordinate plane, it would intersect the graph at $x = -1, \frac{1}{2}$, and 2. Hence, $f(x) = 1 \Rightarrow x = -1, \frac{1}{2}$, 2.

(e) The function is above 1 between $x = -1$ and $x = \frac{1}{2}$, and also to the right of $x = 2$. Hence, $f(x) > 1 \Rightarrow x \in (-1, \frac{1}{2}) \cup (2, 4]$.

17–28   We need to make sure that the radicand { the expression under the radical sign } is greater than or equal to zero and that the denominator is not equal to zero.

17 $f(x) = \sqrt{2x+7}$ • $2x+7 \geq 0 \Rightarrow x \geq -\frac{7}{2} \Leftrightarrow [-\frac{7}{2}, \infty)$

19 $f(x) = \sqrt{9-x^2}$ • $9-x^2 \geq 0 \Rightarrow 9 \geq x^2 \Rightarrow x^2 \leq 9 \Rightarrow |x| \leq 3 \Rightarrow$
$-3 \leq x \leq 3$, or $[-3, 3]$ in interval notation.

21 $f(x) = \dfrac{x+1}{x^3-4x}$ • For this function we must have the denominator not equal to 0.
The denominator is $x^3 - 4x = x(x^2-4) = x(x+2)(x-2)$, so $x \neq 0$, -2, 2. The solution is then all real numbers *except* 0, -2, 2. In interval notation, we have $(-\infty, -2) \cup (-2, 0) \cup (0, 2) \cup (2, \infty)$. We could also denote this solution as $\mathbb{R} - \{ \pm 2, 0 \}$.

$\boxed{23}$ $f(x) = \dfrac{\sqrt{2x-3}}{x^2 - 5x + 4}$ • For this function we must have the radicand greater than or

equal to 0 *and* the denominator not equal to 0. The radicand is greater than or equal to 0 if $2x - 3 \geq 0$, or, equivalently, $x \geq \frac{3}{2}$. The denominator is $(x-1)(x-4)$, so $x \neq 1, 4$. The solution is then all real numbers greater than or equal to $\frac{3}{2}$, excluding 4. In interval notation, we have $[\frac{3}{2}, 4) \cup (4, \infty)$.

$\boxed{25}$ $f(x) = \dfrac{x-4}{\sqrt{x-2}}$ • For this function we must have $x - 2 > 0 \Rightarrow x > 2$.

Note that " $>$ " must be used since the denominator cannot *equal* 0.

In interval notation, we have $(2, \infty)$.

$\boxed{27}$ $f(x) = \sqrt{x+2} + \sqrt{2-x}$ • We must have $x + 2 \geq 0 \Rightarrow x \geq -2$ and as well as $2 - x \geq 0 \Rightarrow x \leq 2$. The domain is the intersection of $x \geq -2$ and $x \leq 2$, that is,
$$[-2, 2].$$

$\boxed{29}$ The graph of the function is increasing on $(-\infty, -3]$ and is decreasing on $[-3, 2]$, so there must be a high point at $x = -3$. Now the graph of the function is increasing on $[2, \infty)$, so there must be a low point at $x = 2$.

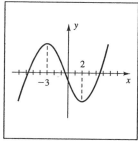

Figure 29

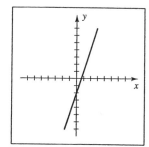

Figure 31

$\boxed{31}$ (a) $f(x) = 3x - 2$ • This is a line with slope 3 and y-intercept 2.

(b) The domain D and the range R are equal to $(-\infty, \infty)$.

(c) f is increasing on its entire domain, that is, $(-\infty, \infty)$.

33 (a) To sketch the graph of $f(x) = 4 - x^2$, we can make use of the symmetry with respect to the y-axis.

x	± 4	± 3	± 2	± 1	0
y	-12	-5	0	3	4

(b) Since we can substitute any number for x, the domain is all real numbers, that is, $D = \mathbb{R}$. By examining *Figure 33*, we see that the values of y are at most 4. Hence, the range of f is all reals less than or equal to 4, that is, $R = (-\infty, 4]$.

(c) A common mistake is to confuse the function values, the y's, with the input values, the x's. We are not interested in the specific y-values for determining if the function is increasing, decreasing, or constant. We are only interested if the y-values are going up, going down, or staying the same. For the function $f(x) = 4 - x^2$, we say f *is increasing on* $(-\infty, 0]$ since the y-values are getting larger as we move from left to right over the x-values from $-\infty$ to 0. Also, f *is decreasing on* $[0, \infty)$ since the y-values are getting smaller as we move from left to right over the x-values from 0 to ∞. Note that this answer would have been the same if the function was $f(x) = 500 - x^2$, $f(x) = -300 - x^2$, or any function of the form $f(x) = a - x^2$, where a is any real number.

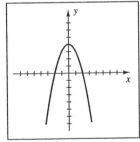

Figure 33

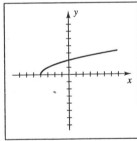

Figure 35

35 (a) $f(x) = \sqrt{x + 4}$ • This is half of a parabola opening to the right.

It has an x-intercept of -4 and a y-intercept of 2.

(b) $x + 4 \geq 0 \Rightarrow x \geq -4$, so the domain is $D = [-4, \infty)$.

The y-values are all positive or zero, so the range is $R = [0, \infty)$.

(c) f is increasing on its entire domain, that is, on $[-4, \infty)$.

$\boxed{37}$ (a) $f(x) = -2$ • This is a horizontal line with y-intercept -2.

(b) The domain is $D = (-\infty, \infty)$ and the range consists of a single value,

so $R = \{-2\}$.

(c) f is constant on its entire domain, that is, $(-\infty, \infty)$.

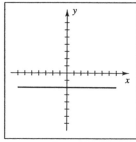

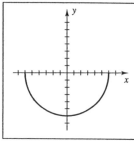

Figure 37   *Figure 39*

$\boxed{39}$ (a) We recognize $y = f(x) = -\sqrt{36 - x^2}$ as the lower half of the circle $x^2 + y^2 = 36$.

(b) To find the domain, we solve $36 - x^2 \geq 0$. $36 - x^2 \geq 0 \Rightarrow x^2 \leq 36 \Rightarrow$

$|x| \leq 6 \Rightarrow D = [-6, 6]$. From *Figure 39*, we see that the y-values vary from

$y = -6$ to $y = 0$. Hence, the range R is $[-6, 0]$.

(c) As we move from left to right, $x = -6$ to $x = 0$, the y-values are decreasing.

From $x = 0$ to $x = 6$, the y-values increase.

Hence, f is decreasing on $[-6, 0]$ and increasing on $[0, 6]$.

$\boxed{41}$ $\dfrac{f(x+h) - f(x)}{h} = \dfrac{[(x+h)^2 + 5] - [x^2 + 5]}{h} = \dfrac{(x^2 + 2xh + h^2 + 5) - (x^2 + 5)}{h} =$

$\dfrac{2xh + h^2}{h} = \dfrac{h(2x + h)}{h} = 2x + h$

$\boxed{43}$ $\dfrac{f(x) - f(a)}{x - a} = \dfrac{\sqrt{x - 3} - \sqrt{a - 3}}{x - a} = \dfrac{\sqrt{x - 3} - \sqrt{a - 3}}{x - a} \cdot \dfrac{\sqrt{x - 3} + \sqrt{a - 3}}{\sqrt{x - 3} + \sqrt{a - 3}} =$

$\dfrac{(x - 3) - (a - 3)}{(x - a)(\sqrt{x - 3} + \sqrt{a - 3})} = \dfrac{x - a}{(x - a)(\sqrt{x - 3} + \sqrt{a - 3})} = \dfrac{1}{\sqrt{x - 3} + \sqrt{a - 3}}$

$\boxed{45}$ As in Example 7, $a = \dfrac{2 - 1}{3 - (-3)} = \dfrac{1}{6}$ and f has the form $f(x) = \frac{1}{6}x + b$.

$f(3) = \frac{1}{6}(3) + b = \frac{1}{2} + b$. But $f(3) = 2$, so $\frac{1}{2} + b = 2 \Rightarrow b = \frac{3}{2}$, and $f(x) = \frac{1}{6}x + \frac{3}{2}$.

$\boxed{47\text{–}56}$ *Note:* A good question to consider is "Given a particular value of x, can a unique value of y be found?" If the answer is yes, the value of y (general formula) is given. If no, two ordered pairs satisfying the relation having x in the first position are given.

$\boxed{47}$ $2y = x^2 + 5 \Rightarrow y = \dfrac{x^2 + 5}{2}$, a function

$\boxed{49}$ $x^2 + y^2 = 4 \Rightarrow y^2 = 4 - x^2 \Rightarrow y = \pm\sqrt{4 - x^2}$, not a function, $(0, \pm 2)$

[51] $y = 3$ is a function since for any x,

$(x,\ 3)$ is the only ordered pair in W having x in the first position.

[53] Any ordered pair with x-coordinate 0 satisfies $xy = 0$.

Two such ordered pairs are $(0,\ 0)$ and $(0,\ 1)$. Not a function

[55] $|y| = |x| \;\Rightarrow\; \pm y = \pm x \;\Rightarrow\; y = \pm x$, not a function, $(1,\ \pm 1)$

[57] $V = lwh = (30 - x - x)(20 - x - x)(x)$

$$= (30 - 2x)(20 - 2x)(x) = 2(15 - x) \cdot 2(10 - x)(x) = 4x(15 - x)(10 - x)$$

[59] (a) The formula for the area of a rectangle is $A = lw$ { Area = length × width }.

$$A = 500 \;\Rightarrow\; xy = 500 \;\Rightarrow\; y = \frac{500}{x}$$

(b) We need to determine the number of linear feet (P) first. There are two walls

of length y, two walls of length $(x - 3)$, and one wall of length x.

Thus, $P = $ Linear feet of wall $= x + 2(y) + 2(x - 3) = 3x + 2\left(\dfrac{500}{x}\right) - 6$.

The cost C is 100 times P, so $C = 100P = 300x + \dfrac{100{,}000}{x} - 600$.

[61] The expression $(h - 25)$ represents the number of feet *above* 25 feet.

$$S(h) = 6(h - 25) + 100 = 6h - 150 + 100 = 6h - 50.$$

[63] (a) Using $(6,\ 48)$ and $(7,\ 50.5)$, we have

$$y - 48 = \frac{50.5 - 48}{7 - 6}(t - 6),\ \text{or}\ y = 2.5t + 33.$$

(b) The slope represents the

yearly increase in height, 2.5 in./yr.

(c) $t = 10 \;\Rightarrow\; y = 2.5(10) + 33 = 58$ in.

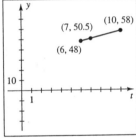

Figure 63

[65] The height of the balloon is $2t$. Using the Pythagorean theorem,

$$d^2 = 100^2 + (2t)^2 = 10{,}000 + 4t^2 \;\Rightarrow\; d = \sqrt{4(2500 + t^2)} \;\Rightarrow\; d = 2\sqrt{t^2 + 2500}.$$

[67] (a) CTP forms a right angle, so the Pythagorean theorem may be applied.

$$(CT)^2 + (PT)^2 = (PC)^2 \;\Rightarrow\; r^2 + y^2 = (h + r)^2 \;\Rightarrow$$

$$r^2 + y^2 = h^2 + 2hr + r^2 \;\Rightarrow\; y^2 = h^2 + 2hr\ \{y > 0\} \;\Rightarrow\; y = \sqrt{h^2 + 2hr}$$

(b) $y = \sqrt{(200)^2 + 2(4000)(200)} = \sqrt{(200)^2(1 + 40)} = 200\sqrt{41} \approx 1280.6$ mi

[69] Form a right triangle with the control booth and the beginning of the runway. Let y

denote the distance from the control booth to the beginning of the runway and apply

the Pythagorean theorem. $y^2 = 300^2 + 20^2 \;\Rightarrow\; y^2 = 90{,}400$. Now form a right

triangle, in a different plane, with sides y and x and hypotenuse d.

Then $d^2 = y^2 + x^2 \;\Rightarrow\; d^2 = 90{,}400 + x^2 \;\Rightarrow\; d = \sqrt{90{,}400 + x^2}$.

71 (b) The maximum y-value of 0.75 occurs when $x \approx 0.55$ and

the minimum y-value of -0.75 occurs when $x \approx -0.55$.

Therefore, the range of f is approximately $[-0.75, 0.75]$.

(c) f is decreasing on $[-2, -0.55]$ and on $[0.55, 2]$. f is increasing on $[-0.55, 0.55]$.

$[-2, 2]$ by $[-2, 2]$ $[-0.7, 1.4]$ by $[-1.1, 1]$

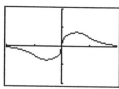   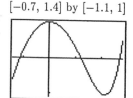

Figure 71   *Figure 73*

73 (b) The maximum y-value of 1 occurs when $x = 0$ and

the minimum y-value of -1.03 occurs when $x \approx 1.06$.

Therefore, the range of f is approximately $[-1.03, 1]$.

(c) f is decreasing on $[0, 1.06]$. f is increasing on $[-0.7, 0]$ and on $[1.06, 1.4]$.

75 For each of (a)–(e), an assignment to Y_1, an appropriate viewing rectangle, and the solution(s) are listed.

(a) $Y_1 = (x^5)^{(1/3)}$,   VR: $[-40, 40, 10]$ by $[-40, 40, 10]$, $x = 8$

(b) $Y_1 = (x^4)^{(1/3)}$,   VR: $[-40, 40, 10]$ by $[-40, 40, 10]$, $x = \pm 8$

$[-40, 40, 10]$ by $[-40, 40, 10]$ $[-40, 40, 10]$ by $[-40, 40, 10]$

   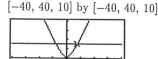

Figure 75(a)   *Figure 75(b)*

(c) $Y_1 = (x^2)^{(1/3)}$,   VR: $[-40, 40, 10]$ by $[-40, 40, 10]$,   no real solutions

(d) $Y_1 = (x^3)^{(1/4)}$,   VR: $[0, 800, 100]$ by $[0, 200, 100]$, $x = 625$

(e) $Y_1 = (x^3)^{(1/2)}$,   VR: $[-30, 30, 10]$ by $[-30, 30, 10]$,   no real solutions

77 (a) There are $95 \times 63 = 5985$ total pixels in the screen.

(b) If a function is graphed in dot mode, only one pixel in each column of pixels on the screen can be darkened. Therefore, there are at most 95 pixels darkened. *Note:* In connected mode this may not be true.

[79] (a) First, we must determine an equation of the line that passes through the points

(1985, 11,450) and (1994, 20,021).

$$y - 11{,}450 = \frac{20{,}021 - 11{,}450}{1994 - 1985}(x - 1985) = \frac{2857}{3}(x - 1985) \;\Rightarrow$$

$$y = \frac{2857}{3}x - \frac{5{,}636{,}795}{3}. \text{ Thus, let } f(x) = \frac{2857}{3}x - \frac{5{,}636{,}795}{3} \text{ and graph } f.$$

(b) The average annual increase in the price paid for a new car is equal to the slope:

$$\frac{2857}{3} \approx \$952.33.$$

(c) Graph $y = \frac{2857}{3}x - \frac{5{,}636{,}795}{3}$ and $y = 25{,}000$ on the same coordinate axes. Their

point of intersection is approximately (1999.2, 25,000). Thus, according to this

model, in the year 1999 the average price paid for a new car was $25,000.

[1984, 2005, 5] by [1E4, 3E4, 1E4]   [1984, 2005, 5] by [1E4, 3E4, 1E4]

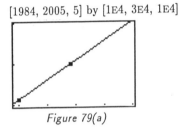   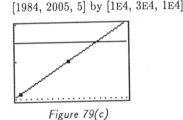

Figure 79(a)   *Figure 79(c)*

2.2 Exercises

[1] $f(x) = 5x^3 + 2x \;\Rightarrow\; f(-x) = 5(-x)^3 + 2(-x) = -5x^3 - 2x = -(5x^3 + 2x) = -f(x).$

Since $f(-x) = -f(x)$, f is odd and its graph is symmetric with respect to the origin.

Note that this means if (a, b) is a point on the graph of f,

then the point $(-a, -b)$ is also on the graph.

[3] $f(x) = 3x^4 + 2x^2 - 5 \;\Rightarrow\; f(-x) = 3(-x)^4 + 2(-x)^2 - 5 = 3x^4 + 2x^2 - 5 = f(x).$

Since $f(-x) = f(x)$, f is even and its graph is symmetric with respect to the y-axis.

Note that this means if (a, b) is a point on the graph of f,

then the point $(-a, b)$ is also on the graph.

[5] $f(x) = 8x^3 - 3x^2 \;\Rightarrow\; f(-x) = 8(-x)^3 - 3(-x)^2 = -8x^3 - 3x^2$

$-f(x) = -1 \cdot f(x) = -1(8x^3 - 3x^2) = -8x^3 + 3x^2$

Since $f(-x) \neq f(x)$ and $f(-x) \neq -f(x)$, f is neither even nor odd.

[7] $f(x) = \sqrt{x^2 + 4} \;\Rightarrow\; f(-x) = \sqrt{(-x)^2 + 4} = \sqrt{x^2 + 4} = f(x).$

Since $f(-x) = f(x)$, f is even and its graph is symmetric with respect to the y-axis.

9. $f(x) = \sqrt[3]{x^3 - x} \Rightarrow$

$f(-x) = \sqrt[3]{(-x)^3 - (-x)} = \sqrt[3]{-x^3 + x} = \sqrt[3]{-1(x^3 - x)} = \sqrt[3]{-1}\sqrt[3]{x^3 - x} = -\sqrt[3]{x^3 - x}.$

$-f(x) = -1 \cdot f(x) = -1 \cdot \sqrt[3]{x^3 - x} = -\sqrt[3]{x^3 - x}.$

Since $f(-x) = -f(x)$, f is odd and its graph is symmetric with respect to the origin.

11. $f(x) = |x| + c$, $c = -3, 1, 3$ • Shift $g(x) = |x|$ {in Figure 1 in the text}

down 3, up 1, and up 3 units, respectively.

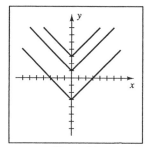

Figure 11

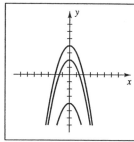

Figure 13

13. $f(x) = -x^2 + c$, $c = -4, 2, 4$ • Shift $g(x) = -x^2$ {in Figure 5 in the text}

down 4, up 2, up 4, respectively.

15. $f(x) = 2\sqrt{x} + c$, $c = -3, 0, 2$ • The graph of $y^2 = x$ is shown in Figure 8 in Section 1.7 in the text. The top half of this graph is the graph of the **square root function**, $h(x) = \sqrt{x}$. The second value of c, 0, gives us the graph of $g(x) = 2\sqrt{x}$, which is a vertical stretching of h by a factor of 2. The effect of *adding* -3 and 2 is to vertically shift g down 3 units and up 2 units, respectively.

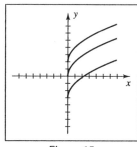

Figure 15

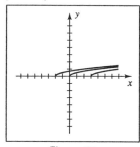

Figure 17

17. $f(x) = \frac{1}{2}\sqrt{x - c}$, $c = -2, 0, 3$ • The graph of $g(x) = \frac{1}{2}\sqrt{x}$ is a vertical compression of the square root function by a factor of $1/(1/2) = 2$. The effect of *subtracting* -2 and 3 from x will be to horizontally shift g left 2 units and right 3 units, respectively. If you forget which way to shift the graph, it is helpful to find the domain of the function. For example, if $h(x) = \sqrt{x - 2}$, then $x - 2$ must be nonnegative. $x - 2 \geq 0 \Rightarrow x \geq 2$, which also indicates a shift of 2 units to the right.

19 $f(x) = c\sqrt{4-x^2}$, $c = -2, 1, 3$ • For $c = 1$, the graph of $g(x) = \sqrt{4-x^2}$ is the upper half of the circle $x^2 + y^2 = 4$. For $c = -2$, reflect g through the x-axis and vertically stretch it by a factor of 2. For $c = 3$, vertically stretch g by a factor of 3.

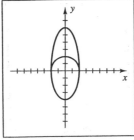

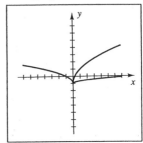

| *Figure 19* | *Figure 21* | *Figure 23* |

21 $f(x) = cx^3$, $c = -\frac{1}{3}, 1, 2$ • For $c = 1$, see a graph of the **cubing function** in Appendix I of the text. For $c = -\frac{1}{3}$, reflect $g(x) = x^3$ through the x-axis and vertically compress it by a factor of $1/(1/3) = 3$. For $c = 2$, vertically stretch g by a factor of 2.

23 $f(x) = \sqrt{cx} - 1$, $c = -1, \frac{1}{9}, 4$ • If $c = 1$, then the graph of $g(x) = \sqrt{x} - 1$ is the graph of the square root function vertically shifted down one unit. For $c = -1$, reflect g through the y-axis. For $c = \frac{1}{9}$, horizontally stretch g by a factor $1/(1/9) = 9$ { x-intercept changes from 1 to 9 }. For $c = 4$, horizontally compress g by a factor 4 { x-intercept changes from 1 to $\frac{1}{4}$ }.

25 You know that $y = f(x+2)$ is the graph of $y = f(x)$ shifted to the left 2 units, so the point $P(0, 5)$ would move to the point $(-2, 5)$. The graph of $y = f(x+2) - 1$ is the graph of $y = f(x+2)$ shifted down 1 unit, so the point $(-2, 5)$ moves to $(-2, 4)$. Summarizing these steps gives us the following:

$$P(0, 5) \quad \{ x+2 \text{ [subtract 2 from the x-coordinate]} \} \quad \rightarrow (-2, 5)$$
$$\{ -1 \text{ [subtract 1 from the y-coordinate]} \} \quad \rightarrow (-2, 4)$$

27 To determine what happens to a point P under this transformation, think of how you would evaluate $y = 2f(x-4) + 1$ for a particular value of x. You would first subtract 4 from x and then put that value into the function, obtaining a corresponding y-value. Next, you would multiply that y-value by 2 and finally, add 1. Summarizing these steps using the given point P, we have the following:

$$P(3, -2) \quad \{ x-4 \text{ [add 4 to the x-coordinate]} \} \quad \rightarrow (7, -2)$$
$$\{ \times 2 \text{ [multiply the y-coordinate by 2]} \} \quad \rightarrow (7, -4)$$
$$\{ +1 \text{ [add 1 to the y-coordinate]} \} \quad \rightarrow (7, -3)$$

29 $P(3, 9)$; $y = \frac{1}{3}f(\frac{1}{2}x) - 1$

$\quad\quad$ { $\frac{1}{2}x$ [multiply the x-coordinate by 2] } $\quad\quad \to (6, 9)$

$\quad\quad$ { $\times\frac{1}{3}$ [multiply the y-coordinate by $\frac{1}{3}$] } $\quad\quad \to (6, 3)$

$\quad\quad$ { -1 [subtract 1 from the y-coordinate] } $\quad \to (6, 2)$

31 For $y = f(x-2) + 3$, the graph of f is shifted 2 units to the right and 3 units up.

33 For $y = f(-x) - 2$, the graph of f is reflected through the y-axis and shifted 2 units down.

35 For $y = -\frac{1}{2}f(x)$, the graph of f is compressed vertically by a factor of 2 and reflected through the x-axis.

37 For $y = -2f(\frac{1}{3}x)$, the graph of f is stretched horizontally by a factor of 3, stretched vertically by a factor of 2, and reflected through the x-axis.

39 (a) $y = f(x+3)$ $\quad \bullet \quad$ shift f left 3 units

$\quad$ (b) $y = f(x-3)$ $\quad \bullet \quad$ shift f right 3 units

$\quad$ (c) $y = f(x) + 3$ $\quad \bullet \quad$ shift f up 3 units

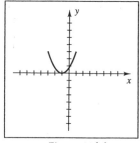 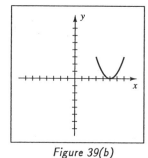 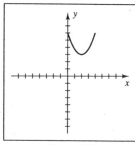

$\quad\quad$ *Figure 39(a)* $\quad\quad\quad\quad\quad$ *Figure 39(b)* $\quad\quad\quad\quad\quad$ *Figure 39(c)*

$\quad$ (d) $y = f(x) - 3$ $\quad \bullet \quad$ shift f down 3 units

$\quad$ (e) $y = -3f(x)$ $\quad \bullet$

$\quad\quad\quad\quad$ reflect f through the x-axis and vertically stretch it by a factor of 3

$\quad$ (f) $y = -\frac{1}{3}f(x)$ $\quad \bullet \quad$ reflect f through the x-axis { the effect of the negative sign

$\quad\quad\quad\quad$ in front of $\frac{1}{3}$ } and vertically compress it by a factor of $1/(1/3) = 3$

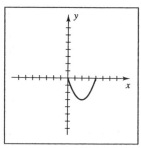 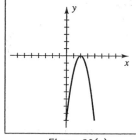 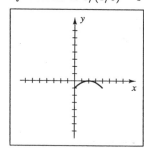

$\quad\quad$ *Figure 39(d)* $\quad\quad\quad\quad\quad$ *Figure 39(e)* $\quad\quad\quad\quad\quad$ *Figure 39(f)*

(g) $y = f(-\frac{1}{2}x)$ •   reflect f through the y-axis { the effect of the negative sign

inside the parentheses } and horizontally stretch it by a factor of $1/(1/2) = 2$

(h) $y = f(2x)$ •   horizontally compress f by a factor of 2

(i) $y = -f(x+2) - 3$ •   shift f left 2 units, reflect it through the x-axis,

and then shift it down 3 units

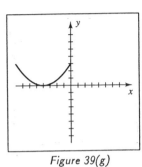

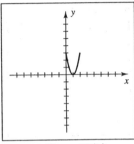

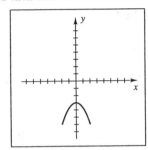

Figure 39(g)   *Figure 39(h)*   *Figure 39(i)*

(j) $y = f(x-2) + 3$ •   shift f right 2 units and up 3

(k) $y = |f(x)|$ •   since no portion of the graph lies below the x-axis,

the graph is unchanged

(l) $y = f(|x|)$ •   include the reflection of the given graph through the y-axis

since all points have positive x-coordinates

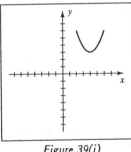

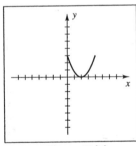

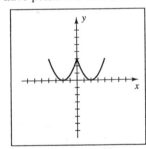

Figure 39(j)   *Figure 39(k)*   *Figure 39(l)*

41 (a) The minimum point on $y = f(x)$ is $(2, -1)$.

On the graph labeled (a), the minimum point is $(-7, 0)$.

It has been shifted left 9 units and up 1. Hence, $y = f(x+9) + 1$.

(b) f is reflected through the x-axis $\Rightarrow$ $y = -f(x)$

(c) f is reflected through the x-axis and shifted left 7 units and down 1 $\Rightarrow$

$$y = -f(x+7) - 1$$

[43] (a) f is shifted left 4 units $\Rightarrow$ $y = f(x+4)$

(b) f is shifted up 1 unit $\Rightarrow$ $y = f(x)+1$

(c) f is reflected through the y-axis $\Rightarrow$ $y = f(-x)$

[45] $f(x) = \begin{cases} 3 & \text{if } x \leq -1 \\ -2 & \text{if } x > -1 \end{cases}$

We can think of f as 2 functions: If $x \leq -1$, then $y = 3$ { include the point $(-1, 3)$ },

and if $x > -1$, then $y = -2$ { exclude the point $(-1, -2)$ }.

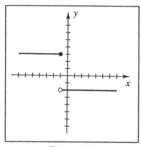

Figure 45

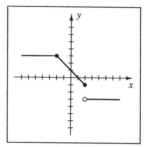

Figure 47

[47] $f(x) = \begin{cases} 3 & \text{if } x < -2 \\ -x+1 & \text{if } |x| \leq 2 \\ -3 & \text{if } x > 2 \end{cases}$

For the second part of the function, we have $|x| \leq 2$, or, equivalently, $-2 \leq x \leq 2$.

On this part of the domain, we want to graph $f(x) = -x+1$, a line with slope -1

and y-intercept 1. Include both endpoints, $(-2, 3)$ and $(2, -1)$.

[49] $f(x) = \begin{cases} x+2 & \text{if } x \leq -1 \\ x^3 & \text{if } |x| < 1 \\ -x+3 & \text{if } x \geq 1 \end{cases}$

If $x \leq -1$, we want the graph of $y = x+2$. To determine the endpoint of this part of

the graph, merely substitute $x = -1$ in $y = x+2$, obtaining $y = 1$. If $|x| < 1$, or,

equivalently, $-1 < x < 1$, we want the graph of $y = x^3$. We do not include the

endpoints $(-1, -1)$ and $(1, 1)$. If $x \geq 1$, we want the graph of $y = -x+3$ and

include its endpoint $(1, 2)$.

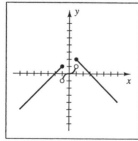

Figure 49

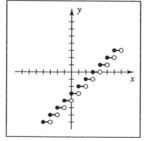

Figure 51(a)

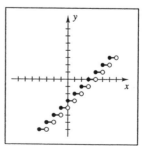

Figure 51(b)

51 (a) $f(x) = [\![x - 3]\!]$ •   shift $g(x) = [\![x]\!]$ right 3 units {see figure on page 63}

(b) $f(x) = [\![x]\!] - 3$ •   shift g down 3 units, which is the same graph as in part (a).

(c) $f(x) = 2[\![x]\!]$ •   vertically stretch g by a factor of 2

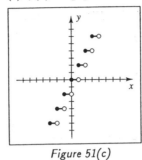 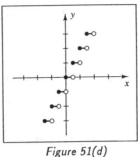 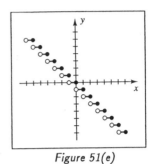

Figure 51(c)   *Figure 51(d)*   *Figure 51(e)*

(d) $f(x) = [\![2x]\!]$ •   horizontally compress g by a factor of 2

Alternatively, we could determine the pattern of "steps" for this function by finding the values of x that make $f(x)$ change from 0 to 1, then from 1 to 2, etc. If $2x = 0$, then $x = 0$, and if $2x = 1$, then $x = \frac{1}{2}$.

Thus, the function will equal 0 from $x = 0$ to $x = \frac{1}{2}$ and then jump to 1 at $x = \frac{1}{2}$.

If $2x = 2$, then $x = 1$. The pattern is established: each step will be $\frac{1}{2}$ unit long.

(e) $f(x) = [\![-x]\!]$ •   reflect g through the y-axis

53 A question you can ask to help determine if a relationship is a function is "If x is a particular value, can I find a unique y-value?" In this case, if x was 16, then $16 = y^2 \Rightarrow y = \pm 4$. Since we cannot find a unique y-value, this is not a function. Graphically, {see Figure 8 in §1.7 in the text} given any x-value greater than 0, there are two points on the graph and a vertical line intersects the graph in more than one point.

55 Reflect each portion of the graph that is below the x-axis through the x-axis.

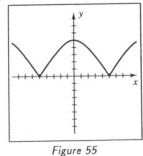 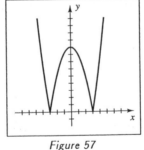 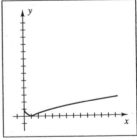

Figure 55   *Figure 57*   *Figure 59*

57 $y = |9 - x^2|$ •   First sketch $y = 9 - x^2$,

then reflect the portions of the graph below the x-axis through the x-axis.

59 $y = |\sqrt{x} - 1|$ • First sketch $y = \sqrt{x} - 1$, which is the graph of the square root function shifted down 1 unit. Then reflect the portions of the graph below the x-axis through the x-axis. See *Figure 59* on the preceding page.

61 (a) For $y = -2f(x)$, multiply the y-coordinates by -2. So the given range, $[-4, 8]$, becomes $[-16, 8]$ {note that we changed the order so that -16 appears first— listing the range as $[8, -16]$ is incorrect}. The domain { x-coordinates} remains the same. $D = [-2, 6], R = [-16, 8]$

(b) For $y = f(\frac{1}{2}x)$, multiply the x-coordinates by 2.

The range { y-coordinates} remains the same. $D = [-4, 12], R = [-4, 8]$

(c) For $y = f(x - 3) + 1$, add 3 to the x-coordinates and add 1 to the y-coordinates.
$$D = [1, 9], R = [-3, 9]$$

(d) For $y = f(x + 2) - 3$, subtract 2 from the x-coordinates and subtract 3 from the y-coordinates. $D = [-4, 4], R = [-7, 5]$

(e) For $y = f(-x)$, multiply all x-coordinates by -1. $D = [-6, 2], R = [-4, 8]$

(f) For $y = -f(x)$, multiply all y-coordinates by -1. $D = [-2, 6], R = [-8, 4]$

(g) $y = f(|x|)$ • Graphically, we can reflect all points with positive x-coordinates through the y-axis, so the domain $[-2, 6]$ becomes $[-6, 6]$. Algebraically, we are replacing x with $|x|$, so $-2 \le x \le 6$ becomes $-2 \le |x| \le 6$, which is equivalent $|x| \le 6$, or, equivalently, $-6 \le x \le 6$. The range stays the same because of the given assumptions: $f(2) = 8$ and $f(6) = -4$; that is, the full range is taken on for $x \ge 0$. Note that the range could not be determined if $f(-2)$ was equal to 8. $D = [-6, 6], R = [-8, 4]$

(h) $y = |f(x)|$ • The points with y-coordinates having values from -4 to 0 will have values from 0 to 4, so the range will be $[0, 8]$. $D = [-2, 6], R = [0, 8]$

63 If $x \le 20{,}000$, then $T(x) = 0.15x$. If $x > 20{,}000$, then the tax is 15% of the first 20,000, which is 3000, plus 20% of the amount over 20,000, which is $(x - 20{,}000)$. We may summarize and simplify as follows:

$$T(x) = \begin{cases} 0.15x & \text{if } x \le 20{,}000 \\ 3000 + 0.20(x - 20{,}000) & \text{if } x > 20{,}000 \end{cases}$$

$$= \begin{cases} 0.15x & \text{if } x \le 20{,}000 \\ 3000 + 0.20x - 4{,}000 & \text{if } x > 20{,}000 \end{cases}$$

$$= \begin{cases} 0.15x & \text{if } x \le 20{,}000 \\ 0.20x - 1000 & \text{if } x > 20{,}000 \end{cases}$$

65 The author receives $1.20 on the first 10,000 copies,

$1.50 on the next 5000, and $1.80 on each additional copy.

$$R(x) = \begin{cases} 1.20x & \text{if } 0 \le x \le 10{,}000 \\ 12{,}000 + 1.50(x - 10{,}000) & \text{if } 10{,}000 < x \le 15{,}000 \\ 19{,}500 + 1.80(x - 15{,}000) & \text{if } x > 15{,}000 \end{cases}$$

$$= \begin{cases} 1.20x & \text{if } 0 \le x \le 10{,}000 \\ 1.50x - 3000 & \text{if } 10{,}000 < x \le 15{,}000 \\ 1.80x - 7500 & \text{if } x > 15{,}000 \end{cases}$$

67 Assign ABS$(1.3x + 2.8)$ to Y_1 and $1.2x + 5$ to Y_2. The viewing rectangle $[-10, 50, 10]$ by $[-10, 50, 10]$ shows intersection points at exactly -3.12 and -22.

The solution of $|1.3x + 2.8| < 1.2x + 5$ is the interval $(-3.12, 22)$.

69 Assign ABS$(1.2x^2 - 10.8)$ to Y_1 and $1.36x + 4.08$ to Y_2. The standard viewing rectangle $[-15, 15]$ by $[-10, 10]$ shows intersection points at exactly -3 and at approximately 1.87 and 4.13. The solution is $(-\infty, -3) \cup (-3, 1.87) \cup (4.13, \infty)$.

71 $f(x) = (0.5x^3 - 4x - 5)$ and $g(x) = (0.5x^3 - 4x - 5) + 4 = f(x) + 4$, so the graph of g can be obtained by shifting the graph of f upward a distance of 4.

$[-12, 12]$ by $[-8, 8]$ $\qquad\qquad$ $[-12, 12]$ by $[-8, 8]$

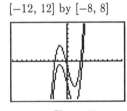 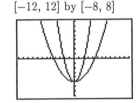

Figure 71 $\qquad\qquad$ Figure 73

73 $f(x) = x^2 - 5$ and $g(x) = (\frac{1}{2}x)^2 - 5$, so $g(x) = f(\frac{1}{2}x)$ and the graph of g can be obtained by stretching the graph of f horizontally by a factor of 2.

75 $f(x) = x^3 - 5x$ and $g(x) = |f(x)|$, so the graph of g is the same as the graph of f if f is nonnegative. If $f(x) < 0$, then the graph of f will be reflected through the x–axis.

$[-12, 12]$ by $[-8, 8]$

Figure 75

77 (a) Option I gives $C_1 = 4(\$29.95) + \$0.25(500 - 200) = 119.80 + 75.00 = \$194.80$.

Option II gives $C_2 = 4(\$39.95) + \$0.15(500) = 159.80 + 75.00 = \$234.80$.

(b) Let x represent the mileage. The cost function for Option I is the piecewise linear function

$$C_1(x) = \begin{cases} 119.80 & \text{if } 0 \le x \le 200 \\ 119.80 + 0.25(x - 200) & \text{if } x > 200 \end{cases}$$

Option II is the linear function $C_2(x) = 159.8 + 0.15x$ for $x \ge 0$.

(c) Let $C_1 = Y_1$ and $C_2 = Y_2$.

Table $Y_1 = 119.80 + 0.25(x - 200) * (x > 200)$ and $Y_2 = 159.80 + 0.15x$

x	Y_1	Y_2	x	Y_1	Y_2
100	119.8	174.8	700	244.8	264.8
200	119.8	189.8	800	269.8	279.8
300	144.8	204.8	900	294.8	294.8
400	169.8	219.8	1000	319.8	309.8
500	194.8	234.8	1100	344.8	324.8
600	219.8	249.8	1200	369.8	339.8

(d) From the table, we see that the options are equal in cost for $x = 900$ miles. Option I is preferable if $x \in [0, 900)$ and Option II is preferable if $x > 900$.

2.3 Exercises

1 Using the standard equation of a parabola with a vertical axis having vertex $V(-3, 1)$, we have $y = a[x - (-3)]^2 + 1$. The coefficient a determines whether the parabola opens upward { a positive } or opens downward { a negative }. If $|a| > 1$, then the parabola is narrower { steeper } than the graph of $y = x^2$; if $|a| < 1$, then the parabola is wider { flatter } than the graph of $y = x^2$. Simplifying the equation gives us $y = a(x + 3)^2 + 1$.

3 We can use the standard equation of a parabola with a vertical axis.

$$V(0, -3) \Rightarrow y = a(x - 0)^2 - 3 \Rightarrow y = ax^2 - 3.$$

5 The approach shown here { like Solution 1 in Example 2 } requires us to *factor out the leading coefficient.*

$$\begin{aligned} f(x) &= -x^2 - 4x - 8 && \{\text{given}\} \\ &= -(x^2 + 4x + \_\_) - 8 + \_\_ && \{\text{factor out } -1 \text{ from } -x^2 - 4x\} \\ &= -(x^2 + 4x + \underline{4}) - 8 + \underline{4} && \{\text{complete the square for } x^2 + 4x\} \\ &= -(x + 2)^2 - 4 && \{\text{equivalent equation}\} \end{aligned}$$

7̄ The approach shown here {like Solution 2 in Example 2} requires us to *divide both sides by the leading coefficient*—remember to multiply both sides by the same coefficient in the end. Either method is fine.

$$f(x) = 2x^2 - 12x + 22 \qquad \{\,\text{given}\,\}$$

$$\tfrac{1}{2}f(x) = (x^2 - 6x + \underline{\ \ }) + 11 - \underline{\ \ } \qquad \{\,\text{divide the equation by 2}\,\}$$

$$= (x^2 - 6x + \underline{\ 9\ }) + 11 - \underline{\ 9\ } \qquad \{\,\text{complete the square for } x^2 - 6x\,\}$$

$$= (x-3)^2 + 2 \qquad \{\,\text{equivalent equation}\,\}$$

$$2 \cdot \tfrac{1}{2}f(x) = 2 \cdot \left[(x-3)^2 + 2\right] \qquad \{\,\text{multiply the equation by 2}\,\}$$

$$f(x) = 2(x-3)^2 + 4 \qquad \{\,\text{the desired standard form}\,\}$$

9̄ $f(x) = -3x^2 - 6x - 5$ {divide by -3 and proceed as in Exercise 7} $\Rightarrow$

$$-\tfrac{1}{3}f(x) = x^2 + 2x + \underline{\ 1\ } + \tfrac{5}{3} - \underline{\ 1\ } = (x+1)^2 + \tfrac{2}{3} \ \Rightarrow$$

$$-3 \cdot -\tfrac{1}{3}f(x) = -3 \cdot \left[(x+1)^2 + \tfrac{2}{3}\right] \ \Rightarrow \ \underline{f(x) = -3(x+1)^2 - 2}$$

11̄ $f(x) = -\tfrac{3}{4}x^2 + 9x - 34$ {divide by $-\tfrac{3}{4}$ and proceed as in Exercise 7} $\Rightarrow$

$$-\tfrac{4}{3}f(x) = x^2 - 12x + \tfrac{136}{3} = x^2 - 12x + \underline{\ 36\ } + \tfrac{136}{3} - \underline{\ 36\ } = (x-6)^2 + \tfrac{28}{3} \ \Rightarrow$$

$$f(x) = (-\tfrac{3}{4})(x-6)^2 + (-\tfrac{3}{4}) \cdot (\tfrac{28}{3}) \ \Rightarrow \ \underline{f(x) = -\tfrac{3}{4}(x-6)^2 - 7}$$

13̄ (a) $x^2 - 4x = 0$ {$a = 1, b = -4, c = 0$} $\Rightarrow$ $x = \dfrac{4 \pm \sqrt{16 - 0}}{2} = \dfrac{4 \pm 4}{2} = 0, 4$

(b) Using the theorem for locating the vertex of a parabola with $y = x^2 - 4x$ gives us x-coordinate $-\dfrac{b}{2a} = -\dfrac{-4}{2(1)} = 2$. The parabola opens upward since $a = 1 > 0$, so there is a minimum value of y. To find the minimum, substitute 2 for x in $f(x) = x^2 - 4x$ to get $f(2) = -4$.

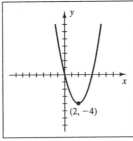

Figure 13

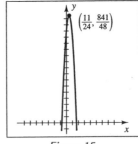

Figure 15

15̄ (a) $f(x) = -12x^2 + 11x + 15 = 0$ $\Rightarrow$ $x = \dfrac{-11 \pm \sqrt{121 + 720}}{-24} = \dfrac{-11 \pm 29}{-24} = -\dfrac{3}{4}, \dfrac{5}{3}$

(b) The x-coordinate of the vertex is given by $x = -\dfrac{b}{2a} = -\dfrac{11}{2(-12)} = \dfrac{11}{24}$.

Note that this value is easily obtained from part (a). The y-coordinate of the vertex is then $f(\tfrac{11}{24}) = \tfrac{841}{48} \approx 17.52$. This is a maximum since $a = -12 < 0$.

$\boxed{17}$ (a) $9x^2 + 24x + 16 = 0 \Rightarrow x = \dfrac{-24 \pm \sqrt{576 - 576}}{18} = \dfrac{-24}{18} = -\dfrac{4}{3}$

(b) $f(x) = 9x^2 + 24x + 16 \Rightarrow -\dfrac{b}{2a} = -\dfrac{24}{2(9)} = -\dfrac{4}{3}.$

$f\left(-\dfrac{4}{3}\right) = 0$ is a minimum since $a > 0$.

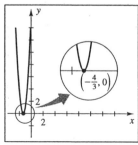   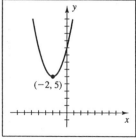   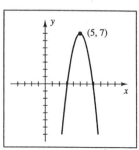

Figure 17   *Figure 19*   *Figure 21*

$\boxed{19}$ (a) $f(x) = x^2 + 4x + 9 = 0 \Rightarrow x = \dfrac{-4 \pm \sqrt{16 - 36}}{2} = \dfrac{-4 \pm \sqrt{-20}}{2} = -2 \pm \sqrt{5}\, i.$

The imaginary part indicates that there are *no* x-intercepts.

(b) $-\dfrac{b}{2a} = -\dfrac{4}{2(1)} = -2.$ $f(-2) = 5$ is a minimum since $a = 1 > 0$.

$\boxed{21}$ (a) $-2x^2 + 20x - 43 = 0 \Rightarrow x = \dfrac{-20 \pm \sqrt{400 - 344}}{-4} = 5 \pm \dfrac{1}{2}\sqrt{14} \approx 6.87,\ 3.13$

(b) $f(x) = -2x^2 + 20x - 43 \Rightarrow -\dfrac{b}{2a} = -\dfrac{20}{2(-2)} = 5.$

$f(5) = 7$ is a maximum since $a < 0$.

$\boxed{23}$ $V(4, -1) \Rightarrow y = a(x - 4)^2 - 1$. Substituting 0 for x and 1 for y gives us

$1 = a(0 - 4)^2 - 1 \Rightarrow 2 = 16a \Rightarrow a = \dfrac{1}{8}$. Hence, $y = \dfrac{1}{8}(x - 4)^2 - 1.$

$\boxed{25}$ $V(-2, 4) \Rightarrow y = a(x + 2)^2 + 4.$ $x = 1,\ y = 0 \Rightarrow 0 = a(1 + 2)^2 + 4 \Rightarrow$

$-4 = 9a \Rightarrow a = -\dfrac{4}{9}$. Hence, $y = -\dfrac{4}{9}(x + 2)^2 + 4.$

$\boxed{27}$ $V(0, -2) \Rightarrow (h, k) = (0, -2).$ $x = 3,\ y = 25 \Rightarrow 25 = a(3 - 0)^2 - 2 \Rightarrow$

$27 = 9a \Rightarrow a = 3$. Hence, $y = 3(x - 0)^2 - 2$, or $y = 3x^2 - 2.$

$\boxed{29}$ $V(3, 5) \Rightarrow y = a(x - 3)^2 + 5$ (∗). If the x-intercept is 0, then the point $(0, 0)$ is on

the parabola. Substituting $x = 0$ and $y = 0$ into (∗) gives us $0 = a(0 - 3)^2 + 5 \Rightarrow$

$-5 = 9a \Rightarrow a = -\dfrac{5}{9}$. Hence, $y = -\dfrac{5}{9}(x - 3)^2 + 5.$

$\boxed{31}$ Since the x-intercepts are -3 and 5, the x-coordinate of the vertex of the parabola is

1 { the average of the x-intercept values }. Since the highest point has y-coordinate 4,

the vertex is $(1, 4)$. $V(1, 4) \Rightarrow y = a(x - 1)^2 + 4$. We can use the point $(-3, 0)$

since there is an x-intercept of -3. $x = -3,\ y = 0 \Rightarrow 0 = a(-3 - 1)^2 + 4 \Rightarrow$

$-4 = 16a \Rightarrow a = -\dfrac{1}{4}$. Hence, $y = -\dfrac{1}{4}(x - 1)^2 + 4.$

33 Let d denote the distance between the parabola and the line.

$$d = (\text{parabola}) - (\text{line}) = (-2x^2 + 4x + 3) - (x - 2) = -2x^2 + 3x + 5.$$

This relation is quadratic and the x-value of its maximum value is

$$-\frac{b}{2a} = -\frac{3}{2(-2)} = \frac{3}{4}. \text{ Thus, maximum } d = -2\left(\frac{3}{4}\right)^2 + 3\left(\frac{3}{4}\right) + 5 = \frac{49}{8} = 6.125.$$

Note that the maximum value *of the parabola* is $f(1) = 5$,

which is not the same as the maximum value of the distance d.

Note: We may find the vertex of a parabola in the following problems using either:

    (1) the complete the square method,

    (2) the formula method, or

    (3) the fact that the vertex lies halfway between the x-intercepts.

35 For $D(h) = -0.058h^2 + 2.867h - 24.239$, the vertex is located at

$$h = \frac{-b}{2a} = \frac{-2.867}{2(-0.058)} \approx 24.72 \text{ km. Since } a < 0, \text{ this will produce a maximum value.}$$

37 Since the x-intercepts of $y = cx(21 - x)$ are 0 and 21, the maximum will occur

halfway between them, that is, when the infant weighs 10.5 lb.

39 (a) $s(t) = -16t^2 + 144t + 100$ will be a maximum when $t = \frac{-b}{2a} = \frac{-144}{2(-16)} = \frac{9}{2}$.

$$s\left(\tfrac{9}{2}\right) = -16\left(\tfrac{9}{2}\right)^2 + 144\left(\tfrac{9}{2}\right) + 100 = -324 + 648 + 100 = 424 \text{ ft.}$$

  (b) When $t = 0$, $s(t) = 100$ ft, which is the height of the building.

41 Let x and $40 - x$ denote the numbers, and their product P is $x(40 - x)$.

P has zeros at 0 and 40 and is a maximum (since $a < 0$) when $x = \frac{0 + 40}{2} = 20$.

The product will be a maximum when both numbers are 20.

43 (a) The 1000 ft of fence is made up of 3 sides of length x and 4 sides of length y.

To express y as a function of x, we need to solve $3x + 4y = 1000$ for y.

$$3x + 4y = 1000 \implies 4y = 1000 - 3x \implies y = 250 - \tfrac{3}{4}x.$$

  (b) Using the value of y from part (a), $A = xy = x(250 - \tfrac{3}{4}x) = -\tfrac{3}{4}x^2 + 250x.$

  (c) A will be a maximum when $x = \frac{-b}{2a} = \frac{-250}{2(-3/4)} = \frac{500}{3} = 166\tfrac{2}{3}$ ft. Using part (a)

to find the corresponding value of y, $y = 250 - \tfrac{3}{4}\left(\tfrac{500}{3}\right) = 250 - 125 = 125$ ft.

45 The parabola has vertex $V\left(\tfrac{9}{2}, 3\right)$. Hence, the equation has the form $y = a\left(x - \tfrac{9}{2}\right)^2 + 3$.

Using the point $(9, 0)$ { or $(0, 0)$ }, we have $0 = a\left(9 - \tfrac{9}{2}\right)^2 + 3 \implies a = -\tfrac{4}{27}$.

Thus, the path may be described by $y = -\tfrac{4}{27}\left(x - \tfrac{9}{2}\right)^2 + 3$.

47 (a) Since the vertex is at $(0, 10)$, an equation for the parabola is $y = ax^2 + 10$.

The points $(200, 90)$ and $(-200, 90)$ are on the parabola.

Substituting $(200, 90)$ for (x, y) yields $90 = a(200)^2 + 10 \implies a = \frac{80}{40{,}000} = \frac{1}{500}$.

Hence, $y = \frac{1}{500}x^2 + 10$.

(b) The cables are spaced 40 ft apart. Using $y = \frac{1}{500}x^2 + 10$ with

$x = 40, 80, 120,$ and 160, we get $y = \frac{66}{5}, \frac{114}{5}, \frac{194}{5},$ and $\frac{306}{5}$, respectively.

There is one cable of length 10 ft and 2 cables of each of the other lengths.

Thus, the total length is $10 + 2\left(\frac{66}{5} + \frac{114}{5} + \frac{194}{5} + \frac{306}{5}\right) = 282$ ft.

49 An equation describing the doorway is $y = ax^2 + 9$. Since the doorway is 6 feet wide

at the base, $x = 3$ when $y = 0 \implies 0 = a(3)^2 + 9 \implies 0 = 9a + 9 \implies 9a = -9 \implies$

$a = -1$. Thus, the equation is $y = -x^2 + 9$. To fit an 8 foot high box through the

doorway, we must find x when $y = 8$. If $y = 8$, then $8 = -x^2 + 9 \implies x^2 = 1 \implies$

$x = \pm 1$. Hence, the box can only be $1 - (-1) = 2$ feet wide.

51 Let x denote the number of pairs of shoes that are ordered. If $x < 50$, then the

amount A of money that the company makes is $40x$. If $50 \le x \le 600$, then each pair

of shoes is discounted $0.04x$, so the price per pair is $40 - 0.04x$, and the amount of

money that the company makes is $(40 - 0.04x)x$. In piecewise form, we have

$$A(x) = \begin{cases} 40x & \text{if } x < 50 \\ (40 - 0.04x)x & \text{if } 50 \le x \le 600 \end{cases}$$

The maximum value of the first part of A is $(\$40)(49) = \$1960$. For the second part

of A, $A = -0.04x^2 + 40x$ has a maximum when $x = \frac{-b}{2a} = \frac{-40}{2(-0.04)} = 500$ pairs.

$A(500) = 10{,}000 > 1960$, so $x = 500$ produces a maximum for both parts of A.

53 (a) Let y denote the number of \$1 decreases in the monthly charge.

$R(y) = (\text{\# of customers})(\text{monthly charge per customer})$

$= (5000 + 500y)(20 - y)$

$= 500(10 + y)(20 - y)$

Now let x denote the monthly charge, which is $20 - y$.

R becomes $500[10 + (20 - x)](x) = 500x(30 - x)$.

(b) R has x-intercepts at 0 and 30, and must have its

vertex halfway between them at $x = 15$.

Note that this gives us $y = 20 - 15 = 5$, and we have

$5000 + 500(5) = 7500$ customers for a revenue of

$500(10 + 5)(20 - 5) = \$112{,}500$.

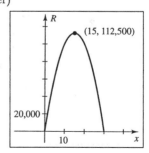

Figure 53

55 From the graph, there are three points of intersection.

Their coordinates are approximately $(-0.57, 0.64)$, $(0.02, -0.27)$, and $(0.81, -0.41)$.

$[-3, 3]$ by $[-2, 2]$ $[-8, 4]$ by $[-1, 7]$

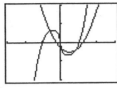   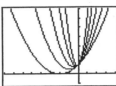

Figure 55   Figure 57

57 Since $a > 0$, all parabolas open upward. From the graph, we can see that smaller values of a result in the parabola opening wider while larger values of a result in the parabola becoming narrower.

59 (a) Let January correspond to 1, February to 2, ..., and December to 12.

(b) Let $f(x) = a(x - h)^2 + k$. The vertex appears to occur near $(7, 0.8)$ Thus, $h = 7$ and $k = 0.8$. Using trial and error, a reasonable value for a is 0.17. Thus, let $f(x) = 0.17(x - 7)^2 + 0.8$. {From the TI-83 Plus, the quadratic regression equation is $y \approx 0.1476x^2 - 1.9951x + 7.8352$.}

(c) $f(4) = 2.33$, compared to the actual value of 2.4 in.

$[0.5, 12]$ by $[0, 8]$ $[-800, 800, 100]$ by $[-100, 200, 100]$

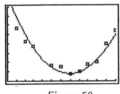   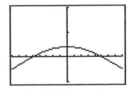

Figure 59   Figure 61

61 (a) The equation of the line passing through $A(-800, -48)$ and $B(-500, 0)$ is $y = \frac{4}{25}x + 80$. The equation of the line passing through $D(500, 0)$ and $E(800, -48)$ is $y = -\frac{4}{25}x + 80$. Let $y = a(x - h)^2 + k$ be the equation of the parabola passing through the points $B(-500, 0)$, $C(0, 40)$, and $D(500, 0)$. The vertex is located at $(0, 40)$ so $y = a(x - 0)^2 + 40$. Since $D(500, 0)$ is on the graph, $0 = a(500 - 0)^2 + 40 \Rightarrow a = -\frac{1}{6250}$ and $y = -\frac{1}{6250}x^2 + 40$. Thus, let

$$f(x) = \begin{cases} \frac{4}{25}x + 80 & \text{if } -800 \leq x < -500 \\ -\frac{1}{6250}x^2 + 40 & \text{if } -500 \leq x \leq 500 \\ -\frac{4}{25}x + 80 & \text{if } 500 < x \leq 800 \end{cases}$$

(b) Graph the equations:

$$Y_1 = (4/25 * x + 80)/(x < -500),$$

$$Y_2 = (-1/6250 * x^2 + 40)/(x \geq -500 \text{ and } x \leq 500),$$

$$Y_3 = (-4/25 * x + 80)/(x > 500)$$

$\boxed{63}$ (a) f must have zeros of 0 and 150. Thus, $f(x) = a(x - 0)(x - 150)$. Also, f will have a maximum of 100 occurring at $x = 75$. (The vertex will be midway between the zeros of f.) $a(75 - 0)(75 - 150) = 100 \Rightarrow a = \frac{100}{(75)(-75)} = -\frac{4}{225}$. $f(x) = -\frac{4}{225}(x)(x - 150) = -\frac{4}{225}x^2 + \frac{8}{3}x$.

(c) The value of k affects both the distance and the height traveled by the object. The distance and height decrease by a factor of $\frac{1}{k}$ when $k > 1$ and increase by a factor of $\frac{1}{k}$ when $0 < k < 1$.

[0, 180, 50] by [0, 120, 50]

[0, 600, 50] by [0, 400, 50]

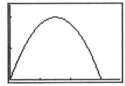

Figure 63(b)

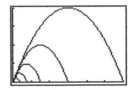

Figure 63(c)

2.4 Exercises

$\boxed{1}$ (a) $(f + g)(3) = f(3) + g(3) = 6 + 9 = 15$

(b) $(f - g)(3) = f(3) - g(3) = 6 - 9 = -3$

(c) $(fg)(3) = f(3) \cdot g(3) = 6 \cdot 9 = 54$

(d) $(f/g)(3) = \dfrac{f(3)}{g(3)} = \dfrac{6}{9} = \dfrac{2}{3}$

$\boxed{3}$ (a) $(f + g)(x) = f(x) + g(x) = (x^2 + 2) + (2x^2 - 1) = 3x^2 + 1;$

$(f - g)(x) = f(x) - g(x) = (x^2 + 2) - (2x^2 - 1) = 3 - x^2;$

$(fg)(x) = f(x) \cdot g(x) = (x^2 + 2) \cdot (2x^2 - 1) = 2x^4 + 3x^2 - 2;$

$\left(\dfrac{f}{g}\right)(x) = \dfrac{f(x)}{g(x)} = \dfrac{x^2 + 2}{2x^2 - 1}$

(b) The domain of $f + g$, $f - g$, and fg is the set of all real numbers, $\mathbb{R}$.

(c) The domain of f/g is the same as in (b), except we must exclude the zeros of g.

$$2x^2 - 1 = 0 \Rightarrow x^2 = \frac{1}{2} \Rightarrow x = \pm\sqrt{\frac{1}{2}} = \pm\frac{\sqrt{2}}{2} \text{ or } \pm\frac{1}{2}\sqrt{2} \Rightarrow$$

Hence, the domain of f/g is all real numbers except $\pm\frac{1}{2}\sqrt{2}$.

[5] (a) $(f+g)(x) = f(x) + g(x) = \sqrt{x+5} + \sqrt{x+5} = 2\sqrt{x+5}$;

$(f-g)(x) = f(x) - g(x) = \sqrt{x+5} - \sqrt{x+5} = 0$;

$(fg)(x) = f(x) \cdot g(x) = \sqrt{x+5} \cdot \sqrt{x+5} = x+5$;

$\left(\dfrac{f}{g}\right)(x) = \dfrac{f(x)}{g(x)} = \dfrac{\sqrt{x+5}}{\sqrt{x+5}} = 1$

(b) The radicand, $x+5$, must be nonnegative; that is $x+5 \geq 0 \;\Rightarrow\; x \geq -5$.

Thus, the domain of $f+g$, $f-g$, and fg is $[-5, \infty)$.

(c) Now the radicand must be positive { can't have zero in the denominator }.

Thus, the domain of f/g is $(-5, \infty)$.

[7] (a) $(f+g)(x) = f(x) + g(x) = \dfrac{2x}{x-4} + \dfrac{x}{x+5} = \dfrac{2x(x+5) + x(x-4)}{(x-4)(x+5)} = \dfrac{3x^2 + 6x}{(x-4)(x+5)}$;

$(f-g)(x) = f(x) - g(x) = \dfrac{2x}{x-4} - \dfrac{x}{x+5} = \dfrac{2x(x+5) - x(x-4)}{(x-4)(x+5)} = \dfrac{x^2 + 14x}{(x-4)(x+5)}$;

$(fg)(x) = f(x) \cdot g(x) = \dfrac{2x}{x-4} \cdot \dfrac{x}{x+5} = \dfrac{2x^2}{(x-4)(x+5)}$;

$\left(\dfrac{f}{g}\right)(x) = \dfrac{f(x)}{g(x)} = \dfrac{2x/(x-4)}{x/(x+5)} = \dfrac{2(x+5)}{x-4}$

(b) The domain of f is $\mathbb{R} - \{4\}$ and the domain of g is $\mathbb{R} - \{-5\}$. The intersection of these two domains, $\mathbb{R} - \{-5, 4\}$, is the domain of the three functions.

(c) To determine the domain of the quotient f/g, we also exclude any values that make the denominator g equal to zero. Hence, we exclude $x = 0$ and the domain of the quotient is all real numbers except -5, 0, and 4, that is, $\mathbb{R} - \{-5, 0, 4\}$.

[9] (a) $(f \circ g)(x) = f(g(x)) = f(-x^2) = 2(-x^2) - 1 = -2x^2 - 1$

(b) $(g \circ f)(x) = g(f(x)) = g(2x - 1) = -(2x - 1)^2 = -(4x^2 - 4x + 1) = -4x^2 + 4x - 1$

(c) $(f \circ f)(x) = f(f(x)) = f(2x - 1) = 2(2x - 1) - 1 = (4x - 2) - 1 = 4x - 3$

(d) $(g \circ g)(x) = g(g(x)) = g(-x^2) = -(-x^2)^2 = -(x^4) = -x^4$

Note: Let $h(x) = (f \circ g)(x) = f(g(x))$ and $k(x) = (g \circ f)(x) = g(f(x))$.

$h(-2)$ and $k(3)$ could be worked two ways, as in Example 3(c) in the text.

[11] (a) $h(x) = f(3x + 7) = 2(3x + 7) - 5 = 6x + 9$

(b) $k(x) = g(2x - 5) = 3(2x - 5) + 7 = 6x - 8$

(c) Using the result from part (a), $h(-2) = 6(-2) + 9 = -12 + 9 = -3$.

(d) Using the result from part (b), $k(3) = 6(3) - 8 = 18 - 8 = 10$.

[13] (a) $h(x) = f(5x) = 3(5x)^2 + 4 = 75x^2 + 4$

(b) $k(x) = g(3x^2 + 4) = 5(3x^2 + 4) = 15x^2 + 20$

(c) $h(-2) = 75(-2)^2 + 4 = 300 + 4 = 304$

(d) $k(3) = 15(3)^2 + 20 = 135 + 20 = 155$

$\boxed{15}$ (a) $h(x) = f(2x - 1) = 2(2x - 1)^2 + 3(2x - 1) - 4 = 8x^2 - 2x - 5$

(b) $k(x) = g(2x^2 + 3x - 4) = 2(2x^2 + 3x - 4) - 1 = 4x^2 + 6x - 9$

(c) $h(-2) = 8(-2)^2 - 2(-2) - 5 = 32 + 4 - 5 = 31$

(d) $k(3) = 4(3)^2 + 6(3) - 9 = 36 + 18 - 9 = 45$

$\boxed{17}$ (a) $h(x) = f(2x^3 - 5x) = 4(2x^3 - 5x) = 8x^3 - 20x$

(b) $k(x) = g(4x) = 2(4x)^3 - 5(4x) = 128x^3 - 20x$

(c) $h(-2) = 8(-2)^3 - 20(-2) = -64 + 40 = -24$

(d) $k(3) = 128(3)^3 - 20(3) = 3456 - 60 = 3396$

$\boxed{19}$ (a) $h(x) = f(-7) = |-7| = 7$ (b) $k(x) = g(|x|) = -7$

(c) $h(-2) = 7$ since h(any value) $= 7$ (d) $k(3) = -7$ since k(any value) $= -7$

$\boxed{21}$ (a) $h(x) = f(\sqrt{x+2}) = (\sqrt{x+2})^2 - 3(\sqrt{x+2}) = x + 2 - 3\sqrt{x+2}$. The domain of $f \circ g$ is the set of all x in the domain of g, $x \ge -2$, such that $g(x)$ is in the domain of f. Since the domain of f is $\mathbb{R}$, any value of $g(x)$ is in its domain. Thus, the domain is all x such that $x \ge -2$. Note that *both $g(x)$ and $f(g(x))$ are* defined for x in $[-2, \infty)$.

(b) $k(x) = g(x^2 - 3x) = \sqrt{(x^2 - 3x) + 2} = \sqrt{x^2 - 3x + 2}$. The domain of $g \circ f$ is the set of all x in the domain of f, $\mathbb{R}$, such that $f(x)$ is in the domain of g. Since the domain of g is $x \ge -2$, we must solve $f(x) \ge -2$.

$$x^2 - 3x \ge -2 \;\Rightarrow\; x^2 - 3x + 2 \ge 0 \;\Rightarrow\; (x - 1)(x - 2) \ge 0$$

Resulting sign:	$\oplus$	$\ominus$	$\oplus$
Sign of $x - 2$:	$-$	$-$	$+$
Sign of $x - 1$:	$-$	$+$	$+$
x values:	1	2	

Diagram 21

From the sign diagram, $(x - 1)(x - 2) \ge 0 \;\Rightarrow\; x \in (-\infty, 1] \cup [2, \infty)$. Thus, the domain is all x such that $x \in (-\infty, 1] \cup [2, \infty)$. Note that *both $f(x)$ and $g(f(x))$* are defined for x in $(-\infty, 1] \cup [2, \infty)$.

$\boxed{23}$ (a) $h(x) = f(\sqrt{3x}) = (\sqrt{3x})^2 - 4 = 3x - 4$.

Domain of $g = [0, \infty)$. Domain of $f = \mathbb{R}$. Since $g(x)$ is always in the domain of f, the domain of $f \circ g$ is the same as the domain of g, $[0, \infty)$. Note that *both $g(x)$ and $f(g(x))$ are* defined for x in $[0, \infty)$.

(b) $k(x) = g(x^2 - 4) = \sqrt{3(x^2 - 4)} = \sqrt{3x^2 - 12}$.

Domain of $f = \mathbb{R}$. Domain of $g = [0, \infty)$.

$f(x) \ge 0 \;\Rightarrow\; x^2 - 4 \ge 0 \;\Rightarrow\; x^2 \ge 4 \;\Rightarrow\; |x| \ge 2 \;\Rightarrow\; x \in (-\infty, -2] \cup [2, \infty)$.

Note that *both $f(x)$ and $g(f(x))$ are* defined for x in $(-\infty, -2] \cup [2, \infty)$.

25 (a) $h(x) = f(\sqrt{x+5}) = \sqrt{\sqrt{x+5} - 2}$.

Domain of $g = [-5, \infty)$. Domain of $f = [2, \infty)$.

$$g(x) \geq 2 \;\Rightarrow\; \sqrt{x+5} \geq 2 \;\Rightarrow\; x+5 \geq 4 \;\Rightarrow\; x \geq -1 \text{ or } x \in [-1, \infty).$$

(b) $k(x) = g(\sqrt{x-2}) = \sqrt{\sqrt{x-2} + 5}$. Domain of $f = [2, \infty)$.

Domain of $g = [-5, \infty)$. $f(x) \geq -5 \;\Rightarrow\; \sqrt{x-2} \geq -5$. This is always true since

the result of a square root is nonnegative. The domain is $[2, \infty)$.

27 (a) $h(x) = f(\sqrt{x^2 - 16}) = \sqrt{3 - \sqrt{x^2 - 16}}$.

Domain of $g = (-\infty, -4] \cup [4, \infty)$. Domain of $f = (-\infty, 3]$.

$$g(x) \leq 3 \;\Rightarrow\; \sqrt{x^2 - 16} \leq 3 \;\Rightarrow\; x^2 - 16 \leq 9 \;\Rightarrow\; x^2 \leq 25 \;\Rightarrow\; x \in [-5, 5].$$

But $|x| \geq 4$ from the domain of g, so the domain of $f \circ g$ is $[-5, -4] \cup [4, 5]$.

(b) $k(x) = g(\sqrt{3-x}) = \sqrt{(\sqrt{3-x})^2 - 16} = \sqrt{3 - x - 16} = \sqrt{-x - 13}$.

Domain of $f = (-\infty, 3]$. Domain of $g = (-\infty, -4] \cup [4, \infty)$.

$f(x) \geq 4 \; \{f(x) \text{ cannot be less than } 0\} \;\Rightarrow$

$$\sqrt{3-x} \geq 4 \;\Rightarrow\; 3 - x \geq 16 \;\Rightarrow\; x \leq -13.$$

29 (a) $h(x) = f\left(\dfrac{2x-5}{3}\right) = \dfrac{3\left(\dfrac{2x-5}{3}\right) + 5}{2} = \dfrac{2x - 5 + 5}{2} = \dfrac{2x}{2} = x$.

Domain of $g = \mathbb{R}$. Domain of $f = \mathbb{R}$. All values of $g(x)$ are in the domain of f.

Hence, the domain of $f \circ g$ is $\mathbb{R}$.

(b) $k(x) = g\left(\dfrac{3x+5}{2}\right) = \dfrac{2\left(\dfrac{3x+5}{2}\right) - 5}{3} = \dfrac{3x + 5 - 5}{3} = \dfrac{3x}{3} = x$.

Domain of $f = \mathbb{R}$. Domain of $g = \mathbb{R}$. All values of $f(x)$ are in the domain of g.

Hence, the domain of $g \circ f$ is $\mathbb{R}$.

31 (a) $h(x) = f\left(\dfrac{1}{x^3}\right) = \left(\dfrac{1}{x^3}\right)^2 = \dfrac{1}{x^6}$. Domain of $g = \mathbb{R} - \{0\}$. Domain of $f = \mathbb{R}$.

All values of $g(x)$ are in the domain of f. Hence, the domain of $f \circ g$ is $\mathbb{R} - \{0\}$.

(b) $k(x) = g(x^2) = \dfrac{1}{(x^2)^3} = \dfrac{1}{x^6}$. Domain of $f = \mathbb{R}$. Domain of $g = \mathbb{R} - \{0\}$.

All values of $f(x)$ are in the domain of g except for 0.

Since f is 0 when x is 0, the domain of $f \circ g$ is $\mathbb{R} - \{0\}$.

33 (a) $h(x) = f\left(\dfrac{x-3}{x-4}\right) = \dfrac{\dfrac{x-3}{x-4} - 1}{\dfrac{x-3}{x-4} - 2} \cdot \dfrac{x-4}{x-4} = \dfrac{x - 3 - 1(x-4)}{x - 3 - 2(x-4)} = \dfrac{1}{5-x}$.

Domain of $g = \mathbb{R} - \{4\}$. Domain of $f = \mathbb{R} - \{2\}$. $g(x) \neq 2 \;\Rightarrow$

$$\dfrac{x-3}{x-4} \neq 2 \;\Rightarrow\; x - 3 \neq 2x - 8 \;\Rightarrow\; x \neq 5. \text{ The domain is } \mathbb{R} - \{4, 5\}.$$

(b) $k(x) = g\left(\dfrac{x-1}{x-2}\right) = \dfrac{\dfrac{x-1}{x-2} - 3}{\dfrac{x-1}{x-2} - 4} \cdot \dfrac{x-2}{x-2} = \dfrac{x-1-3(x-2)}{x-1-4(x-2)} = \dfrac{-2x+5}{-3x+7}.$

Domain of $f = \mathbb{R} - \{2\}$. Domain of $g = \mathbb{R} - \{4\}$. $f(x) \ne 4 \Rightarrow$

$\dfrac{x-1}{x-2} \ne 4 \Rightarrow x - 1 \ne 4x - 8 \Rightarrow x \ne \frac{7}{3}$. The domain is $\mathbb{R} - \{2, \frac{7}{3}\}$.

$\boxed{35}$ $(f \circ g)(x) = f(g(x)) = f(x+3) = (x+3)^2 - 2.$

$(f \circ g)(x) = 0 \Rightarrow (x+3)^2 - 2 = 0 \Rightarrow (x+3)^2 = 2 \Rightarrow x + 3 = \pm\sqrt{2} \Rightarrow$

$$x = -3 \pm \sqrt{2}$$

$\boxed{37}$ (a) $(f \circ g)(6) = f(g(6)) = f(8) = 5$ (b) $(g \circ f)(6) = g(f(6)) = g(7) = 6$

     (c) $(f \circ f)(6) = f(f(6)) = f(7) = 6$ (d) $(g \circ g)(6) = g(g(6)) = g(8) = 5$

$\boxed{39}$ $(D \circ R)(x) = D(R(x)) = D(20x) =$

$$\sqrt{400 + (20x)^2} = \sqrt{400 + 400x^2} = \sqrt{400(1 + x^2)} = \sqrt{400}\,\sqrt{x^2 + 1} = 20\sqrt{x^2 + 1}$$

$\boxed{41}$ We need to examine $(fg)(-x)$—if we obtain $(fg)(x)$, then fg is an even function,

                 whereas if we obtain $-(fg)(x)$, then fg is an odd function.

$(fg)(-x) = f(-x)g(-x)$ { definition of the product of two functions }

           $= -f(x)g(x)$ { f is odd, so $f(-x) = -f(x)$; g is even, so $g(-x) = g(x)$ }

           $= -(fg)(x)$ { definition of the product of two functions }

Since $(fg)(-x) = -(fg)(x)$, fg is an odd function.

$\boxed{43}$ $(\text{ROUND2} \circ \text{SSTAX})(437.21) = \text{ROUND2}(\text{SSTAX}(437.21))$

                                $= \text{ROUND2}(0.0715 \cdot 437.21)$

                                $= \text{ROUND2}(31.260515) = 31.26$

$\boxed{45}$ $r = 6t$ and $A = \pi r^2 \Rightarrow A = \pi(6t)^2 = 36\pi t^2$ ft².

$\boxed{47}$ The formula for the volume of a cone is $V = \frac{1}{3}\pi r^2 h = \frac{1}{3}\pi r^3$ { since $h = r$ } $\Rightarrow$

     $r^3 = \dfrac{3V}{\pi} \Rightarrow r = \sqrt[3]{\dfrac{3V}{\pi}}.$ $V = 243\pi t \Rightarrow r = \sqrt[3]{\dfrac{3 \cdot 243\pi t}{\pi}} = \sqrt[3]{729t} = 9\sqrt[3]{t}$ ft.

$\boxed{49}$ Let l denote the length of the rope. At $t = 0$, $l = 20$. At $t = 1$, $l = 25$.

At time t, $l = 20 + 5t$, *not* just $5t$. We have a right triangle with sides 20, h, and l.

$h^2 + 20^2 = l^2 \Rightarrow$

$$h = \sqrt{(20 + 5t)^2 - 20^2} = \sqrt{25t^2 + 200t} = \sqrt{25(t^2 + 8t)} = 5\sqrt{t^2 + 8t}.$$

$\boxed{51}$ From Exercise 69 of Section 2.1, $d(x) = \sqrt{90{,}400 + x^2}$. The distance x of the plane

from the control tower is 500 feet plus 150 feet per second, that is, $x(t) = 500 + 150t$.

Thus, $d(t) = \sqrt{90{,}400 + (500 + 150t)^2} = \sqrt{90{,}400 + 250{,}000 + 150{,}000t + 22{,}500t^2}$

        $= \sqrt{22{,}500t^2 + 150{,}000t + 340{,}400} = 10\sqrt{225t^2 + 1500t + 3404}.$

---

53 $y = (x^2 + 3x)^{1/3}$ • Suppose you were to find the value of y if x was equal to 3. Using a calculator, you might compute the value of $x^2 + 3x$ first, and then raise that result to the $\frac{1}{3}$ power. Thus, we would choose $y = u^{1/3}$ and $u = x^2 + 3x$.

55 For $y = \dfrac{1}{(x-3)^4}$, choose $u = x - 3$ and $y = 1/u^4 = u^{-4}$.

57 For $y = (x^4 - 2x^2 + 5)^5$, choose $u = x^4 - 2x^2 + 5$ and $y = u^5$.

59 For $y = \dfrac{\sqrt{x+4} - 2}{\sqrt{x+4} + 2}$, there is not a "simple" choice for y as in previous exercises.

One choice for u is $u = x + 4$. Then y would be $\dfrac{\sqrt{u} - 2}{\sqrt{u} + 2}$.

Another choice for u is $u = \sqrt{x+4}$. Then y would be $\dfrac{u - 2}{u + 2}$.

61 $(f \circ g)(x) = f(g(x)) = f(x^3 + 1) = \sqrt{x^3 + 1} - 1$. We will multiply this expression by

$\dfrac{\sqrt{x^3 + 1} + 1}{\sqrt{x^3 + 1} + 1}$, treating it as though it was one factor of a difference of two squares.

$$\left(\sqrt{x^3 + 1} - 1\right) \times \frac{\sqrt{x^3 + 1} + 1}{\sqrt{x^3 + 1} + 1} = \frac{\left(\sqrt{x^3 + 1}\right)^2 - 1^2}{\sqrt{x^3 + 1} + 1} = \frac{x^3}{\sqrt{x^3 + 1} + 1}$$

Thus, $(f \circ g)(0.0001) = f(g(10^{-4})) \approx \dfrac{(10^{-4})^3}{2} = 5 \times 10^{-13}$.

63 *Note:* If we relate this problem to the composite function $g(f(u))$, then Y_2 can be considered to be f, Y_1 {which can be considered to be u} is the function of x we are substituting into f, and Y_3 {which can be considered to be g} is accepting Y_2's output values as its input values. Hence, Y_3 is a function of a function of a function.

(a) $y = -2f(x)$; $Y_1 = x$, $Y_2 = 3\sqrt{(Y_1 + 2)(6 - Y_1)} - 4$, graph $Y_3 = -2Y_2$

{turn off Y_1 and Y_2, leaving only Y_3 on}; $D = [-2, 6]$, $R = [-16, 8]$

Note: The graphs often do not show the correct endpoints—you need to change the viewing rectangle or zoom in to actually view them on the screen.

[−12, 12, 2] by [−16, 8, 2]   [−12, 12, 2] by [−16, 8, 2]

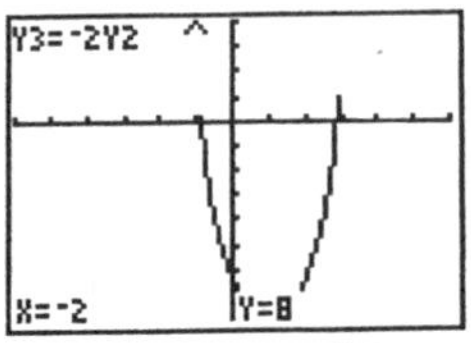

Figure 63(a)

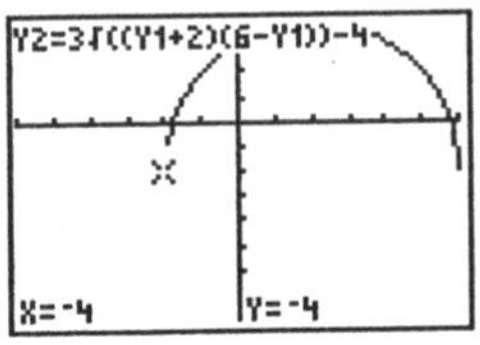

Figure 63(b)

(b) $y = f(\tfrac{1}{2}x)$; $Y_1 = 0.5x$, graph Y_2; $D = [-4, 12]$, $R = [-4, 8]$

(c) $y = f(x - 3) + 1$; $Y_1 = x - 3$, graph $Y_3 = Y_2 + 1$; $D = [1, 9]$, $R = [-3, 9]$

[−12, 12, 2] by [−6, 10, 2]

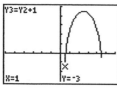

Figure 63(c)

[−12, 12, 2] by [−8, 8, 2]

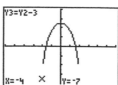

Figure 63(d)

(d) $y = f(x + 2) - 3$; $Y_1 = x + 2$, graph $Y_3 = Y_2 - 3$; $D = [-4, 4]$, $R = [-7, 5]$

(e) $y = f(-x)$; $Y_1 = -x$, graph Y_2; $D = [-6, 2]$, $R = [-4, 8]$

[−12, 12, 2] by [−8, 8, 2]

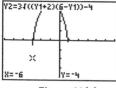

Figure 63(e)

[−12, 12, 2] by [−8, 8, 2]

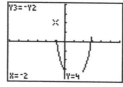

Figure 63(f)

(f) $y = -f(x)$; $Y_1 = x$, graph $Y_3 = -Y_2$; $D = [-2, 6]$, $R = [-8, 4]$

(g) $y = f(\,|\,x\,|\,)$; $Y_1 = \text{abs } x$, graph Y_2; $D = [-6, 6]$, $R = [-4, 8]$

[−12, 12, 2] by [−6, 10, 2]

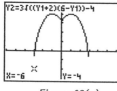

Figure 63(g)

[−4, 8] by [−2, 10]

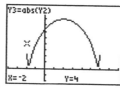

Figure 63(h)

(h) $y = |\,f(x)\,|$; $Y_1 = x$, graph $Y_3 = \text{abs } Y_2$; $D = [-2, 6]$, $R = [0, 8]$

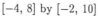

2.5 Exercises

[1] (a) f is one-to-one and $f(4) = 5$, so $f^{-1}(5) = 4$.

(b) g is *not* one-to-one since $g(1) = g(5) = 6$, so we cannot find $g^{-1}(6)$. Not possible

Note: To help determine if you should try to prove the function is one-to-one or look for a counterexample to show that it is not one-to-one, consider the question: "If y was a particular value, could I find a unique x?" If the answer is yes, try to prove the function is one-to-one. Also, consider the Horizontal Line Test listed on page 236 in the text.

[3] If y was a particular value, say 5, we would have $5 = 3x - 7$. Trying to solve for x would yield $12 = 3x \Rightarrow x = 4$. Since we could find a unique x, we will try to prove that the function is one-to-one. **Proof** Suppose that $f(a) = f(b)$ for some numbers a and b in the domain. This gives us $3a - 7 = 3b - 7 \Rightarrow 3a = 3b \Rightarrow a = b$. Since $f(a) = f(b)$ implies that $a = b$, we conclude that f is one-to-one.

[5] If y was a particular value, say 7, we would have $7 = x^2 - 9$. Trying to solve for x would yield $16 = x^2 \Rightarrow x = \pm 4$. Since we could not find a *unique* x, we will show that two different numbers have the same function value. Using the information already obtained, $f(4) = 7$ and $f(-4) = 7$, but $4 \neq -4$ and hence, f is *not* one-to-one.

[7] Suppose $f(a) = f(b)$ with $a, b \geq 0$. $\sqrt{a} = \sqrt{b} \Rightarrow (\sqrt{a})^2 = (\sqrt{b})^2 \Rightarrow a = b$.

f is one-to-one.

[9] For $f(x) = |x|$, $f(-1) = 1 = f(1)$. f is *not* one-to-one.

[11] For $f(x) = \sqrt{4 - x^2}$, $f(-1) = \sqrt{3} = f(1)$. f is *not* one-to-one.

[13] Suppose $f(a) = f(b)$. $\frac{1}{a} = \frac{1}{b} \Rightarrow a = b$. f is one-to-one.

Note: For Exercises 15–18, we need to show that $f(g(x)) = x = g(f(x))$.

[15] $f(g(x)) = 3\left(\frac{x+2}{3}\right) - 2 = x + 2 - 2 = x$. $g(f(x)) = \frac{(3x - 2) + 2}{3} = \frac{3x}{3} = x$.

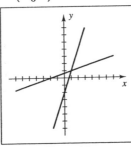

Figure 15

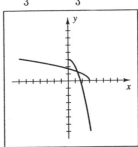

Figure 17

[17] $f(g(x)) = -(\sqrt{3 - x})^2 + 3 = -(3 - x) + 3 = x$.

$g(f(x)) = \sqrt{3 - (-x^2 + 3)} = \sqrt{x^2} = |x| = x$ { since $x \geq 0$ }.

[19] We need to solve for x. $y = 3x + 5 \Rightarrow y - 5 = 3x \Rightarrow x = \frac{y - 5}{3}$.

Now exchange x and y. $y = f^{-1}(x) = \frac{x - 5}{3}$

[21] $f(x) = \frac{1}{3x - 2} \Rightarrow y(3x - 2) = 1 \Rightarrow 3xy - 2y = 1 \Rightarrow 3xy = 2y + 1 \Rightarrow$

$$x = \frac{2y + 1}{3y} \Rightarrow f^{-1}(x) = \frac{2x + 1}{3x}$$

[23] $f(x) = \frac{3x + 2}{2x - 5} \Rightarrow 2xy - 5y = 3x + 2 \Rightarrow 2xy - 3x = 5y + 2 \Rightarrow$

$$x(2y - 3) = 5y + 2 \Rightarrow x = \frac{5y + 2}{2y - 3} \Rightarrow f^{-1}(x) = \frac{5x + 2}{2x - 3}$$

$\boxed{25}$ $f(x) = 2 - 3x^2$, $x \le 0$ $\Rightarrow$ $y + 3x^2 = 2$ $\Rightarrow$

$$x^2 = \frac{2-y}{3} \;\Rightarrow\; x = \pm\sqrt{\frac{2-y}{3}} \;\{\text{choose minus since } x \le 0\} \;\Rightarrow\; f^{-1}(x) = -\sqrt{\frac{2-x}{3}}$$

$\boxed{27}$ $f(x) = 2x^3 - 5$ $\Rightarrow$ $\dfrac{y+5}{2} = x^3$ $\Rightarrow$ $x = \sqrt[3]{\dfrac{y+5}{2}}$ $\Rightarrow$ $f^{-1}(x) = \sqrt[3]{\dfrac{x+5}{2}}$

$\boxed{29}$ $f(x) = \sqrt{3-x}$ $\Rightarrow$ $y^2 = 3 - x$ $\Rightarrow$

$\qquad\qquad x = 3 - y^2$ $\{\text{Since } y \ge 0 \text{ for } f,\ x \ge 0 \text{ for } f^{-1}.\}$ $\Rightarrow$ $f^{-1}(x) = 3 - x^2,\ x \ge 0$

$\boxed{31}$ $f(x) = \sqrt[3]{x} + 1$ $\Rightarrow$ $y - 1 = \sqrt[3]{x}$ $\Rightarrow$ $x = (y-1)^3$ $\Rightarrow$ $f^{-1}(x) = (x-1)^3$

$\boxed{33}$ $f(x) = x$ $\Rightarrow$ $y = x$ $\Rightarrow$ $x = y$ $\Rightarrow$ $f^{-1}(x) = x$

$\boxed{35}$ $f(x) = x^2 - 6x$, $x \ge 3$ $\Rightarrow$ $y = x^2 - 6x$ $\Rightarrow$ $x^2 - 6x - y = 0$.

This is a quadratic equation in x with $a = 1$, $b = -6$, and $c = -y$.

Solving with the quadratic formula gives us

$$x = \frac{-(-6) \pm \sqrt{(-6)^2 - 4(1)(-y)}}{2(1)} = \frac{6 \pm \sqrt{36 + 4y}}{2} = \frac{6 \pm 2\sqrt{9+y}}{2} = 3 \pm \sqrt{9+y}.$$

Since $x \ge 3$, we choose the "+" and obtain $f^{-1}(x) = 3 + \sqrt{x+9}$.

$\boxed{37}$ (a) $(g^{-1} \circ f^{-1})(2) = g^{-1}(f^{-1}(2))$

$\qquad\qquad\qquad\qquad\quad = g^{-1}(5) \qquad \{\text{since } f(5) = 2\}$

$\qquad\qquad\qquad\qquad\quad = 3 \qquad\qquad \{\text{since } g(3) = 5\}$

$\quad$ (b) $(g^{-1} \circ h)(3) = g^{-1}(h(3)) = g^{-1}(1) = -1$ $\{\text{since } g(-1) = 1\}$

$\quad$ (c) $(h^{-1} \circ f \circ g^{-1})(3) = h^{-1}(f(g^{-1}(3)))$

$\qquad\qquad\qquad\qquad\qquad\quad = h^{-1}(f(2))$ $\{\text{since } g(2) = 3\}$

$\qquad\qquad\qquad\qquad\qquad\quad = h^{-1}(-1)$

$\qquad\qquad\qquad\qquad\qquad\quad = 5 \qquad\qquad \{\text{since } -1 = 4 - x \;\Rightarrow\; x = 5\}$

$\boxed{39}$ (a) Remember that the domain of f is the range of f^{-1} and that the range of f is the domain of f^{-1}.

$\quad$ (b) $D = [-1, 2]$; $R = [\frac{1}{2}, 4]$ $\qquad\qquad$ (c) $D_1 = R = [\frac{1}{2}, 4]$; $R_1 = D = [-1, 2]$

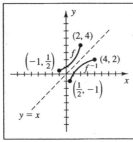

Figure 39

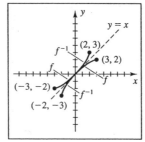

Figure 41

[41] (a) See *Figure 41* on the preceding page.

(b) $D = [-3, 3]; R = [-2, 2]$ (c) $D_1 = R = [-2, 2]; R_1 = D = [-3, 3]$

[43] (a) Since f is one-to-one, an inverse exists.

$$\text{If } f(x) = ax + b, \text{ then } f^{-1}(x) = \frac{x-b}{a} \text{ for } a \neq 0.$$

(b) No, because a constant function is not one-to-one.

[45] (a) $f(x) = -x + b \Rightarrow y = -x + b \Rightarrow x = -y + b,$ or $f^{-1}(x) = -x + b.$

(b) $f(x) = \frac{ax+b}{cx-a}$ for $c \neq 0 \Rightarrow y = \frac{ax+b}{cx-a} \Rightarrow cyx - ya = ax + b \Rightarrow$

$$cyx - ax = ay + b \Rightarrow x(cy - a) = ay + b \Rightarrow x = \frac{ay+b}{cy-a}, \text{ or } f^{-1}(x) = \frac{ax+b}{cx-a}.$$

(c) The graph of f is symmetric about the line $y = x$. Thus, $f(x) = f^{-1}(x)$.

[47] From *Figure 47*, we see that f is always increasing. Thus, f is one-to-one.

$[-6, 6]$ by $[-4, 4]$ $[-1, 2]$ by $[-1, 4]$

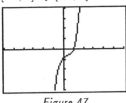   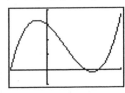

Figure 47   *Figure 49*

[49] (a) The high point is approximately $(-0.27, 3.31)$ and the low point is $(1.22, -0.20)$.

Thus, f decreases on $[-0.27, 1.22]$.

(b) The function g is the same as f with a limited domain. The domain of g^{-1} is the same as the range of g, that is, $[-0.20, 3.31]$. The range of g^{-1} is the same as the domain of g, that is, $[-0.27, 1.22]$.

[51] The graph of f will be reflected through the line $y = x$. $[-12, 12]$ by $[-8, 8]$

$y = \sqrt[3]{x-1} \Rightarrow y^3 = x - 1 \Rightarrow$

$x = y^3 + 1 \Rightarrow f^{-1}(x) = x^3 + 1.$

Graph $Y_1 = \sqrt[3]{x-1},\ Y_2 = x^3 + 1,$ and $Y_3 = x.$

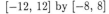
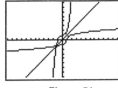

Figure 51

[53] (a) $V(23) = 35(23) = 805 \text{ ft}^3/\text{min}$

(b) $V^{-1}(x) = \frac{1}{35}x.$ Given an air circulation of x cubic feet per minute, $V^{-1}(x)$ computes the maximum number of people that should be in the restaurant at one time.

(c) $V^{-1}(2350) = \frac{1}{35}(2350) \approx 67.1 \Rightarrow$ the maximum number of people is 67.

2.6 Exercises

1 u is directly proportional to v $\Rightarrow$ $u = kv$. If $v = 30$, then $u = 12$ $\Rightarrow$ $12 = k(30)$.

Solving for the constant of proportionality, k, we have $12 = 30k$ $\Rightarrow$ $k = \frac{2}{5}$.

3 r varies directly as s and inversely as t $\Rightarrow$ $r = k\frac{s}{t}$.

If $s = -2$ and $t = 4$, then $r = 7$ $\Rightarrow$ $7 = \frac{k(-2)}{4}$.

Solving for the constant of proportionality, k, we have $7 = \frac{k(-2)}{4}$ $\Rightarrow$ $k = -14$.

5 y is directly proportional to the square of x and inversely proportional to the cube of

z $\Rightarrow$ $y = k\frac{x^2}{z^3}$. Substituting $x = 5$, $z = 3$, and $y = 25$ gives us $25 = \frac{k(25)}{27}$.

Solving for the constant of proportionality, k, we have $k = 27$.

7 z is directly proportional to the product of the square of x and the cube of y $\Rightarrow$

$z = kx^2y^3$. Substituting $x = 7$, $y = -2$, and $z = 16$ gives us $16 = k(49)(-8)$.

Solving for the constant of proportionality, k, we have $k = -\frac{2}{49}$.

9 y is directly proportional to x and inversely proportional to the square of z $\Rightarrow$

$y = k\frac{x}{z^2}$. Substituting $x = 4$, $z = 3$, and $y = 16$ gives us $16 = \frac{k(4)}{9}$.

Solving for the constant of proportionality, k, we have $k = 36$.

11 y is directly proportional to the square root of x and inversely proportional to the

cube of z $\Rightarrow$ $y = k\frac{\sqrt{x}}{z^3}$. Substituting $x = 9$, $z = 2$, and $y = 5$ gives us $5 = \frac{k(3)}{8}$.

Solving for the constant of proportionality, k, we have $k = \frac{40}{3}$.

13   (a) $P = kd$ (b) $118 = k(2)$ $\Rightarrow$ $k = 59$

      (c) $P = 59(5) = 295$ lb/ft^2

15   (a) $R = k\frac{l}{d^2} = \frac{kl}{d^2}$ (b) $25 = \frac{k(100)}{(0.01)^2}$ $\Rightarrow$ $k = \frac{1}{40{,}000}$

      (c) $R = \frac{50}{(40{,}000)(0.015)^2} = \frac{50}{9}$ ohms

17   (a) $P = k\sqrt{l}$ (b) $1.5 = k\sqrt{2}$ $\Rightarrow$ $k = \frac{3}{4}\sqrt{2}$

      (c) $P = \frac{3}{4}\sqrt{2}(\sqrt{6}) = \frac{3}{2}\sqrt{3}$ sec

19   (a) $T = kd^{3/2}$ (b) $365 = k(93)^{3/2}$ $\Rightarrow$ $k = \frac{365}{(93)^{3/2}}$

      (c) $T = \frac{365}{(93)^{3/2}} \cdot (67)^{3/2} \approx 223.2$ days

21 (a) $V = k\sqrt{L}$ (b) $35 = k\sqrt{50}$ $\Rightarrow$ $k = \frac{7}{2}\sqrt{2}$

   (c) $V = \frac{7}{2}\sqrt{2}(\sqrt{150}) = 35\sqrt{3} \approx 60.6$ mi/hr

23 (a) $W = kh^3$ (b) $200 = k(6)^3$ $\Rightarrow$ $k = \frac{25}{27}$

   (c) 5 feet 6 inches is $\frac{11}{2}$ feet. Hence, $W = \frac{25}{27}(\frac{11}{2})^3 \approx 154.1$ lb, or 154 lb.

25 (a) $F = kPr^4$ $\Rightarrow$ $P = \dfrac{F}{kr^4}$ under normal conditions.

   (b) "Normal flow rates triple" means that we will use $3F$ for the flow rate F.

      "Radius increases by 10%" means that we will use $(r + 10\%r) = 1.1r$ for the

      radius. Hence, $3F = kP(1.1r)^4$ $\Rightarrow$ $P = \dfrac{3F}{(1.1)^4 kr^4} \approx 2.05\left(\dfrac{F}{kr^4}\right)$,

                                  or about 2.05 times as hard as normal.

27 $C = \dfrac{kDE}{Vt}$ $\Rightarrow$ $D = \left(\dfrac{Ct}{k}\right)\dfrac{V}{E}$, where $\dfrac{Ct}{k}$ is constant. If V is twice its original value

   and E is 0.8 of its original value (reduced by 20%), then D becomes $D_1 = \left(\dfrac{Ct}{k}\right)\dfrac{2V}{0.8E}$.

   Comparing D_1 to D, we have $\dfrac{D_1}{D} = \dfrac{\left(\dfrac{Ct}{k}\right)\dfrac{2V}{0.8E}}{\left(\dfrac{Ct}{k}\right)\dfrac{V}{E}} = \dfrac{2}{0.8} = 2.5 = 250\%$ of its original

                                   value. Thus, D increases by 250%.

29 The square of the distance from the origin to the point (x, y) is $x^2 + y^2$. $d = \dfrac{k}{x^2 + y^2}$.

   If (x_1, y_1) is the new point that has density d_1, then

$$d_1 = \frac{k}{x_1^2 + y_1^2} = \frac{k}{(\frac{1}{3}x)^2 + (\frac{1}{3}y)^2} = \frac{k}{\frac{1}{9}x^2 + \frac{1}{9}y^2} = \frac{k}{\frac{1}{9}(x^2 + y^2)} = 9 \cdot \frac{k}{x^2 + y^2} = 9d.$$

                                 The density d is multiplied by 9.

31 We need to examine y/x for the given set of data points.

$$\frac{y}{x} = \frac{0.72}{0.6} = \frac{1.44}{1.2} = \frac{5.04}{4.2} = \frac{8.52}{7.1} = \frac{11.16}{9.3} = 1.2 \;\Rightarrow\; y = 1.2x.$$

           Thus, y varies directly as x with constant of variation $k = 1.2$.

33 $x^2y = -10.1$ for each data point.

    Thus, y varies inversely as x^2 with constant of variation $k = -10.1$, and $y = -\dfrac{10.1}{x^2}$.

35 (a) $D = kS^{2.3}$ $\Rightarrow$ $k = \dfrac{D}{S^{2.3}}$. Using the 6 data points: $\dfrac{33}{20^{2.3}} \approx 0.0336$;

      $\dfrac{86}{30^{2.3}} \approx 0.0344$; $\dfrac{167}{40^{2.3}} \approx 0.0345$; $\dfrac{278}{50^{2.3}} \approx 0.0344$; $\dfrac{414}{60^{2.3}} \approx 0.0337$; $\dfrac{593}{70^{2.3}} \approx 0.0338$.

                         Let $k = 0.034$. Thus, $D = 0.034S^{2.3}$.

(b) Graph the data together with $Y_1 = 0.034x^{2.3}$.

$[0, 75, 10]$ by $[0, 600, 100]$

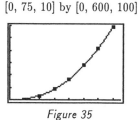

Figure 35

Chapter 2 Review Exercises

$\boxed{1}$ (a) $f(x) = \dfrac{x}{\sqrt{x+3}} \Rightarrow f(1) = \dfrac{1}{\sqrt{4}} = \dfrac{1}{2}$ (b) $f(-1) = -\dfrac{1}{\sqrt{2}}$ (c) $f(0) = \dfrac{0}{\sqrt{3}} = 0$

(d) $f(-x) = \dfrac{-x}{\sqrt{-x+3}} = -\dfrac{x}{\sqrt{3-x}}$ (e) $-f(x) = -1 \cdot f(x) = -\dfrac{x}{\sqrt{x+3}}$

(f) $f(x^2) = \dfrac{x^2}{\sqrt{x^2+3}}$ (g) $[f(x)]^2 = \left(\dfrac{x}{\sqrt{x+3}}\right)^2 = \dfrac{x^2}{x+3}$

$\boxed{3}$ $f(x) = \dfrac{-2(x^2-20)(5-x)}{(6-x^2)^{4/3}} \Rightarrow f(4) = \dfrac{(-)(-)(+)}{(+)} = +.$ $f(4)$ is positive.

Note that the factor -2 always contributes one negative sign to the quotient and that the denominator is always positive if $x \neq \pm\sqrt{6}$.

$\boxed{4}$ (a) $3x - 4 \geq 0 \Rightarrow x \geq \frac{4}{3}$; $D = [\frac{4}{3}, \infty)$.

Since y is the result of a square root, $y \geq 0$; $R = [0, \infty)$.

(b) $D = $ All real numbers except -3.

Since y is the square of the nonzero term $\dfrac{1}{x+3}$, $y > 0$; $R = (0, \infty)$.

$\boxed{5}$ $\dfrac{f(a+h) - f(a)}{h} = \dfrac{[-(a+h)^2 + (a+h) + 5] - [-a^2 + a + 5]}{h}$

$= \dfrac{-a^2 - 2ah - h^2 + a + h + 5 + a^2 - a - 5}{h}$

$= \dfrac{-2ah - h^2 + h}{h} = \dfrac{h(-2a - h + 1)}{h} = -2a - h + 1$

$\boxed{7}$ $f(x) = ax + b$ is the desired form. $a = \text{slope} = \dfrac{7-2}{3-1} = \dfrac{5}{2}$. $f(x) = \frac{5}{2}x + b \Rightarrow$

$f(1) = \frac{5}{2} + b$, but $f(1) = 2$, so $\frac{5}{2} + b = 2$, and $b = -\frac{1}{2}$. Thus, $f(x) = \frac{5}{2}x - \frac{1}{2}$.

9 $y = (x-3)^2 - 2$ is a parabola that opens upward and has

vertex $(3, -2)$.

x-intercepts: $y = 0 \;\Rightarrow\; 0 = (x-3)^2 - 2 \;\Rightarrow\;$

$(x-3)^2 = 2 \;\Rightarrow\; x - 3 = \pm\sqrt{2} \;\Rightarrow\; x = 3 \pm \sqrt{2}$

y-intercept: $x = 0 \;\Rightarrow\; y = (0-3)^2 - 2 = 9 - 2 = 7$

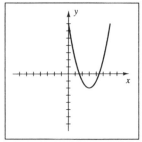

Figure 9

11 $P(-1, 4);\; y = -2\,f(x-3) + 1$

$\quad\{\,x - 3 \text{ [add 3 to the } x\text{-coordinate]}\,\}$ $\qquad\qquad \to (2, 4)$

$\quad\{\,\times -2 \text{ [multiply the } y\text{-coordinate by } -2\,]\,\}$ $\quad \to (2, -8)$

$\quad\{\,+1 \text{ [add 1 to the } y\text{-coordinate]}\,\}$ $\qquad\quad \to (2, -7)$

13 (a) The graph of $f(x) = \dfrac{1 - 3x}{2} = -\dfrac{3}{2}x + \dfrac{1}{2}$ is a line with slope $-\dfrac{3}{2}$ and y-intercept $\dfrac{1}{2}$.

(b) The function is defined for all x, so the domain D is the set of all real numbers.

The range is the set of all real numbers, so $D = \mathbb{R}$ and $R = \mathbb{R}$.

(c) The function f is decreasing on $(-\infty, \infty)$.

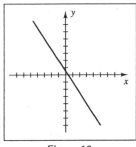

Figure 13

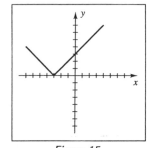

Figure 15

15 (a) The graph of $f(x) = |\,x + 3\,|$ can be thought of as

the graph of $g(x) = |\,x\,|$ shifted left 3 units.

(b) The function is defined for all x, so the domain D is the set of all real numbers.

The range is the set of all nonnegative numbers, that is, $R = [0, \infty)$.

(c) The function f is decreasing on $(-\infty, -3]$ and is increasing on $[-3, \infty)$.

17 (a) $f(x) = 1 - \sqrt{x+1} = -\sqrt{x+1} + 1$ • We can think of this graph as the graph

of $y = \sqrt{x}$ shifted left 1 unit, reflected through the x-axis, and shifted up 1 unit.

See *Figure 17* on the next page.

(b) For f to be defined, we must have $x + 1 \geq 0 \;\Rightarrow\; x \geq -1$, so $D = [-1, \infty)$.

The y-values of f are all the values less than or equal to 1, so $R = (-\infty, 1]$.

(c) The function f is decreasing on $[-1, \infty)$.

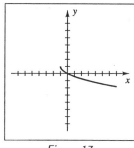

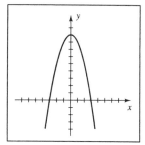

Figure 17   *Figure 19*

19 (a) $f(x) = 9 - x^2 = -x^2 + 9$ • We can think of this graph as the graph of $y = x^2$

reflected through the x-axis and shifted up 9 units.

(b) The function is defined for all x, so the domain D is the set of all real numbers.

The y-values of f are all the values less than or equal to 9, so $R = (-\infty, 9]$.

(c) The function f is increasing on $(-\infty, 0]$ and is decreasing on $[0, \infty)$.

21 (a) $f(x) = \begin{cases} x^2 & \text{if } x < 0 \\ 3x & \text{if } 0 \le x < 2 \\ 6 & \text{if } x \ge 2 \end{cases}$

If $x < 0$, we want the graph of the parabola $y = x^2$. The endpoint of this part of

the graph is $(0, 0)$, but it is not included. If $0 \le x < 2$, we want the graph of

$y = 3x$, a line with slope 3 and y-intercept 0. Now we do include the endpoint

$(0, 0)$, but we don't include the endpoint $(2, 6)$. If $x \ge 2$, we want the graph of

the horizontal line $y = 6$ and include its endpoint $(2, 6)$, so there are no open

endpoints on the graph of f.

(b) The function is defined for all x, so the domain D is the set of all real numbers.

The y-values of f are all the values greater than or equal to 0, so $R = [0, \infty)$.

(c) f is decreasing on $(-\infty, 0]$, increasing on $[0, 2]$, and constant on $[2, \infty)$.

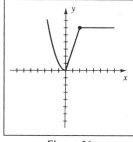

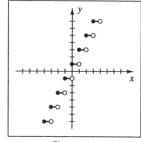

Figure 21   *Figure 22*

22 (a) The "2" in front of $[\![x]\!]$ has the effect of doubling all the y-values of $g(x) = [\![x]\!]$.

The "+1" has the effect of vertically shifting the graph of $h(x) = 2[\![x]\!]$ up 1 unit.

(continued)

(b) The function is defined for all x, so the domain D is the set of all real numbers. The range of $g(x) = [\![x]\!]$ is the set of integers, that is, $\{\ldots, -2, -1, 0, 1, 2, \ldots\}$. The range of $h(x) = 2[\![x]\!]$ is the set of even integers since we are doubling the values of g—that is, $\{\ldots, -4, -2, 0, 2, 4, \ldots\}$. Since $f(x) = 1 + h(x)$, the range R of f is $\{\ldots, -3, -1, 1, 3, \ldots\}$.

(c) The function f is constant on intervals such as $[0, 1)$, $[1, 2)$, and $[2, 3)$. In general, f is constant on $[n, n+1)$, where n is any integer.

$\boxed{23}$ (a) $y = \sqrt{x}$ • This is the square root function, whose graph is a half-parabola.

(b) $y = \sqrt{x+4}$ • shift $y = \sqrt{x}$ left 4 units

(c) $y = \sqrt{x} + 4$ • shift $y = \sqrt{x}$ up 4 units

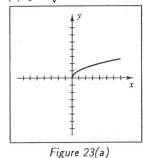 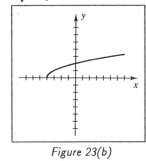 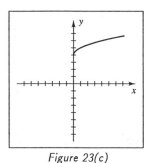

| Figure 23(a) | Figure 23(b) | Figure 23(c) |

(d) $y = 4\sqrt{x}$ • vertically stretch the graph of $y = \sqrt{x}$ by a factor of 4

(e) $y = \frac{1}{4}\sqrt{x}$ • vertically compress the graph of $y = \sqrt{x}$ by a factor of $1/\frac{1}{4} = 4$

(f) $y = -\sqrt{x}$ • reflect the graph of $y = \sqrt{x}$ through the x-axis

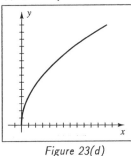 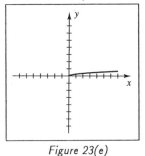 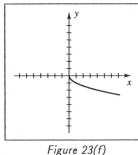

| Figure 23(d) | Figure 23(e) | Figure 23(f) |

$\boxed{25}$ $V(2, -4)$ and $P(-2, 4)$ with $y = a(x-h)^2 + k \;\Rightarrow\; 4 = a(-2-2)^2 - 4 \;\Rightarrow$
$$8 = 16a \;\Rightarrow\; a = \tfrac{1}{2}. \text{ An equation is } y = \tfrac{1}{2}(x-2)^2 - 4.$$

$\boxed{26}$ The graph could be made by taking the graph of $y = |x|$, reflecting it through the x-axis $\{y = -|x|\}$, then shifting that graph to the right 2 units $\{y = -|x-2|\}$, and then shifting that graph down 1 unit, resulting in the graph of the equation $y = -|x-2| - 1$.

$\boxed{27}$ $f(x) = 5x^2 + 30x + 49 \Rightarrow -\dfrac{b}{2a} = -\dfrac{30}{2(5)} = -3.$

$f(-3) = 4$ is a minimum since $a = 5 > 0.$

$\boxed{29}$ $f(x) = -12(x+1)^2 - 37$ is in the standard form. $f(-1) = -37$ is a maximum.

$\boxed{30}$ $f(x) = 3(x+2)(x-10)$ has x-intercepts at -2 and 10. The vertex is halfway

between them at $x = 4$. $f(4) = 3 \cdot 6 \cdot (-6) = -108$ is a minimum.

$\boxed{31}$ $f(x) = -2x^2 + 12x - 14 = -2(x^2 - 6x + \underline{}) - 14 + \underline{}$ { complete the square }

$= -2(x^2 - 6x + \underline{9}) - 14 + 18 = -2(x-3)^2 + 4.$

$\boxed{33}$ The domain of $f(x) = \sqrt{4-x^2}$ is $[-2, 2]$. The domain of $g(x) = \sqrt{x}$ is $[0, \infty)$.

(a) The domain of fg is the intersection of those two domains, $[0, 2]$.

(b) The domain of f/g is the same as that of fg,

excluding any values that make g equal to 0. Thus, the domain of f/g is $(0, 2]$.

$\boxed{35}$ (a) $(f \circ g)(x) = f(g(x)) = 2(3x+2)^2 - 5(3x+2) + 1 = 18x^2 + 9x - 1$

(b) $(g \circ f)(x) = g(f(x)) = 3(2x^2 - 5x + 1) + 2 = 6x^2 - 15x + 5$

$\boxed{37}$ (a) $(f \circ g)(x) = f(g(x)) = f(\sqrt{x-3}) = \sqrt{25 - (\sqrt{x-3})^2}$

$= \sqrt{25 - (x-3)} = \sqrt{28 - x}.$

Domain of $g(x) = \sqrt{x-3} = [3, \infty)$. Domain of $f(x) = \sqrt{25 - x^2} = [-5, 5]$.

$g(x) \le 5$ { $g(x)$ cannot be less than 0 } $\Rightarrow$

$\sqrt{x-3} \le 5 \Rightarrow x - 3 \le 25 \Rightarrow x \le 28.$ $[3, \infty) \cap (-\infty, 28] = [3, 28]$

(b) $(g \circ f)(x) = g(f(x)) = g(\sqrt{25 - x^2}) = \sqrt{\sqrt{25 - x^2} - 3}.$

Domain of $f = [-5, 5]$. Domain of $g = [3, \infty)$.

$f(x) \ge 3 \Rightarrow \sqrt{25 - x^2} \ge 3 \Rightarrow 25 - x^2 \ge 9 \Rightarrow x^2 \le 16 \Rightarrow x \in [-4, 4].$

$\boxed{39}$ For $y = \sqrt[3]{x^2 - 5x}$, choose $u = x^2 - 5x$ and $y = \sqrt[3]{u}$.

$\boxed{41}$ $f(x) = 10 - 15x \Rightarrow 15x = 10 - y \Rightarrow$

$x = \dfrac{10 - y}{15} \Rightarrow f^{-1}(x) = \dfrac{10 - x}{15}$

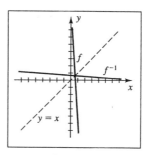

Figure 41

43 (a) The point $(1, 2)$ is on the graph, so $f(1) = 2$.

   (b) $(f \circ f)(1) = f(f(1)) = f(2) = 4$.

   (c) $f(2) = 4$ and f is one-to-one $\Rightarrow$ $f^{-1}(4) = 2$.

   (d) $y = 4$ when $x = 2$, so $f(x) = 4$ $\Rightarrow$ $x = 2$.

   (e) $y > 4$ when $x > 2$, so $f(x) > 4$ $\Rightarrow$ $x > 2$.

44 Since f and g are one-to-one functions, we know that $f(2) = 7$, $f(4) = 2$, and $g(2) = 5$ imply that $f^{-1}(7) = 2$, $f^{-1}(2) = 4$, and $g^{-1}(5) = 2$, respectively. Part (c) is included—try part (d) and explain why you *can't* find the value.

   (c) $(f^{-1} \circ g^{-1})(5) = f^{-1}(g^{-1}(5)) = f^{-1}(2) = 4$

45 $y = \dfrac{k\sqrt[3]{x}}{z^2}$ $\Rightarrow$ $6 = \dfrac{k\sqrt[3]{8}}{3^2}$ $\Rightarrow$ $6 = \dfrac{2k}{9}$ $\Rightarrow$ $k = \dfrac{54}{2} = 27$

47 (a) $F = aC + b$ is the desired form. $F = 32$ when $C = 0$ $\Rightarrow$ $F = aC + 32$. $F = 212$ when $C = 100$ $\Rightarrow$ $212 = 100a + 32$ $\Rightarrow$ $a = \dfrac{180}{100} = \dfrac{9}{5}$ and hence, $F = \dfrac{9}{5}C + 32$.

   (b) If C increases $1°$, F increases $\left(\dfrac{9}{5}\right)°$, or $1.8°$.

49 $V = \pi r^2 h$ and $V = 24\pi \Rightarrow h = \dfrac{24}{r^2}$. $S = \pi r^2 + 2\pi r h = \pi r^2 + 2\pi r \cdot \dfrac{24}{r^2} = \pi r^2 + \dfrac{48\pi}{r}$.

$$C = (0.30)(\pi r^2) + (0.10)\left(\dfrac{48\pi}{r}\right) = \dfrac{3\pi r^2}{10} + \dfrac{48\pi}{10r} = \dfrac{3\pi(r^3 + 16)}{10r}.$$

51 (a) Using similar triangles, $\dfrac{r}{x} = \dfrac{2}{4} \Rightarrow r = \dfrac{1}{2}x$.

   (b) Volume$_{\text{cone}}$ + Volume$_{\text{cup}}$ = Volume$_{\text{total}}$ $\Rightarrow \dfrac{1}{3}\pi r^2 h + \pi r^2 h = 5 \Rightarrow$

$$\dfrac{1}{3}\pi\left(\dfrac{1}{2}x\right)^2(x) + \pi(2)^2(y) = 5 \Rightarrow 5 - \dfrac{\pi}{12}x^3 = 4\pi y \Rightarrow y = \dfrac{5}{4\pi} - \dfrac{1}{48}x^3$$

53 Let t denote the time (in hr) after 1:00 P.M. If the starting point for ship B is the origin, then the locations of A and B are $-30 + 15t$ and $-10t$, respectively. Using the Pythagorean theorem, $d^2 = (-30 + 15t)^2 + (-10t)^2 = 325t^2 - 900t + 900$. The time at which the distance between the ships is minimal is the same as the time at which the square of the distance between the ships is minimal.

$$\text{Thus, } t = -\dfrac{b}{2a} = -\dfrac{-900}{2(325)} = \dfrac{18}{13}, \text{ or about 2:23 P.M.}$$

55 (a) $g = 32 \Rightarrow f(t) = -16t^2 + 16t$. Solving $f(t) = 0$ gives us $-16t(t - 1) \Rightarrow t = 0, 1$.

                        The player is in the air for 1 second.

   (b) $t = -\dfrac{b}{2a} = -\dfrac{16}{2(-16)} = \dfrac{1}{2}$. $f\left(\dfrac{1}{2}\right) = 4 \Rightarrow$ the player jumps 4 feet high.

   (c) $g = \dfrac{32}{6} \Rightarrow f(t) = -\dfrac{8}{3}t^2 + 16t$. Solving $f(t) = 0$ yields $t = 0$ or 6.

   The player would be in the air for 6 seconds on the moon.

$$t = -\dfrac{b}{2a} = -\dfrac{16}{2(-8/3)} = 3. \quad f(3) = 24 \Rightarrow \text{the player jumps 24 feet high.}$$

57 $C = \dfrac{kP_1P_2}{d^2}$; $2000 = \dfrac{k(10,000)(5000)}{(25)^2} \Rightarrow k = \dfrac{1}{40}$; $C = \dfrac{(10,000)(15,000)}{40(100)^2} = 375$

Chapter 2 Discussion Exercises

1 The expression $2x + h + 6$ represents the slope of the line between the points $P(x,\ f(x))$ and $Q(x+h,\ f(x+h))$. If $h = 0$, then $2x + 6$ represents the slope of the tangent line at the point $P(x,\ f(x))$.

3 For the graph of $g(x) = \sqrt{f(x)}$, where $f(x) = ax^2 + bx + c$, consider 2 cases:

(1) $(a > 0)$ If f has 0 or 1 x-intercept(s), the domain of g is $\mathbb{R}$ and its range is $[\sqrt{k},\ \infty)$, where k is the y-value of the vertex of f. If f has 2 x-intercepts (say x_1 and x_2 with $x_1 < x_2$), then the domain of g is $(-\infty,\ x_1] \cup [x_2,\ \infty)$ and its range is $[0,\ \infty)$. The general shape is similar to the v-shape of the graph of $y = \sqrt{a}\,|x|$.

(2) $(a < 0)$ If f has no x-intercepts, there is no graph of g. If f has 1 x-intercept, the graph of g consists of that point. If f has 2 x-intercepts, the domain of g is $[x_1,\ x_2]$ and the range is $[0,\ \sqrt{k}]$. The shape of g is that of the top half of an oval.

The main advantage of graphing g as a composition (say $Y_1 = f$ and $Y_2 = \sqrt{Y_1}$ on a graphing calculator) is to observe the relationship between the range of f and the domain of g.

$\boxed{5}$ The values of the x-intercepts (if they exist) are found by using the quadratic formula $\left(x = -\dfrac{b}{2a} \pm \dfrac{\sqrt{b^2 - 4ac}}{2a} \right)$. Hence, the distance d from the axis of symmetry, $x = -\dfrac{b}{2a}$, to either x-intercept is $d = \dfrac{\sqrt{b^2 - 4ac}}{2 \, |a|}$ and $d^2 = \dfrac{b^2 - 4ac}{4a^2}$. From page 173, the y-coordinate of the vertex is $h = c - \dfrac{b^2}{4a} = \dfrac{4ac - b^2}{4a}$, so

$$\frac{h}{d^2} = \frac{\dfrac{4ac - b^2}{4a}}{\dfrac{b^2 - 4ac}{4a^2}} = -\frac{4a^2(4ac - b^2)}{4a(b^2 - 4ac)} = -a.$$

Thus, $h = -ad^2$. Note that this relationship also reveals a connection between the discriminant D and the y-coordinate of the vertex, namely $h = -D^2/(4a)$.

$\boxed{7}$ $D = 0.0833x^2 - 0.4996x + 3.5491 \;\Rightarrow\; 0.0833x^2 - 0.4996x + (3.5491 - D) = 0.$

Solving for x with the quadratic formula yields

$$x = \frac{0.4996 \pm \sqrt{(-0.4996)^2 - 4(0.0833)(3.5491 - D)}}{2(0.0833)}, \text{ or, equivalently,}$$

$$x = \frac{4996 \pm \sqrt{33{,}320{,}000D - 93{,}295{,}996}}{1666}.$$

From *Figure 7* (a graph of D), we see that $3 \le x \le 15$ corresponds to the right half of the parabola. Hence, we choose the plus sign in the equation for x.

[−20, 20, 2] by [0, 40, 2]   [−15, 15] by [−10, 10]

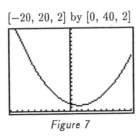

Figure 7

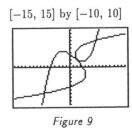

Figure 9

$\boxed{9}$ (a) $f(x) = 0.0346x^3 - 0.0731x^2 - 0.7154x + 3.4462$

(b) If f is increasing, then f^{-1} moves from left to right.

If f is decreasing, then f^{-1} moves from right to left.

(c) They can only occur on the line $y = x$.

Chapter 3: Polynomial and Rational Functions

☐1 $f(x) = 2x^3 + c$ • The graph of $y = g(x) = 2x^3$ is the graph of $y = x^3$ stretched vertically by a factor of 2.

(a) The effect of the "3" for c is to shift the graph of g *up* 3 units.

(b) The effect of the "−3" for c is to shift the graph of g *down* 3 units.

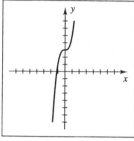

Figure 1(a)

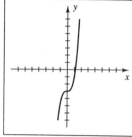

Figure 1(b)

☐3 $f(x) = ax^3 + 2$ • The "+2" will shift each graph up 2 units.

(a) The effect of the "2" for a is to vertically stretch $g(x) = x^3 + 2$ by a factor of 2 and make it appear "steeper."

(b) The effect of the "$\frac{1}{3}$" for a is to vertically compress $g(x) = x^3$ by a factor of $1/(1/3) = 3$, making it appear "flatter." The "−" reflects the graph of $h(x) = \frac{1}{3}x^3$ through the x-axis.

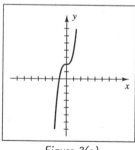

Figure 3(a)

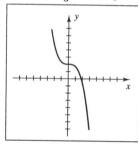

Figure 3(b)

☐5 We need to show that there is a sign change between $f(3)$ and $f(4)$ for $f(x) = x^3 - 4x^2 + 3x - 2$.

$$f(3) = 27 - 36 + 9 - 2 = -2 \quad \text{and} \quad f(4) = 64 - 64 + 12 - 2 = 10.$$

Since $f(3) = -2 < 0$ and $f(4) = 10 > 0$, the intermediate value theorem for polynomial functions assures us that f takes on every value between -2 and 10 in the interval $[3, 4]$, namely, 0.

$\boxed{7}$ As in Exercise 5, with $f(x) = -x^4 + 3x^3 - 2x + 1$, $a = 2$, $b = 3$,

we have $f(2) = 5 > 0$ and $f(3) = -5 < 0$.

$\boxed{9}$ As in Exercise 5, with $f(x) = x^5 + x^3 + x^2 + x + 1$, $a = -\frac{1}{2}$, $b = -1$,

we have $f(-\frac{1}{2}) = \frac{19}{32} > 0$ and $f(-1) = -1 < 0$.

$\boxed{11}$ $f(x) = \frac{1}{4}x^3 - 2 = \frac{1}{4}(x^3 - 8) = \frac{1}{4}(x - 2)(x^2 + 2x + 4)$. The general shape of the graph is

that of $g(x) = x^3$. To find the x-intercepts, set y $\{f(x)\}$ equal to 0. This means

either $x - 2 = 0$ or $x^2 + 2x + 4 = 0$. $x - 2 = 0$ $\Rightarrow$ $x = 2$ and $x^2 + 2x + 4 = 0$ $\Rightarrow$

$x = -1 \pm \sqrt{3}\,i$. The imaginary solutions mean that we have no x-intercepts from the

factor $x^2 + 2x + 4$ and the only x-intercept is 2. To find the y-intercept, set x equal

to 0. $x = 0$ $\Rightarrow$ $y = -2$.

The graph of f lies *above* the x-axis for all values of x greater than 2, so

$f(x) > 0$ if $x > 2$. The graph of f lies *below* the x-axis for all values of x less than 2,

so $f(x) < 0$ if $x < 2$.

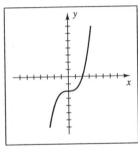

Figure 11

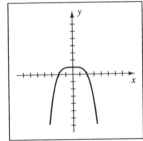

Figure 13

$\boxed{13}$ $f(x) = -\frac{1}{16}x^4 + 1 = -\frac{1}{16}(x^4 - 16) = -\frac{1}{16}(x^2 + 4)(x + 2)(x - 2)$. The general shape of

the graph is that of $g(x) = -x^4$. The x-intercepts are -2 and 2. The y-intercept is 1.

The graph of f lies *above* the x-axis for all values of x such that $-2 < x < 2$, so we

can write $f(x) > 0$ if $|x| < 2$. The graph of f lies *below* the x-axis for all values of x

such that $x < -2$ or $x > 2$, so we can write $f(x) < 0$ if $|x| > 2$.

$\boxed{15}$ $f(x) = x^4 - 4x^2 = x^2(x + 2)(x - 2)$. The general shape of the graph is that of

$g(x) = x^4$. See *Figure 15* on the next page. The x-intercepts are -2, 0, and 2. The

y-intercept is 0. The graph of f lies *above* the x-axis for all values of x such that

$x < -2$ or $x > 2$, so we can write $f(x) > 0$ if $|x| > 2$. The graph of f lies *below* the

x-axis for all values of x such that $-2 < x < 0$ or $0 < x < 2$, so we can write $f(x) < 0$

if $0 < |x| < 2$. Note that $x = 0$ is excluded in the last inequality since $f(x)$ *is equal*

to 0 if $x = 0$.

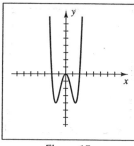

Figure 15

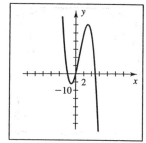

Figure 17

17 $f(x) = -x^3 + 3x^2 + 10x = -x(x^2 - 3x - 10) = -x(x+2)(x-5)$. The general shape of the graph is that of $g(x) = -x^3$. The x-intercepts are -2, 0, and 5. The y-intercept is 0. The graph of f lies *above* the x-axis for all values of x such that $x < -2$ or $0 < x < 5$, so $f(x) > 0$ if $x < -2$ or $0 < x < 5$. The graph of f lies *below* the x-axis for all values of x such that $-2 < x < 0$ or $x > 5$, so $f(x) < 0$ if $-2 < x < 0$ or $x > 5$.

19 $f(x) = \frac{1}{6}(x+2)(x-3)(x-4)$ •   If we were to multiply the terms of f, we would obtain a polynomial of the form $f(x) = \frac{1}{6}x^3 + (\text{other terms})$, so the general shape of the graph is that of $g(x) = x^3$. The x-intercepts are -2, 3, and 4. The y-intercept is $f(0) = \frac{1}{6}(0+2)(0-3)(0-4) = \frac{1}{6}(2)(-3)(-4) = 4$.

To determine the intervals on which $f(x) > 0$ and $f(x) < 0$, you may want to refer back to the concept of a sign diagram. Below is a sign diagram for this function. Note how the sign of each region correlates to the sign of $f(x)$.

Resulting sign:	$\ominus$	$\oplus$	$\ominus$	$\oplus$
Sign of $x - 4$:	$-$	$-$	$-$	$+$
Sign of $x - 3$:	$-$	$-$	$+$	$+$
Sign of $x + 2$:	$-$	$+$	$+$	$+$
x values:		-2 3		4

Diagram 19

The graph of f lies *above* the x-axis for all values of x such that $-2 < x < 3$ or $x > 4$, so $f(x) > 0$ if $-2 < x < 3$ or $x > 4$. The graph of f lies *below* the x-axis for all values of x such that $x < -2$ or $3 < x < 4$, so $f(x) < 0$ if $x < -2$ or $3 < x < 4$.

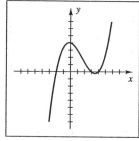

Figure 19

21 $f(x) = x^3 + 2x^2 - 4x - 8 = x^2(x+2) - 4(x+2)$ { factor by grouping }

$$= (x^2 - 4)(x + 2) \qquad \text{\{ factor out } (x+2) \text{\}}$$

$$= (x+2)(x-2)(x+2) = (x+2)^2(x-2).$$

The general shape of the graph is that of $g(x) = x^3$. The x-intercepts are -2 and 2. The y-intercept is -8.

If an intercept value has a corresponding *linear* factor { a factor with exponent 1 }, the graph will go through that intercept—that is, there will be a sign change from positive to negative or negative to positive in the function. Thus, at $x = 2$ the graph of the function "goes through" the point. If an intercept value has a corresponding quadratic factor (as in the case of -2 and $(x+2)^2$), then the function will not change sign. Hence, at $x = -2$ the graph of the function "touches the point and turns around."

The above discussion can be generalized as follows:

(1) If an intercept value has a corresponding factor raised to an <u>odd</u> power, then the function <u>will</u> change sign at that point.

(2) If an intercept value has a corresponding factor raised to an <u>even</u> power, then the function <u>will not</u> change sign at that point.

The graph of f lies *above* the x-axis for all values of x such that $x > 2$, so $f(x) > 0$ if $x > 2$. The graph of f lies *below* the x-axis for all values of x such that $x < -2$ or $-2 < x < 2$, so we write $f(x) < 0$ if $x < -2$ or $|x| < 2$.

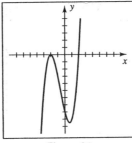

Figure 21   *Figure 23*

23 $f(x) = x^4 - 6x^2 + 8 = (x^2 - 2)(x+2)(x-2)$. The general shape of the graph is that of $g(x) = x^4$. The x-intercepts are ± 2 and $\pm\sqrt{2}$. The y-intercept is 8. The graph of f lies *above* the x-axis for all values of x such that $x < -2$ or $-\sqrt{2} < x < \sqrt{2}$ or $x > 2$, so we can write $f(x) > 0$ if $|x| > 2$ or $|x| < \sqrt{2}$. The graph of f lies *below* the x-axis for all values of x such that $-2 < x < -\sqrt{2}$ or $\sqrt{2} < x < 2$, so we can write $f(x) < 0$ if $\sqrt{2} < |x| < 2$.

25 $f(x) = x^2(x+2)(x-1)^2(x-2)$ • If we were to multiply the terms of f, we would obtain a polynomial of the form $f(x) = x^6 +$ (other terms), so the general shape of the graph is that of $g(x) = x^6$. The x-intercepts are -2, 0, 1, and 2. The y-intercept is $f(0) = 0$.

The graph of f lies *above* the x-axis for all values of x such that $x < -2$ or $x > 2$, so we can write $f(x) > 0$ if $|x| > 2$. The graph of f lies *below* the x-axis for all values of x such that $-2 < x < 0$ or $0 < x < 1$ or $1 < x < 2$, so we can write $f(x) < 0$ if $|x| < 2$, $x \neq 0$, $x \neq 1$.

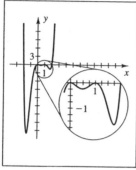

Figure 25

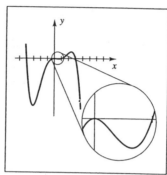

Figure 27

27 The sign of $f(x)$ is positive on $(-\infty, -4)$, so the graph of f must be above the x-axis on that interval. The sign of $f(x)$ is negative on $(-4, 0)$, so the graph of f must cross the x-axis at $x = -4$ and be below the x-axis on the interval $(-4, 0)$. The sign of $f(x)$ is negative on $(0, 1)$, so the graph of f must touch the x-axis at $x = 0$ and then fall below the x-axis on $(0, 1)$. The sign of $f(x)$ is positive on $(1, 3)$, so the graph of f must cross the x-axis at $x = 1$ and be above the x-axis on that interval. The sign of $f(x)$ is negative on $(3, \infty)$, so the graph of f must cross the x-axis at $x = 3$ and be below the x-axis on the interval $(3, \infty)$.

29 (a) The graph of $f(x) = (x-a)(x-b)(x-c)$, where $a < 0 < b < c$, must have one negative zero, a, and two positive zeros, b and c.

The general shape is that of a cubic polynomial.

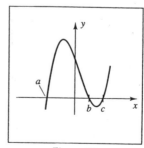

Figure 29

(b) The y-intercept is $f(0) = (-a)(-b)(-c) = -abc$.

(c) The solution to $f(x) < 0$ is $(-\infty, a) \cup (b, c)$;

that is, where the graph of f is below the x-axis.

(d) The solution to $f(x) \geq 0$ is $[a, b] \cup [c, \infty)$;

that is, where the graph of f is above or on the x-axis.

31 If $f(x)$ is a polynomial such that the coefficient of every odd power of x is 0, then $f(x)$ consists of only even-powered terms. If n is even, then $(-x)^n = x^n$ and hence $f(-x) = f(x)$, which is the condition that must be true for f to be an even function. Thus, f is an even function.

33 If a graph contains the point $(-1, 4)$, then $f(-1) = 4$.

For this function, $f(x) = 3x^3 - kx^2 + x - 5k$,

$$f(-1) = 3(-1)^3 - k(-1)^2 + (-1) - 5k = -3 - k - 1 - 5k = -4 - 6k.$$

Thus, $-4 - 6k$ must equal 4. Solving for k yields $-4 - 6k = 4 \implies k = -\frac{4}{3}$.

35 If one zero of $f(x) = x^3 - 2x^2 - 16x + 16k$ is 2, then $f(2) = 0$. But $f(2) = 16k - 32$, so $16k - 32 = 0 \implies 16k = 32 \implies k = 2$. Thus,

$$\begin{aligned} f(x) &= x^3 - 2x^2 - 16x + 32 && \{\text{let } k = 2\} \\ &= x^2(x - 2) - 16(x - 2) && \{\text{factor by grouping}\} \\ &= (x^2 - 16)(x - 2) && \{\text{factor out } (x - 2)\} \\ &= (x + 4)(x - 4)(x - 2). \end{aligned}$$

So the other two zeros of f are ± 4.

37 $P(x) = \frac{1}{2}(5x^3 - 3x) = \frac{1}{2}x(5x^2 - 3)$.

$P(x) = 0 \implies x = 0$ or $x^2 = \frac{3}{5} \iff x = \pm\sqrt{\frac{3}{5}} = \pm\sqrt{\frac{3 \cdot 5}{5 \cdot 5}} = \pm\frac{1}{5}\sqrt{15}$.

$P(x) > 0$ on $(-\frac{1}{5}\sqrt{15}, 0)$ and $(\frac{1}{5}\sqrt{15}, \infty)$.

$P(x) < 0$ on $(-\infty, -\frac{1}{5}\sqrt{15})$ and $(0, \frac{1}{5}\sqrt{15})$.

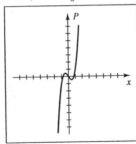

Figure 37

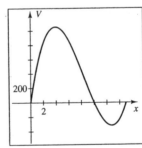

Figure 39

39 (a) $V(x) = lwh = (30 - x - x)(20 - x - x)x = x(20 - 2x)(30 - 2x)$.

(b) $V(x) = x(20 - 2x)(30 - 2x) = x(-2)(x - 10)(-2)(x - 15) = 4x(x - 10)(x - 15)$.

The x-intercepts of V are 0, 10, and 15. $V(x) > 0$ on $(0, 10)$ and $(15, \infty)$.

Allowable values for x are in $(0, 10)$.

41 We will work part (b) first.

(b) $T = \frac{1}{20}t(t-12)(t-24)$ has the general shape of a cubic polynomial with zeros at

0, 12, and 24. Its sign pattern is negative, positive, negative, positive.

(a) $T = \frac{1}{20}t(t-12)(t-24) = 0 \Rightarrow t = 0, 12, 24$. $T > 0$ for

$0 < t < 12$ { 6 A.M. to 6 P.M. }; $T < 0$ for $12 < t < 24$ { 6 P.M. to 6 A.M. }.

(c) 12 noon corresponds to $t = 6$, $T(6) = 32.4 > 32\,°F$ and $T(7) = 29.75 < 32\,°F$

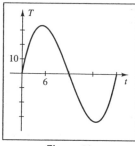

Figure 41

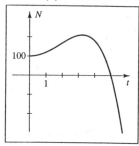

Figure 43

43 (a) $N(t) = -t^4 + 21t^2 + 100$

$\qquad = -(t^4 - 21t^2 - 100)$ { factor out -1 }

$\qquad = -(t^2 - 25)(t^2 + 4)$ { factor as a quadratic in t^2 }

$\qquad = -(t+5)(t-5)(t^2+4)$ { difference of two squares }

N has the general shape of a quartic polynomial reflected through the t-axis.

If $t > 0$, then $N(t) > 0$ for $0 < t < 5$.

(b) The population becomes extinct when $N = 0$. This occurs after 5 years.

45 (a) $f(x) = 2x^4$, $g(x) = 2x^4 - 5x^2 + 1$, $h(x) = 2x^4 + 5x^2 - 1$, $k(x) = 2x^4 - x^3 + 2x$

x	$f(x)$	$g(x)$	$h(x)$	$k(x)$
-60	25,920,000	25,902,001	25,937,999	26,135,880
-40	5,120,000	5,112,001	5,127,999	5,183,920
-20	320,000	318,001	321,999	327,960
20	320,000	318,001	321,999	312,040
40	5,120,000	5,112,001	5,127,999	5,056,080
60	25,920,000	25,902,001	25,937,999	25,704,120

(b) As $|x|$ becomes large, the function values become similar.

(c) The term with the highest power of x: $2x^4$.

47 (a) [-9, 9] by [-6, 6]   [-9, 9] by [-6, 6]

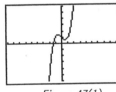

Figure 47(1)

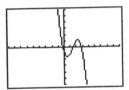

Figure 47(2)

[-9, 9] by [-6, 6]   [-9, 9] by [-6, 6]

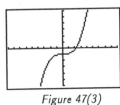

Figure 47(3)

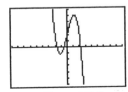

Figure 47(4)

(b) (1) $f(x) = x^3 - x + 1$ •   As x approaches ∞, $f(x)$ approaches ∞; as x approaches $-\infty$, $f(x)$ approaches $-\infty$

(2) $f(x) = -x^3 + 4x^2 - 3x - 1$ •   As x approaches ∞, $f(x)$ approaches $-\infty$; as x approaches $-\infty$, $f(x)$ approaches ∞

(3) $f(x) = 0.1x^3 - 1$ •   As x approaches ∞, $f(x)$ approaches ∞; as x approaches $-\infty$, $f(x)$ approaches $-\infty$

(4) $f(x) = -x^3 + 4x + 2$ •   As x approaches ∞, $f(x)$ approaches $-\infty$; as x approaches $-\infty$, $f(x)$ approaches ∞

(c) For the cubic function $f(x) = ax^3 + bx^2 + cx + d$ with $a > 0$, $f(x)$ approaches ∞ as x approaches ∞ and $f(x)$ approaches $-\infty$ as x approaches $-\infty$. With $a < 0$, $f(x)$ approaches $-\infty$ as x approaches ∞ and $f(x)$ approaches ∞ as x approaches $-\infty$.

49 From the graph, $f(x) = x^3 + 0.2x^2 - 2.6x + 1.1$ has three zeros.

They are approximately -1.89, 0.49, and 1.20.

[-4.5, 4.5] by [-3, 3]   [-4.5, 4.5] by [-3, 3]

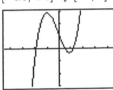

Figure 49

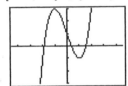

Figure 51

51 From the graph, $f(x) = x^3 - 3x + 1$ has three zeros.

They are approximately -1.88, 0.35, and 1.53.

53 If $f(x) = x^3 + 5x - 2$ and $k = 1$, then $f(x) > k$ on $(0.56, \infty)$.

Use an intersect feature to find the value 0.56.

[−4.5, 4.5] by [−3, 3]   [−4.5, 4.5] by [−3, 3]

   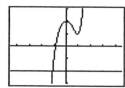

Figure 53   *Figure 55*

55 If $f(x) = x^5 - 2x^2 + 2$ and $k = -2$, then $f(x) > k$ on $(-1.10, \infty)$.

57 From the graph, there are three points of intersection.

Their coordinates are approximately $(-1.29, -0.77)$, $(0.085, 2.66)$, and $(1.36, -0.42)$.

[−4.5, 4.5] by [−2, 4]   [1975, 1995] by [20, 40]

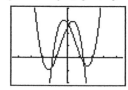   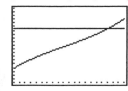

Figure 57   *Figure 59*

59 (a) To enter f in your graphing calculator, let $Y_1 = X - 1975$ and then assign

$$0.0014Y_1{}^3 - 0.0388Y_1{}^2 + 0.8783Y_1 + 23.82 \quad \text{to} \quad Y_2.$$

From the graph, we determine that there were approximately 23.82 million recipients in 1975 and 37.066 million recipients in 1995. The number of recipients is increasing quite rapidly.

(b) Graph $Y_3 = 34.4$ (shown in *Figure 59*). The graphs of Y_2 and Y_3 intersect when $x \approx 1991.9792 \approx 1992$. There were 34.4 million recipients in 1992.

3.2 Exercises

Note: Refer to the illustration on page 231. We will walk through the steps to obtain the quotient and remainder for this example.

(1) Determine the quotient of the first terms of the dividend $(x^4 - 16)$ and the divisor $(x^2 + 3x + 1)$. $\frac{x^4}{x^2} = x^2$—this is the first term of the resulting quotient.

(2) Multiply the result in step (1), x^2, by the divisor, $x^2 + 3x + 1$. This product is $x^4 + 3x^3 + x^2$.

(3) Subtract the result in step (2) from the dividend. This result is $-3x^3 - x^2$ (there is no need to worry about the lower degree terms here, namely, -16). A common mistake is to forget to subtract <u>all</u> of the terms and obtain (in this case) $3x^3 + x^2$.

(4) Repeat steps (1) through (3) for the remaining dividend until the degree of the remainder is less than the degree of the divisor.

$\boxed{1}$ $f(x) = 2x^4 - x^3 - 3x^2 + 7x - 12$; $p(x) = x^2 - 3$ •

$$
\begin{array}{r}
2x^2 - \quad x + \quad 3 \\
x^2 - 3 \overline{\smash{\big)}\, 2x^4 - \quad x^3 - \quad 3x^2 + \quad 7x - \quad 12} \\
2x^4 \qquad\quad - \quad 6x^2 \\
\hline
-x^3 + \quad 3x^2 \\
-x^3 \qquad\quad + \quad 3x \\
\hline
3x^2 + \quad 4x \\
3x^2 \qquad\quad - \quad 9 \\
\hline
4x - \quad 3
\end{array}
$$

First step: $\dfrac{2x^4}{x^2} = 2x^2$

★ $q(x) = 2x^2 - x + 3$; $r(x) = 4x - 3$

$\boxed{3}$ $f(x) = 3x^3 + 2x - 4$; $p(x) = 2x^2 + 1$ •

$$
\begin{array}{r}
\frac{3}{2}x \\
2x^2 + 1 \overline{\smash{\big)}\, 3x^3 - \quad 0x^2 + \quad 2x \quad - \quad 4} \\
3x^3 \qquad\qquad + \quad \frac{3}{2}x \\
\hline
\frac{1}{2}x \quad - \quad 4
\end{array}
$$

First step: $\dfrac{3x^3}{2x^2} = \dfrac{3}{2}x$

★ $q(x) = \frac{3}{2}x$; $r(x) = \frac{1}{2}x - 4$

$\boxed{5}$ Note that we can't divide $f(x) = 7x + 2$ by $p(x) = 2x^2 - x - 4$ since the degree of the dividend is already less than the degree of the divisor. So the quotient is 0 and the remainder is $7x + 2$.

$\boxed{7}$ $f(x) = 9x + 4$; $p(x) = 2x - 5$ •

$$
\begin{array}{r}
\frac{9}{2} \\
2x - 5 \overline{\smash{\big)}\, 9x \quad + \quad 4} \\
9x \quad - \quad \frac{45}{2} \\
\hline
\frac{53}{2}
\end{array}
$$

First step: $\dfrac{9x}{2x} = \dfrac{9}{2}$

★ $q(x) = \frac{9}{2}$; $r(x) = \frac{53}{2}$

9 We wish to find $f(2)$ by using the remainder theorem. Thus, we need to divide $f(x) = 3x^3 - x^2 + 5x - 4$ by $x - 2$ using either long division or synthetic division. Using synthetic division, we have

$$
\begin{array}{r|rrrr}
2 & 3 & -1 & 5 & -4 \\
  &   & 6 & 10 & 30 \\
\hline
  & 3 & 5 & 15 & 26
\end{array}
$$

The last number in the third row, 26, is our remainder. Hence, $f(2) = 26$.

11 We divide by $x + 3$ or $x - (-3)$. Note that we include a 0 for the missing x^3 term in $f(x) = x^4 - 6x^2 + 4x - 8$.

$$
\begin{array}{r|rrrrr}
-3 & 1 & 0 & -6 & 4 & -8 \\
   &   & -3 & 9 & -9 & 15 \\
\hline
   & 1 & -3 & 3 & -5 & 7
\end{array}
$$

Hence, $f(-3) = 7$.

13 To show that $x + 3$ is a factor of $f(x) = x^3 + x^2 - 2x + 12$, we must show that $f(-3) = 0$. $f(-3) = -27 + 9 + 6 + 12 = 0$, and, hence, $x + 3$ is a factor of $f(x)$.

15 To show that $x + 2$ is a factor of $f(x) = x^{12} - 4096$, we must show that $f(-2) = 0$.
$f(-2) = (-2)^{12} - 4096 = 4096 - 4096 = 0$, and, hence, $x + 2$ is a factor of $f(x)$.

17 f has degree 3 with zeros $-2, 0, 5$ $\Rightarrow$
$$f(x) = a\big[x - (-2)\big](x - 0)(x - 5) \;\{\text{let } a = 1\}$$
$$= x(x + 2)(x - 5) = x(x^2 - 3x - 10) = x^3 - 3x^2 - 10x$$

19 f has degree 4 with zeros $-2, \pm 1, 4$ $\Rightarrow$
$$f(x) = a(x + 2)(x + 1)(x - 1)(x - 4) \;\{\text{let } a = 1\}$$
$$= (x^2 - 1)(x^2 - 2x - 8) = x^4 - 2x^3 - 9x^2 + 2x + 8$$

21
$$
\begin{array}{r|rrrr}
2 & 2 & -3 & 4 & -5 \\
  &   & 4 & 2 & 12 \\
\hline
  & 2 & 1 & 6 & 7
\end{array}
$$
The synthetic division indicates that the quotient is $2x^2 + x + 6$ and the remainder is 7.

23
$$
\begin{array}{r|rrrr}
-3 & 1 & 0 & -8 & -5 \\
   &   & -3 & 9 & -3 \\
\hline
   & 1 & -3 & 1 & -8
\end{array}
$$
The synthetic division indicates that the quotient is $x^2 - 3x + 1$ and the remainder is -8.

25 $3x^5 + 6x^2 + 7 = \underline{3}\,x^5 + \underline{0}\,x^4 + \underline{0}\,x^3 + \underline{6}\,x^2 + \underline{0}\,x + \underline{7}$

$$
\begin{array}{r|rrrrrr}
-2 & 3 & 0 & 0 & 6   & 0 & 7 \\
   &   & -6 & 12 & -24 & 36 & -72 \\
\hline
   & 3 & -6 & 12 & -18 & 36 & -65
\end{array}
$$
The synthetic division indicates that the quotient is
$3x^4 - 6x^3 + 12x^2 - 18x + 36$
and the remainder is -65.

27 $\frac{1}{2}$ | 4   0   −5   0   1

              2   1   −2   −1

          4   2   −4   −2   0

The synthetic division indicates that the quotient is

$$4x^3 + 2x^2 - 4x - 2$$

and the remainder is 0.

29 3 | 2   3   −4   4

           6   27   69

      2   9   23   73

The synthetic division indicates that $f(3) = 73$.

31 −0.2 | 0.3   0   0.04   −0.034

            −0.06   0.012   −0.0104

      0.3   −0.06   0.052   −0.0444

The remainder is −0.0444 and the remainder theorem indicates that this is $f(-0.2)$.

33 $2 + \sqrt{3}$ | 1   3   −5

            $2 + \sqrt{3}$ $13 + 7\sqrt{3}$

      1    $5 + \sqrt{3}$ $8 + 7\sqrt{3}$

The remainder is $8 + 7\sqrt{3}$ and the remainder theorem indicates that this is the value of $f(2 + \sqrt{3})$.

35 −2 is a zero of $f(x)$ if we can show that $f(-2) = 0$.

        −2 | 3   8   −2   −10   4

                   −6   −4   12   −4

           3   2   −6   2   0

Hence, $f(-2) = 0$ and −2 is a zero of $f(x)$.

37 $\frac{1}{2}$ | 4   −6   8   −3

             2   −2   3

      4   −4   6   0

Hence, $f(\frac{1}{2}) = 0$ and $\frac{1}{2}$ is a zero of $f(x)$.

39 $f(x) = kx^3 + x^2 + k^2 x + 3k^2 + 11$, so

$$f(-2) = k(-2)^3 + (-2)^2 + k^2(-2) + 3k^2 + 11$$
$$= -8k + 4 - 2k^2 + 3k^2 + 11$$
$$= k^2 - 8k + 15$$

The remainder, $k^2 - 8k + 15$, must be zero if $f(x)$ is to be divisible by $x + 2$.

$$k^2 - 8k + 15 = 0 \implies (k - 3)(k - 5) = 0 \implies k = 3, 5.$$

[41] $x - c$ will not be a factor of $f(x) = 3x^4 + x^2 + 5$ for any real number c if the remainder of dividing $f(x)$ by $x - c$ is not zero for any real number c. The remainder is $f(c)$, or $3c^4 + c^2 + 5$. Since $3c^4 + c^2 + 5$ is greater than or equal to 5 for any real number c, it is never zero and hence, $x - c$ is never a factor of $f(x)$.

[43] $f(x) = 3x^{100} + 5x^{85} - 4x^{38} + 2x^{17} - 6 \Rightarrow f(-1) = 3 - 5 - 4 - 2 - 6 = -14$.

{ change sign on coefficients of odd powers of x }

[45] If $f(x) = x^n - y^n$ and n is even, then $f(-y) = (-y)^n - (y)^n = y^n - y^n = 0$.

Hence, $x + y$ is a factor of $x^n - y^n$.

[47] (a) The radius of the resulting cylinder is x and the height is y, that is, $6 - x$.

$$V = \pi r^2 h = \pi x^2 (6 - x)$$

(b) The volume of the cylinder of radius 1 and altitude 5 is $\pi (1)^2 5 = 5\pi$. To determine another value of x which would result in the same volume, we need to solve the equation $5\pi = \pi x^2 (6 - x)$. $5\pi = \pi x^2 (6 - x) \Rightarrow 5 = 6x^2 - x^3 \Rightarrow$ $x^3 - 6x^2 + 5 = 0$.   We know that $x = 1$ is a solution to this equation. Synthetically dividing, we obtain

$$\begin{array}{r|rrrr} 1 & 1 & -6 & 0 & 5 \\ & & 1 & -5 & -5 \\ \hline & 1 & -5 & -5 & 0 \end{array}$$

The quotient is $x^2 - 5x - 5$. Using the quadratic formula, we see that the solutions are $x = \dfrac{5 \pm \sqrt{45}}{2}$. Since x must be positive in the first quadrant, $\dfrac{5 + \sqrt{45}}{2} \approx 5.85$ would be an allowable value of x. If $x = \frac{1}{2}(5 + \sqrt{45})$, then

$y = 6 - x = 6 - \frac{1}{2}(5 + \sqrt{45}) = \frac{12}{2} - \frac{5}{2} - \frac{1}{2}\sqrt{45} = \frac{1}{2}(7 - \sqrt{45})$.

The point P is $P(x, y) = \left(\frac{1}{2}(5 + \sqrt{45}), \frac{1}{2}(7 - \sqrt{45}) \right) \approx (5.85, 0.15)$.

[49] (a) $A = lw = (2x)(y) = 2x(4 - x^2) = 8x - 2x^3$

(b) This solution is similar to the solution in Exercise 47(b).

$A = 6 \Rightarrow 6 = 8x - 2x^3 \Rightarrow x^3 - 4x + 3 = 0 \Rightarrow (x - 1)(x^2 + x - 3) = 0 \Rightarrow$

$x = 1$ or $x^2 + x - 3 = 0$. Solving the second equation, we get $x = \dfrac{-1 \pm \sqrt{13}}{2}$.

The value $\dfrac{\sqrt{13} - 1}{2}$ would be an allowable value of x.

The base, $2x$, would then be $\sqrt{13} - 1 \approx 2.61$.

51 If $f(x) = x^8 - 7.9x^5 - 0.8x^4 + x^3 + 1.2x - 9.81$,

then the remainder is $f(0.21) \approx -9.55$.

[-1, 1] by [-20, 0, 2]   [-9, 9] by [-3, 9]

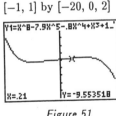

Figure 51   Figure 53

53 $f(1.6) = -2k^4 + 2.56k^3 + 3.2k + 4.096$. Graph $y = -2k^4 + 2.56k^3 + 3.2k + 4.096$ (that

is, $y = -2x^4 + 2.56x^3 + 3.2x + 4.096$). From the graph, we see that $y = 0$ when

$k \approx -0.75$, 1.96. Thus, if k assumes either of these values, $f(1.6) = 0$ and f will be

divisible by $x - 1.6$ by the factor theorem.

3.3 Exercises

1 The polynomial is of the form $f(x) = a(x - (-1))(x - 2)(x - 3)$.

$f(-2) = a(-1)(-4)(-5) = 80 \Rightarrow -20a = 80 \Rightarrow a = -4$.

Hence, the polynomial is $-4(x + 1)(x - 2)(x - 3)$, or $-4x^3 + 16x^2 - 4x - 24$.

3 The polynomial is of the form $f(x) = a(x - (-4))(x - 3)(x - 0)$.

$f(2) = a(6)(-1)(2) = -36 \Rightarrow -12a = -36 \Rightarrow a = 3$.

Hence, the polynomial is $3(x + 4)(x - 3)(x)$, or $3x^3 + 3x^2 - 36x$.

5 The polynomial is of the form $f(x) = a(x - (-2i))(x - 2i)(x - 3)$.

$f(1) = a(1 + 2i)(1 - 2i)(-2) = 20 \Rightarrow -10a = 20 \Rightarrow a = -2$.

Hence, the polynomial is $-2(x + 2i)(x - 2i)(x - 3) = -2(x^2 + 4)(x - 3)$,

or $-2x^3 + 6x^2 - 8x + 24$.

7 $f(x) = a(x + 4)^2(x - 3)^2 = (x^2 + x - 12)^2 \{a = 1\} = x^4 + 2x^3 - 23x^2 - 24x + 144$

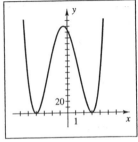

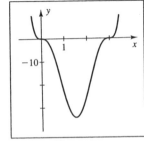

Figure 7   Figure 9

9 $f(x) = a(x)^3(x - 3)^3$ so $f(2) = a(8)(-1) = -8a$. But $f(2) = -24$, so $-8a = -24$, or,

$a = 3$. $f(x) = 3(x)^3(x^3 - 9x^2 + 27x - 27) = 3x^6 - 27x^5 + 81x^4 - 81x^3$.

11 The graph has x-intercepts at -1, $\frac{3}{2}$, 3, and $f(0) = \frac{7}{2}$.

$$f(x) = a(x+1)(x-\tfrac{3}{2})(x-3); \; f(0) = a(1)(-\tfrac{3}{2})(-3) = \tfrac{7}{2} \; \Rightarrow \; \tfrac{9}{2}a = \tfrac{7}{2} \; \Rightarrow \; a = \tfrac{7}{9}.$$

$$f(x) = \tfrac{7}{9}(x+1)(x-\tfrac{3}{2})(x-3)$$

13 3 is a zero of multiplicity one, 1 is a zero of multiplicity two, and $f(0) = 3$.

$$f(x) = a(x-1)^2(x-3); \; f(0) = a(1)(-3) = 3 \; \Rightarrow \; a = -1. \;\; f(x) = -1(x-1)^2(x-3).$$

15 The values of x that make $f(x) = x^2(3x+2)(2x-5)^3$ equal to zero are 0, $-\frac{2}{3}$, and $\frac{5}{2}$.

The exponents associated with these zeros are 2, 1, and 3, respectively. Thus, 0 is a

root of multiplicity 2, $-\frac{2}{3}$ is a root of multiplicity 1, and $\frac{5}{2}$ is a root of multiplicity 3.

17 $f(x) = 4x^5 + 12x^4 + 9x^3 = x^3(4x^2 + 12x + 9) = x^3(2x+3)^2$, so 0 is a root of

multiplicity 3 and $-\frac{3}{2}$ is a root of multiplicity 2.

19 $f(x) = (x^2 + x - 12)^3(x^2 - 9)^2 = \left[(x+4)(x-3)\right]^3\left[(x+3)(x-3)\right]^2$

$$= (x+4)^3(x-3)^3(x+3)^2(x-3)^2 = (x+4)^3(x+3)^2(x-3)^5$$

★ -4 (multiplicity 3); -3 (multiplicity 2); 3 (multiplicity 5)

21 $f(x) = x^4 + 7x^2 - 144 = (x^2 + 16)(x^2 - 9) = (x^2 + 16)(x+3)(x-3)$

★ $\pm 4i$, ± 3 (each of multiplicity 1)

23 Using synthetic division,

$$
\begin{array}{r|rrrrr}
-3 & 1 & 7 & 13 & -3 & -18 \\
   &   & -3 & -12 & -3 & 18 \\
\hline
   & 1 & 4 & 1 & -6 & 0 \\
\end{array}
$$

$$
\begin{array}{r|rrrr}
-3 & 1 & 4 & 1 & -6 \\
   &   & -3 & -3 & 6 \\
\hline
   & 1 & 1 & -2 & 0 \\
\end{array}
$$

The remaining polynomial, $x^2 + x - 2$, can be factored as $(x+2)(x-1)$.

Hence, $f(x) = (x+3)^2(x+2)(x-1)$.

25 *Note:* If the sum of the coefficients of the polynomial is 0, then 1 is a zero of the

polynomial. Synthetically dividing 1 five times, we obtain the remaining polynomial,

$x+1$. Hence, $f(x) = x^6 - 4x^5 + 5x^4 - 5x^2 + 4x - 1 = (x-1)^5(x+1)$.

Note: For the following exercises, let $f(x)$ denote the polynomial, P, the number of sign changes in $f(x)$, and N, the number of sign changes in $f(-x)$. The types of possible solutions are listed in the order positive, negative, nonreal complex.

27 $f(x) = 4x^3 - 6x^2 + x - 3$. The sign pattern for coefficients is $+, -, +, -$. Since there are 3 sign changes in $f(x)$, P = 3. $f(-x) = -4x^3 - 6x^2 - x - 3$. *Note:* Simply change the sign of the coefficients of the odd-powered terms of $f(x)$ to find $f(-x)$. The sign pattern for coefficients is $-, -, -, -$. Since there are no sign changes in $f(-x)$, N = 0. If there are 3 positive solutions, then there are no negative or nonreal complex. If there is 1 positive solution, there are no negative and 2 nonreal complex solutions.

29 $f(x) = 4x^3 + 2x^2 + 1 \Rightarrow$ P = 0. $f(-x) = -4x^3 + 2x^2 + 1 \Rightarrow$ N = 1. ★ 0, 1, 2

31 $f(x) = 3x^4 + 2x^3 - 4x + 2 \Rightarrow$ P = 2.
$f(-x) = 3x^4 - 2x^3 + 4x + 2 \Rightarrow$ N = 2. ★ 2, 2, 0; 2, 0, 2; 0, 2, 2; 0, 0, 4

33 $f(x) = x^5 + 4x^4 + 3x^3 - 4x + 2 \Rightarrow$ P = 2.
$f(-x) = -x^5 + 4x^4 - 3x^3 + 4x + 2 \Rightarrow$ N = 3. ★ 2, 3, 0; 2, 1, 2; 0, 3, 2; 0, 1, 4

35 Synthetically dividing 5 into the polynomial yields a bottom row consisting of all nonnegative numbers (see below). Synthetically dividing -2 into the polynomial yields a bottom row that alternates in sign (see below).

5 │	1	-4	-5	7
		5	5	0
	1	1	0	7

-2 │	1	-4	-5	7
		-2	12	-14
	1	-6	7	-7

These results indicate that the upper bound is 5 and the lower bound is -2. From the graph of $f(x) = x^3 - 4x^2 - 5x + 7$, we see that the bounds given by the theorem (5 and -2) are indeed the smallest and largest integers that are upper and lower bounds.

37 Synthetically dividing 2 into the polynomial yields a bottom row consisting of all nonnegative numbers (see below). Synthetically dividing -2 into the polynomial yields a bottom row that alternates in sign (see below).

2 │	1	-1	-2	3	6
		2	2	0	6
	1	1	0	3	12

-2 │	1	-1	-2	3	6
		-2	6	-8	10
	1	-3	4	-5	16

These results indicate that the upper bound is 2 and the lower bound is -2.

(continued)

From the graph of $f(x) = x^4 - x^3 - 2x^2 + 3x + 6$, we see that there are *no* real zeros. Remember, the theorem only gives us upper positive and lower negative bounds for the zeros of a polynomial—it doesn't guarantee that there are any zeros.

39 Similar to Exercise 35 with $f(x) = 2x^5 - 13x^3 + 2x - 5$, upper bound 3, and lower bound −3.

$$
\begin{array}{r|rrrrrr}
3 & 2 & 0 & -13 & 0 & 2 & -5 \\
  &   & 6 & 18 & 15 & 45 & 141 \\
\hline
  & 2 & 6 & 5 & 15 & 47 & 136
\end{array}
$$

$$
\begin{array}{r|rrrrrr}
-3 & 2 & 0 & -13 & 0 & 2 & -5 \\
   &   & -6 & 18 & -15 & 45 & -141 \\
\hline
   & 2 & -6 & 5 & -15 & 47 & -146
\end{array}
$$

41 The zero at $x = -1$ must be of even multiplicity since the graph does not cross the x-axis. Since we want f to have minimal degree, we will let the associated factor be $(x + 1)^2$. The graph goes through the zero at $x = 1$ without "flattening out," so the associated factor is $(x - 1)^1$. Because the graph flattens out at the zero at $x = 2$, we will let its associated factor be $(x - 2)^3$. Thus, f has the form

$$f(x) = a(x + 1)^2(x - 1)(x - 2)^3.$$

Since the y-intercept is $(0, -2)$, we have $f(0) = a(1)(-1)(-8) = 8a$ and $f(0) = -2 \Rightarrow 8a = -2 \Rightarrow a = -\frac{1}{4}$. Hence, $f(x) = -\frac{1}{4}(x + 1)^2(x - 1)(x - 2)^3$.

43 (a) Similar to the discussion in the solution of Exercise 41, we have

$$f(x) = a(x + 3)^3(x + 1)(x - 2)^2.$$

(b) $a = 1$ and $x = 0 \Rightarrow f(0) = 1(0 + 3)^3(0 + 1)(0 - 2)^2 = 1(3)^3(1)(-2)^2 = 108.$

45 From the graph, we see that $f(x) = x^5 - 16.75x^3 + 12.75x^2 + 49.5x - 54$ has zeros of −4, −2, 1.5, and 3. There is a double root at 1.5. Since the leading coefficient of f is 1, we have $f(x) = 1(x + 4)(x + 2)(x - 1.5)^2(x - 3)$.

$$[-5, 5] \text{ by } [-150, 150, 25]$$

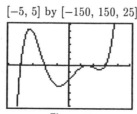

Figure 45

47 Since the zeros are -2, 1, 2, and 3, the polynomial must have the form

$$f(x) = a(x + 2)(x - 1)(x - 2)(x - 3).$$

Now, $f(0) = a(2)(-1)(-2)(-3) = -12a$ and $f(0) = -24 \Rightarrow -12a = -24 \Rightarrow a = 2$.
Let $f(x) = 2(x + 2)(x - 1)(x - 2)(x - 3)$. Since f has been completely determined, we
must check the remaining data point(s). $f(-1) = 2(1)(-2)(-3)(-4) = -48 \neq -52$.

Thus, a fourth-degree polynomial *does not* fit the data points.

49 Since the zeros are 2, 5.2, and 10.1, the polynomial must have the form

$$f(x) = a(x - 2)(x - 5.2)(x - 10.1).$$

Now, $f(1.1) = a(-33.21) = -49.815 \Rightarrow a = 1.5$.
Let $f(x) = 1.5(x - 2)(x - 5.2)(x - 10.1)$. Since f has been completely determined, we
must check the remaining data points. $f(3.5) = 25.245$ and $f(6.4) = -29.304$.

Thus, a third-degree polynomial *does* fit the data points.

51 The zeros are 0, 5, 19, 24, and $f(12) = 10$. $f(t) = a(t)(t - 5)(t - 19)(t - 24)$;
$f(12) = a(12)(7)(-7)(-12) = 10 \Rightarrow 7056a = 10 \Rightarrow a = \frac{10}{7056} = \frac{5}{3528}$.

$$f(t) = \frac{5}{3528}t(t - 5)(t - 19)(t - 24)$$

53 The graph of f does not cross the x-axis at a zero of even multiplicity, but does cross
the x-axis at a zero of odd multiplicity. The higher the multiplicity of a zero, the
more horizontal the graph of f is near that zero.

$[-3, 3]$ by $[-2, 2]$ $[-3, 3]$ by $[-3, 1]$

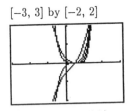   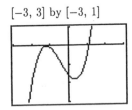

Figure 53   *Figure 55*

55 From the graph of f there are two zeros. They are -1.2 and 1.1. The zero at -1.2
has even multiplicity and the zero at 1.1 has odd multiplicity. Since f has degree 3,
the zero at -1.2 must have multiplicity 2 and the zero at 1.1 has multiplicity 1.

57 From the graph of $A(t) = -\frac{1}{2400}t^3 + \frac{1}{20}t^2 + \frac{7}{6}t + 340$, we see that $A = 400$ when $t \approx 27.1$. Thus, the carbon dioxide concentration will be 400 in $1980 + 27.1 = 2007.1$, or, during the year 2007.

[0, 60, 10] by [0, 600, 100]   [0.5, 12.5] by [−30, 50, 10]

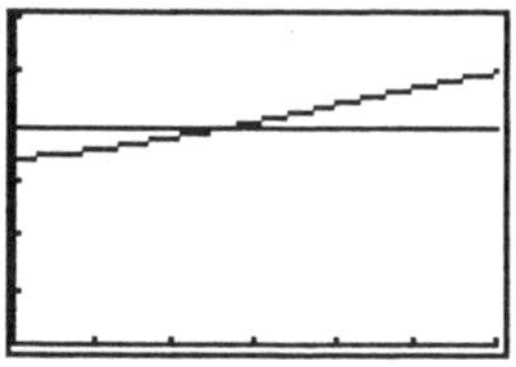

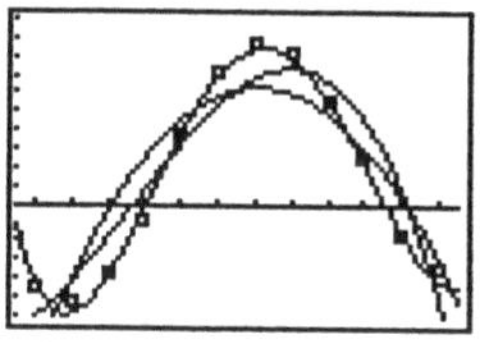

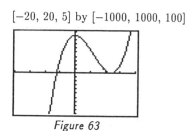

Figure 57   *Figure 59*

59 (a) Graphing the data and the functions show that the best fit is $h(x)$.

(b) Since the temperature changes sign between April and May and between October and November, an average temperature of 0°F occurs when $4 \le x \le 5$ and $10 \le x \le 11$.

(c) Finding the zeros of h between 1 and 12 gives us $x \approx 4.02, 10.53$.

61 Let $r = 6$ and $k = 0.7$. Graph $Y_1 = \frac{4k}{3}\pi r^3 - \pi x^2 r + \frac{1}{3}\pi x^3 = \frac{604.8\pi}{3} - 6\pi x^2 + \frac{1}{3}\pi x^3$ and determine the positive zeros. There are two zeros located at $x \approx 7.64, 15.47$. Since the sphere floats, it will not sink deeper than twice the radius, which is 12 centimeters. Thus, the pine sphere will sink approximately 7.64 centimeters into the water.

[−20, 20, 5] by [−800, 800, 100]   [−20, 20, 5] by [−1000, 1000, 100]

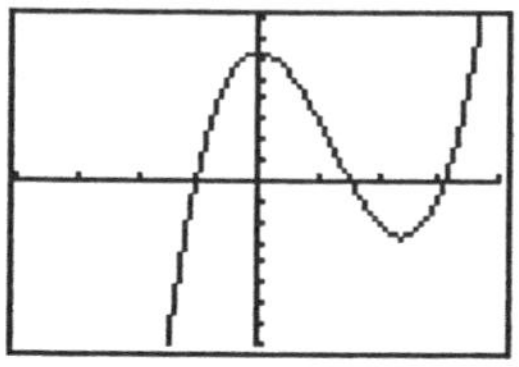

Figure 61   *Figure 63*

63 Let $r = 6$ and $k = 1$. Graph $Y_1 = \frac{4k}{3}\pi r^3 - \pi x^2 r + \frac{1}{3}\pi x^3 = 288\pi - 6\pi x^2 + \frac{1}{3}\pi x^3$ and determine the positive zero. There is a zero at $x = 12$. This means that the entire sphere is just submerged. The sphere has the same density as water and neither sinks nor floats, much like a balloon filled with water.

Note: It is helpful to remember that if $a + bi$ is a zero, then $\left[x^2 - 2ax + (a^2 + b^2) \right]$

is the associated quadratic factor for the following exercises. (page 254 in the text)

[1] Since $3 + 2i$ is a root, so is $3 - 2i$. As in the preceding note, $a = 3$ and $b = -2$, and

hence, $-2a = -6$ and $a^2 + b^2 = 13$.   Thus, the polynomial is of the form

$\left[x - (3 + 2i) \right]\left[x - (3 - 2i) \right] = x^2 - 6x + 13$.

[3] Since 2 is a zero, $x - 2$ is a factor. The associated quadratic factor for $-2 \pm 5i$ is

$x^2 - 2(-2)x + (-2)^2 + (5)^2$, or, equivalently, $x^2 + 4x + 29$. Hence, the polynomial is

of the form $(x - 2)(x^2 + 4x + 29)$.

[5] Since -1 and 0 are zeros, $x + 1$ and x are factors. The associated quadratic factor for

$3 \pm i$ is $x^2 - 2(3)x + (3)^2 + (1)^2$, or, equivalently, $x^2 - 6x + 10$. Hence, the polynomial

is of the form $(x)(x + 1)(x^2 - 6x + 10)$.

[7] Since $4 + 3i$ is a root, so is $4 - 3i$. The associated quadratic factor for $4 \pm 3i$ is

$$x^2 - 2(4)x + (4)^2 + (3)^2, \text{ or, equivalently, } x^2 - 8x + 25.$$

Since $-2 + i$ is a root, so is $-2 - i$. The associated quadratic factor for $-2 \pm i$ is

$$x^2 - 2(-2)x + (-2)^2 + (1)^2, \text{ or, equivalently, } x^2 + 4x + 5.$$

Hence, the polynomial is of the form $(x^2 - 8x + 25)(x^2 + 4x + 5)$.

[9] If $-2i$ is a zero, then $2i$ is a zero. Thus, $(x - 2i)(x - (-2i)) = x^2 - 4i^2 = x^2 + 4$ is the

associated factor. Hence, the polynomial is of the form $(x)(x^2 + 4)(x^2 - 2x + 2)$.

[11] The constant term, 6, has positive integer divisors 1, 2, 3, 6. The leading coefficient,

1, has integer divisors ± 1. By the theorem on rational zeros of a polynomial, any

rational root will be of the form

$$\frac{\text{positive divisors of 6}}{\text{divisors of 1}} = \frac{1, \, 2, \, 3, \, 6}{\pm 1} \longrightarrow \pm 1, \, \pm 2, \, \pm 3, \, \pm 6.$$

Note that this quotient gives us the same possible rational roots as the one on page

256:

$$\frac{\text{factors of } a_0}{\text{factors of } a_n} = \frac{\pm 1, \, \pm 2, \, \pm 3, \, \pm 6}{\pm 1} \longrightarrow \pm 1, \, \pm 2, \, \pm 3, \, \pm 6.$$

Since ± 1 are always potential solutions, they are as good as any other values to

begin with. If the coefficients of the polynomial ($x^3 + 3x^2 - 4x + 6$ in this exercise)

sum to 0, then 1 is a zero. In this case, the sum is $1 + 3 + (-4) + 6 = 6 \neq 0$. Thus, 1

is eliminated and we will try -1. If $f(x)$ is the polynomial, then -1 will be a zero if

the coefficients of $f(-x)$ sum to 0. In this case, $f(-x) = -x^3 + 3x^2 - 10x + 6$ and the

sum of the coefficients is $(-1) + 3 + (-10) + 6 = -2 \neq 0$.   (continued)

Thus, -1 is not a zero. We now use synthetic division to show that ± 2, ± 3, and ± 6 are not zeros of $f(x)$, and thus, the equation has no rational root.

$\boxed{13}$ For $x^5 - 3x^3 + 4x^2 + x - 2 = 0$, $a_0 = -2$ and $a_n = 1$.

$$\frac{\text{factors of } a_0}{\text{factors of } a_n} = \frac{\pm 1, \ \pm 2}{\pm 1} \longrightarrow \pm 1, \ \pm 2$$

Show that ± 1 and ± 2 are not zeros of $f(x)$.

$\boxed{15}$ The constant term, -8, has positive integer divisors 1, 2, 4, 8. The leading coefficient has integer divisors ± 1. Hence, any rational root will be of the form

$$\frac{\text{positive divisors of } -8}{\text{divisors of } 1} = \frac{1, \ 2, \ 4, \ 8}{\pm 1} \longrightarrow \pm 1, \ \pm 2, \ \pm 4, \ \pm 8.$$

We now need to try to find one solution of the equation. Once we have one solution, the resulting polynomial will be a quadratic and then we can use the quadratic formula to find the other 2 solutions. As in the solution to Exercise 11, we will try ± 1 first. After determining that 1 is not a solution, we try -1. In this case, $f(-x) = -x^3 - x^2 + 10x - 8$ and the sum of the coefficients is $(-1) + (-1) + 10 + (-8) = 0$. Thus, -1 is a zero and we will use synthetic division to find the remaining polynomial.

$$
\begin{array}{r|rrrr}
-1 & 1 & -1 & -10 & -8 \\
   &   & -1 & 2 & 8 \\
\hline
   & 1 & -2 & -8 & 0
\end{array}
$$

The remaining polynomial is $(x^2 - 2x - 8) = (x-4)(x+2)$.

Hence, the solutions are -2, -1, and 4.

$\boxed{17}$ As previously noted, we first check ± 1. Neither of these values are solutions so we list the possible rational roots. The values listed below are obtained by taking quotients of each number in the numerator with ± 1, and then each number in the numerator with ± 2, while discarding any repeat choices.

$$\frac{1, 2, 3, 5, 6, 10, 15, 30}{\pm 1, \ \pm 2} \longrightarrow$$

$$\pm 1, \ \pm 2, \ \pm 3, \ \pm 5, \ \pm 6, \ \pm 10, \ \pm 15, \ \pm 30, \ \pm \tfrac{1}{2}, \ \pm \tfrac{3}{2}, \ \pm \tfrac{5}{2}, \ \pm \tfrac{15}{2}$$

Trying 2 we obtain

$$
\begin{array}{r|rrrr}
2 & 2 & -3 & -17 & 30 \\
  &   & 4 & 2 & -30 \\
\hline
  & 2 & 1 & -15 & 0
\end{array}
$$

The remaining polynomial is $(2x^2 + x - 15) = (2x - 5)(x + 3)$.

Hence, the solutions are -3, 2, and $\tfrac{5}{2}$.

$\boxed{19}$ $\dfrac{1,\ 2,\ 4,\ 7,\ 8,\ 14,\ 28,\ 56}{\pm 1} \longrightarrow \pm 1,\ \pm 2,\ \pm 4,\ \pm 7,\ \pm 8,\ \pm 14,\ \pm 28,\ \pm 56.$ Again, ± 1

are not solutions. Trying 4 and then -7 (using a slightly different format) we obtain

$$
\begin{array}{r|rrrrr}
4 & 1 & 3 & -30 & -6 & 56 \\
  &   & 4 & 28 & -8 & -56 \\
\hline
-7 & 1 & 7 & -2 & -14 & 0 \\
  &   & -7 & 0 & 14 & \\
\hline
  & 1 & 0 & -2 & 0 &
\end{array}
$$

The remaining polynomial is $x^2 - 2$. Its solutions are $\pm\sqrt{2}$.

Hence, the solutions are $-7,\ \pm\sqrt{2}$, and 4.

$\boxed{21}$ We first factor out x^2, leaving us with the equation $x^2(6x^3 + 19x^2 + x - 6) = 0$. The x^2 factor indicates that the number 0 is a zero of multiplicity two. We can now concentrate on solving the equation $6x^3 + 19x^2 + x - 6 = 0$.

$$
\dfrac{1,\ 2,\ 3,\ 6}{\pm 1,\ \pm 2,\ \pm 3,\ \pm 6} \longrightarrow \pm 1,\ \pm 2,\ \pm 3,\ \pm 6,\ \pm\tfrac{1}{2},\ \pm\tfrac{3}{2},\ \pm\tfrac{1}{3},\ \pm\tfrac{2}{3},\ \pm\tfrac{1}{6}
$$

$$
\begin{array}{r|rrrr}
-3 & 6 & 19 & 1 & -6 \\
   &   & -18 & -3 & 6 \\
\hline
   & 6 & 1 & -2 & 0
\end{array}
$$

$6x^2 + x - 2 = (3x + 2)(2x - 1)$ $\qquad\qquad\qquad$ $\bigstar$ $-3,\ -\tfrac{2}{3},\ \tfrac{1}{2}$

$\boxed{23}$ $\dfrac{1,\ 3,\ 9,\ 27}{\pm 1,\ \pm 2,\ \pm 4,\ \pm 8} \longrightarrow$

$\pm 1,\ \pm 3,\ \pm 9,\ \pm 27,\ \pm\tfrac{1}{2},\ \pm\tfrac{1}{4},\ \pm\tfrac{1}{8},\ \pm\tfrac{3}{2},\ \pm\tfrac{3}{4},\ \pm\tfrac{3}{8},\ \pm\tfrac{9}{2},\ \pm\tfrac{9}{4},\ \pm\tfrac{9}{8},\ \pm\tfrac{27}{2},\ \pm\tfrac{27}{4},\ \pm\tfrac{27}{8}$

This is a tough one because only $-\tfrac{3}{4}$ is a solution.

$$
\begin{array}{r|rrrr}
-\tfrac{3}{4} & 8 & 18 & 45 & 27 \\
   &   & -6 & -9 & -27 \\
\hline
   & 8 & 12 & 36 & 0
\end{array}
$$

$8x^2 + 12x + 36 = 0 \ \{\text{divide by } 4\} \ \Rightarrow \ 2x^2 + 3x + 9 = 0 \ \Rightarrow$

$x = \dfrac{-3 \pm \sqrt{9 - 72}}{4} = -\tfrac{3}{4} \pm \tfrac{1}{4}\sqrt{63}\,i = -\tfrac{3}{4} \pm \tfrac{3}{4}\sqrt{7}\,i$ $\qquad$ $\bigstar$ $-\tfrac{3}{4},\ -\tfrac{3}{4} \pm \tfrac{3}{4}\sqrt{7}\,i$

$\boxed{25}$ $f(x) = 6x^5 - 23x^4 + 24x^3 + x^2 - 12x + 4$ has zeros at $-\tfrac{2}{3},\ \tfrac{1}{2},\ 1$ (mult. 2), and 2.

Thus, $f(x) = 6(x + \tfrac{2}{3})(x - \tfrac{1}{2})(x - 1)^2(x - 2) = (3x + 2)(2x - 1)(x - 1)^2(x - 2)$.

[27] From the graph, we see that $f(x) = 2x^3 - 25.4x^2 + 3.02x + 24.75$ has zeros of approximately -0.9, 1.1, and 12.5. Since the leading coefficient of f is 2, we have $f(x) = 2(x + 0.9)(x - 1.1)(x - 12.5)$.

$$[-5, \, 15, \, 5] \text{ by } [-600, \, 100, \, 100]$$

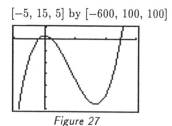

Figure 27

[29] No. If i is a root, then $-i$ is also a root. Hence, the polynomial would have factors $x - 1$, $x + 1$, $x - i$, $x + i$ and therefore would be of degree greater than 3.

[31] Since n is odd and nonreal complex zeros occur in conjugate pairs for polynomials with real coefficients, there must be at least one real zero.

[33] (a) From Exercise 39 of Section 3.1, the allowable range was $0 < x < 10$ and
$V(x) = 4x(x - 10)(x - 15)$. $V = 1000 \Rightarrow 4x(x - 10)(x - 15) = 1000 \Rightarrow$
$x(x - 10)(x - 15) = 250 \Rightarrow x(x^2 - 25x + 150) - 250 = 0 \Rightarrow$
$x^3 - 25x^2 + 150x - 250 = 0$. We can show that 5 is a root, so the last equation
can be written as $(x - 5)(x^2 - 20x + 50) = 0$. Using the quadratic equation gives
us $(x - 5)\left[x - (10 - 5\sqrt{2})\right]\left[x - (10 + 5\sqrt{2})\right] = 0$. Discard $10 + 5\sqrt{2}$ $\{ > 10 \}$.

The dimensions of the box are x, $20 - 2x$, and $30 - 2x$.

So the two boxes $[\![A]\!]$ and $[\![B]\!]$ having volume 1000 in^3 have dimensions

$\quad [\![A]\!]$: $5 \times \underline{20 - 2(5)} \times \underline{30 - 2(5)} = 5 \times 10 \times 20$ and

$\quad [\![B]\!]$: $(10 - 5\sqrt{2}) \times \underline{20 - 2(10 - 5\sqrt{2})} \times \underline{30 - 2(10 - 5\sqrt{2})} =$
$\qquad\qquad\qquad\qquad\qquad (10 - 5\sqrt{2}) \times (10\sqrt{2}) \times (10 + 10\sqrt{2})$.

(b) The surface area function is

$S(x) = (20 - 2x)(30 - 2x) + 2(x)(20 - 2x) + 2(x)(30 - 2x) = -4x^2 + 600$.

$S(5) = 500$ and $S(10 - 5\sqrt{2}) = 400\sqrt{2} \approx 565.7$, so box $[\![A]\!]$ has less surface area.

[35] (a) Let x denote one of the sides of the triangle and $x + 1$ its hypotenuse.

Using the Pythagorean theorem, the third side y is given by

$$x^2 + y^2 = (x + 1)^2 \Rightarrow y^2 = (x^2 + 2x + 1) - x^2 \Rightarrow y = \sqrt{2x + 1}.$$

Hence, the sides of the triangle are x, $\sqrt{2x + 1}$, and $x + 1$.

$A = \frac{1}{2}bh \Rightarrow 30 = \frac{1}{2}x\sqrt{2x + 1} \Rightarrow 60 = x\sqrt{2x + 1} \Rightarrow 60^2 = x^2(2x + 1) \Rightarrow$
$\qquad\qquad\qquad\qquad\qquad 3600 = 2x^3 + x^2 \Rightarrow 2x^3 + x^2 - 3600 = 0$.

(b) There is one sign change in $f(x) = 2x^3 + x^2 - 3600$.

By Descartes' rule of signs there is one positive real root.

Synthetically dividing 13 into f, we obtain

$$
\begin{array}{r|rrrr}
13 & 2 & 1 & 0 & -3600 \\
   &   & 26 & 351 & 4563 \\
\hline
   & 2 & 27 & 351 & 963
\end{array}
$$

The numbers in the third row are nonnegative,

so 13 is an upper bound for the zeros of f.

(c) $2x^3 + x^2 - 3600 = 0 \Leftrightarrow (x - 12)(2x^2 + 25x + 300) = 0$. The solutions of

$2x^2 + 25x + 300 = 0$ are $x = -\frac{25}{4} \pm \frac{5}{4}\sqrt{71}\,i$, and hence, $x = 12$ is the only real

solution. The legs of the triangle are 12 ft and 5 ft, and the hypotenuse is 13 ft.

[37] (a) Volume$_{\text{total}}$ = Volume$_{\text{cube}}$ + Volume$_{\text{roof}}$

$\quad = x^3 + \frac{1}{2}bhx = x^3 + \frac{1}{2}(x)(6 - x)(x) = x^3 + \frac{1}{2}x^2(6 - x)$.

(b) Volume = 80 $\Rightarrow$ $x^3 + \frac{1}{2}x^2(6 - x) = 80$ $\Rightarrow$ $x^3 + 3x^2 - \frac{1}{2}x^3 = 80$ $\Rightarrow$

$\frac{1}{2}x^3 + 3x^2 = 80$ $\Rightarrow$ $x^3 + 6x^2 - 160 = 0$ $\Rightarrow$ $(x - 4)(x^2 + 10x + 40) = 0$.

The length of the side is 4 ft.

[39] $x^5 + 1.1x^4 - 3.21x^3 - 2.835x^2 + 2.7x + 0.62 = -1$ $\Leftrightarrow$

$x^5 + 1.1x^4 - 3.21x^3 - 2.835x^2 + 2.7x + 1.62 = 0$.

The graph of $y = x^5 + 1.1x^4 - 3.21x^3 - 2.835x^2 + 2.7x + 1.62$ intersects the x-axis

three times. The zeros at -1.5 and 1.2 have even multiplicity (since the graph is

tangent to the x-axis at these points) and the zero at -0.5 has odd multiplicity (since

the graph crosses the x-axis at this point). Since the equation has degree 5, the only

possibility is that the zeros at -1.5 and 1.2 have multiplicity 2 and the zero at -0.5

has multiplicity 1. Thus, the equation has no nonreal solutions.

[−4.5, 4.5] by [−3, 3]   [−4.5, 4.5] by [−3, 3]

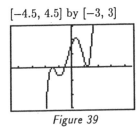   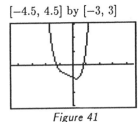

Figure 39   Figure 41

[41] Graph $y = x^4 + 1.4x^3 + 0.44x^2 - 0.56x - 0.96$.

From the graph, zeros are located at -1.2 and 0.8. Using synthetic division,

$\dfrac{x^4 + 1.4x^3 + 0.44x^2 - 0.56x - 0.96}{x + 1.2} = x^3 + 0.2x^2 + 0.2x - 0.8$ and

$\dfrac{x^3 + 0.2x^2 + 0.2x - 0.8}{x - 0.8} = x^2 + x + 1.$ The zeros of $x^2 + x + 1$ are $-\frac{1}{2} \pm \frac{\sqrt{3}}{2}i$.

Thus, the solutions to the equation are $-1.2,\ 0.8,\ -\frac{1}{2} \pm \frac{\sqrt{3}}{2}i$.

$\boxed{43}$ From the graph, we see that $D(h) = 0.4$ when $h \approx 10{,}200$.

Thus, the density of the atmosphere is 0.4 kg/m^3 at 10,200 m.

$[0, 30{,}000, 2000]$ by $[0, 1.2, 0.2]$

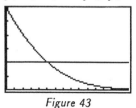

Figure 43

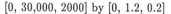

3.5 Exercises

Note: We will use the guidelines listed in the text on page 267.

Below is a summary of these guidelines.

Let $f(x) = \dfrac{a_n x^n + a_{n-1} x^{n-1} + \cdots + a_1 x + a_0}{b_k x^k + b_{k-1} x^{k-1} + \cdots + b_1 x + b_0}$, where $a_n \neq 0$ and $b_k \neq 0$.

(1) Find the x-intercepts { the zeros of the numerator }.

(2) Find the vertical asymptotes { the zeros of the denominator }.

(3) Find the y-intercept { the ratio of constant terms, a_0/b_0 }.

(4) Find the horizontal or oblique asymptote.

    (a) If $n < k$, then $y = 0$ { the x-axis } is the horizontal asymptote.

    (b) If $n = k$,

        then $y = a_n/b_k$ { the ratio of leading coefficients } is the horizontal asymptote.

    (c) If $n > k$, then the asymptote is found by long division. The graph of f is

        asymptotic to $y = q(x)$, where $q(x)$ is the quotient of the division process.

(5) Find the intersection points of the function and the asymptote found in step 3. This will help us decide *how* the function approaches the asymptote.

(6) Sketch the graph by regions, where the regions are determined by the vertical asymptotes. We may use the sign of a particular function value to help us determine if the function is positive or negative. We will not plot any points since the purpose in this section is to determine the general shape of the graph and understand the general principles involved with asymptotes.

boxed{1} $f(x) = 4/x$ •

  (a) (1) There are no x-intercepts since the numerator is never equal to zero.

      (2) There is a vertical asymptote at $x = 0$ {the y-axis}.

      (3) There is no y-intercept since the function is undefined for $x = 0$.

      (4) The degree of the numerator is 0, which is less than the degree of the denominator, 1, so the horizontal asymptote is $y = 0$ {the x-axis}.

      (5) Setting the function equal to the value of the asymptote found in step 4 gives us $4/x = 0$, which has no solutions.

      (6) There is one vertical asymptote, and it separates the plane into 2 regions. For the region $x < 0$, $f(x) = 4/x < 0$—that is, the y-values are negative. This indicates that the graph is under the y-axis as in *Figure 1*. For the region $x > 0$, $f(x) > 0$, and the graph is above the x-axis.

  (b) The domain D is the set of all nonzero real numbers—that is, $\mathbb{R} - \{0\}$. The range R is equal to the same set of numbers.

  (c) The function is decreasing on $(-\infty, 0)$ and also on $(0, \infty)$.

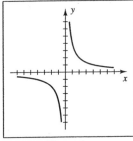

Figure 1

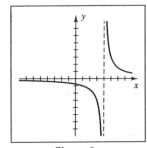

Figure 3

boxed{3} $f(x) = \dfrac{3}{x - 4} = 3\left(\dfrac{1}{x - 4}\right)$ •

      Rather than use the guidelines, we will think of this graph in terms of the shifting and stretching properties presented in Section 2.2. Consider the graph of $g(x) = 1/x$, which is similar to the graph in Exercise 1. Now $h(x) = 1/(x - 4)$ is just $g(x - 4)$, which shifts the graph of g to the right by 4 units, making $x = 4$ the vertical asymptote. The effect of the 3 in the numerator is to vertically stretch the graph of h by a factor of 3.

5 $f(x) = \dfrac{-3x}{x+2}$ •

(1) Numerator $= 0 \;\Rightarrow\; -3x = 0 \;\Rightarrow\; x = 0$, the <u>only</u> x-intercept.

(2) Denominator $= 0 \;\Rightarrow\; x+2 = 0 \;\Rightarrow\; x = -2$, the only vertical asymptote.

(3) $f(0) = \frac{0}{2} = 0 \;\Rightarrow\;$ 0 is the y-intercept. { We already knew this from step 1. }

(4) Degree of numerator $= 1 =$ degree of denominator $\;\Rightarrow$

$$y = \tfrac{-3}{1} \; \{\text{ratio of leading coefficients}\} = -3 \text{ is the horizontal asymptote.}$$

(5) Function $=$ horizontal asymptote value $\;\Rightarrow\; f(x) = -3 \;\Rightarrow\; \dfrac{-3x}{x+2} = -3 \;\Rightarrow$

$-3x = -3(x+2) \;\Rightarrow\; -3x = -3x - 6 \;\Rightarrow\; 0 = -6$. This is a contradiction and

indicates that there are <u>no</u> intersection points on the horizontal asymptote.

Remember, the function can not intersect a vertical asymptote, but can intersect

the horizontal asymptote.

(6) For the region $x < -2$, there are no x-intercepts and the graph must be below the

horizontal asymptote $y = -3$. { If it was above $y = -3$, it would have to

intersect the x-axis at some point, but from (1), we know that 0 is the *only*

x-intercept. }

For the region $x > -2$, there is an x-intercept at 0 and the graph is above the

horizontal asymptote. A common mistake is to confuse the x-axis with the

horizontal asymptote. Remember, as $x \to \infty$ or $x \to -\infty$, $f(x)$ will get close

to the horizontal or oblique asymptote, not the x-axis { unless the x-axis is

the horizontal asymptote }.

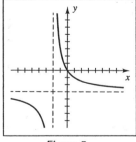

Figure 5

$\boxed{7}$ $f(x) = \frac{4x-1}{2x+3}$ •

(1) $4x - 1 = 0 \Rightarrow 4x = 1 \Rightarrow x = \frac{1}{4}$, the <u>only</u> x-intercept.

(2) $2x + 3 = 0 \Rightarrow 2x = -3 \Rightarrow x = -\frac{3}{2}$, the only vertical asymptote.

(3) $f(0) = \frac{-1}{3} = -\frac{1}{3} \Rightarrow -\frac{1}{3}$ is the y-intercept.

(4) Degree of numerator $= 1 =$ degree of denominator $\Rightarrow$

$$y = \frac{4}{2} \{\text{ratio of leading coefficients}\} = 2 \text{ is the horizontal asymptote.}$$

(5) Function = horizontal asymptote value $\Rightarrow$ $f(x) = 2 \Rightarrow \frac{4x-1}{2x+3} = 2 \Rightarrow$

$4x - 1 = 2(2x + 3) \Rightarrow 4x - 1 = 4x + 6 \Rightarrow -1 = 6$.   This is a contradiction
and indicates that there are <u>no</u> intersection points on the horizontal asymptote.
Remember, the function can not intersect a vertical asymptote, but can intersect
the horizontal asymptote.

(6) For the region $x < -\frac{3}{2}$, there are no x-intercepts and the graph must be above the
horizontal asymptote $y = 2$.   {If it was below $y = 2$, it would have to
intersect the x-axis at some point, but from (1), we know that $\frac{1}{4}$ is the *only*
x-intercept.}   As x approaches $-\frac{3}{2}$ from the left, y approaches ∞.   As x
approaches $-\infty$, y approaches 2 through values larger than 2.

For the region $x > -\frac{3}{2}$, there is a y-intercept at $-\frac{1}{3}$ and an x-intercept at $\frac{1}{4}$, so the
graph is below the horizontal asymptote.   As x approaches $-\frac{3}{2}$ from the
right, y approaches $-\infty$.   As x approaches ∞, y approaches 2 through values
smaller than 2.

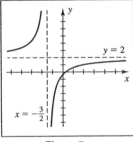

Figure 7

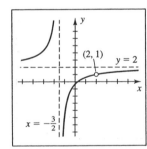

Figure 9

$\boxed{9}$ $f(x) = \frac{(4x-1)(x-2)}{(2x+3)(x-2)} = \frac{4x-1}{2x+3}$ for $x \neq 2$.   The reduced function is the same as the

function in Exercise 7.   The only difference is that this function has a hole in the

graph at $x = 2$.   To determine the value of y when $x = 2$, substitute 2 into $\frac{4x-1}{2x+3}$ to

get $\frac{7}{7} = 1$.   There is a hole in the graph at $(2, 1)$.

11 $f(x) = \dfrac{x-2}{x^2 - x - 6} = \dfrac{x-2}{(x+2)(x-3)}$ •

(1) $x - 2 = 0 \Rightarrow x = 2$, the only x-intercept

(2) $(x+2)(x-3) = 0 \Rightarrow x = -2$ and $x = 3$, the vertical asymptotes

(3) $f(0) = \frac{-2}{-6} = \frac{1}{3} \Rightarrow \frac{1}{3}$ is the y-intercept

(4) Degree of numerator $= 1 < 2 =$ degree of denominator $\Rightarrow$

$y = 0$ is the horizontal asymptote.

(5) We have already solved $f(x) = 0$ in step 1.

Hence, we know that the function will cross the horizontal asymptote at $(2, 0)$.

(6) For $x < -2$, we have no information and will examine $f(-5)$ $\{-5$ is to the left of
$-2\}$. Using the factored form of f and only the signs of the values, we

obtain $f(-5) = \dfrac{(-)}{(-)(-)} = \{$ some negative number $\} < 0$ since the combination

of 3 negative signs will be negative. Hence, the graph will be below the

horizontal asymptote.

For $-2 < x < 3$, we have an x-intercept at 2 and need to know what the function

does on each side of 2. Choosing 0 and 2.5 for test points, we check the sign

of $f(0)$ and $f(2.5)$ as above. $f(0) = \dfrac{(-)}{(+)(-)} > 0$ and $f(2.5) = \dfrac{(+)}{(+)(-)} < 0$.

Hence, the function will change signs from positive to negative as it passes

through 2.

For $x > 3$, $f(5) = \dfrac{(+)}{(+)(+)} > 0$ and the function is above the horizontal asymptote.

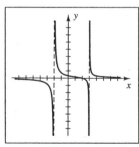

Figure 11

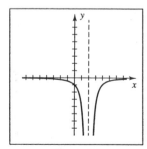

Figure 13

13 $f(x) = \dfrac{-4}{(x-2)^2}$ • Note that the function is always negative since it is the quotient

of a negative and a positive, provided $x \neq 2$. The y-intercept is $f(0) = -1$, the

vertical asymptote is $x = 2$, and the horizontal asymptote is $y = 0$.

$\boxed{15}$ $f(x) = \dfrac{x-3}{x^2-1} = \dfrac{x-3}{(x+1)(x-1)}$ •

(1) $x - 3 = 0 \;\Rightarrow\; x = 3$, the only x-intercept

(2) $(x+1)(x-1) = 0 \;\Rightarrow\; x = -1$ and $x = 1$, the vertical asymptotes

(3) $f(0) = \frac{-3}{-1} = 3 \;\Rightarrow\; 3$ is the y-intercept

(4) Degree of numerator $= 1 < 2 =$ degree of denominator $\;\Rightarrow\;$

$y = 0$ is the horizontal asymptote.

(5) We have already solved $f(x) = 0$ in step 1.

Hence, we know that the function will cross the horizontal asymptote at $(3,\ 0)$.

(6) For $-1 < x < 1$, we have the y-intercept at 3. Since the function cannot intersect the x-axis { the only x-intercept is at 3 }, the function remains positive and we have $f(x) \to \infty$ as $x \to 1^-$ and $f(x) \to \infty$ as $x \to -1^+$.

For $x > 1$, we have the x-intercept at 3 as a point to work with. The function will change sign at 3 since 3 is a zero of odd multiplicity { 1 }. If the function was going to merely touch the point $(3,\ 0)$ and "turn around", 3 would have to be a zero of even multiplicity. Hence, it suffices to find only one sign. We choose 2 as our test point. $f(2) = \dfrac{(-)}{(+)(+)} < 0$. The function values are negative for $1 < x < 3$ and positive for $x > 3$. Remember, eventually $f(x)$ will be extremely close to 0 as $x \to \infty$. Make sure your sketch reflects that fact.

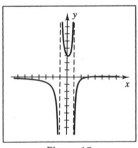

Figure 15

$\boxed{17}$ $f(x) = \dfrac{2x^2 - 2x - 4}{x^2 + x - 12} = \dfrac{2(x^2 - x - 2)}{(x+4)(x-3)} = \dfrac{2(x+1)(x-2)}{(x+4)(x-3)}$ • As a generalization, you

should always factor the function first. For steps 1–6 of the guidelines, it is convenient to use the original function for steps 3, 4, and 5, and to use the factored form for steps 1, 2, and 6. After this problem, we will concentrate on steps 5 and 6. You should be able to pick up the information for steps 1–4 by merely looking at both forms of the function {except for the oblique asymptote case}.

(1) $2(x+1)(x-2) = 0 \Rightarrow x = -1, 2$; the x-intercepts

(2) $(x+4)(x-3) = 0 \Rightarrow x = -4, 3$; the vertical asymptotes

(3) $f(0) = \frac{-4}{-12} = \frac{1}{3} \Rightarrow \frac{1}{3}$ is the y-intercept

(4) Degree of numerator $= 2 =$ degree of denominator $\Rightarrow$

$$y = \tfrac{2}{1} = 2 \text{ is the horizontal asymptote.}$$

(5) $\dfrac{2x^2 - 2x - 4}{x^2 + x - 12} = 2 \Rightarrow 2x^2 - 2x - 4 = 2(x^2 + x - 12) \Rightarrow$

$2x^2 - 2x - 4 = 2x^2 + 2x - 24 \Rightarrow 20 = 4x \Rightarrow x = 5$. Hence, the function will

intersect the horizontal asymptote <u>only</u> at the point (5, 2).

(6) For $x < -4$, there are no x-intercepts so the function is above the horizontal asymptote.

For $-4 < x < 3$, the function must go through $(-1, 0)$, $(0, \frac{1}{3})$, and $(2, 0)$. It must go "down" by both vertical asymptotes because if it went "up", it would have to cross the horizontal asymptote. But we know this only occurs at the point (5, 2) and not in this region.

For $x > 3$, we know that the graph goes up as we get close to 3 because if it went down, there would have to be an x-intercept, but there isn't one. The function goes through (5, 2), but then turns around {before touching the x-axis} and gets very close to 2 as x increases.

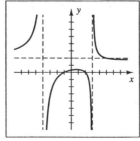

Figure 17

Note: We will let $I(x, y)$ denote the intersection point found in step 5.

19 $f(x) = \dfrac{-x^2 - x + 6}{x^2 + 3x - 4} = \dfrac{-1(x^2 + x - 6)}{(x+4)(x-1)} = \dfrac{-1(x+3)(x-2)}{(x+4)(x-1)}$ •

(5) $\dfrac{-x^2 - x + 6}{x^2 + 3x - 4} = -1 \;\Rightarrow\; -x^2 - x + 6 = -x^2 - 3x + 4 \;\Rightarrow$

$$2x = -2 \;\Rightarrow\; x = -1,\; I = (-1, -1)$$

(6) For $x < -4$, the function is below $y = -1$ since there are no x-intercepts.

For $-4 < x < 1$, the function passes through $(-3, 0)$, $(-1, -1)$, and $(0, -\frac{3}{2})$.
  • These points are known from finding information in steps 1–5.

For $x > 1$, the function passes through $(2, 0)$ and is always above the horizontal asymptote since it only intersects the horizontal asymptote at $(-1, -1)$.

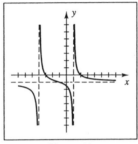

Figure 19

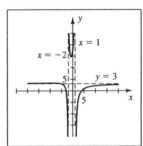

Figure 21

21 $f(x) = \dfrac{3x^2 - 3x - 36}{x^2 + x - 2} = \dfrac{3(x^2 - x - 12)}{(x+2)(x-1)} = \dfrac{3(x+3)(x-4)}{(x+2)(x-1)}$ •

(5) $\dfrac{3x^2 - 3x - 36}{x^2 + x - 2} = 3 \;\Rightarrow\; 3x^2 - 3x - 36 = 3x^2 + 3x - 6 \;\Rightarrow$

$$-30 = 6x \;\Rightarrow\; x = -5,\; I = (-5, 3)$$

(6) For $x < -2$, the function passes through $(-3, 0)$ and $(-5, 3)$ and stays above the horizontal asymptote, $y = 3$, as $x \to -\infty$.

For $-2 < x < 1$, we have the y-intercept at 18. Since the graph doesn't intersect the horizontal asymptote in this region, the function goes up as it gets close to each vertical asymptote.

For $x > 1$, we have the x-intercept 4 and the function stays below the horizontal asymptote.

$\boxed{23}$ $f(x) = \dfrac{-2x^2 + 10x - 12}{x^2 + x} = \dfrac{-2(x^2 - 5x + 6)}{(x+1)(x)} = \dfrac{-2(x-2)(x-3)}{(x+1)(x)}$ •

(2) Since $(x+1)(x) = 0 \Rightarrow x = -1$ and 0,

we have the y-axis as a vertical asymptote, and hence, there is no y-intercept.

(5) $\dfrac{-2x^2 + 10x - 12}{x^2 + x} = -2 \Rightarrow -2x^2 + 10x - 12 = -2x^2 - 2x \Rightarrow$

$12x = 12 \Rightarrow x = 1, I = (1, -2)$

(6) For $-1 < x < 0$, $f(-\frac{1}{2}) = \dfrac{(-)(-)(-)}{(+)(-)} > 0$ and the function is above the x-axis.

For $x > 0$, the function goes through $(1, -2)$, $(2, 0)$, and $(3, 0)$ and gradually gets

closer to $y = -3$ as $x \to \infty$.

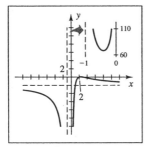

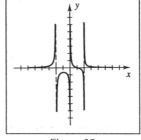

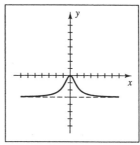

| *Figure 23* | *Figure 25* | *Figure 27* |

$\boxed{25}$ $f(x) = \dfrac{x-1}{x^3 - 4x} = \dfrac{x-1}{(x)(x^2 - 4)} = \dfrac{x-1}{(x+2)(x)(x-2)}$ •

(5) The horizontal asymptote is $y = 0$ {the x-axis} and the x-intercept is 1, so we

know that the function will cross the horizontal asymptote at $(1, 0)$.

(6) For $-2 < x < 0$, $f(-1) = \dfrac{(-)}{(+)(-)(-)} < 0$ and the function is below the x-axis.

For $0 < x < 1$, $f(\frac{1}{2}) = \dfrac{(-)}{(+)(+)(-)} > 0$ and the function is above the x-axis.

For $1 < x < 2$, $f(\frac{3}{2}) = \dfrac{(+)}{(+)(+)(-)} < 0$ and the function is below the x-axis.

$\boxed{27}$ $f(x) = \dfrac{-3x^2}{x^2 + 1}$ • f is an even function

(2) There are no vertical asymptotes since $x^2 + 1 \neq 0$ for any real number.

(5) $\dfrac{-3x^2}{x^2 + 1} = -3 \Rightarrow -3x^2 = -3x^2 - 3 \Rightarrow 0 = -3 \Rightarrow$

there are no intersection points of the horizontal asymptote and the function.

(6) We have the x-intercept at 0 and the function must get close to the

horizontal asymptote, $y = -3$, as $x \to \infty$ or $x \to -\infty$.

29 $f(x) = \dfrac{x^2 - x - 6}{x + 1} = \dfrac{(x + 2)(x - 3)}{x + 1}$ •

(4) We could use long division to find the oblique asymptote, but since the denominator is of the from $x - c$, we will use synthetic division.

$$
\begin{array}{r|rrr}
-1 & 1 & -1 & -6 \\
   &   & -1 & 2 \\
\hline
   & 1 & -2 & -4
\end{array}
$$

The third row indicates that $f(x) = x - 2 - \dfrac{4}{x + 1}$. The expression $\dfrac{4}{x + 1} \to 0$ as $x \to \pm\infty$, so $y = x - 2$ is an oblique asymptote for f.

(5) $\dfrac{x^2 - x - 6}{x + 1} = x - 2 \Rightarrow x^2 - x - 6 = (x - 2)(x + 1) \Rightarrow$

$x^2 - x - 6 = x^2 - x - 2 \Rightarrow -6 = -2 \Rightarrow$ there are no intersection points of the oblique asymptote and the function.

(6) For $x < -1$, we have the x-intercept at -2 and the function is above the oblique asymptote.

For $x > -1$, we have the points $(0, -6)$ and $(3, 0)$ and the function is below the oblique asymptote.

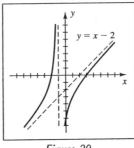

Figure 29

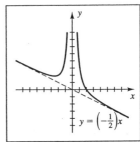

Figure 31

31 $f(x) = \dfrac{8 - x^3}{2x^2} = \dfrac{(2 - x)(4 + 2x + x^2)}{2x^2}$ •

(4) By dividing, we can write f in the form $f(x) = -\dfrac{1}{2}x + \dfrac{8}{2x^2}$. The oblique asymptote is $y = -\dfrac{1}{2}x$.

(5) Function = oblique asymptote $\Rightarrow \dfrac{8 - x^3}{2x^2} = -\dfrac{1}{2}x \Rightarrow 8 - x^3 = -x^3 \Rightarrow$

$8 = 0 \Rightarrow$ there are no intersection points of the asymptote and the function.

(6) For $x < 0$, there are no x-intercepts, so the function is above the oblique asymptote.

For $x > 0$, we have the x-intercept 2, and the function is above the asymptote.

33 $f(x) = \dfrac{2x^2 + x - 6}{x^2 + 3x + 2} = \dfrac{(x+2)(2x-3)}{(x+2)(x+1)} = \dfrac{2x-3}{x+1}$ for $x \neq -2$. We must reduce the

function first and then *sketch the reduced function*. After sketching the reduced

function, we need to remember to put a hole in the graph where the original function

was undefined. In this case, the reduced function is $f(x) = \dfrac{2x-3}{x+1}$. We sketch this as

we did the other rational functions. To determine the value of y when $x = -2$,

substitute -2 into $\dfrac{2x-3}{x+1}$ to get 7. There is a hole in the graph at $(-2, 7)$.

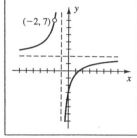

Figure 33

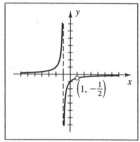

Figure 35

35 $f(x) = \dfrac{x-1}{1-x^2} = \dfrac{x-1}{(1+x)(1-x)} = \dfrac{-1(1-x)}{(1+x)(1-x)} = \dfrac{-1}{x+1}$ for $x \neq 1$.

Substituting 1 for x in $\dfrac{-1}{x+1}$ gives us a hole in the graph at $\left(1, -\dfrac{1}{2}\right)$.

37 $f(x) = \dfrac{x^2 + x - 2}{x+2} = \dfrac{(x+2)(x-1)}{x+2} = x - 1$ for $x \neq -2$.

Substituting -2 for x in $x - 1$ gives us a hole in the graph $\{$a line$\}$ at $(-2, -3)$.

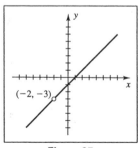

Figure 37

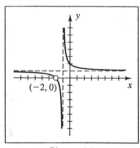

Figure 39

39 $f(x) = \dfrac{x^2 + 4x + 4}{x^2 + 3x + 2} = \dfrac{(x+2)^2}{(x+1)(x+2)} = \dfrac{x+2}{x+1}$ for $x \neq -2$. Substituting -2 for x in

$\dfrac{x+2}{x+1}$ gives us 0, so there is a hole in the graph $\{$on the x-axis$\}$ at $(-2, 0)$.

41 The vertical asymptote of $x = 4$ implies that $x - 4$ is a factor in the denominator. Since 3 is an x-intercept, $x - 3$ is a factor in the numerator. The ratio of leading coefficients must be -1 because the horizontal asymptote is $y = -1$, and hence, we must have a factor of -1 in the numerator. Therefore, f has the form

$$f(x) = \frac{-1(x - 3)}{x - 4}, \text{ or equivalently, } \frac{3 - x}{x - 4}.$$

43 The vertical asymptotes of $x = -3$ and $x = 1$ imply that $x + 3$ and $x - 1$ are factors in the denominator. Since -1 is an x-intercept, $x + 1$ is a factor in the numerator. The hole at $x = 2$ indicates that $x - 2$ is a factor in the numerator and in the denominator. If we let a denote an arbitrary coefficient, then f has the form

$$f(x) = \frac{a(x + 1)(x - 2)}{(x - 1)(x + 3)(x - 2)}.$$

Note that $y = 0$ is the horizontal asymptote since the degree of the numerator is less than the degree of the denominator. To determine the value of a, we use the fact that $f(0) = -2$.

$$f(0) = -2 \text{ and } f(0) = \frac{a(1)}{(-1)(3)} = \frac{a}{-3} \;\Rightarrow\; \frac{a}{-3} = -2 \;\Rightarrow\; a = 6.$$

Therefore, f has the form $f(x) = \dfrac{6(x + 1)(x - 2)}{(x - 1)(x + 3)(x - 2)}$, or equivalently, $\dfrac{6x^2 - 6x - 12}{x^3 - 7x + 6}$.

45 $f(x) = -\dfrac{2}{x - 1}$ • The domain of f is $\mathbb{R} - \{1\}$. The horizontal asymptote of f is $y = 0$ and the range of f is $\mathbb{R} - \{0\}$, or equivalently, $(-\infty, 0) \cup (0, \infty)$. The domain of f^{-1} is equal to the range of f; that is, $\mathbb{R} - \{0\}$. The range of f^{-1} is equal to the domain of f; that is, $\mathbb{R} - \{1\}$, or equivalently, $(-\infty, 1) \cup (1, \infty)$.

47 $f(x) = \dfrac{4x + 5}{3x - 8}$ • The domain of f is $\mathbb{R} - \{\frac{8}{3}\}$. The horizontal asymptote of f is $y = \frac{4}{3}$ and the range of f is $\mathbb{R} - \{\frac{4}{3}\}$, or equivalently, $(-\infty, \frac{4}{3}) \cup (\frac{4}{3}, \infty)$. The domain of f^{-1} is equal to the range of f; that is, $\mathbb{R} - \{\frac{4}{3}\}$. The range of f^{-1} is equal to the domain of f; that is, $\mathbb{R} - \{\frac{8}{3}\}$, or equivalently, $(-\infty, \frac{8}{3}) \cup (\frac{8}{3}, \infty)$.

49 (a) The radius of the outside cylinder is $(r + 0.5)$ ft and its height is $(h + 1)$ ft. Since the volume is 16π ft^3 and the volume is $\pi(\text{radius})^2(\text{height})$, we have

$$16\pi = \pi(r + 0.5)^2(h + 1) \;\Rightarrow\; h + 1 = \frac{16\pi}{\pi(r + 0.5)^2} \;\Rightarrow\; h = \frac{16}{(r + 0.5)^2} - 1.$$

(b) Substituting for h from part (a), $V(r) = \pi r^2 h = \pi r^2 \left[\dfrac{16}{(r + 0.5)^2} - 1 \right]$.

(c) r and h must both be positive. $h > 0 \Rightarrow$

$$\frac{16}{(r+0.5)^2} - 1 > 0 \Rightarrow \frac{16}{(r+0.5)^2} > 1 \Rightarrow 16 > (r+0.5)^2 \Rightarrow$$

$$(r+0.5)^2 < 16 \Rightarrow \sqrt{(r+0.5)^2} < \sqrt{16} \Rightarrow |r+0.5| < 4 \Rightarrow$$

$-4 < r + 0.5 < 4 \Rightarrow -4.5 < r < 3.5$. The last inequality combined with the

fact that $r > 0$ means that the excluded values are $r \le 0$ and $r \ge 3.5$.

51 (a) Since 5 gallons of water flow into the tank each minute, $V(t) = 50 + 5t$. Since

each additional gallon of water contains 0.1 lb of salt, $A(t) = 5(0.1)t = 0.5t$.

(b) $c(t) = \dfrac{A(t)}{V(t)} = \dfrac{0.5t}{50+5t} \cdot \dfrac{2}{2} = \dfrac{t}{10t+100}$ lb/gal

(c) As $t \to \infty$, $c(t) \to 0.1$ lb. of salt per gal.

53 (a) $R > S \Rightarrow \dfrac{4500\,S}{S+500} > S \Rightarrow \dfrac{4500\,S}{S+500} - S > 0 \Rightarrow$

$$\dfrac{4500\,S}{S+500} - \dfrac{S(S+500)}{S+500} > 0 \Rightarrow \dfrac{4500\,S - S^2 - 500\,S}{S+500} > 0 \Rightarrow$$

$\dfrac{4000\,S - S^2}{S+500} > 0 \Rightarrow \dfrac{S^2 - 4000\,S}{S+500} < 0$ {multiply both sides by -1 and reverse

the direction of the inequality} $\Rightarrow \dfrac{S(S-4000)}{S+500} < 0$. From *Diagram 53* and

using the fact that $S > 0$, we conclude that $R > S$ when $0 < S < 4000$.

Resulting sign:	$\ominus$	$\oplus$	$\ominus$	$\oplus$
$S - 4000$:	−	−	−	+
S:	−	−	+	+
$S + 500$:	−	+	+	+
S values:	−500	0	4000	

Diagram 53

(b) The greatest possible number of offspring that survive to maturity is 4500, the

horizontal asymptote value. 90% of 4500 is 4050.

$R = 4050 \Rightarrow 4050 = \dfrac{4500\,S}{S+500} \Rightarrow 4050S + 2{,}025{,}000 = 4500S \Rightarrow$

$$2{,}025{,}000 = 450S \Rightarrow S = 4500.$$

(c) 80% of 4500 is 3600. $R = 3600 \Rightarrow 3600 = \dfrac{4500\,S}{S+500} \Rightarrow$

$$3600S + 1{,}800{,}000 = 4500S \Rightarrow 1{,}800{,}000 = 900S \Rightarrow S = 2000.$$

(d) A 125% increase $\left\{\dfrac{4500-2000}{2000} \times 100\right\}$ in the number S of spawners produces only

a 12.5% increase $\left\{\dfrac{4050-3600}{3600} \times 100\right\}$ in the number R of offspring surviving to

maturity.

55 Assign $20x^2 + 80x + 72$ to Y_1, $10x^2 + 40x + 41$ to Y_2, and Y_1/Y_2 to Y_3. Zoom-in around $(-2, -8)$ to confirm that this a low point and that there is not a vertical asymptote of $x = -2$. To determine the vertical asymptotes, graph Y_2 only { turn off Y_3 }, and look for its zeros. If these values are not zeros of the numerator, then they are the values of the vertical asymptotes. No vertical asymptotes in this case.

$[-9, 3]$ by $[-9, 3]$ $[0.7, 1.3, 0.1]$ by $[0.8, 1.2, 0.1]$

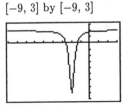   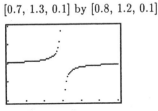

Figure 55   Figure 57

57 $f(x) = \dfrac{(x-1)^2}{(x-0.999)^2}$ • Note that the standard viewing rectangle gives the horizontal line $y = 1$. *Figure 57* was obtained by using Dot mode. If Connected mode is used, the calculator will draw a near-vertical line at $x = 0.999$. An equation of the vertical asymptote is $x = 0.999$.

59 (a) The graph of g is the horizontal line $y = 1$ with holes at $x = 0$, ± 1, ± 2, ± 3.

The TI-85/86 shows a small break in the line to indicate a hole.

(b) The graph of h is the graph of p with holes at $x = 0$, ± 1, ± 2, ± 3.

61 (a) GPA with additional credits = desired GPA $\Rightarrow$ $\dfrac{48(2.75) + y(4.0)}{48 + y} = x$ $\Rightarrow$

$132 + 4y = 48x + xy$ $\Rightarrow$ $132 - 48x = xy - 4y$ $\Rightarrow$ $132 - 48x = y(x - 4)$ $\Rightarrow$

$$y = \frac{132 - 48x}{x - 4}$$

(b)   (c) $[2, 4]$ by $[0, 1000, 100]$

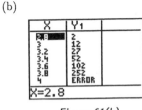   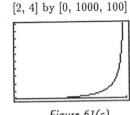

Figure 61(b)   Figure 61(c)

(d) The vertical asymptote of the graph is $x = 4$.

(e) Regardless of the number of additional credit hours obtained at 4.0, a cumulative GPA of 4.0 is not attainable.

Chapter 3 Review Exercises

$\boxed{1}$ To graph $f(x) = (x+2)^3$, shift $y = x^3$ left 2 units.

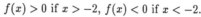

$f(x) > 0$ if $x > -2$, $f(x) < 0$ if $x < -2$.

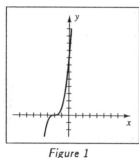
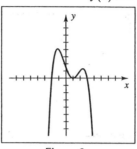
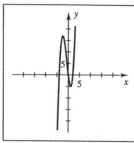

Figure 1	Figure 3	Figure 5

$\boxed{3}$ $f(x) = -\frac{1}{4}(x+2)(x-1)^2(x-3)$ has zeros at -2, 1 (multiplicity 2), and 3.

The graph of f has the general shape of $y = -x^4$.

$$f(x) > 0 \text{ if } -2 < x < 1 \text{ or } 1 < x < 3, \ f(x) < 0 \text{ if } x < -2 \text{ or } x > 3.$$

$\boxed{5}$ $f(x) = x^3 + 2x^2 - 8x = x(x^2 + 2x - 8) = x(x+4)(x-2)$.

f has zeros at -4, 0, and 2. It has the general shape of $y = x^3$.

$$f(x) > 0 \text{ if } -4 < x < 0 \text{ or } x > 2, \ f(x) < 0 \text{ if } x < -4 \text{ or } 0 < x < 2.$$

$\boxed{7}$ $f(0) = -9 < 100$ and $f(10) = 561 > 100$. By the intermediate value theorem for polynomial functions, f takes on every value between -9 and 561. Hence, there is at least one real number a in $[0, 10]$ such that $f(a) = 100$.

$\boxed{9}$ $f(x) = 3x^5 - 4x^3 + x + 5$; $p(x) = x^3 - 2x + 7$ •

$$
\begin{array}{r}
3x^2 \qquad\qquad +2 \\
x^3 - 2x + 7 \overline{\smash{\big)}\ 3x^5 + 0x^4 - 4x^3 + 0x^2 + x \ +5} \\
\underline{3x^5 \qquad\quad - 6x^3 + 21x^2 \qquad\qquad} \\
+2x^3 - 21x^2 \\
\underline{+2x^3 \qquad - 4x + 14} \\
-21x^2 + 5x \ - 9
\end{array}
$$

First step: $\dfrac{3x^5}{x^3} = 3x^2$

★ $q(x) = 3x^2 + 2$; $r(x) = -21x^2 + 5x - 9$

$\boxed{11}$ Synthetically dividing $f(x) = -4x^4 + 3x^3 - 5x^2 + 7x - 10$ by $x + 2$, we have

$$
\begin{array}{r|rrrrr}
-2 & -4 & 3 & -5 & 7 & -10 \\
   &   & 8 & -22 & 54 & -122 \\
\hline
   & -4 & 11 & -27 & 61 & -132
\end{array}
$$

By the remainder theorem, $f(-2) = -132$.

[13] Synthetically dividing $f(x) = 6x^5 - 4x^2 + 8$ by $x + 2$, we have

$$
\begin{array}{r|rrrrrr}
-2 & 6 & 0 & 0 & -4 & 0 & 8 \\
   &   & -12 & 24 & -48 & 104 & -208 \\
\hline
   & 6 & -12 & 24 & -52 & 104 & -200
\end{array}
$$

The synthetic division indicates that the quotient is

$6x^4 - 12x^3 + 24x^2 - 52x + 104$

and the remainder is -200.

[15] Since $-3 + 5i$ is a zero, so is $-3 - 5i$.

$f(x) = a[x - (-3 + 5i)][x - (-3 - 5i)](x + 1) = a(x^2 + 6x + 34)(x + 1)$.

$f(1) = a(41)(2)$ and $f(1) = 4 \Rightarrow 82a = 4 \Rightarrow a = \frac{2}{41}$.

Hence, $f(x) = \frac{2}{41}(x^2 + 6x + 34)(x + 1)$.

[17] $f(x) = x^5(x + 3)^2$

   { leading coefficient is 1 }

   { 0 is a zero of multiplicity 5 }

   { −3 is a zero of multiplicity 2 }

   $= x^5(x^2 + 6x + 9)$

   $= x^7 + 6x^6 + 9x^5$

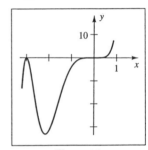

Figure 17

[19] $f(x) = (x^2 - 2x + 1)^2(x^2 + 2x - 3) = [(x - 1)^2]^2(x + 3)(x - 1) = (x + 3)(x - 1)^5$

★ 1 (multiplicity 5); −3 (multiplicity 1)

[21] (a) Let $f(x) = 2x^4 - 4x^3 + 2x^2 - 5x - 7$. Since there are 3 sign changes in $f(x)$ and 1 sign change in $f(-x) = 2x^4 + 4x^3 + 2x^2 + 5x - 7$, there are either 3 positive and 1 negative solution or 1 positive, 1 negative, and 2 nonreal complex solutions.

(b) Upper bound is 3, lower bound is −1

[23] Since there are only even powers, $7x^6 + 2x^4 + 3x^2 + 10 \geq 10$ for every real number x.

[25] $\dfrac{\text{positive divisors of 3}}{\text{divisors of 16}} = \dfrac{1,\ 3}{\pm 1,\ \pm 2,\ \pm 4,\ \pm 8,\ \pm 16} \longrightarrow$

$\pm 1,\ \pm 3,\ \pm\frac{1}{2},\ \pm\frac{3}{2},\ \pm\frac{1}{4},\ \pm\frac{3}{4},\ \pm\frac{1}{8},\ \pm\frac{3}{8},\ \pm\frac{1}{16},\ \pm\frac{3}{16}$

After a few attempts, we try $-\frac{1}{2}$.

$$
\begin{array}{r|rrrr}
-\frac{1}{2} & 16 & -20 & -8 & 3 \\
   &   & -8 & 14 & -3 \\
\hline
   & 16 & -28 & 6 & 0
\end{array}
$$

$16x^2 - 28x + 6 = 0 \Rightarrow 8x^2 - 14x + 3 = 0 \Rightarrow (4x - 1)(2x - 3) = 0 \Rightarrow x = \frac{1}{4}, \frac{3}{2}$.

Thus, the solutions of $16x^3 - 20x^2 - 8x + 3 = 0$ are $x = -\frac{1}{2}, \frac{1}{4}, \frac{3}{2}$.

27 $f(x) = \dfrac{-2}{(x+1)^2}$ •   There is a vertical asymptote of $x = -1$. Since the numerator

is negative and the denominator is positive, the functions values are negative and so

f is always below the x-axis.

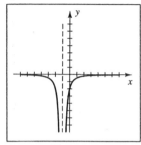

Figure 27

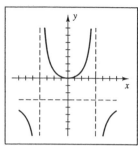

Figure 29

29 $f(x) = \dfrac{3x^2}{16 - x^2} = \dfrac{3x^2}{(4+x)(4-x)}$. $f(x) = -3 \Rightarrow 3x^2 = 3x^2 - 48 \Rightarrow 0 = -48$,

a contradiction $\Rightarrow$ the function does not intersect the horizontal asymptote.

31 $f(x) = \dfrac{x^3 - 2x^2 - 8x}{-x^2 + 2x} = \dfrac{x(x^2 - 2x - 8)}{x(2-x)} = \dfrac{(x-4)(x+2)}{2-x}$ for $x \neq 0$.

Substituting $x = 0$ into $\dfrac{(x-4)(x+2)}{2-x}$ gives us -4. Thus, we have a hole at $(0, -4)$.

By long division, $f(x) = -x + \dfrac{-8x}{-x^2 + 2x}$, so $y = -x$ is an oblique asymptote.

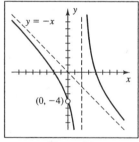

Figure 31

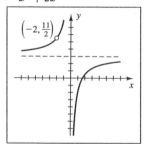

Figure 33

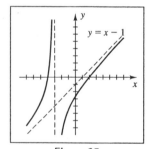

Figure 35

33 $f(x) = \dfrac{3x^2 + x - 10}{x^2 + 2x} = \dfrac{(3x-5)(x+2)}{x(x+2)} = \dfrac{3x-5}{x}$ if $x \neq -2$. The horizontal asymptote

is $y = 3$. The x-intercept is $\frac{5}{3}$ and there is no y-intercept since $x = 0$ is a vertical

asymptote. Substituting $x = -2$ into $\dfrac{3x-5}{x}$ gives us $\frac{11}{2}$. Thus, we have a hole at

$(-2, \frac{11}{2})$.

35 $f(x) = \dfrac{x^2 + 2x - 8}{x + 3} = \dfrac{(x+4)(x-2)}{x+3} = x - 1 - \dfrac{5}{x+3}$.   The vertical asymptote is

$x = -3$. The x-intercepts are -4 and 2. $y = x - 1$ is an oblique asymptote.

37 $y = cx^2(x^2 - 4lx + 6l^2)$ •

(a) $l = 10$, $x = 10$, and $y = 2$ $\Rightarrow$ $2 = c(10)^2(10^2 - 4 \cdot 10 \cdot 10 + 6 \cdot 10^2)$ $\Rightarrow$

$2 = c(100)(100 - 400 + 600)$ $\Rightarrow$ $2 = c(100)(300)$ $\Rightarrow$

$$2 = 30{,}000c \Rightarrow c = \tfrac{1}{15{,}000}$$

(b) $x = 6.1$ $\Rightarrow$ $y \approx 0.9754 < 1$ and $x = 6.2$ $\Rightarrow$ $y \approx 1.0006 > 1$.

39 $T = \frac{1}{20}t(t - 12)(t - 24)$ and $T = 32$ $\Rightarrow$ $\frac{1}{20}t(t - 12)(t - 24) = 32$ $\Rightarrow$

$t(t^2 - 36t + 288) = 640$ $\Rightarrow$ $t^3 - 36t^2 + 288t - 640 = 0$.

Synthetically dividing $t^3 - 36t^2 + 288t - 640$ by $t - 4$ gives us

$$
\begin{array}{r|rrrr}
4 & 1 & -36 & 288 & -640 \\
  &   & 4 & -128 & 640 \\
\hline
  & 1 & -32 & 160 & 0 \\
\end{array}
$$

Now use the quadratic formula to solve $t^2 - 32t + 160 = 0$ for t.

$$t = \frac{32 \pm \sqrt{32^2 - 4(1)(160)}}{2(1)} = \frac{32 \pm \sqrt{384}}{2} = \frac{32 \pm 8\sqrt{6}}{2} = 16 \pm 4\sqrt{6}. \text{ Since } 0 \le t \le 24,$$

$t = 4$ (10:00 A.M.) and $t = 16 - 4\sqrt{6} \approx 6.2020$ (12:12 P.M.) are the times when the

temperature was $32\,°\mathrm{F}$.

41 (a) S is the value that is changing. As S gets large, the value of $R = \dfrac{kS^n}{S^n + a^n}$ will

approach the ratio of leading coefficients, $k/1$. Hence, an equation of the

horizontal asymptote is $R = k$.

(b) k is the maximum rate at which the liver can remove alcohol from the

bloodstream.

42 $C(x) = \dfrac{20x}{101 - x}$ •

(a) $C(100) = \$2{,}000{,}000.00$ and

$C(90) \approx \$163{,}636.36$

(b) Notice the great difference in cost as x approaches 100.

$C(50)$ is only $\$19{,}607.84$ and $C(80)$ is $\$76{,}190.48$.

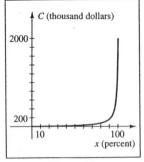

Figure 42

Chapter 3 Discussion Exercises

[1] **For even-degreed polynomials:** the domain is $\mathbb{R}$ and the number of x-intercepts ranges from *zero* to the degree of the polynomial; if the leading coefficient is positive, the range is of the form $[c, \infty)$ and the general shape has $y \to \infty$ as $|x| \to \infty$; if the leading coefficient is negative, the range is of the form $(-\infty, c]$ and the general shape has $y \to -\infty$ as $|x| \to \infty$.

For odd-degreed polynomials: the domain is $\mathbb{R}$, the range is $\mathbb{R}$, and the number of x-intercepts ranges from *one* to the degree of the polynomial; if the leading coefficient is positive, then $y \to \infty$ as $x \to \infty$ and $y \to -\infty$ as $x \to \infty$, if the leading coefficient is negative, then $y \to -\infty$ as $x \to \infty$ and $y \to \infty$ as $x \to -\infty$.

[3] By long division, we obtain the quotient $2x^2 - 7x + 5$ with remainder -6. By synthetic division with $k = -3/2$, we obtain a bottom row of 4 -14 10 -6. The first three numbers are twice the coefficients of the quotient and the last number is the remainder. For the factor $ax + b$, we can use synthetic division with $k = -b/a$, and obtain a times the quotient and the remainder in the bottom row.

[5] From your previous experience, you know that 2 points specify a first-degree polynomial (a line) and that 3 points specify a second-degree polynomial (a parabola). After working Discussion Exercise 4, you should guess that 4 points specify a third-degree polynomial, so a logical conclusion is that $n + 1$ points specify an n-degree polynomial.

[7] If the common factor is never equal to zero for any real number, then it can be canceled and has no effect on its graph. Such a factor is $x^2 + 1$, and an example of a function is $f(x) = \dfrac{(x^2 + 1)(x - 1)}{(x^2 + 1)(x - 2)}$.

⑨ $y = \dfrac{9x}{\sqrt{x^2+1}} \Rightarrow y\sqrt{x^2+1} = 9x \Rightarrow y^2(x^2+1) = 81x^2 \Rightarrow$

$y^2x^2 + y^2 = 81x^2 \Rightarrow y^2x^2 - 81x^2 = -y^2 \Rightarrow x^2(y^2-81) = -y^2 \Rightarrow$

$x^2 = \dfrac{y^2}{81-y^2} \Rightarrow x = \pm\sqrt{\dfrac{y^2}{81-y^2}} \Rightarrow x = \pm\dfrac{|y|}{\sqrt{81-y^2}} \Rightarrow x = \pm\dfrac{y}{\sqrt{81-y^2}}.$

From the original equation, we see that x and y are always both positive or both

negative, so the last equation can be simplified to $x = y/\sqrt{81-y^2}$. Hence,

$f^{-1}(y) = y/\sqrt{81-y^2}$ or, equivalently, $f^{-1}(x) = x/\sqrt{81-x^2}$. The denominator of

f^{-1} is zero for $x = \pm 9$, which are the vertical asymptotes. They are related to the

horizontal asymptotes of f, which are $y = \pm 9$.

Chapter 4: Exponential and Logarithmic Functions

$\boxed{1}$ $7^{x+6} = 7^{3x-4}$ $\Rightarrow$ $x + 6 = 3x - 4$ $\Rightarrow$ $10 = 2x$ $\Rightarrow$ $x = 5$

$\boxed{3}$ $3^{2x+3} = 3^{(x^2)}$ $\Rightarrow$ $2x + 3 = x^2$ $\Rightarrow$ $x^2 - 2x - 3 = 0$ $\Rightarrow$ $(x-3)(x+1) = 0$ $\Rightarrow$

$$x = -1, \ 3$$

$\boxed{5}$ We need to obtain the same base on each side of the equals sign, then we can apply

part (2) of the theorem about exponential functions being one-to-one.

$2^{-100x} = (0.5)^{x-4}$ $\Rightarrow$ $(2^{-1})^{100x} = \left(\frac{1}{2}\right)^{x-4}$ $\Rightarrow$ $\left(\frac{1}{2}\right)^{100x} = \left(\frac{1}{2}\right)^{x-4}$ $\Rightarrow$

$$100x = x - 4 \ \Rightarrow \ 99x = -4 \ \Rightarrow \ x = -\frac{4}{99}$$

$\boxed{7}$ $4^{x-3} = 8^{4-x}$ $\Rightarrow$ $(2^2)^{x-3} = (2^3)^{4-x}$ $\Rightarrow$ $2^{2x-6} = 2^{12-3x}$ $\Rightarrow$

$$2x - 6 = 12 - 3x \ \Rightarrow \ 5x = 18 \ \Rightarrow \ x = \frac{18}{5}$$

$\boxed{9}$ $4^x \cdot \left(\frac{1}{2}\right)^{3-2x} = 8 \cdot (2^x)^2$ $\Rightarrow$ $2^{2x} \cdot (2^{-1})^{3-2x} = 2^3 \cdot 2^{2x}$ $\Rightarrow$

$2^{2x-3+2x} = 2^{3+2x}$ $\Rightarrow$ $2^{4x-3} = 2^{2x+3}$ $\Rightarrow$

$$4x - 3 = 2x + 3 \ \Rightarrow \ 2x = 6 \ \Rightarrow \ x = 3$$

$\boxed{11}$ (a) Let $g = f(x) = 2^x$ for reference purposes.

The graph of g goes through the points $\left(-1, \frac{1}{2}\right)$, $(0, 1)$, and $(1, 2)$.

(b) $f(x) = -2^x$ •

Reflect g through the x-axis since f is just $-1 \cdot 2^x$. Do not confuse this function

with $(-2)^x$ — remember, the base is positive for exponential functions.

(c) $f(x) = 3 \cdot 2^x$ • vertically stretch g by a factor of 3

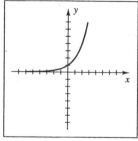 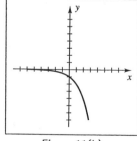 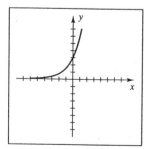

Figure 11(a)   Figure 11(b)   Figure 11(c)

(d) $f(x) = 2^{x+3}$ • shift g left 3 units since f is $g(x+3)$

(e) $f(x) = 2^x + 3$ • shift g up 3 units

(f) $f(x) = 2^{x-3}$ • shift g right 3 units since f is $g(x-3)$

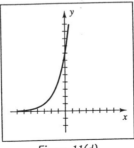

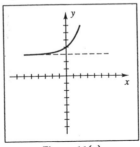

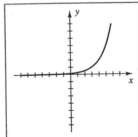

Figure 11(d)   Figure 11(e)   Figure 11(f)

(g) $f(x) = 2^x - 3$ • shift g down 3 units

(h) $f(x) = 2^{-x}$ • reflect g through the y-axis since f is $g(-x)$

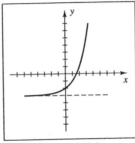

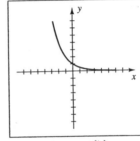

Figure 11(g)   Figure 11(h)

(i) $f(x) = (\frac{1}{2})^x$ • $(\frac{1}{2})^x = (2^{-1})^x = 2^{-x}$, same graph as in part (h)

(j) $f(x) = 2^{3-x}$ • $2^{3-x} = 2^{-(x-3)}$, shift g right 3 units and reflect through the

line $x = 3$. Alternatively, $2^{3-x} = 2^3 2^{-x} = 8(\frac{1}{2})^x$, vertically stretch $y = (\frac{1}{2})^x$ (the

graph in part (i)) by a factor of 8.

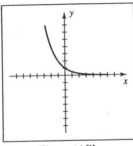

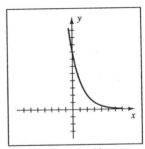

Figure 11(i)   Figure 11(j)

13 $f(x) = \left(\frac{2}{5}\right)^{-x} = \left[\left(\frac{2}{5}\right)^{-1}\right]^x = \left(\frac{5}{2}\right)^x$ •   goes through $\left(-1, \frac{2}{5}\right)$, $(0, 1)$, and $\left(1, \frac{5}{2}\right)$

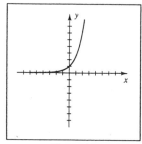

Figure 13

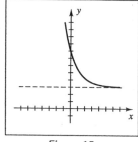

Figure 15

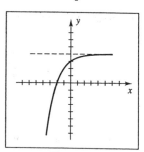

Figure 17

15 $f(x) = 5\left(\frac{1}{2}\right)^x + 3$ •   vertically stretch $y = \left(\frac{1}{2}\right)^x$ by a factor of 5 and shift up 3 units

17 $f(x) = -\left(\frac{1}{2}\right)^x + 4$ •   reflect $y = \left(\frac{1}{2}\right)^x$ through the x-axis and shift up 4 units

19 $f(x) = 2^{|x|} = \begin{cases} 2^x & \text{if } x \geq 0 \\ 2^{-x} & \text{if } x < 0 \end{cases} = \begin{cases} 2^x & \text{if } x \geq 0 \\ \left(\frac{1}{2}\right)^x & \text{if } x < 0 \end{cases}$

Use the portion of $y = 2^x$ with $x \geq 0$ and reflect it through the y-axis since f is even.

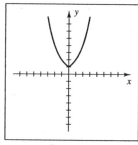

Figure 19

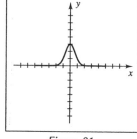

Figure 21

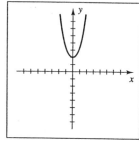

Figure 23

Note: For Exercises 21, 22, and 5 of the review exercises, refer to Example 6 in the text

for the basic graph of $y = a^{-x^2} = (a^{-1})^{(x^2)} = \left(\frac{1}{a}\right)^{(x^2)}$, where $a > 1$.

21 $f(x) = 3^{1-x^2} = 3^1 3^{-x^2} = 3\left(\frac{1}{3}\right)^{(x^2)}$ •   stretch $y = \left(\frac{1}{3}\right)^{(x^2)}$ by a factor of 3

23 $f(x) = 3^x + 3^{-x}$ •   Adding the functions $g(x) = 3^x$ and $h(x) = 3^{-x} = \left(\frac{1}{3}\right)^x$
together, we see that the y-intercept will be 2. If $x > 0$, f looks like $y = 3^x$ since 3^x
dominates 3^{-x} (3^{-x} gets close to 0 and 3^x grows very large). If $x < 0$, f looks like
$y = 3^{-x}$ since 3^{-x} dominates 3^x.

25 Since the y-intercept is 8, we must have $f(0) = 8$. $f(x) = ba^x$,
so $f(0) = 8 \Rightarrow 8 = ba^0 \Rightarrow 8 = b(1) \Rightarrow 8 = b$, and hence, $f(x) = 8a^x$.
Since the point $P(3, 1)$ is on the graph, $f(3) = 1$.
Thus, $1 = 8a^3 \Rightarrow a^3 = \frac{1}{8} \Rightarrow a = \frac{1}{2}$, and the function is $f(x) = 8\left(\frac{1}{2}\right)^x$.

27 The horizontal asymptote is $y = 32$, so $f(x) = ba^{-x} + c$ has the form

$f(x) = ba^{-x} + 32$. The y-intercept is 212, so $f(0) = 212 \Rightarrow 212 = ba^{-0} + 32 \Rightarrow$

$212 = b(1) + 32 \Rightarrow 180 = b \Rightarrow f(x) = 180a^{-x} + 32$. The function passes through

the point $P(2, 112)$, so $f(2) = 112 \Rightarrow 112 = 180a^{-2} + 32 \Rightarrow 80 = 180a^{-2} \Rightarrow$

$80 = \dfrac{180}{a^2} \Rightarrow a^2 = \dfrac{180}{80} \Rightarrow a^2 = \dfrac{9}{4} \Rightarrow a = \dfrac{3}{2}$ {since a must be positive, $a = -\dfrac{3}{2}$ is

not allowed} $\Rightarrow f(x) = 180(1.5)^{-x} + 32$.

29 (a) $N(t) = 100(0.9)^t \Rightarrow N(1) = 100(0.9)^1 = 90$

   (b) $N(5) = 100(0.9)^5 \approx 59$ (c) $N(10) = 100(0.9)^{10} \approx 35$

31 $f(t) = 600(3)^{t/2}$ •

   (a) 8:00 A.M. corresponds to $t = 1$ and $f(1) = 600\sqrt{3} \approx 1039$.

   10:00 A.M. corresponds to $t = 3$ and $f(3) = 600(3\sqrt{3}) = 1800\sqrt{3} \approx 3118$.

   11:00 A.M. corresponds to $t = 4$ and $f(4) = 600(9) = 5400$.

   (b) The graph of f is an increasing exponential that passes through $(0, 600)$ and the

points in part (a).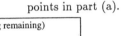

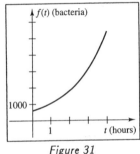

Figure 31

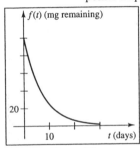

Figure 33

33 (a) $f(t) = 100(2)^{-t/5}$, so $f(5) = 100(2)^{-1} = 50$ mg, $f(10) = 100(2)^{-2} = 25$ mg, and

$$f(12.5) = 100(2)^{-2.5} = \frac{100}{4\sqrt{2}} = \frac{25}{2}\sqrt{2} \approx 17.7 \text{ mg}.$$

   (b) The endpoints of this decreasing exponential are $(0, 100)$ and $(30, 1.5625)$.

35 A half-life of 1600 years means that when $t = 1600$, the amount remaining, $q(t)$, will

be one-half the original amount—that is, $\frac{1}{2}q_0$. $q(t) = \frac{1}{2}q_0$ when $t = 1600 \Rightarrow$

$\frac{1}{2}q_0 = q_0 2^{k(1600)} \Rightarrow 2^{-1} = 2^{1600k} \Rightarrow 1600k = -1 \Rightarrow k = -\frac{1}{1600}$.

37 Using $A = P\left(1 + \frac{r}{n}\right)^{nt}$, we have $P = 1000$, $r = 0.12$, and $n = 12$.

Consider A to be a function of t, that is, $A(t) = 1000\left(1 + \frac{0.12}{12}\right)^{12t} = 1000(1.01)^{12t}$.

   (a) $A\left(\frac{1}{12}\right) = \$1010.00$ (b) $A\left(\frac{6}{12}\right) \approx \$1061.52$

   (c) $A(1) \approx \$1126.83$ (d) $A(20) \approx \$10{,}892.55$

39 $C = 10{,}000 \Rightarrow V(t) = 0.78(10{,}000)(0.85)^{t-1} = 7800(0.85)^{t-1}$

(a) $V(1) = \$7800$ (b) $V(4) \approx \$4790.18$, or \$4790   (c) $V(7) \approx \$2941.77$, or \$2942

41 $t = 2006 - 1626 = 380;\ A = \$24(1 + 0.06/4)^{4 \cdot 380} = \$161{,}657{,}351{,}965.80.$

That's right—\$161 billion!

43 (a) Examine the pattern formed by the value y in the year n.

year (n)	value (y)
0	y_0
1	$(1-a)y_0 = y_1$
2	$(1-a)y_1 = (1-a)\big[(1-a)y_0\big] = (1-a)^2 y_0 = y_2$
3	$(1-a)y_2 = (1-a)\big[(1-a)^2 y_0\big] = (1-a)^3 y_0 = y_3$

(b) $s = (1-a)^T y_0 \Rightarrow (1-a)^T = s/y_0 \Rightarrow 1 - a = \sqrt[T]{s/y_0} \Rightarrow a = 1 - \sqrt[T]{s/y_0}$

45 (a) $r = 0.12,\ t = 30,\ L = 90{,}000 \Rightarrow$

$$k = \left(1 + \frac{0.12}{12}\right)^{12 \cdot 30} \approx 35.95 \quad \text{and} \quad M = \frac{90{,}000(0.12)k}{12(k-1)} \approx 925.75.$$

(b) Total interest paid = total of payments − original loan amount

$$= (12 \cdot 30 \text{ payments}) \times \$925.75 - \$90{,}000 = \$243{,}270$$

47 $r = 0.15,\ t = 3,\ M = 220 \Rightarrow k = \left(1 + \frac{0.15}{12}\right)^{12 \cdot 3} \approx 1.56.$

$$M = \frac{Lrk}{12(k-1)} \Rightarrow L = \frac{12M(k-1)}{rk} = \frac{12(220)(k-1)}{(0.15)k} \approx \$6{,}346.40$$

49 (a) $f(3) = 13^{\sqrt{3+1.1}} \approx 180.1206$ (b) $g(1.43) = \left(\frac{5}{42}\right)^{-1.43} \approx 20.9758$

(c) $h(1.06) = (2^{1.06} + 2^{-1.06})^{2(1.06)} \approx 7.3639$

51 Part (b) may be interpreted as doubling an investment at 8.5%.

(a) If $y = (1.085)^x$ and $x = 40$, then $y \approx 26.13$.   (b) If $y = 2$, then $x \approx 8.50$.

[0, 60, 5] by [0, 40, 5]   [−10.5, 10.5] by [−7, 7]

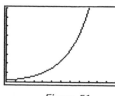

Figure 51

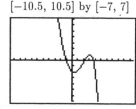

Figure 53

53 First graph $y = 1.4x^2 - 2.2^x - 1$. The x-intercepts are the roots of the equation.

Using a zero or root feature, the roots are $x \approx -1.02$, 2.14, and 3.62.

55 (a) f is not one-to-one since

the horizontal line $y = -0.1$ intersects the graph of f more than once.

(b) The only zero of f is $x = 0$.

[−3, 3] by [−2, 2]

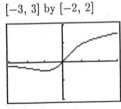

Figure 55

[−4, 1] by [−2, 3]

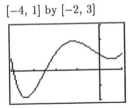

Figure 57

57 (a) The low points are about $(-3.37, -1.79)$ and $(0.52, 0.71)$. The high point is

approximately $(-1.19, 1.94)$. Thus, f is increasing on $[-3.37, -1.19]$ and

$[0.52, 1]$. f is decreasing on $[-4, -3.37]$ and $[-1.19, 0.52]$.

(b) The range of f on $[-4, 1]$ is approximately $[-1.79, 1.94]$.

59 *Figure 59* is a graph of $N(t) = 1000(0.9)^t$.

By using an intersect feature, we can determine that $N = 500$ when $t \approx 6.58$ yr.

[0, 10] by [0, 1000, 100]

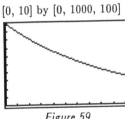

Figure 59

[0, 7.5] by [0, 5]

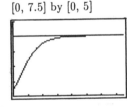

Figure 61

61 Graph $y = 4(0.125)^{(0.25^x)}$. The line $y = k = 4$ is a horizontal asymptote for the

Gompertz function. The maximum number of sales of the product approaches k.

63 Using an intersect feature, we determine that

$$A = \frac{100\left(1 + \frac{0.05}{12}\right)\left[\left(1 + \frac{0.05}{12}\right)^{12n} - 1\right]}{\frac{0.05}{12}} \text{ and } A = 100,000 \text{ intersect when } n \approx 32.8.$$

[0, 40, 10] by [0, 200,000, 50,000]

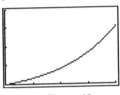

Figure 63

65 (a) Let $x = 0$ correspond to 1910, $x = 20$ to 1930, ..., and $x = 85$ to 1995.

Graph the data together with the functions

(1) $f(x) = 0.809(1.094)^x$ and **(2)** $g(x) = 0.375x^2 - 18.4x + 88.1$.

$[-10, 90, 10]$ by $[-200, 1500, 100]$ $\qquad$ $[-10, 90, 10]$ by $[-200, 1500, 100]$

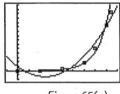 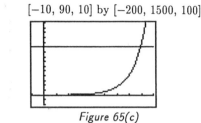

Figure 65(a) $\qquad\qquad\qquad$ Figure 65(c)

(b) The exponential function f best models the data.

(c) Graph $Y_1 = f(x)$ and $Y_2 = 1000$. The graphs intersect at $x \approx 79$, or in 1989.

67 (a) $t = 1999 - 1974 = 25$ and r/n is 0.0025, so $A = P\left(1 + \frac{r}{n}\right)^{nt} =$

$$\$353{,}022(1 + 0.0025)^{12 \cdot 25} = \$746{,}648.43; \text{ westegg gives } \$1{,}286{,}896.$$

(b) Let r denote the annual interest rate. $\$6{,}616{,}585 = \$353{,}022\left(1 + \frac{r}{1}\right)^{1 \cdot 25} \Rightarrow$

$$\frac{6{,}616{,}585}{353{,}022} = (1 + r)^{25} \Rightarrow 1 + r = \sqrt[25]{\frac{6{,}616{,}585}{353{,}022}} \Rightarrow r = \sqrt[25]{\frac{6{,}616{,}585}{353{,}022}} - 1 \approx$$

$$0.1244, \text{ so } r \text{ is about } 12.44\%.$$

(c) In part (a), the base is constant and the variable is in the exponent, so this is an *exponential* function. In part (b), the base is variable and the exponent is constant, so this is a *polynomial* function.

4.2 Exercises

Note: Examine Figure 1 in this section to reinforce the idea that $y = e^x$ is just a special case of $y = a^x$ with $a > 1$.

1 (a) $f(x) = e^{-x}$ • reflect $y = e^x$ through the y-axis

(b) $f(x) = -e^x$ • reflect $y = e^x$ through the x-axis

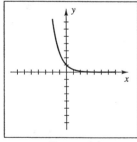 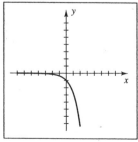

Figure 1(a) $\qquad\qquad\qquad$ Figure 1(b)

$\boxed{3}$ (a) $f(x) = e^{x+4}$ • shift $y = e^x$ left 4 units

(b) $f(x) = e^x + 4$ • shift $y = e^x$ up 4 units

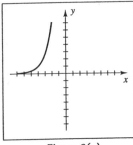

Figure 3(a)

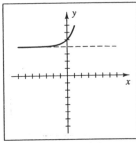

Figure 3(b)

$\boxed{5}$ $A = Pe^{rt} = 1000e^{(0.0825)(5)} \approx \$1510.59$

$\boxed{7}$ $100{,}000 = Pe^{(0.11)(18)} \Rightarrow P = \dfrac{100{,}000}{e^{1.98}} \approx \$13{,}806.92$

$\boxed{9}$ $13{,}464 = 1000e^{(r)(20)} \Rightarrow e^{20r} = 13.464$. Using a trial and error approach on a scientific calculator or tracing and zooming or an intersect feature on a graphing calculator, we determine that $e^x \approx 13.464$ if $x \approx 2.6$. Thus, $20r = 2.6$ and $r = 0.13$ or 13%.

$\boxed{11}$ $e^{(x^2)} = e^{7x-12} \Rightarrow x^2 = 7x - 12 \Rightarrow x^2 - 7x + 12 = 0 \Rightarrow$
$$(x-3)(x-4) = 0 \Rightarrow x = 3, 4$$

$\boxed{13}$ $xe^x + e^x = 0 \Rightarrow e^x(x+1) = 0 \Rightarrow x = -1$ { Note that $e^x \neq 0$. }

$\boxed{15}$ $x^3(4e^{4x}) + 3x^2e^{4x} = 0 \Rightarrow x^2e^{4x}(4x+3) = 0 \Rightarrow x = -\frac{3}{4}, 0$. Note that $e^{4x} \neq 0$.

$\boxed{17}$ When we multiply $(e^x + e^{-x})$ times $(e^x + e^{-x})$, we get
$$e^{x+x} + e^{x+(-x)} + e^{(-x)+x} + e^{(-x)+(-x)} = e^{2x} + e^0 + e^0 + e^{-2x} = e^{2x} + 2 + e^{-2x}.$$

$$\frac{(e^x + e^{-x})(e^x + e^{-x}) - (e^x - e^{-x})(e^x - e^{-x})}{(e^x + e^{-x})^2} = \frac{(e^{2x} + 2 + e^{-2x}) - (e^{2x} - 2 + e^{-2x})}{(e^x + e^{-x})^2}$$

$$= \frac{4}{(e^x + e^{-x})^2}$$

$\boxed{19}$ $W(t) = W_0e^{kt}$; $t = 30 \Rightarrow W(30) = 68e^{(0.2)(30)} \approx 27{,}433$ mg, or 27.43 grams

$\boxed{21}$ The year 2010 corresponds to $t = 2010 - 1980 = 30$. Using the law of growth formula on page 300 with $q_0 = 227$ and $r = 0.007$, we have $N(t) = 227e^{0.007t}$.
$$\text{Thus, } N(30) = 227e^{(0.007)(30)} \approx 280.0 \text{ million.}$$

$\boxed{23}$ $N(t) = N_0e^{-0.2t} \Rightarrow N(10) = N_0e^{-2}$. The percentage of the original number still alive after 10 years is $\left(\dfrac{N(10)}{N_0}\right) \times 100 = \left(\dfrac{N_0e^{-2}}{N_0}\right) \times 100 = 100e^{-2} \approx 13.5\%$.

25 The year 2010 corresponds to $t = 2010 - 1978 = 32$.

$$N(32) = 5000e^{(0.0036)(32)} = 5000e^{(0.1152)} \approx 5610.$$

27 $h = 40,000 \Rightarrow p = 29e^{(-0.000034)(40,000)} = 29e^{(-1.36)} \approx 7.44$ in.

29 $x = 1 \Rightarrow y = 79.041 + 6.39(1) - e^{3.261 - 0.993(1)} \approx 75.77$ cm.

$x = 1 \Rightarrow R = 6.39 + 0.993e^{3.261 - 0.993(1)} \approx 15.98$ cm/yr.

31 $2010 - 1971 = 39 \Rightarrow t = 39$ years. Using the continuously compounded interest

formula with $P = 1.60$ and $r = 0.05$, we have $A = 1.60e^{(0.05)(39)} \approx \$11.25$ per hour.

33 (a) Note here that the amount of money invested is not of interest and that we are

only concerned with the percent of growth.

$$\left(1 + \frac{0.07}{4}\right)^{4 \cdot 1} \approx 1.0719. \quad (1.0719 - 1) \times 100\% = 7.19\%$$

(b) $e^{(0.07)(1)} \approx 1.0725.$ $(1.0725 - 1) \times 100\% = 7.25\%.$ The results indicate that we

would receive an extra 0.06% in interest by investing our money in an account

that is compounded continuously rather than quarterly. This is only an extra 6

cents on a \$100 investment, but \$600 extra on a \$1,000,000 investment (actually

\$649.15 if the computations are carried beyond 0.01%).

35 It may be of interest to compare this graph with the graph of $y = (1.085)^x$ in

Exercise 51 of §4.1. Both are compounding functions with $r = 8.5\%$.

Note that $e^{0.085x} = (e^{0.085})^x \approx (1.0887)^x > (1.085)^x$ for $x > 0$.

(a) If $y = e^{0.085x}$ and $x = 40$, then $y \approx 29.96$.   (b) If $y = 2$, then $x \approx 8.15$.

[0, 60, 5] by [0, 40, 5]

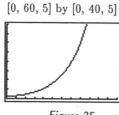

Figure 35

37 (a) As $x \to \infty$, $e^{-x} \to 0$ and f will resemble $\frac{1}{2}e^x$.

As $x \to -\infty$, $e^x \to 0$ and f will resemble $-\frac{1}{2}e^x$.

See *Figure 37(a)* on the next page.

(b) At $x = 0$, $f(x) = 0$, and g will have a vertical asymptote since g is undefined (division by 0). As $x \to \infty$, $f(x) \to \infty$, and since the reciprocal of a large positive number is a small positive number, we have $g(x) \to 0$. As $x \to -\infty$, $f(x) \to -\infty$, and since the reciprocal of a large negative number is a small negative number, we have $g(x) \to 0$.

$[-7.5, 7.5]$ by $[-5, 5]$

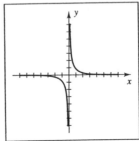

Figure 37(a)

Figure 37(b)

39 (a) $f(x) = \dfrac{e^x - e^{-x}}{e^x + e^{-x}} = \dfrac{e^x - 1/e^x}{e^x + 1/e^x} \cdot \dfrac{e^x}{e^x} = \dfrac{e^{2x} - 1}{e^{2x} + 1}$. At $x = 0$, $f(x) = 0$. As $x \to \infty$,

$f(x) \to 1$ since the numerator and denominator are nearly the same number. As $x \to -\infty$, $f(x) \to \dfrac{0 - 1}{0 + 1} = -1$.

(b) At $x = 0$, we will have a vertical asymptote. As $x \to \infty$, $f(x) \to 1$, and since g is the reciprocal of f, $g(x) \to 1$. Similarly, as $x \to -\infty$, $g(x) \to -1$.

$[-4.5, 4.5]$ by $[-3, 3]$

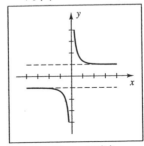

Figure 39(a)

Figure 39(b)

41 The approximate coordinates of the points where the graphs of f and g intersect are $(-1.04, -0.92)$, $(2.11, 2.44)$, and $(8.51, 70.42)$. The region near the origin in *Figure 41(a)* is enhanced in *Figure 41(b)*. Thus, the solutions are $x \approx -1.04$, 2.11, and 8.51.

$[-3, 11]$ by $[-10, 80, 10]$ $[-2.26, 3.34]$ by $[-7.14, 8.57, 10]$

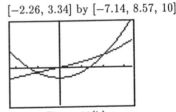

Figure 41(a)   *Figure 41(b)*

43 $f(x) = x + 1$ is a more accurate approximation to e^x near $x = 0$, whereas $g(x) = 1.72x + 1$ is a more accurate approximation to e^x near $x = 1$.

$[0, 4.5]$ by $[0, 3]$

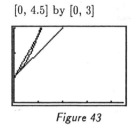

Figure 43

45 From the graph, we see that f has zeros at $x \approx 0.11$, 0.79, and 1.13.

$[-2, 2.5]$ by $[-1, 2]$

Figure 45

$[0, 200, 50]$ by $[0, 8]$

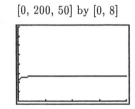

Figure 47

47 From the graph, there is a horizontal asymptote of $y \approx 2.71$.

f is approaching the value of e asymptotically.

49 $Y_1 = e^{-x}$ and $Y_2 = x$ intersect when $x \approx 0.567$.

$[-4.5, 4.5]$ by $[-3, 3]$

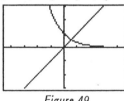

Figure 49

$[-5.5, 5]$ by $[-2, 5]$

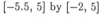
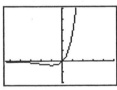

Figure 51

51 f is increasing on $[-1, \infty)$ and f is decreasing on $(-\infty, -1]$.

53 (a) When $y = 0$ and $z = 0$, the equation becomes $C = \dfrac{2Q}{\pi vab} e^{-h^2/(2b^2)}$. As h increases, the exponent becomes a larger *negative* value, and hence the concentration C decreases.

(b) When $z = 0$, the equation becomes $C = \dfrac{2Q}{\pi vab} e^{-y^2/(2a^2)} e^{-h^2/(2b^2)}$.

As y increases, the concentration C decreases.

55 (a) Choose two arbitrary points that appear to lie on the curve such as (0, 1.225)
and (10,000, 0.414).

$$f(0) = Ce^0 = C = 1.225 \quad \text{and} \quad f(10,000) = 1.225e^{10,000k} = 0.414.$$

To solve the last equation, graph $Y_1 = 1.225e^{10,000x}$ and $Y_2 = 0.414$. The graph
of Y_1 increases so rapidly that it nearly "covers" the positive y-axis, so a viewing
rectangle such as $[-0.001, 0]$ by $[0, 1]$ is needed to see the point of intersection.
The graphs intersect at $x \approx -0.0001085$. Thus, $f(x) = 1.225e^{-0.0001085x}$.

(b) $f(3000) \approx 0.885$ { actual $= 0.909$ } and $f(9000) \approx 0.461$ { actual $= 0.467$ }.

$$[-1000, 10,100, 1000] \text{ by } [0, 1.5, 0.5]$$

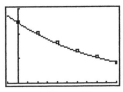

Figure 55

4.3 Exercises

Note: Exercises 1–4 are designed to familiarize the reader with the definition of $\log_a$ in
this section. It is very important that you can generalize your understanding of
this definition to the following case:

$$\log_{\text{base}}(\text{argument}) = \text{exponent} \quad \textit{is equivalent to} \quad (\text{base})^{\text{exponent}} = \text{argument}$$

Later in this section, you will also want to be able to use the following two special
cases with ease:

$$\log(\text{argument}) = \text{exponent} \quad \textit{is equivalent to} \quad (10)^{\text{exponent}} = \text{argument}$$

$$\ln(\text{argument}) = \text{exponent} \quad \textit{is equivalent to} \quad (e)^{\text{exponent}} = \text{argument}$$

1 (a) In this case, the *base* is 4, the *exponent* is 3, and the *argument* is 64. Thus,

$$4^3 = 64 \quad \text{is equivalent to} \quad \log_4 64 = 3.$$

(b) $4^{-3} = \frac{1}{64} \iff \log_4 \frac{1}{64} = -3$

(c) $t^r = s \iff \log_t s = r$

(d) $3^x = 4 - t \iff \log_3(4 - t) = x$

(e) In this case, the *base* is 5, the *exponent* is $7t$, and the *argument* is $\frac{a+b}{a}$. Thus,

$$5^{7t} = \frac{a+b}{a} \quad \text{is equivalent to} \quad \log_5 \frac{a+b}{a} = 7t.$$

(f) $(0.7)^t = 5.3 \quad \Leftrightarrow \quad \log_{0.7}(5.3) = t$

$\boxed{3}$ (a) In this case, the *base* is 2, the *argument* is 32, and the *exponent* is 5. Thus,

$$\log_2 32 = 5 \quad \text{is equivalent to} \quad 2^5 = 32.$$

(b) $\log_3 \frac{1}{243} = -5 \quad \Leftrightarrow \quad 3^{-5} = \frac{1}{243}$

(c) $\log_t r = p \quad \Leftrightarrow \quad t^p = r$

(d) $\log_3(x+2) = 5 \quad \Leftrightarrow \quad 3^5 = (x+2)$

(e) In this case, the *base* is 2, the *argument* is m, and the *exponent* is $3x + 4$. Thus,

$$\log_2 m = 3x + 4 \quad \text{is equivalent to} \quad 2^{3x+4} = m.$$

(f) $\log_b 512 = \frac{3}{2} \quad \Leftrightarrow \quad b^{3/2} = 512$

$\boxed{5}$ $2a^{t/3} = 5 \quad \Leftrightarrow \quad a^{t/3} = \frac{5}{2} \quad \Leftrightarrow \quad t/3 = \log_a \frac{5}{2} \quad \Leftrightarrow \quad t = 3\log_a \frac{5}{2}$

$\boxed{7}$ In order to solve for t, we must isolate the expression containing t—in this case, that expression is the exponential a^{Ct}. $A = Ba^{Ct} + D \Rightarrow A - D = Ba^{Ct} \Rightarrow$

$$\frac{A-D}{B} = a^{Ct} \Rightarrow Ct = \log_a\left(\frac{A-D}{B}\right) \Rightarrow t = \frac{1}{C}\log_a\left(\frac{A-D}{B}\right).$$

The confusing step to most students in the above solution is $\frac{A-D}{B} = a^{Ct} \Rightarrow Ct = \log_a\left(\frac{A-D}{B}\right)$. This is similar to $y = a^x \Rightarrow \log_a y = x$, except x and y are more complicated expressions.

$\boxed{9}$ (a) Changing $10^5 = 100{,}000$ to logarithmic form gives us $\log_{10} 100{,}000 = 5$.

Since this is a common logarithm, we denote it as $\log 100{,}000 = 5$.

(b) $10^{-3} = 0.001 \quad \Leftrightarrow \quad \log 0.001 = -3$

(c) $10^x = y + 1 \quad \Leftrightarrow \quad \log(y+1) = x$

(d) Changing $e^7 = p$ to logarithmic form gives us $\log_e p = 7$.

Since this is a natural logarithm, we denote it as $\ln p = 7$.

(e) Changing $e^{2t} = 3 - x$ to logarithmic form gives us $\log_e(3-x) = 2t$.

Since this is a natural logarithm, we denote it as $\ln(3-x) = 2t$.

[11] (a) Remember that $\log x = 50$ is the same as $\log_{10} x = 50$.

Changing to exponential form, we have $10^{50} = x$.

(b) $\log x = 20t$ $\{\log_{10} x = 20t\}$ $\Leftrightarrow$ $10^{20t} = x$

(c) Remember that $\ln x = 0.1$ is the same as $\log_e x = 0.1$.

Changing to exponential form, we have $e^{0.1} = x$.

(d) $\ln w = 4 + 3x$ $\{\log_e w = 4 + 3x\}$ $\Leftrightarrow$ $e^{4+3x} = w$

(e) $\ln(z - 2) = \frac{1}{6}$ $\{\log_e(z - 2) = \frac{1}{6}\}$ $\Leftrightarrow$ $e^{1/6} = z - 2$

[13] (a) $\log_5 1 = 0$ since $5^0 = 1$ $\{\log_a 1 = 0$ for any allowable base $a\}$

(b) $\log_3 3 = 1$ $\{\log_a a = 1$ since $a^1 = a\}$

(c) Remember that you cannot take the logarithm, any base, of a negative number.

Hence, $\log_4(-2)$ is undefined.

(d) $\log_7 7^2 = 2$ $\{\log_a a^x = x\}$

(e) $3^{\log_3 8} = 8$ $\{a^{\log_a x} = x\}$

(f) $\log_5 125 = \log_5 5^3 = 3$, as in part (d)

(g) We will change the form of $\frac{1}{16}$ so that it can be written as an exponential expression with the same base as the logarithm—in this case, that base is 4.

$$\log_4 \tfrac{1}{16} = \log_4 4^{-2} = -2$$

[15] Parts (a)–(f) are direct applications of the properties in the chart on page 314.

(a) $10^{\log 3} = 3$ (b) $\log 10^5 = 5$ (c) $\log 100 = \log 10^2 = 2$

(d) $\log 0.0001 = \log 10^{-4} = -4$ (e) $e^{\ln 2} = 2$ (f) $\ln e^{-3} = -3$

For part (g), we use a property of exponents that will enable us to use the property $e^{\ln x} = x$. (g) $e^{2 + \ln 3} = e^2 e^{\ln 3} = e^2(3) = 3e^2$

[17] $\log_4 x = \log_4(8 - x)$ $\Rightarrow$ $x = 8 - x$ {since the logarithm function is one-to-one} $\Rightarrow$ $2x = 8$ $\Rightarrow$ $x = 4$. We must check to make sure that all proposed solutions do not make any of the original expressions undefined. Checking $x = 4$ in the original equation gives us $\log_4 4 = \log_4(8 - 4)$, which is a true statement, so $x = 4$ is a valid solution.

[19] $\log_5(x - 2) = \log_5(3x + 7)$ $\Rightarrow$ $x - 2 = 3x + 7$ $\Rightarrow$ $2x = -9$ $\Rightarrow$ $x = -\frac{9}{2}$. The value $x = -\frac{9}{2}$ is extraneous since it makes either of the original logarithm expressions undefined. Hence, there is no solution.

[21] $\log x^2 = \log(-3x - 2)$ $\Rightarrow$ $x^2 = -3x - 2$ $\Rightarrow$ $x^2 + 3x + 2 = 0$ $\Rightarrow$ $(x + 1)(x + 2) = 0$ $\Rightarrow$ $x = -1, -2$.

Checking -1 and -2, we find that both are valid solutions.

$\boxed{23}$ $\log_3(x-4) = 2$ $\Rightarrow$ $x-4 = 3^2$ $\Rightarrow$ $x-4 = 9$ $\Rightarrow$ $x = 13$

$\boxed{25}$ $\log_9 x = \frac{3}{2}$ $\Rightarrow$ $x = 9^{3/2} = (9^{1/2})^3$ { remember, root first, power second } $= 3^3 = 27$

$\boxed{27}$ $\ln x^2 = -2$ $\Rightarrow$ $x^2 = e^{-2}$ $\Rightarrow$ $x^2 = \frac{1}{e^2}$ $\Rightarrow$ $x = \pm\sqrt{\frac{1}{e^2}}$ $\Rightarrow$ $x = \pm\frac{1}{e}$

$\boxed{29}$ $e^{2\ln x} = 9$ $\Rightarrow$ $(e^{\ln x})^2 = 9$ $\Rightarrow$ $x^2 = 9$ $\Rightarrow$ $x = \pm 3$; -3 is extraneous

$\boxed{31}$ $e^{x\ln 3} = 27$ $\Rightarrow$ $(e^{\ln 3})^x = 27$ $\Rightarrow$ $3^x = 3^3$ $\Rightarrow$ $x = 3$

$\boxed{33}$ (a) $f(x) = \log_4 x$ • This graph has a vertical asymptote of $x = 0$ and

goes through $(\frac{1}{4}, -1)$, $(1, 0)$, and $(4, 1)$. For reference purposes, call this $F(x)$.

(b) $f(x) = -\log_4 x$ • reflect F through the x-axis

(c) $f(x) = 2\log_4 x$ • vertically stretch F by a factor of 2

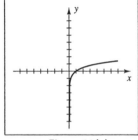 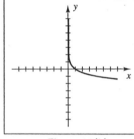 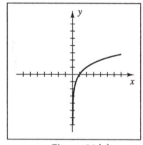

  Figure 33(a)   *Figure 33(b)*   *Figure 33(c)*

(d) $f(x) = \log_4(x+2)$ • shift F left 2 units

(e) $f(x) = (\log_4 x) + 2$ • shift F up 2 units

(f) $f(x) = \log_4(x-2)$ • shift F right 2 units

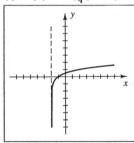 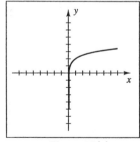 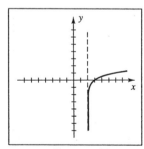

  Figure 33(d)   *Figure 33(e)*   *Figure 33(f)*

(g) $f(x) = (\log_4 x) - 2$ •   shift F down 2 units

(h) $f(x) = \log_4 |x|$ •   include the reflection of F through the y-axis since x may

be positive or negative, but $\log_4 |x|$ will give the same result

(i) $f(x) = \log_4(-x)$ • x must be negative so that $-x$ is positive,

reflect F through the y-axis

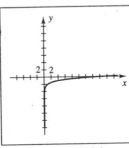

Figure 33(g)

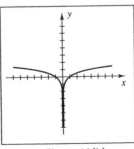

Figure 33(h)

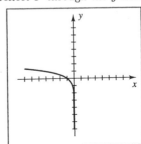

Figure 33(i)

(j) $f(x) = \log_4(3 - x) = \log_4[-(x - 3)]$ •   Shift F right 3 units and reflect
through the line $x = 3$. It may be helpful to determine the domain of this
function. We know that $3 - x$ must be positive for the function to be defined.
Thus, $3 - x > 0 \Rightarrow 3 > x$, or, equivalently, $x < 3$.

(k) $f(x) = |\log_4 x|$ •

reflect points with negative y-coordinates through the x-axis

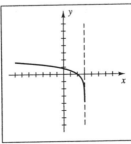

Figure 33(j)

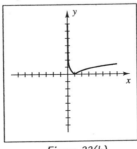

Figure 33(k)

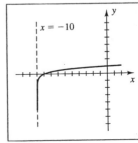

Figure 35

35   $f(x) = \log(x + 10)$ •   This is the graph of $g(x) = \log x$ shifted 10 units to the left.
There is a vertical asymptote of $x = -10$. g goes through $(\frac{1}{10}, -1)$, $(1, 0)$, and
$(10, 1)$, so f goes through $(\frac{1}{10} - 10, -1)$, $(1 - 10, 0)$, and $(10 - 10, 1)$,
that is, $(-9.9, -1)$, $(-9, 0)$, and $(0, 1)$.

$\boxed{37}$ $f(x) = \ln|x|$ • This is the graph of $g(x) = \ln x$ and its reflection through the y-axis. There is a vertical asymptote of $x = 0$. The graph of g goes through $(\frac{1}{e}, -1)$, $(1, 0)$, and $(e, 1)$, so the graph of f goes through $(\pm\frac{1}{e}, -1)$, $(\pm 1, 0)$, and $(\pm e, 1)$.

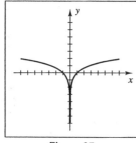

Figure 37

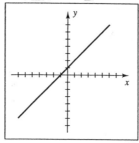

Figure 39

$\boxed{39}$ $f(x) = \ln e + x = 1 + x$. This is the graph of a line with slope 1 and y-intercept 1.

$\boxed{41}$ This is the graph of F reflected through the x-axis.   ★ $f(x) = -F(x)$

$\boxed{43}$ This is the graph of F shifted right 2 units.   ★ $f(x) = F(x - 2)$

$\boxed{45}$ This is the graph of F shifted up 1 unit.   ★ $f(x) = F(x) + 1$

$\boxed{47}$ (a) $\log x = 3.6274 \Rightarrow x = 10^{3.6274} \approx 4240.333$, or 4240 to three significant figures

(b) $\log x = 0.9469 \Rightarrow x = 10^{0.9469} \approx 8.849$, or 8.85

(c) $\log x = -1.6253 \Rightarrow x = 10^{-1.6253} \approx 0.023697$, or 0.0237

(d) $\ln x = 2.3 \Rightarrow x = e^{2.3} \approx 9.974$, or 9.97

(e) $\ln x = 0.05 \Rightarrow x = e^{0.05} \approx 1.051$, or 1.05

(f) $\ln x = -1.6 \Rightarrow x = e^{-1.6} \approx 0.2019$, or 0.202

$\boxed{49}$ $f(x) = 1000(1.05)^x = 1000e^{x \ln 1.05}$ (see the illustration on page 314).

This is approximately $1000e^{x\,(0.0487901642)} \approx 1000e^{0.0488x}$,

so the growth rate of f is about 4.88%.

$\boxed{51}$ $q = q_0(2)^{-t/1600} \Rightarrow \frac{q}{q_0} = 2^{-t/1600}$ { change to logarithm form } $\Rightarrow$

$$-\frac{t}{1600} = \log_2\left(\frac{q}{q_0}\right) \text{ { multiply by } } -1600 \text{ } \} \Rightarrow t = -1600\log_2\left(\frac{q}{q_0}\right)$$

$\boxed{53}$ $I = 20e^{-Rt/L} \Rightarrow \frac{I}{20} = e^{-Rt/L} \Rightarrow \ln\left(\frac{I}{20}\right) = -\frac{Rt}{L} \Rightarrow t = -\frac{L}{R}\ln\left(\frac{I}{20}\right)$

$\boxed{55}$ $I = 10^a I_0 \Rightarrow R = \log\left(\frac{I}{I_0}\right) = \log\left(\frac{10^a I_0}{I_0}\right) = \log 10^a = a$.

Hence, for $10^2 I_0$, $10^4 I_0$, and $10^5 I_0$, the answers are: (a) 2   (b) 4   (c) 5

$\boxed{57}$ We will find a general formula for α first.

$$I = 10^a I_0 \Rightarrow \alpha = 10\log\left(\frac{I}{I_0}\right) = 10\log\left(\frac{10^a I_0}{I_0}\right) = 10\,(\log 10^a) = 10(a) = 10a.$$

Hence, for $10^1 I_0$, $10^3 I_0$, and $10^4 I_0$, the answers are: (a) 10   (b) 30   (c) 40

59 1980 corresponds to $t = 0$ and $N(0) = 227$ million. We need to find t when

$$N = 2 \cdot 227. \quad 2 \cdot 227 = 227 e^{0.007t} \quad \Rightarrow \quad 2 = e^{0.007t} \quad \Rightarrow \quad \ln 2 = 0.007t \quad \Rightarrow$$

$$t = \frac{\ln 2}{0.007} \quad \Rightarrow \quad t \approx 99, \text{ which corresponds to the year } 1980 + 99 = 2079.$$

Alternatively, using the doubling time formula from page 316,

$$t = (\ln 2)/r = (\ln 2)/0.007 \approx 99.$$

61 (a) $\ln W = \ln 2.4 + (1.84)h$ { change to exponential form } $\Rightarrow$

$$W = e^{[\ln 2.4 + (1.84)h]} \quad \Rightarrow \quad W = e^{\ln 2.4} e^{1.84h} \text{ { since } } e^x e^y = e^{x+y} \text{ } \Rightarrow$$

$$W = 2.4 e^{1.84h}$$

(b) $h = 1.5 \quad \Rightarrow \quad W = 2.4 e^{(1.84)(1.5)} = 2.4 e^{2.76} \approx 37.92$ kg

63 (a) $p(h) = 10 \quad \Rightarrow \quad 10 = 14.7 e^{-0.0000385h} \quad \Rightarrow \quad \frac{10}{14.7} = e^{-0.0000385h} \quad \Rightarrow$

$$\ln\left(\tfrac{10}{14.7}\right) = -0.0000385h \quad \Rightarrow \quad h = -\frac{1}{0.0000385} \ln\left(\tfrac{10}{14.7}\right) \approx 10{,}007 \text{ ft.}$$

(b) At sea level, $h = 0$, and $p(0) = 14.7$. Setting $p(h)$ equal to $\frac{1}{2}(14.7)$,

and solving as in part (a), we have $h = -\frac{1}{0.0000385} \ln\left(\tfrac{1}{2}\right) \approx 18{,}004$ ft.

65 (a) $t = 0 \quad \Rightarrow \quad W = 2600\left(1 - 0.51 e^{-0.075(0)}\right)^3 = 2600(0.49)^3 \approx 305.9$ kg

(b) (1) From the graph, if $W = 1800$, t appears to be about 20.

(2) Solving the equation for t, we have $1800 = 2600\left(1 - 0.51 e^{-0.075t}\right)^3 \quad \Rightarrow$

$$\tfrac{1800}{2600} = \left(1 - 0.51 e^{-0.075t}\right)^3 \quad \Rightarrow \quad \sqrt[3]{\tfrac{9}{13}} = 1 - 0.51 e^{-0.075t} \quad \Rightarrow$$

$$\tfrac{51}{100} e^{-0.075t} = 1 - \sqrt[3]{\tfrac{9}{13}} \quad \Rightarrow \quad e^{-0.075t} = (\tfrac{100}{51})(1 - \sqrt[3]{\tfrac{9}{13}}) \quad \Rightarrow$$

$$-0.075t = \ln\left[(\tfrac{100}{51})(1 - \sqrt[3]{\tfrac{9}{13}})\right] \quad \Rightarrow \quad t = \frac{\ln\left[(\tfrac{100}{51})(1 - \sqrt[3]{\tfrac{9}{13}})\right]}{-0.075} \approx 19.8 \text{ yr.}$$

67 $D = ae^{-bx}$ with $a = 5.5$ and $b = 0.10$ •   The population density equals 2000 per

square mile means that $D = 2$. $D = 2 \quad \Rightarrow \quad 5.5 e^{-0.1x} = 2 \quad \Rightarrow \quad e^{-0.1x} = \tfrac{4}{11} \quad \Rightarrow$

$$-0.1x = \ln \tfrac{4}{11} \quad \Rightarrow \quad x = -10 \ln \tfrac{4}{11} \text{ mi} \approx 10.1 \text{ mi}$$

69 Since the half-life is eight days, $A(t) = \frac{1}{2} A_0$ when $t = 8$.

Thus, $\frac{1}{2} A_0 = A_0 a^{-8} \quad \Rightarrow \quad a^{-8} = \tfrac{1}{2} \quad \Rightarrow \quad \tfrac{1}{a^8} = \tfrac{1}{2} \quad \Rightarrow$

$$a^8 = 2 \quad \Rightarrow \quad a = 2^{1/8} \text{ { take the eighth root } } \approx 1.09.$$

71 (a) Since $\log P$ is an increasing function, increasing the population increases the

walking speed. Pedestrians have faster average walking speeds in large cities.

(b) $S = 5 \Rightarrow 5 = 0.05 + 0.86 \log P \Rightarrow 4.95 = 0.86 \log P \Rightarrow$

$$\frac{4.95}{0.86} = \log P \Rightarrow P = 10^{4.95/0.86} \approx 570{,}000$$

[73] (a) $f(x) = \ln(x+1) + e^x \Rightarrow f(2) = \ln(2+1) + e^2 \approx 8.4877$

(b) $g(x) = \dfrac{(\log x)^2 - \log x}{4} \Rightarrow g(3.97) = \dfrac{(\log 3.97)^2 - \log 3.97}{4} \approx -0.0601$

[75] *Figure 75* shows a graph of $Y_1 = x \ln x$ and $Y_2 = 1$.

By using an intersect feature, we determine that

$x \ln x = 1$ when $x \approx 1.763$. Alternatively, we could

graph $Y_1 = x \ln x - 1$, and find the zero of that graph.

[0, 4] by [−1, 1.67]

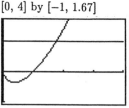

Figure 75

[77] The domain of $f(x) = 2.2 \log(x+2)$ is $x > -2$. The domain of $g(x) = \ln x$ is $x > 0$.

From *Figure 77*, we determine that f intersects g at about 14.90. Thus, $f(x) \ge g(x)$

on approximately (0, 14.90], not (−2, 14.90] since g is not defined if $x \le 0$. In

general, the larger the base of the logarithm, the slower its graph will rise. In this

case, we have base 10 and base $e \approx 2.72$, so we know that g will eventually intersect

f even if we don't see this intersection in our first window.

[−2, 16] by [−4, 8]   [3, 5, 0.5] by [0, 1, 0.5]

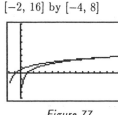   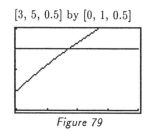

Figure 77   *Figure 79*

[79] (a) $R = 2.07 \ln x - 2.04 = 2.07 \ln \frac{C}{H} - 2.04 = 2.07 \ln \frac{242}{78} - 2.04 \approx 0.3037 \approx 30\%$.

(b) Graph $Y_1 = 2.07 \ln x - 2.04$ and $Y_2 = 0.75$.

From the graph, $R \approx 0.75$ when $x = \frac{C}{H} \approx 3.85$.

4.4 Exercises

[1] (a) $\log_4(xz) = \log_4 x + \log_4 z$ (b) $\log_4(y/x) = \log_4 y - \log_4 x$

(c) $\log_4 \sqrt[3]{z} = \log_4 z^{1/3} = \frac{1}{3} \log_4 z$

[3] $\log_a \dfrac{x^3 w}{y^2 z^4} = \log_a(x^3 w) - \log_a(y^2 z^4) = (\log_a x^3 + \log_a w) - (\log_a y^2 + \log_a z^4)$

$$= 3 \log_a x + \log_a w - 2 \log_a y - 4 \log_a z$$

The most common mistake is to not have the minus sign in front of $4 \log_a z$.

This error results from not having the parentheses in the correct place.

[5] $\log \dfrac{\sqrt[3]{z}}{x\sqrt{y}} = \log \sqrt[3]{z} - \log x\sqrt{y} = \log z^{1/3} - \log x - \log y^{1/2} = \frac{1}{3}\log z - \log x - \frac{1}{2}\log y$

[7] $\ln \sqrt[4]{\dfrac{x^7}{y^5 z}} = \ln x^{7/4} - \ln y^{5/4} z^{1/4} = \ln x^{7/4} - \ln y^{5/4} - \ln z^{1/4} = \frac{7}{4}\ln x - \frac{5}{4}\ln y - \frac{1}{4}\ln z$

As a generalization for exercises similar to those in 1–8, if the exponents on the variables are positive, then the sign in front of the individual logarithms will be positive if the variable was originally in the numerator and negative if the variable was originally in the denominator.

[9] (a) $\log_3 x + \log_3 (5y) = \log_3 (x \cdot 5y) = \log_3 (5xy)$

    (b) $\log_3 (2z) - \log_3 x = \log_3 (2z/x)$ (c) $5\log_3 y = \log_3 y^5$

[11] $2\log_a x + \frac{1}{3}\log_a (x-2) - 5\log_a (2x+3)$

$\quad = \log_a x^2 + \log_a (x-2)^{1/3} - \log_a (2x+3)^5$ { logarithm law 3 }

$\quad = \log_a x^2 \sqrt[3]{x-2} - \log_a (2x+3)^5$ { logarithm law 1 }

$\quad = \log_a \dfrac{x^2 \sqrt[3]{x-2}}{(2x+3)^5}$ { logarithm law 2 }

[13] $\log (x^3 y^2) - 2\log x\sqrt[3]{y} - 3\log \left(\frac{x}{y}\right) = \log (x^3 y^2) - \left[\log \left(x\sqrt[3]{y}\right)^2 + \log \left(\frac{x}{y}\right)^3\right]$

$\quad\quad\quad\quad\quad\quad\quad\quad\quad\quad = \log (x^3 y^2) - \left[\log \left(x^2 y^{2/3} \cdot (x^3/y^3)\right)\right]$

$\quad\quad\quad\quad\quad\quad\quad\quad\quad\quad = \log \dfrac{x^3 y^2}{x^5 y^{-7/3}} = \log \dfrac{y^{13/3}}{x^2}$

[15] $\ln y^3 + \frac{1}{3}\ln (x^3 y^6) - 5\ln y = \ln y^3 + \ln (xy^2) - \ln y^5 = \ln\left[(xy^5)/y^5\right] = \ln x$

[17] $\log_6 (2x-3) = \log_6 12 - \log_6 3 \;\Rightarrow\; \log_6 (2x-3) = \log_6 \frac{12}{3} \;\Rightarrow\; 2x-3 = 4 \;\Rightarrow$

$\quad\quad\quad\quad\quad\quad\quad\quad\quad\quad\quad\quad\quad\quad\quad\quad 2x = 7 \;\Rightarrow\; x = \frac{7}{2}$

[19] $2\log_3 x = 3\log_3 5 \;\Rightarrow\; \log_3 x^2 = \log_3 5^3 \;\Rightarrow\; x^2 = 125 \;\Rightarrow\; x = \pm\sqrt{125} = \pm 5\sqrt{5};$

$\quad\quad\quad\quad -5\sqrt{5}$ is extraneous since it would make $\log_3 x$ undefined

[21] $\log x - \log (x+1) = 3\log 4 \;\Rightarrow\; \log \dfrac{x}{x+1} = \log 4^3 \;\Rightarrow\; \dfrac{x}{x+1} = 64 \;\Rightarrow$

$\quad x = 64x + 64 \;\Rightarrow\; -63x = 64 \;\Rightarrow\; x = -\frac{64}{63}$. But $-\frac{64}{63}$ is extraneous since it makes

an argument negative, so there is no solution for the equation.

[23] $\ln (-4-x) + \ln 3 = \ln (2-x) \;\Rightarrow\; \ln\left[(-4-x) \cdot 3\right] = \ln (2-x) \;\Rightarrow$

$\ln (-12-3x) = \ln (2-x) \;\Rightarrow\; -12-3x = 2-x \;\Rightarrow\; 2x = -14 \;\Rightarrow\; x = -7.$

Remember, the solution of a logarithmic equation may be negative—you must examine what happens to the original logarithm expressions. In this case, we have $\ln 3 + \ln 3 = \ln 9$, which is true.

$\boxed{25}$ $\log_2(x+7) + \log_2 x = 3 \Rightarrow \log_2(x^2 + 7x) = 3 \Rightarrow x^2 + 7x = 2^3 \Rightarrow$

$\quad x^2 + 7x - 8 = 0 \Rightarrow (x+8)(x-1) = 0 \Rightarrow x = -8, 1;\ -8$ is extraneous

$\boxed{27}$ $\log_3(x+3) + \log_3(x+5) = 1 \Rightarrow \log_3\big[(x+3)(x+5)\big] = 1 \Rightarrow$

$\quad \log_3(x^2 + 8x + 15) = 1 \Rightarrow x^2 + 8x + 15 = 3 \Rightarrow$

$\quad x^2 + 8x + 12 = 0 \Rightarrow (x+2)(x+6) = 0 \Rightarrow x = -6, -2;\ -6$ is extraneous

$\boxed{29}$ $\log(x+3) = 1 - \log(x-2) \Rightarrow \log(x+3) + \log(x-2) = 1 \Rightarrow$

$\quad \log\big[(x+3)(x-2)\big] = 1 \Rightarrow x^2 + x - 6 = 10^1 \Rightarrow$

$\quad x^2 + x - 16 = 0 \Rightarrow x = \dfrac{-1 + \sqrt{65}}{2} \approx 3.53;\ \dfrac{-1 - \sqrt{65}}{2} \approx -4.53$ is extraneous

$\boxed{31}$ $\ln x = 1 - \ln(x+2) \Rightarrow \ln x + \ln(x+2) = 1 \Rightarrow$

$\quad \ln\big[x(x+2)\big] = 1 \Rightarrow x^2 + 2x = e^1 \Rightarrow x^2 + 2x - e = 0 \Rightarrow$

$\quad x = \dfrac{-2 \pm \sqrt{4+4e}}{2} = \dfrac{-2 \pm 2\sqrt{1+e}}{2} = -1 \pm \sqrt{1+e}.$

$\quad x = -1 + \sqrt{1+e} \approx 0.93$ is a valid solution,

$\qquad\qquad\qquad$ but $x = -1 - \sqrt{1+e} \approx -2.93$ is extraneous.

$\boxed{33}$

$\log_3(x-2) = \log_3 27 - \log_3(x-4) - 5^{\log_5 1}$	given
$\log_3(x-2) + \log_3(x-4) = \log_3 3^3 - 1$	$5^{\log_5 1} = 1$
$\log_3\big[(x-2)(x-4)\big] = 3 - 1$	$\log_3 3^3 = 3$
$\log_3(x^2 - 6x + 8) = 2$	simplify
$x^2 - 6x + 8 = 3^2$	exponential form
$x^2 - 6x - 1 = 0$	subtract 9

Now solve $x^2 - 6x - 1 = 0$ using the quadratic formula to get $x = (6 \pm \sqrt{40})/2 =$
$3 \pm \sqrt{10}$. Since $3 - \sqrt{10}$ makes the arguments of the logarithms negative,

$\qquad\qquad\qquad\qquad$ the only solution is $3 + \sqrt{10}$.

$\boxed{35}$ $f(x) = \log_3(3x) = \log_3 3 + \log_3 x = \log_3 x + 1 \quad \bullet \quad$ shift $y = \log_3 x$ up 1 unit

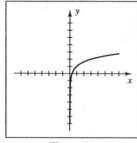

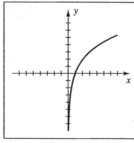

$\qquad$ Figure 35 $\qquad\qquad\qquad\qquad$ Figure 37

$\boxed{37}$ $f(x) = 3\log_3 x \quad \bullet \quad$ vertically stretch $y = \log_3 x$ by a factor of 3

39 $f(x) = \log_3(x^2) = 2\log_3 x$ • Vertically stretch $y = \log_3 x$ by a factor of 2 and include its reflection through the y-axis since the domain of the original function, $f(x) = \log_3(x^2)$, is $\mathbb{R} - \{0\}$. Keep in mind that the laws of logarithms are established for positive real numbers, so that when we make the step $\log_3(x^2) = 2\log_3 x$, it is only for positive x.

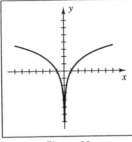

Figure 39

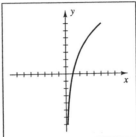

Figure 41

41 $f(x) = \log_2(x^3) = 3\log_2 x$ • vertically stretch $y = \log_2 x$ by a factor of 3

43 $f(x) = \log_2 \sqrt{x} = \log_2 x^{1/2} = \frac{1}{2}\log_2 x$ •

vertically compress $y = \log_2 x$ by a factor of $1/(1/2) = 2$

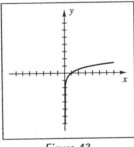

Figure 43

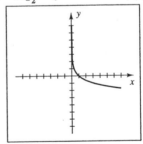

Figure 45

45 $f(x) = \log_3\left(\frac{1}{x}\right) = \log_3 x^{-1} = -\log_3 x$ • reflect $y = \log_3 x$ through the x-axis

47 The values of $F(x) = \log_2 x$ are doubled and the reflection of F through the y-axis is included. The domain of f is $\mathbb{R} - \{0\}$ and $f(x) = \log_2 x^2$.

49 $F(x) = \log_2 x$ is shifted up 3 units since $(1, 0)$ on F is $(1, 3)$ on the graph.

Hence, $f(x) = 3 + \log_2 x = \log_2 2^3 + \log_2 x = \log_2(8x)$.

51 $V_1 = 2$ and $V_2 = 4.5$ $\Rightarrow$ db $= 20\log\dfrac{V_2}{V_1} = 20\log\dfrac{4.5}{2} \approx 7.04$, or $+7$.

53 $\log y = \log b - k\log x$ $\Rightarrow$ $\log y = \log b - \log x^k$ $\Rightarrow$ $\log y = \log\dfrac{b}{x^k}$ $\Rightarrow$ $y = \dfrac{b}{x^k}$

55 $c = 0.5$ and $z_0 = 0.1$ $\Rightarrow$

$v = c \ln(z/z_0)$

$= (0.5) \ln(z/0.1)$

$= \frac{1}{2} \ln(10z)$

$= \frac{1}{2}(\ln 10 + \ln z)$

$= \frac{1}{2} \ln 10 + \frac{1}{2} \ln z \approx \frac{1}{2} \ln z + 1.15$.

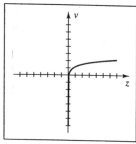

Figure 55

57 (a) $R(x) = a \log\left(\frac{x}{x_0}\right)$ $\Rightarrow$ $R(x_0) = a \log\left(\frac{x_0}{x_0}\right) = a \log 1 = a \cdot 0 = 0$

(b) $R(2x) = a \log\left(\frac{2x}{x_0}\right) = a \log\left(2 \cdot \frac{x}{x_0}\right) = a\left[\log 2 + \log\left(\frac{x}{x_0}\right)\right]$

$= a \log 2 + a \log\left(\frac{x}{x_0}\right) = R(x) + a \log 2$

59 $\ln I_0 - \ln I = kx$ $\Rightarrow$ $\ln\frac{I_0}{I} = kx$ $\Rightarrow$ $x = \frac{1}{k} \ln\frac{I_0}{I} = \frac{1}{0.39} \ln 1.12 \approx 0.29$ cm.

61 From the graph, the coordinates of the points of intersection are approximately

$(1.02, 0.48)$ and $(2.40, 0.86)$. $f(x) \geq g(x)$ on the intervals $(0, 1.02]$ and $[2.40, \infty)$.

[0, 6] by [−1, 3]

Figure 61

[0, 8] by [−1.67, 3.67]

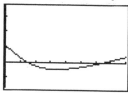

Figure 63

63 Graph $y = e^{-x} - 2 \log(1 + x^2) + 0.5x$ and estimate any x-intercepts. From the graph,

we see that the roots of the equation are approximately $x \approx 1.41, 6.59$.

65 (a) f is increasing on $[0.2, 0.63]$ and $[6.87, 16]$. f is decreasing on $[0.63, 6.87]$.

(b) The maximum value of f is 4.61 when $x = 16$.

The minimum value of f is approximately -3.31 when $x \approx 6.87$.

[0.2, 16, 2] by [−4.77, 5.77]

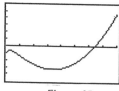

Figure 65

[−5, 10] by [−2, 8]

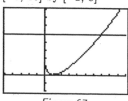

Figure 67

67 Graph $Y_1 = x \log x - \log x$ and $Y_2 = 5$. The graphs intersect at $x \approx 6.94$.

[69] Let $d = x$. Graph $Y_1 = I_0 - 20 \log x - kx = 70 - 20 \log x - 0.076x$ and $Y_2 = 20$.

At the point of intersection, $x \approx 115.3$. The distance is approximately 115 meters.

4.5 Exercises

[1]   (a) $5^x = 8 \;\Rightarrow\; \log 5^x = \log 8 \;\Rightarrow\; x \log 5 = \log 8 \;\Rightarrow\; x = \dfrac{\log 8}{\log 5} \approx 1.29$

    (b) $5^x = 8 \;\Rightarrow\; x = \log_5 8 = \dfrac{\log 8}{\log 5}$ { by the change of base formula } ≈ 1.29

[3]   (a) $3^{4-x} = 5 \;\Rightarrow\; \log(3^{4-x}) = \log 5 \;\Rightarrow\; (4-x)\log 3 = \log 5 \;\Rightarrow$

       $4 - x = \dfrac{\log 5}{\log 3} \;\Rightarrow\; x = 4 - \dfrac{\log 5}{\log 3} \approx 2.54$. *Note:* The answer could also

                be written as $4 - \dfrac{\log 5}{\log 3} = \dfrac{4\log 3 - \log 5}{\log 3} = \dfrac{\log 81 - \log 5}{\log 3} = \dfrac{\log \frac{81}{5}}{\log 3}$.

    (b) $3^{4-x} = 5 \;\Rightarrow\; 4 - x = \log_3 5 \;\Rightarrow\; x = 4 - \dfrac{\log 5}{\log 3} \approx 2.54$.

[5] $\log_5 6 = \dfrac{\log 6}{\log 5} \left\{ \text{or } \dfrac{\ln 6}{\ln 5} \right\} \approx 1.1133$

[7] $\log_9 0.2 = \dfrac{\log 0.2}{\log 9} \left\{ \text{or, equivalently, } \dfrac{\ln 0.2}{\ln 9} \right\} \approx -0.7325$.

     Note that either log or ln can be used here. You should get comfortable using both.

[9] $\dfrac{\log_5 16}{\log_5 4} = \log_4 16 = \log_4 4^2 = 2$

[11] The steps are similar to those in Example 4. $3^{x+4} = 2^{1-3x} \;\Rightarrow$

$\log(3^{x+4}) = \log(2^{1-3x}) \;\Rightarrow\; (x+4)\log 3 = (1 - 3x)\log 2 \;\Rightarrow$

$x \log 3 + 4 \log 3 = \log 2 - 3x \log 2 \;\Rightarrow\; x \log 3 + 3x \log 2 = \log 2 - 4\log 3 \;\Rightarrow$

$x(\log 3 + 3\log 2) = \log 2 - 4\log 3 \;\Rightarrow\; x = \dfrac{\log 2 - \log 81}{\log 3 + \log 8} \;\Rightarrow\; x = \dfrac{\log \frac{2}{81}}{\log 24} \approx -1.16$

[13] $2^{2x-3} = 5^{x-2} \;\Rightarrow\; \log(2^{2x-3}) = \log(5^{x-2}) \;\Rightarrow\; (2x-3)\log 2 = (x-2)\log 5 \;\Rightarrow$

$2x \log 2 - 3\log 2 = x \log 5 - 2\log 5 \;\Rightarrow\; 2x \log 2 - x \log 5 = 3\log 2 - 2\log 5 \;\Rightarrow$

$x(2\log 2 - \log 5) = \log 2^3 - \log 5^2 \;\Rightarrow\; x = \dfrac{\log 8 - \log 25}{\log 4 - \log 5} \;\Rightarrow\; x = \dfrac{\log \frac{8}{25}}{\log \frac{4}{5}} \approx 5.11$

[15] $2^{-x} = 8 \;\Rightarrow\; 2^{-x} = 2^3 \;\Rightarrow\; -x = 3 \;\Rightarrow\; x = -3$

[17] $\log x = 1 - \log(x-3) \;\Rightarrow\; \log x + \log(x-3) = 1 \;\Rightarrow\; \log(x^2 - 3x) = 1 \;\Rightarrow$

$x^2 - 3x = 10^1 \;\Rightarrow\; x^2 - 3x - 10 = 0 \;\Rightarrow\; (x-5)(x+2) = 0 \;\Rightarrow\; x = 5, -2$.

                      The solution $x = 5$ checks, but -2 is extraneous.

19 $\log(x^2 + 4) - \log(x + 2) = 2 + \log(x - 2) \Rightarrow \log\left(\dfrac{x^2 + 4}{x + 2}\right) - \log(x - 2) = 2 \Rightarrow$

$\log\left(\dfrac{x^2 + 4}{x^2 - 4}\right) = 2 \Rightarrow \dfrac{x^2 + 4}{x^2 - 4} = 10^2 \Rightarrow x^2 + 4 = 100(x^2 - 4) \Rightarrow$

$x^2 + 4 = 100x^2 - 400 \Rightarrow 404 = 99x^2 \Rightarrow x^2 = \frac{404}{99} \Rightarrow$

$x = \pm\sqrt{\dfrac{4 \cdot 101}{9 \cdot 11}} = \pm\frac{2}{3}\sqrt{\dfrac{101}{11}} \approx \pm 2.02;\ -\frac{2}{3}\sqrt{\dfrac{101}{11}}$ is extraneous

21 See Example 5 for more detail concerning this type of exercise.

$5^x + 125(5^{-x}) = 30 \left\{\text{multiply by } 5^x\right\} \Rightarrow (5^x)^2 + 125 = 30(5^x) \Rightarrow$

$(5^x)^2 - 30(5^x) + 125 = 0 \left\{\text{recognize as a quadratic in } 5^x \text{ and factor}\right\} \Rightarrow$

$(5^x - 5)(5^x - 25) = 0 \Rightarrow 5^x = 5, 25 \Rightarrow 5^x = 5^1, 5^2 \Rightarrow x = 1, 2$

23 $4^x - 3(4^{-x}) = 8 \left\{\text{multiply by } 4^x\right\} \Rightarrow (4^x)^2 - 3 = 8(4^x) \Rightarrow$

$(4^x)^2 - 8(4^x) - 3 = 0 \left\{\text{recognize as a quadratic in } 4^x \text{ and use the quadratic}\right.$

formula $\left.\right\} \Rightarrow 4^x = \dfrac{-(-8) \pm \sqrt{(-8)^2 - 4(1)(-3)}}{2(1)} = \dfrac{8 \pm \sqrt{76}}{2} = \dfrac{8 \pm 2\sqrt{19}}{2} = 4 \pm \sqrt{19}.$

Since 4^x is positive and $4 - \sqrt{19}$ is negative, $4 - \sqrt{19}$ is discarded.

Continuing, $4^x = 4 + \sqrt{19} \Rightarrow x = \log_4(4 + \sqrt{19}) = \dfrac{\log(4 + \sqrt{19})}{\log 4} \left\{\text{use the change}\right.$

of base formula to approximate $\left.\right\} \approx 1.53$

25 $\log(x^2) = (\log x)^2 \Rightarrow 2\log x = (\log x)^2 \Rightarrow (\log x)^2 - 2\log x = 0 \Rightarrow$

$(\log x)(\log x - 2) = 0 \Rightarrow \log x = 0, 2 \Rightarrow x = 10^0, 10^2 \Rightarrow x = 1 \text{ or } 100$

27 Don't confuse $\log(\log x)$ with $(\log x)(\log x)$. The first expression is

the log of the log of x, whereas the second expression is the log of x times itself.

$\log(\log x) = 2 \Rightarrow \log x = 10^2 = 100 \Rightarrow x = 10^{100}$

29 $x^{\sqrt{\log x}} = 10^8 \left\{\text{take the log of both sides}\right\} \Rightarrow \log\left(x^{\sqrt{\log x}}\right) = \log 10^8 \Rightarrow$

$\sqrt{\log x}\,(\log x) = 8 \Rightarrow (\log x)^{1/2}(\log x)^1 = 8 \Rightarrow (\log x)^{3/2} = 8 \Rightarrow$

$\left[(\log x)^{3/2}\right]^{2/3} = (8)^{2/3} \Rightarrow \log x = (\sqrt[3]{8})^2 = 4 \Rightarrow x = 10{,}000$

31 Since $e^{2x} = (e^x)^2$, we recognize $e^{2x} + 2e^x - 15$ as a quadratic in e^x and factor it.

$e^{2x} + 2e^x - 15 = 0 \Rightarrow (e^x + 5)(e^x - 3) = 0 \Rightarrow e^x = -5, 3.$

But $e^x > 0$, so $e^x = 3 \Rightarrow x = \ln 3.$

Note: For Exercises 33–40 and 41–42 of the Chapter Review Exercises, let D denote the domain of the function determined by the original equation, and R its range. These are then the range and domain, respectively, of the equation listed in the answer.

$\boxed{33}$ $y = \dfrac{10^x + 10^{-x}}{2} \ \{D = \mathbb{R}, \ R = [1, \infty)\} \ \Rightarrow \ 2y = 10^x + 10^{-x}$

$\left\{ \text{since } 10^{-x} = \dfrac{1}{10^x}, \text{ multiply by } 10^x \text{ to eliminate denominator} \right\} \ \Rightarrow$

$10^{2x} - 2y\,10^x + 1 = 0 \ \{\text{treat as a quadratic in } 10^x\} \ \Rightarrow$

$$10^x = \frac{2y \pm \sqrt{4y^2 - 4}}{2} = y \pm \sqrt{y^2 - 1} \ \Rightarrow \ x = \log\left(y \pm \sqrt{y^2 - 1}\right)$$

$\boxed{35}$ $y = \dfrac{10^x - 10^{-x}}{10^x + 10^{-x}} \ \{D = \mathbb{R}, \ R = (-1, 1)\} \ \Rightarrow \ y\,10^x + y\,10^{-x} = 10^x - 10^{-x} \ \Rightarrow$

$y\,10^{2x} + y = 10^{2x} - 1 \ \Rightarrow \ (y - 1)\,10^{2x} = -1 - y \ \Rightarrow$

$$10^{2x} = \frac{-1 - y}{y - 1} \ \Rightarrow \ 2x = \log\left(\frac{1 + y}{1 - y}\right) \ \Rightarrow \ x = \tfrac{1}{2}\log\left(\frac{1 + y}{1 - y}\right)$$

$\boxed{37}$ $y = \dfrac{e^x - e^{-x}}{2} \ \{D = R = \mathbb{R}\} \ \Rightarrow \ 2y = e^x - e^{-x} \ \Rightarrow$

$$e^{2x} - 2y\,e^x - 1 = 0 \ \Rightarrow \ e^x = \frac{2y \pm \sqrt{4y^2 + 4}}{2} = y \pm \sqrt{y^2 + 1};$$

$\sqrt{y^2 + 1} > y$, so $y - \sqrt{y^2 + 1} < 0$, but $e^x > 0$ and thus, $x = \ln\left(y + \sqrt{y^2 + 1}\right)$

$\boxed{39}$ $y = \dfrac{e^x + e^{-x}}{e^x - e^{-x}} \ \{D = \mathbb{R} - \{0\}, \ R = (-\infty, -1) \cup (1, \infty)\} \ \Rightarrow$

$ye^x - ye^{-x} = e^x + e^{-x} \ \Rightarrow \ ye^{2x} - y = e^{2x} + 1 \ \Rightarrow$

$$(y - 1)e^{2x} = y + 1 \ \Rightarrow \ e^{2x} = \frac{y + 1}{y - 1} \ \Rightarrow \ 2x = \ln\left(\frac{y + 1}{y - 1}\right) \ \Rightarrow \ x = \tfrac{1}{2}\ln\left(\frac{y + 1}{y - 1}\right)$$

$\boxed{41}$ $f(x) = \log_2(x + 3) \quad \bullet \quad x = 0 \ \Rightarrow \ y\text{-intercept} = \log_2 3 = \dfrac{\log 3}{\log 2} \approx 1.5850$

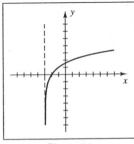

Figure 41

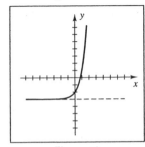

Figure 43

$\boxed{43}$ $f(x) = 4^x - 3 \quad \bullet \quad y = 0 \ \Rightarrow \ 4^x = 3 \ \Rightarrow \ x\text{-intercept} = \log_4 3 = \dfrac{\log 3}{\log 4} \approx 0.7925$

$\boxed{45}$ (a) vinegar: pH $\approx -\log(6.3 \times 10^{-3}) = -(\log 6.3 + \log 10^{-3}) = -(\log 6.3 - 3) =$

$$3 - \log 6.3 \approx 2.2$$

(b) carrots: pH $\approx -\log(1.0 \times 10^{-5}) = 5 - \log 1.0 = 5$

(c) sea water: pH $\approx -\log(5.0 \times 10^{-9}) = 9 - \log 5.0 \approx 8.3$

$\boxed{47}$ $[H^+] < 10^{-7} \Rightarrow \log[H^+] < \log 10^{-7}$ {since log is an increasing function} $\Rightarrow$

$\log[H^+] < -7 \Rightarrow -\log[H^+] > -(-7) \Rightarrow$ pH > 7 for basic solutions;

similarly, pH < 7 for acidic solutions.

$\boxed{49}$ Solving $A = P(1 + \frac{r}{n})^{nt}$ for t with $A = 2P$, $r = 0.06$, and $n = 12$ yields

$2P = P(1 + \frac{0.06}{12})^{12t}$ {divide by P} $\Rightarrow$

$2 = (1.005)^{12t}$ {take the ln of both sides} $\Rightarrow$

$\ln 2 = \ln(1.005)^{12t} \Rightarrow \ln 2 = 12t \ln(1.005) \Rightarrow$

$$t = \frac{\ln 2}{12 \ln(1.005)} \approx 11.58 \text{ yr, or about 11 years and 7 months.}$$

$\boxed{51}$ 50% of the light reaching a depth of 13 meters corresponds to the equation

$\frac{1}{2}I_0 = I_0 c^{13}$. Solving for c, we have $c^{13} = \frac{1}{2} \Rightarrow c = \sqrt[13]{\frac{1}{2}} = 2^{-1/13}$.

Now letting $I = 0.01 I_0$, $c = 2^{-1/13}$, and using the formula from Example 8,

$$x = \frac{\log(I/I_0)}{\log c} = \frac{\log[(0.01 I_0)/I_0]}{\log 2^{-1/13}} = \frac{\log 10^{-2}}{-\frac{1}{13}\log 2} = \frac{-2}{-\frac{1}{13}\log 2} = \frac{26}{\log 2} \approx 86.4 \text{ m.}$$

$\boxed{53}$ (a) The graph of $A = 100[1 - (0.9)^t]$ is an

increasing exponential that passes through

$(0, 0)$, $(5, \approx 41)$, and $(10, \approx 65)$.

(b) $A = 50 \Rightarrow 50 = 100[1 - (0.9)^t] \Rightarrow$

$1 - (0.9)^t = \frac{50}{100} \Rightarrow (0.9)^t = 0.5 \Rightarrow$

$t = \log_{0.9}(0.5) = \frac{\log 0.5}{\log 0.9} \approx 6.58 \text{ min.}$

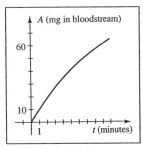

Figure 53

$\boxed{55}$ (a) $F = F_0(1 - m)^t \Rightarrow (1 - m)^t = \frac{F}{F_0} \Rightarrow \log(1 - m)^t = \log\left(\frac{F}{F_0}\right) \Rightarrow$

$$t \log(1 - m) = \log\left(\frac{F}{F_0}\right) \Rightarrow t = \frac{\log(F/F_0)}{\log(1 - m)}$$

(b) Using part (a) with $F = \frac{1}{2}F_0$ and $m = 0.00005$,

$$t = \frac{\log(F/F_0)}{\log(1 - m)} = \frac{\log(\frac{1}{2}F_0/F_0)}{\log(1 - 0.00005)} = \frac{\log \frac{1}{2}}{\log 0.99995} \approx 13{,}863 \text{ generations.}$$

57 (a) $t = 10 \Rightarrow h = \dfrac{120}{1 + 200e^{-0.2(10)}} = \dfrac{120}{1 + 200e^{-2}} \approx 4.28$ ft

(b) $h = 50 \Rightarrow 50 = \dfrac{120}{1 + 200e^{-0.2t}} \Rightarrow 1 + 200e^{-0.2t} = \dfrac{120}{50} \Rightarrow$

$200e^{-0.2t} = \dfrac{12}{5} - 1 \Rightarrow e^{-0.2t} = \dfrac{7}{5} \cdot \dfrac{1}{200} \Rightarrow e^{-0.2t} = 0.007 \Rightarrow$

$-0.2t = \ln 0.007 \Rightarrow t = \dfrac{\ln 0.007}{-0.2} \approx 24.8$ yr

59 $\dfrac{v_0}{v_1} = \left(\dfrac{h_0}{h_1}\right)^P \Rightarrow \ln\dfrac{v_0}{v_1} = P\ln\dfrac{h_0}{h_1} \Rightarrow P = \dfrac{\ln(v_0/v_1)}{\ln(h_0/h_1)} = \dfrac{\ln(25/6)}{\ln(200/35)} \approx 0.82$

61 If $y = c2^{kx}$ and $x = 0$, then $y = c2^0 = c = 4$. Thus, $y = 4(2)^{kx}$. Similarly,

$x = 1 \Rightarrow y = 4(2)^k = 3.249 \Rightarrow 2^k = \dfrac{3.249}{4} \Rightarrow k = \log_2\left(\dfrac{3.249}{4}\right) \approx -0.300$. Thus,

$y = 4(2)^{-0.3x}$. Checking the remaining two points, we see that $x = 2 \Rightarrow y \approx 2.639$ and $x = 3 \Rightarrow y \approx 2.144$. The four points lie on the graph of $y = 4(2)^{-0.3x}$

to within three-decimal-place accuracy.

63 If $y = c\log(kx + 10)$ and $x = 0$, then $y = c\log 10 = c = 1.5$.

Thus, $y = 1.5\log(kx + 10)$. Similarly, $x = 1 \Rightarrow y = 1.5\log(k + 10) = 1.619 \Rightarrow$

$\dfrac{1.619}{1.5} = \log(k + 10) \Rightarrow k + 10 = 10^{1.619/1.5} \Rightarrow k = 10^{1.619/1.5} - 10 \approx 2.004$. Thus,

$y = 1.5\log(2.004x + 10)$.   Checking the remaining two points, we see that $x = 2 \Rightarrow y \approx 1.720$, and $x = 3 \Rightarrow y \approx 1.807$, not 1.997. The points do not lie on the graph of $y = c\log(kx + 10)$ to within three-decimal-place accuracy.

65 $h(5.3) = \log_4 5.3 - 2\log_8(1.2 \times 5.3) = \dfrac{\ln 5.3}{\ln 4} - \dfrac{2\ln 6.36}{\ln 8} \approx -0.5764$

67 The minimum point on the graph of $y = x - \ln(0.3x) - 3\log_3 x$ is about $(3.73, 0.023)$.

So there are *no* x-intercepts, and hence, no roots of the equation on $(0, 9)$.

[0, 9] by [−1, 5]   [−1, 17] by [−1, 11]

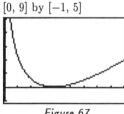
Figure 67

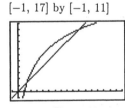
Figure 69

69 The graphs of $f(x) = x$ and $g(x) = 3\log_2 x$ intersect at approximately $(1.37, 1.37)$ and $(9.94, 9.94)$. Hence, the solutions of the equation $f(x) = g(x)$ are 1.37 and 9.94.

71 From the graph, we see that the graphs of f and g intersect at three points. Their coordinates are approximately $(-0.32, 0.50)$, $(1.52, -1.33)$, and $(6.84, -6.65)$. The region near the origin in *Figure 71(a)* is enhanced in *Figure 71(b)*. Thus, $f(x) > g(x)$ on $(-\infty, -0.32)$ and $(1.52, 6.84)$.

$[-5, 10]$ by $[-8, 2]$

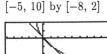
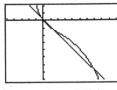

Figure 71(a)

$[-1.53, 2.26]$ by $[-2.92, 1.05]$

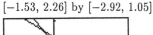
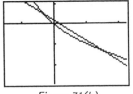

Figure 71(b)

73 (1) The graph of $n(t) = 85e^{t/3}$ is increasing rapidly and soon becomes greater than 100. It is doubtful that the average score would improve dramatically without any review.

$[0, 5]$ by $[0, 200, 20]$

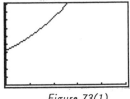

Figure 73(1)

$[0, 5]$ by $[0, 100, 10]$

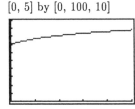

Figure 73(2)

(2) The graph of $n(t) = 70 + 10\ln(t+1)$ is also increasing. It is incorrect because $n(0) = 70 \neq 85$. Also, one would not expect the average score to improve without any review.

(3) The graph of $n(t) = 86 - e^t$ decreases rapidly to zero. The *average* score probably would not be zero after 5 weeks.

$[0, 5]$ by $[0, 100, 10]$

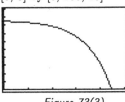

Figure 73(3)

$[0, 5]$ by $[0, 100, 10]$

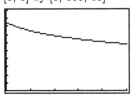

Figure 73(4)

(4) The graph of $n(t) = 85 - 15\ln(t+1)$ is decreasing. During the first weeks it decreases most rapidly and then starts to level off. Of the four functions, **this function seems to best model the situation.**

75 (a) The ozone level is decreasing by 11% per year. The fraction of ozone remaining x
   years after April 1992 is given by the function $f(x) = (0.89)^x$ {decreasing by 11%
   is the same as retaining 89%}. We must approximate x when $f(x) = 0.5$. Using
   a table, $f(x) = 0.5$ when $x \approx 6$. Thus, in 1998 the ozone level would be 50% of
   its normal amount.

x	1	2	3	4	5	6	7
$f(x)$	0.89	0.79	0.70	0.63	0.56	0.50	0.44

(b) $(0.89)^t = 0.5 \Rightarrow \ln(0.89)^t = \ln 0.5 \Rightarrow t \ln 0.89 = \ln 0.5 \Rightarrow$

$$t = \frac{\ln 0.5}{\ln 0.89} \approx 5.948, \text{ or in 1998.}$$

Chapter 4 Review Exercises

1 $f(x) = 3^{x+2}$ • shift $y = 3^x$ left 2 units

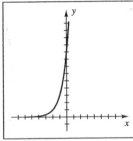

Figure 1

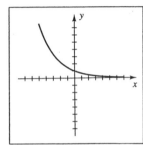

Figure 3

3 $f(x) = \left(\frac{3}{2}\right)^{-x} = \left(\frac{2}{3}\right)^x$ • goes through $(-1, \frac{3}{2})$, $(0, 1)$, and $(1, \frac{2}{3})$

5 $f(x) = 3^{-x^2} = (3^{-1})^{(x^2)} = \left(\frac{1}{3}\right)^{(x^2)}$ • see the note in §4.1 before Exercise 21

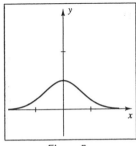

Figure 5

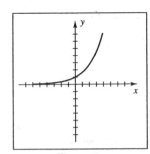

Figure 7

7 $f(x) = e^{x/2} = (e^{1/2})^x \approx (1.65)^x$ • goes through $(-1, 1/\sqrt{e})$, $(0, 1)$, and $(1, \sqrt{e})$;
                                                    or approximately $(-1, 0.61)$, $(0, 1)$, and $(1, 1.65)$

⑨ $f(x) = e^{x-2}$ • shift $y = e^x$ right 2 units

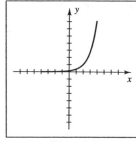

Figure 9

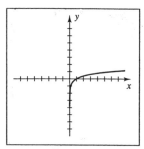

Figure 11

⑪ $f(x) = \log_6 x$ • goes through $(\frac{1}{6}, -1)$, $(1, 0)$, and $(6, 1)$

⑬ $f(x) = \log_4 (x^2) = 2\log_4 x$ • stretch $y = \log_4 x$ by a factor of 2 and include its

reflection through the y-axis since the domain of the original function is $\mathbb{R} - \{0\}$

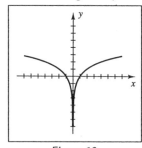

Figure 13

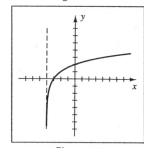

Figure 15

⑮ $f(x) = \log_2 (x + 4)$ • shift $y = \log_2 x$ left 4 units

⑰ (a) $\log_2 \frac{1}{16} = \log_2 2^{-4} = -4$ (b) $\log_\pi 1 = 0$ (c) $\ln e = 1$

(d) $6^{\log_6 4} = 4$ (e) $\log 1{,}000{,}000 = \log 10^6 = 6$

(f) $10^{3\log 2} = 10^{\log 2^3} = 2^3 = 8$ (g) $\log_4 2 = \log_4 4^{1/2} = \frac{1}{2}$

⑲ $2^{3x-1} = \frac{1}{2} \Rightarrow 2^{3x-1} = 2^{-1} \Rightarrow 3x - 1 = -1 \Rightarrow 3x = 0 \Rightarrow x = 0$

㉑ $\log \sqrt{x} = \log(x - 6) \Rightarrow \sqrt{x} = x - 6 \Rightarrow (\sqrt{x})^2 = (x-6)^2 \Rightarrow$

$x = x^2 - 12x + 36 \Rightarrow x^2 - 13x + 36 = 0 \Rightarrow (x-4)(x-9) = 0 \Rightarrow$

$$x = 4, 9; 4 \text{ is extraneous}$$

㉓ $\log_4 (x + 1) = 2 + \log_4 (3x - 2) \Rightarrow \log_4 (x+1) - \log_4 (3x - 2) = 2 \Rightarrow$

$\log_4\left(\frac{x+1}{3x-2}\right) = 2 \Rightarrow \frac{x+1}{3x-2} = 4^2 \Rightarrow x + 1 = 16(x - 2) \Rightarrow$

$$x + 1 = 48x - 32 \Rightarrow 33 = 47x \Rightarrow x = \frac{33}{47}$$

㉕ $\ln(x + 2) = \ln e^{\ln 2} - \ln x \Rightarrow \ln(x + 2) = \ln 2 - \ln x \Rightarrow$

$\ln(x + 2) + \ln x = \ln 2 \Rightarrow \ln[x(x+2)] = \ln 2 \Rightarrow x^2 + 2x = 2 \Rightarrow$

$x^2 + 2x - 2 = 0 \Rightarrow x = \dfrac{-2 \pm \sqrt{4 + 8}}{2} = \dfrac{-2 \pm \sqrt{12}}{2} = \dfrac{-2 \pm 2\sqrt{3}}{2} = -1 + \sqrt{3} \; \{x > 0\}$

27 $2^{5-x} = 6 \ \Rightarrow \ 5 - x = \log_2 6 \ \Rightarrow \ 5 - \log_2 6 = x \ \Rightarrow \ x = 5 - \dfrac{\log 6}{\log 2},$

$$\text{or equivalently,} \ \frac{5\log 2 - \log 6}{\log 2} = \frac{\log 2^5 - \log 6}{\log 2} = \frac{\log \frac{32}{6}}{\log 2} = \frac{\log \frac{16}{3}}{\log 2} \approx 2.415$$

29 $2^{5x+3} = 3^{2x+1} \ \Rightarrow \ \log\left(2^{5x+3}\right) = \log\left(3^{2x+1}\right) \ \Rightarrow$

$(5x + 3)\log 2 = (2x + 1)\log 3 \ \Rightarrow \ 5x\log 2 + 3\log 2 = 2x\log 3 + \log 3 \ \Rightarrow$

$5x\log 2 - 2x\log 3 = \log 3 - 3\log 2 \ \Rightarrow \ x(5\log 2 - 2\log 3) = \log 3 - \log 2^3 \ \Rightarrow$

$$x = \frac{\log 3 - \log 8}{\log 32 - \log 9} = \frac{\log \frac{3}{8}}{\log \frac{32}{9}} \approx -0.773$$

31 $\log_4 x = \sqrt[3]{\log_4 x} \ \Rightarrow \ \log_4 x = \left(\log_4 x\right)^{1/3} \ \Rightarrow \ \left(\log_4 x\right)^3 = \log_4 x \ \Rightarrow$

$\left(\log_4 x\right)^3 - \log_4 x = 0 \ \Rightarrow \ \log_4 x\left[\left(\log_4 x\right)^2 - 1\right] = 0 \ \Rightarrow$

$$\log_4 x = 0 \ \text{or} \ \log_4 x = \pm 1 \ \Rightarrow \ x = 1 \ \text{or} \ x = 4, \tfrac{1}{4}$$

33 $10^{2\log x} = 5 \ \Rightarrow \ 10^{\log x^2} = 5 \ \Rightarrow \ x^2 = 5 \ \Rightarrow \ x = \pm\sqrt{5}; \ -\sqrt{5} \ \text{is extraneous}$

35 $x^2\left(-2xe^{-x^2}\right) + 2xe^{-x^2} = 0 \ \Rightarrow \ 2xe^{-x^2}\left(-x^2 + 1\right) = 0 \ \Rightarrow \ x = 0, \ \pm 1 \left\{ e^{-x^2} \neq 0 \right\}$

37 (a) $\log x^2 = \log(6 - x) \ \Rightarrow \ x^2 = 6 - x \ \Rightarrow \ x^2 + x - 6 = 0 \ \Rightarrow$

$$(x + 3)(x - 2) = 0 \ \Rightarrow \ x = -3, \ 2$$

(b) $2\log x = \log(6 - x) \ \Rightarrow \ \log x^2 = \log(6 - x)$, which is the equation in part (a).

This equation has the same solutions provided they are in the domain.

$$\text{But} -3 \ \text{is extraneous, so 2 is the only solution.}$$

38 (a) $\ln\left(e^x\right)^2 = 16 \ \Rightarrow \ 2\ln e^x = 16 \ \Rightarrow \ 2x = 16 \ \Rightarrow \ x = 8$

(b) $\ln e^{\left(x^2\right)} = 16 \ \Rightarrow \ x^2 = 16 \ \Rightarrow \ x = \pm 4$

39 $\log x^4 \sqrt[3]{y^2/z} = \log\left(x^4 y^{2/3} z^{-1/3}\right) = \log x^4 + \log y^{2/3} + \log z^{-1/3}$

$$= 4\log x + \tfrac{2}{3}\log y - \tfrac{1}{3}\log z$$

40 $\log(x^2/y^3) + 4\log y - 6\log\sqrt{xy} = \log\left(\dfrac{x^2}{y^3}\right) + \log y^4 - \log\left[(xy)^{1/2}\right]^6$

$$= \log\left(\frac{x^2 \cdot y^4}{y^3}\right) - \log\left(x^3 y^3\right)$$

$$= \log\left(\frac{x^2 y}{x^3 y^3}\right) = \log\left(\frac{1}{xy^2}\right)$$

$$= \log 1 - \log\left(xy^2\right)$$

$$= 0 - \log\left(xy^2\right)$$

$$= -\log\left(xy^2\right)$$

41 $y = \dfrac{1}{10^x + 10^{-x}} \left\{ D = \mathbb{R},\ R = (0, \tfrac{1}{2}] \right\} \;\Rightarrow\; y\,10^x + y\,10^{-x} = 1 \;\Rightarrow$

$y\,10^{2x} + y = 10^x \;\Rightarrow\; y\,10^{2x} - 10^x + y = 0 \;\Rightarrow\; 10^x = \dfrac{1 \pm \sqrt{1 - 4y^2}}{2y} \;\Rightarrow$

$$x = \log\!\left(\dfrac{1 \pm \sqrt{1 - 4y^2}}{2y} \right)$$

43 (a) $x = \ln 6.6 \approx 1.89$ (b) $\log x = 1.8938 \;\Rightarrow\; x = 10^{1.8938} \approx 78.3$

(c) $\ln x = -0.75 \;\Rightarrow\; x = e^{-0.75} \approx 0.472$

45 (a) For $y = \log_2 (x + 1)$, $D = (-1, \infty)$ and $R = \mathbb{R}$.

(b) $y = \log_2 (x + 1) \;\Rightarrow\; x = \log_2 (y + 1) \;\Rightarrow\; 2^x = y + 1 \;\Rightarrow\; y = 2^x - 1$,

$$D = \mathbb{R},\ R = (-1, \infty)$$

47 (a) $Q(t) = 2(3^t) \;\Rightarrow\; Q(0) = 2(3^0) = 2$ { in thousands }, or 2000

(b) $Q(\tfrac{10}{60}) = 2000(3^{1/6}) \approx 2401$; $Q(\tfrac{30}{60}) = 2000(3^{1/2}) \approx 3464$; $Q(1) = 2(3) = 6000$

49 (a) $N = 64(0.5)^{t/8} = 64\!\left[(0.5)^{1/8}\right]^t \approx 64(0.917)^t$

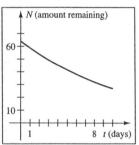

(b) $N = \tfrac{1}{2}N_0 \;\Rightarrow\; \tfrac{1}{2}N_0 = N_0(0.5)^{t/8} \;\Rightarrow$

$(\tfrac{1}{2})^1 = (\tfrac{1}{2})^{t/8} \;\Rightarrow\; 1 = t/8 \;\Rightarrow\; t = 8$ days

Figure 49

51 (a) Using $A = Pe^{rt}$ with $A = \$35{,}000$, $P = \$10{,}000$, and $r = 11\%$, we have

$35{,}000 = 10{,}000e^{0.11t} \;\Rightarrow\; e^{0.11t} = 3.5 \;\Rightarrow\; 0.11t = \ln 3.5 \;\Rightarrow$

$$t = \tfrac{1}{0.11}\ln 3.5 \approx 11.39 \text{ yr.}$$

(b) $2 \cdot 10{,}000 = 10{,}000e^{0.11t} \;\Rightarrow\; e^{0.11t} = 2 \;\Rightarrow\; 0.11t = \ln 2 \;\Rightarrow\; t \approx 6.30$ yr

Alternatively, using the doubling time formula, $t = (\ln 2)/r = (\ln 2)/0.11 \approx 6.30$.

53 (a) $\alpha = 10\log\!\left(\dfrac{I}{I_0}\right) \;\Rightarrow\; \dfrac{\alpha}{10} = \log\!\left(\dfrac{I}{I_0}\right) \;\Rightarrow\; 10^{\alpha/10} = \dfrac{I}{I_0} \;\Rightarrow\; I = I_0 10^{\alpha/10}$

(b) Let $I(\alpha)$ be the intensity corresponding to α decibels.

$I(\alpha + 1) = I_0 10^{(\alpha + 1)/10} = I_0 10^{\alpha/10} 10^{1/10} = I(\alpha)\,10^{1/10} \approx 1.26\,I(\alpha)$,

which represents a 26% increase in $I(\alpha)$.

55 $R = 2.3 \log (A + 3000) - 5.1 \Rightarrow R + 5.1 = 2.3 \log (A + 3000) \Rightarrow$

$\dfrac{R + 5.1}{2.3} = \log (A + 3000) \Rightarrow$

$A + 3000 = 10^{(R + 5.1)/2.3}$ { change to exponential form } $\Rightarrow$

$$A = 10^{(R + 5.1)/2.3} - 3000$$

57 $R = 4 \Rightarrow 2.3 \log (A + 14{,}000) - 6.6 = 4 \Rightarrow \log (A + 14{,}000) = \frac{10.6}{2.3} \Rightarrow$

$$A = 10^{106/23} - 14{,}000 \approx 26{,}615.9 \text{ mi}^2.$$

59 Substituting $v = 0$ and $m = m_1 + m_2$ in $v = -a \ln m + b$ yields

$0 = -a \ln (m_1 + m_2) + b$. Thus, $b = a \ln (m_1 + m_2)$. At burnout, $m = m_1$,

and hence, $v = -a \ln m_1 + b = -a \ln m_1 + a \ln (m_1 + m_2)$

$$= a[\ln (m_1 + m_2) - \ln m_1] \Rightarrow v = a \ln \left(\dfrac{m_1 + m_2}{m_1} \right)$$

61 (a) $\log E = 11.4 + (1.5)R \Rightarrow E = 10^{11.4 + 1.5R}$ { merely change the form }

(b) $R = 8.4 \Rightarrow E = 10^{11.4 + 1.5(8.4)} = 10^{24}$ ergs

63 $h = 70.228 + 5.104t + 9.222 \ln t$ and $R = 5.104 + (9.222/t)$.

$$t = 2 \Rightarrow h \approx 86.8 \text{ cm and } R = 9.715 \text{ cm/yr.}$$

65 (a) $x = 4\% \Rightarrow T = -8310 \ln (0.04) \approx 26{,}749$ yr.

(b) $T = 10{,}000 \Rightarrow 10{,}000 = -8310 \ln x \Rightarrow -\dfrac{10{,}000}{8310} = \ln x \Rightarrow$

$$x = e^{-1000/831} \approx 0.30, \text{ or } 30\%.$$

67 $N(t) = \frac{1}{2}N_0 \Rightarrow \frac{1}{2}N_0 = N_0(0.805)^t \Rightarrow 2^{-1} = (0.805)^t \Rightarrow$

$\ln (2^{-1}) = \ln (0.805)^t \Rightarrow -\ln 2 = t \ln (0.805) \Rightarrow$

$$t = -\dfrac{\ln 2}{\ln (0.805)} \approx 3.196 \text{ millennia, or } 3196 \text{ yr}$$

Chapter 4 Discussion Exercises

1 (a) The y-intercept is a, so it increases as a does. The graph flattens out as a increases.

(b) Graph $Y_1 = \frac{a}{2}(e^{x/a} + e^{-x/a}) + (30 - a)$ on $[-20, 20]$ by $[30, 32]$ and check for $Y_1 < 32$ at $x = 20$. In this case it turns out that $a = 101$ is the smallest integer value that satisfies the conditions, so an equation is

$$y = \tfrac{101}{2}(e^{x/101} + e^{-x/101}) - 71.$$

$\boxed{3}$ (a) $x^y = y^x \Rightarrow \ln(x^y) = \ln(y^x) \Rightarrow y \ln x = x \ln y \Rightarrow \dfrac{\ln x}{x} = \dfrac{\ln y}{y}$

(b) Once you find two values of $(\ln x)/x$ that are the same (such as 0.36652), you know that the corresponding x-values, x_1 and x_2, satisfy the relationship $(x_1)^{x_2} \approx (x_2)^{x_1}$. In particular, when $(\ln x)/x \approx 0.36652$, we find that $x_1 \approx 2.50$ and $x_2 \approx 2.97$. Note that $2.50^{2.97} \approx 2.97^{2.50} \approx 15.20$.

(c) Note that $f(e) = \dfrac{1}{e}$. Any horizontal line $y = k$, with $0 < k < \dfrac{1}{e}$, will intersect the graph at the two points $\left(x_1, \dfrac{\ln x_1}{x_1} \right)$ and $\left(x_2, \dfrac{\ln x_2}{x_2} \right)$, where $1 < x_1 < e$ and $x_2 > e$.

$\boxed{5}$ Logarithm law 3 states that it is valid only for positive real numbers, so $y = \log_3(x^2)$ is equivalent to $y = 2\log_3 x$ only for $x > 0$. The domain of $y = \log_3(x^2)$ is $\mathbb{R} - \{0\}$, whereas the domain of $y = 2\log_3 x$ is $x > 0$.

$\boxed{7}$ There are 4 points of intersection. Listed in order of difficulty to find we have: $(-0.9999011, 0.00999001)$, $(-0.0001, 0.01)$, $(100, 0.01105111)$, and $(36{,}102.844, 4.6928 \times 10^{13})$. Exponential function values (with base > 1) are greater than polynomial function values (with leading term positive) for very large values of x.

$\boxed{9}$ $60{,}000 = 40{,}000b^5 \Rightarrow b = \sqrt[5]{1.5} \approx 1.0844717712$, or 8.44717712%. Mentally—it would take $70/8.5 \approx 8^+$ years to double and there would be about $40/8 = 5$ doubling periods, $2^5 = 32$ and $32 \cdot \$40{,}000 = \$1{,}280{,}000$.

$$\text{Actual} = \$40{,}000(1 + 0.0844717712)^{40} = \$1{,}025{,}156.25.$$

$\boxed{11}$ On the TI-83 Plus, enter the sum of the days, $\{0, 5168, 6728, 8136, 8407, 8735, 8857, 9010, 9274, 9631, 9666\}$, and the averages, $\{1003.16, 2002.25, 3004.46, 4003.33, 5023.55, 6010.00, 7022.44, 8038.88, 9033.23, 10{,}006.78, 11{,}014.69\}$, in L_1 and L_2, respectively. Use ExpReg under STAT CALC to obtain $y = ab^x$, where $a = 753.76507865478$ and $b = 1.0002468662739$. Plot the data along with the exponential function and the line $y = 20{,}000$. The functions intersect at approximately $13{,}282$. This value corresponds to March 27, 2009. *Note:* The TI-83 has a convenient "day between dates" function for problems of this nature.

Using the first and last milestone figures and the continuously compounded interest formula gives us $A = Pe^{rt} \Leftrightarrow 11{,}014.69 = 1003.16e^{r(9666)} \Rightarrow r \approx 0.0002478869045$ (daily) or about 9.05% annually. *Note:* The Dow was 100.25 on $1/12/1906$. You may want to include this information and examine the differences it makes in any type of prediction.

A discussion of practical considerations should lead to mention of crashes, corrections, and the validity of any model over too long of a period of time.

13 (a) [−10, 110, 10] by [0, 1E10, 1E9]

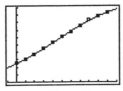

Figure 13

(b) Based on the shape of the curve in part (a), a logistic model is definitely more
appropriate.

(c) Using the logistic model feature, we obtain $y = c/(1 + ae^{-bx})$ with

$$a = 3.567015872, \ b = 0.0281317818, \text{ and } c = 1.1355368 \times 10^{10}.$$

(d) As x gets large, the term ae^{-bx} gets closer to zero, so y approaches $c/1 = c$.

Chapter 5: The Trigonometric Functions

Note: Exercises 1 and 3: The answers listed are the smallest (in magnitude) two positive coterminal angles and two negative coterminal angles.

1 (a) $120° + 1(360°) = 480°,$ $\qquad 120° + 2(360°) = 840°;$

$\qquad 120° - 1(360°) = -240°,$ $\qquad 120° - 2(360°) = -600°$

(b) $135° + 1(360°) = 495°,$ $\qquad 135° + 2(360°) = 855°;$

$\qquad 135° - 1(360°) = -225°,$ $\qquad 135° - 2(360°) = -585°$

(c) $-30° + 1(360°) = 330°,$ $\qquad -30° + 2(360°) = 690°;$

$\qquad -30° - 1(360°) = -390°,$ $\qquad -30° - 2(360°) = -750°$

3 (a) $620° - 1(360°) = 260°,$ $\qquad 620° + 1(360°) = 980°;$

$\qquad 620° - 2(360°) = -100°,$ $\qquad 620° - 3(360°) = -460°$

(b) $\frac{5\pi}{6} + 1(2\pi) = \frac{5\pi}{6} + \frac{12\pi}{6} = \frac{17\pi}{6},$ $\qquad \frac{5\pi}{6} + 2(2\pi) = \frac{5\pi}{6} + \frac{24\pi}{6} = \frac{29\pi}{6};$

$\qquad \frac{5\pi}{6} - 1(2\pi) = \frac{5\pi}{6} - \frac{12\pi}{6} = -\frac{7\pi}{6},$ $\qquad \frac{5\pi}{6} - 2(2\pi) = \frac{5\pi}{6} - \frac{24\pi}{6} = -\frac{19\pi}{6}$

(c) $-\frac{\pi}{4} + 1(2\pi) = -\frac{\pi}{4} + \frac{8\pi}{4} = \frac{7\pi}{4},$ $\qquad -\frac{\pi}{4} + 2(2\pi) = -\frac{\pi}{4} + \frac{16\pi}{4} = \frac{15\pi}{4};$

$\qquad -\frac{\pi}{4} - 1(2\pi) = -\frac{\pi}{4} - \frac{8\pi}{4} = -\frac{9\pi}{4},$ $\qquad -\frac{\pi}{4} - 2(2\pi) = -\frac{\pi}{4} - \frac{16\pi}{4} = -\frac{17\pi}{4}$

5 (a) $90° - 5°17'34'' = 84°42'26''$ $\qquad$ (b) $90° - 32.5° = 57.5°$

7 (a) $180° - 48°51'37'' = 131°8'23''$ $\qquad$ (b) $180° - 136.42° = 43.58°$

Note: Multiply each degree measure by $\frac{\pi}{180}$ to obtain the listed radian measure.

9 (a) $150° \cdot \frac{\pi}{180} = \frac{5 \cdot 30\pi}{6 \cdot 30} = \frac{5\pi}{6}$ $\qquad\qquad$ (b) $-60° \cdot \frac{\pi}{180} = -\frac{60\pi}{3 \cdot 60} = -\frac{\pi}{3}$

(c) $225° \cdot \frac{\pi}{180} = \frac{5 \cdot 45\pi}{4 \cdot 45} = \frac{5\pi}{4}$

11 (a) $450° \cdot \frac{\pi}{180} = \frac{5 \cdot 90\pi}{2 \cdot 90} = \frac{5\pi}{2}$ $\qquad\qquad$ (b) $72° \cdot \frac{\pi}{180} = \frac{2 \cdot 36\pi}{5 \cdot 36} = \frac{2\pi}{5}$

(c) $100° \cdot \frac{\pi}{180} = \frac{5 \cdot 20\pi}{9 \cdot 20} = \frac{5\pi}{9}$

Note: Multiply each radian measure by $\frac{180}{\pi}$ to obtain the listed degree measure.

13 (a) $\frac{2\pi}{3} \cdot \left(\frac{180}{\pi}\right)° = \left(\frac{2 \cdot 3 \cdot 60\pi}{3\pi}\right)° = 120°$ $\qquad$ (b) $\frac{11\pi}{6} \cdot \left(\frac{180}{\pi}\right)° = \left(\frac{11 \cdot 30 \cdot 6\pi}{6\pi}\right)° = 330°$

(c) $\frac{3\pi}{4} \cdot \left(\frac{180}{\pi}\right)° = \left(\frac{3 \cdot 45 \cdot 4\pi}{4\pi}\right)° = 135°$

15 (a) $-\frac{7\pi}{2} \cdot \left(\frac{180}{\pi}\right)° = -\left(\frac{7 \cdot 90 \cdot 2\pi}{2\pi}\right)° = -630°$ $\qquad$ (b) $7\pi \cdot \left(\frac{180}{\pi}\right)° = (7 \cdot 180)° = 1260°$

(c) $\frac{\pi}{9} \cdot \left(\frac{180}{\pi}\right)° = \left(\frac{20 \cdot 9\pi}{9\pi}\right)° = 20°$

17 We first convert 2 radians to degrees. $2 \cdot \left(\frac{180}{\pi}\right)° \approx 114.59156° = 114° + 0.59156°.$

We now use the decimal portion, $0.59156°$, and convert it to minutes.

Since $60' = 1°$, we have $0.59156° = 0.59156 \,(60') = 35.4936'.$

Using the decimal portion, $0.4936'$, we convert it to seconds.

Since $60'' = 1'$, we have $0.4936' = 0.4936 \,(60'') \approx 30''.$ $\qquad$ ∴ 2 radians $\approx 114°35'30''$

19 $5 \cdot \left(\frac{180}{\pi}\right)^{\circ} \approx 286.4789^{\circ};\ 0.4789\,(60') = 28.734';\ 0.734\,(60'') \approx 44''.$

$$\therefore\ 5 \text{ radians} \approx 286^{\circ}28'44''$$

21 Since $1' = \left(\frac{1}{60}\right)^{\circ}$, $41' = \left(\frac{41}{60}\right)^{\circ}$. Thus, $37^{\circ}41' = \left(37 + \frac{41}{60}\right)^{\circ} \approx 37.6833^{\circ}.$

23 Since $1'' = \left(\frac{1}{3600}\right)^{\circ}$, $27'' = \left(\frac{27}{3600}\right)^{\circ}$. Thus, $115^{\circ}26'27'' = \left(115 + \frac{26}{60} + \frac{27}{3600}\right)^{\circ} \approx 115.4408^{\circ}.$

25 We have 63° and a portion of one more degree. Since $1^{\circ} = 60'$,

$0.169^{\circ} = 0.169\,(60') = 10.14'$. We now have $10'$ and a portion of one more minute.

Since $1' = 60''$, $0.14' = 0.14\,(60'') = 8.4''$. $\therefore\ 63.169^{\circ} \approx 63^{\circ}10'8''$

27 $0.6215\,(60') = 37.29';\ 0.29\,(60'') = 17.4'';\ \therefore\ 310.6215^{\circ} \approx 310^{\circ}37'17''$

29 We will use the formula for the length of a circular arc.

$$s = r\theta\ \Rightarrow\ r = \frac{s}{\theta} = \frac{10}{4} = 2.5 \text{ cm.}$$

31 (a) $s = r\theta = 8 \cdot \left(45 \cdot \frac{\pi}{180}\right) = 8 \cdot \frac{\pi}{4} = 2\pi \approx 6.28$ cm

　　(b) $A = \frac{1}{2}r^2\theta = \frac{1}{2}(8)^2\left(\frac{\pi}{4}\right) = 8\pi \approx 25.13$ cm^2

33 (a) Remember that θ is measured in radians. $s = r\theta\ \Rightarrow\ \theta = \frac{s}{r} = \frac{7}{4} = 1.75$ radians.

　　　　Converting to degrees, we have $\frac{7}{4} \cdot \left(\frac{180}{\pi}\right)^{\circ} = \left(\frac{315}{\pi}\right)^{\circ} \approx 100.27^{\circ}.$

　　(b) $A = \frac{1}{2}r^2\theta = \frac{1}{2}(4)^2\left(\frac{7}{4}\right) = 14$ cm^2

35 (a) A measure of 50° is equivalent to $\left(50 \cdot \frac{\pi}{180}\right)$ radians. The radius is one-half of the

　　　　diameter. Thus, $s = r\theta = \left(\frac{1}{2} \cdot 16\right)\left(50 \cdot \frac{\pi}{180}\right) = 8 \cdot \frac{5\pi}{18} = \frac{20\pi}{9} \approx 6.98$ m.

　　(b) $A = \frac{1}{2}r^2\theta = \frac{1}{2}(8)^2\left(\frac{5\pi}{18}\right) = \frac{80\pi}{9} \approx 27.93$ m^2

37 radius $= \frac{1}{2} \cdot 8000$ miles $= 4000$ miles

　　(a) $s = r\theta = 4000\left(60 \cdot \frac{\pi}{180}\right) = \frac{4000\pi}{3} \approx 4189$ miles

　　(b) $s = r\theta = 4000\left(45 \cdot \frac{\pi}{180}\right) = 1000\pi \approx 3142$ miles

　　(c) $s = r\theta = 4000\left(30 \cdot \frac{\pi}{180}\right) = \frac{2000\pi}{3} \approx 2094$ miles

　　(d) $s = r\theta = 4000\left(10 \cdot \frac{\pi}{180}\right) = \frac{2000\pi}{9} \approx 698$ miles

　　(e) $s = r\theta = 4000\left(1 \cdot \frac{\pi}{180}\right) = \frac{200\pi}{9} \approx 70$ miles

39 $\theta = \frac{s}{r} = \frac{500}{4000} = \frac{1}{8}$ radian; $\left(\frac{1}{8}\right)\left(\frac{180}{\pi}\right)^{\circ} = \left(\frac{45}{2\pi}\right)^{\circ} \approx 7^{\circ}10'$

41 23 hours, 56 minutes, and 4 seconds $= 23(60)^2 + 56(60) + 4 = 86,164$ sec.

Since the earth turns through 2π radians in $86,164$ seconds,

it rotates through $\frac{2\pi}{86,164} \approx 7.29 \times 10^{-5}$ radians in one second.

43 (a) $\left(40\ \dfrac{\text{revolutions}}{\text{minute}}\right)\left(2\pi\ \dfrac{\text{radians}}{\text{revolution}}\right) = 80\pi\ \dfrac{\text{radians}}{\text{minute}}.$

Note:　Remember to write out and "cancel" the units if you are unsure about

　　　　what units your answer is measured in.

(b) The distance that a point on the circumference travels is
$$s = r\theta = (5 \text{ in.}) \cdot 80\pi = 400\pi \text{ in.}$$

Hence, its linear speed is
$$400\pi \text{ in./min} = \frac{100\pi}{3} \text{ ft/min } \{400\pi \cdot \frac{1}{12}\} \approx 104.72 \text{ ft/min.}$$

$\boxed{45}$ (a) $200 \text{ rpm} = 200 \frac{\text{rev}}{\text{min}} \cdot \frac{2\pi \text{ rad}}{1 \text{ rev}} = 400\pi \frac{\text{rad}}{\text{min}}$

(b) $s = r\theta = (5.7)(400\pi) = 2280\pi$ cm. Linear speed $= 2280\pi$ cm/min.

Divide by 60 to obtain cm/sec: $\frac{2280\pi}{60} = 38\pi \approx 119$ cm/sec

(c) Let x be the desired angular speed. The linear speed at 5.7 cm must equal the linear speed at 3 cm, so $2280\pi = 3x \Rightarrow x = 760\pi$ rad/min. Divide by 2π to obtain revolutions per minute: $\frac{760\pi}{2\pi} = 380$ rpm

(d) One approach is to simply return to part (c), replace 3 with r, and divide by 2π to create the function. $2280\pi = rx \Rightarrow x = \frac{2280\pi}{r}$, so $S(r) = \frac{2280\pi}{r(2\pi)} = \frac{1140}{r}$. As the radius increases, the drive motor speed decreases. Thus, S and r vary *inversely*.

$\boxed{47}$ (a) The distance that the cargo is lifted is equal to the arc length that the cable is moved through. $s = r\theta = (\frac{1}{2} \cdot 3)(\frac{7\pi}{4}) = \frac{21\pi}{8} \approx 8.25$ ft.

(b) $s = r\theta \Rightarrow d = (\frac{1}{2} \cdot 3)\theta \Rightarrow \theta = (\frac{2}{3}d)$ radians. For example, to lift the cargo 6 feet, the winch must rotate $\frac{2}{3} \cdot 6 = 4$ radians, or about 229°.

$\boxed{49}$ $\text{Area}_{\text{small}} = \frac{1}{2}r^2\theta = \frac{1}{2}(\frac{1}{2} \cdot 18)^2 \cdot (\frac{2\pi}{6}) = \frac{27\pi}{2}$. $\text{Area}_{\text{large}} = \frac{1}{2}(\frac{1}{2} \cdot 26)^2 \cdot (\frac{2\pi}{8}) = \frac{169\pi}{8}$.

$\text{Cost}_{\text{small}} = \text{Area}_{\text{small}} \div \text{Cost} = \frac{27\pi}{2} \div 2 \approx 21.21 \text{ in}^2/\text{dollar.}$

$\text{Cost}_{\text{large}} = \text{Area}_{\text{large}} \div \text{Cost} = \frac{169\pi}{8} \div 3 \approx 22.12 \text{ in}^2/\text{dollar.}$

The large slice provides slightly more pizza per dollar.

$\boxed{51}$ $\frac{40 \text{ miles}}{\text{hour}} = \frac{40 \text{ miles}}{\text{hour}} \cdot \frac{1 \text{ hour}}{3600 \text{ seconds}} \cdot \frac{5280 \text{ feet}}{\text{mile}} \cdot \frac{12 \text{ inches}}{\text{foot}} = \frac{704 \text{ inches}}{\text{second}}.$

The circumference of the wheel is $2\pi(14)$ inches. The back sprocket then rotates

$\frac{704 \text{ inches}}{\text{second}} \cdot \frac{1 \text{ revolution}}{28\pi \text{ inches}} = \frac{704 \text{ revolutions}}{28\pi \text{ second}}$ or $\frac{704}{28\pi} \cdot 2\pi = \frac{352}{7}$ radians per second.

The front sprocket's angular speed is given by { from Exercise 50 }

$$\theta_1 = \frac{r_2\theta_2}{r_1} = \frac{2 \cdot \frac{352}{7}}{5} = \frac{704}{35} \approx 20.114 \text{ radians per second}$$

or 3.2 revolutions per second or 192.08 revolutions per minute.

5.2 Exercises

Note: Answers are in the order *sin, cos, tan, cot, sec, csc* for any exercises that require the values of the six trigonometric functions.

1. Using the definition of the trigonometric functions,

$$\sin\theta = \frac{\text{opp}}{\text{hyp}} = \frac{4}{5}, \quad \cos\theta = \frac{\text{adj}}{\text{hyp}} = \frac{3}{5}, \quad \text{and} \quad \tan\theta = \frac{\text{opp}}{\text{adj}} = \frac{4}{3}.$$

We now use the reciprocal identities to find the values of the other trigonometric functions:

$$\cot\theta = \frac{1}{\tan\theta} = \frac{3}{4}, \quad \sec\theta = \frac{1}{\cos\theta} = \frac{5}{3}, \quad \text{and} \quad \csc\theta = \frac{1}{\sin\theta} = \frac{5}{4}.$$

3. Using the Pythagorean theorem, $(\text{adj})^2 + (\text{opp})^2 = (\text{hyp})^2 \Rightarrow$

$$(\text{adj})^2 = (\text{hyp})^2 - (\text{opp})^2 \Rightarrow \text{adj} = \sqrt{(\text{hyp})^2 - (\text{opp})^2} = \sqrt{5^2 - 2^2} = \sqrt{21}.$$

$$\star \ \frac{2}{5}, \frac{\sqrt{21}}{5}, \frac{2}{\sqrt{21}}, \frac{\sqrt{21}}{2}, \frac{5}{\sqrt{21}}, \frac{5}{2}$$

5. Using the Pythagorean theorem, $\text{hyp} = \sqrt{(\text{adj})^2 + (\text{opp})^2} = \sqrt{b^2 + a^2} = \sqrt{a^2 + b^2}.$

$$\star \ \frac{a}{\sqrt{a^2 + b^2}}, \frac{b}{\sqrt{a^2 + b^2}}, \frac{a}{b}, \frac{b}{a}, \frac{\sqrt{a^2 + b^2}}{b}, \frac{\sqrt{a^2 + b^2}}{a}$$

7. Using the Pythagorean theorem, $(\text{adj})^2 + (\text{opp})^2 = (\text{hyp})^2 \Rightarrow$

$$(\text{adj})^2 = (\text{hyp})^2 - (\text{opp})^2 \Rightarrow \text{adj} = \sqrt{(\text{hyp})^2 - (\text{opp})^2} = \sqrt{c^2 - b^2}.$$

$$\star \ \frac{b}{c}, \frac{\sqrt{c^2 - b^2}}{c}, \frac{b}{\sqrt{c^2 - b^2}}, \frac{\sqrt{c^2 - b^2}}{b}, \frac{c}{\sqrt{c^2 - b^2}}, \frac{c}{b}$$

9. Since we want to find the value of the hypotenuse x, we need to use a trigonometric function which relates x to two given parts of the triangle—in this case, the angle $30°$ and the opposite side of length 4. The sine function relates the opposite side and the hypotenuse. Hence, $\sin 30° = \frac{4}{x} \Rightarrow \frac{1}{2} = \frac{4}{x} \Rightarrow x = 8$. The tangent function relates the opposite and the adjacent side. Hence, $\tan 30° = \frac{4}{y} \Rightarrow \frac{\sqrt{3}}{3} = \frac{4}{y} \Rightarrow y = 4\sqrt{3}.$

11. The easy way to find the values of x and y is to remember that the lengths of the legs in a $45°$–$45°$–$90°$ triangle must be the same; so $y = 7$. Also, the hypotenuse is $\sqrt{2}$ times the length of either leg; that is, $x = 7\sqrt{2}$. We can also use the trigonometric relationships as follows:

$$\sin 45° = \frac{7}{x} \Rightarrow \frac{\sqrt{2}}{2} = \frac{7}{x} \Rightarrow x = 7\sqrt{2}; \ \tan 45° = \frac{7}{y} \Rightarrow 1 = \frac{7}{y} \Rightarrow y = 7$$

13. $\sin 60° = \frac{x}{8} \Rightarrow \frac{\sqrt{3}}{2} = \frac{x}{8} \Rightarrow x = 4\sqrt{3}$ and $\cos 60° = \frac{y}{8} \Rightarrow \frac{1}{2} = \frac{y}{8} \Rightarrow y = 4.$

Note: It may help to sketch a triangle as shown for Exercises 15 and 19.

Use the Pythagorean theorem to find the remaining side.

$\boxed{15}$ $(\text{adj})^2 + (\text{opp})^2 = (\text{hyp})^2$ $\Rightarrow$ $(\text{adj})^2 + 3^2 = 5^2$ $\Rightarrow$ $\text{adj} = \sqrt{25-9} = 4.$

$\star\ \dfrac{3}{5}, \dfrac{4}{5}, \dfrac{3}{4}, \dfrac{4}{3}, \dfrac{5}{4}, \dfrac{5}{3}$

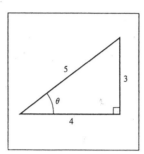

Figure 15

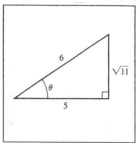

Figure 19

$\boxed{17}$ $12^2 + 5^2 = (\text{hyp})^2$ $\Rightarrow$ $\text{hyp} = \sqrt{144+25} = 13.$

$\star\ \dfrac{5}{13}, \dfrac{12}{13}, \dfrac{5}{12}, \dfrac{12}{5}, \dfrac{13}{12}, \dfrac{13}{5}$

$\boxed{19}$ $5^2 + (\text{opp})^2 = 6^2$ $\Rightarrow$ $\text{opp} = \sqrt{36-25} = \sqrt{11}.$

$\star\ \dfrac{\sqrt{11}}{6}, \dfrac{5}{6}, \dfrac{\sqrt{11}}{5}, \dfrac{5}{\sqrt{11}}, \dfrac{6}{5}, \dfrac{6}{\sqrt{11}}$

$\boxed{21}$ Let h denote the height of the tree.

$$\tan\theta = \frac{\text{opp}}{\text{adj}} \ \Rightarrow \ \tan 60° = \frac{h}{200} \ \Rightarrow \ h = 200\tan 60° = 200\sqrt{3} \approx 346.4 \text{ ft.}$$

$\boxed{23}$ Let d be the distance that the stone was moved. The side opposite $9°$ is 30 feet and

the hypotenuse is d. Thus, $\sin 9° = \dfrac{30}{d} \ \Rightarrow \ d = \dfrac{30}{\sin 9°} \approx 192 \text{ ft.}$

$\boxed{25}$ $\sin\theta = \dfrac{1.22\lambda}{D} \ \Rightarrow \ D = \dfrac{1.22\lambda}{\sin\theta} = \dfrac{1.22 \times 550 \times 10^{-9}}{\sin 0.00003769°} \approx 1.02 \text{ meters}$

$\boxed{27}$ *Note:* Be sure that your calculator is in degree mode.

(a) $\sin 42° \approx 0.6691$ (b) $\cos 77° \approx 0.2250$

(c) $\csc 123° = \dfrac{1}{\sin 123°} \approx 1.1924$ (d) $\sec(-190°) = \dfrac{1}{\cos(-190°)} \approx -1.0154$

$\boxed{29}$ *Note:* Be sure that your calculator is in radian mode.

(a) $\cot\dfrac{\pi}{13} = \dfrac{1}{\tan(\pi/13)} \approx 4.0572$ (b) $\csc 1.32 = \dfrac{1}{\sin 1.32} \approx 1.0323$

(c) $\cos(-8.54) \approx -0.6335$ (d) $\tan\dfrac{3\pi}{7} \approx 4.3813$

$\boxed{31}$ (a) Since $1 + \tan^2 4\beta = \sec^2 4\beta$, $\tan^2 4\beta - \sec^2 4\beta = -1.$

(b) $4\tan^2\beta - 4\sec^2\beta = 4(\tan^2\beta - \sec^2\beta) = 4(-1) = -4$

$\boxed{33}$ (a) $5\sin^2\theta + 5\cos^2\theta = 5(\sin^2\theta + \cos^2\theta) = 5(1) = 5$

(b) $5\sin^2(\theta/4) + 5\cos^2(\theta/4) = 5\left[\sin^2(\theta/4) + \cos^2(\theta/4)\right] = 5(1) = 5$

35 $\dfrac{\sin^3\theta + \cos^3\theta}{\sin\theta + \cos\theta} = \dfrac{(\sin\theta + \cos\theta)(\sin^2\theta - \sin\theta\cos\theta + \cos^2\theta)}{\sin\theta + \cos\theta}$ {factor, sum of cubes}

$\qquad\qquad\qquad = \sin^2\theta - \sin\theta\cos\theta + \cos^2\theta$ {cancel $(\sin\theta + \cos\theta)$}

$\qquad\qquad\qquad = (\sin^2\theta + \cos^2\theta) - \sin\theta\cos\theta$ {group terms}

$\qquad\qquad\qquad = 1 - \sin\theta\cos\theta$ {Pythagorean identity}

37 $\dfrac{2 - \tan\theta}{2\csc\theta - \sec\theta} = \dfrac{2 - \dfrac{\sin\theta}{\cos\theta}}{2 \cdot \dfrac{1}{\sin\theta} - \dfrac{1}{\cos\theta}}$ {tangent and reciprocal identities}

$\qquad\qquad\qquad = \dfrac{\dfrac{2\cos\theta - \sin\theta}{\cos\theta}}{\dfrac{2\cos\theta - \sin\theta}{\sin\theta\cos\theta}}$ {combine terms}

$\qquad\qquad\qquad = \dfrac{\dfrac{1}{1}}{\dfrac{1}{\sin\theta}}$ {cancel like terms} $= \sin\theta$

39 $\cot\theta = \dfrac{\cos\theta}{\sin\theta}$ {cotangent identity} $= \dfrac{\sqrt{1 - \sin^2\theta}}{\sin\theta}$ {$\sin^2\theta + \cos^2\theta = 1$}

41 $\sec\theta = \dfrac{1}{\cos\theta}$ {reciprocal identity} $= \dfrac{1}{\sqrt{1 - \sin^2\theta}}$

43 One solution is $\sin\theta = \sqrt{1 - \cos^2\theta} = \sqrt{1 - \dfrac{1}{\sec^2\theta}} = \dfrac{\sqrt{\sec^2\theta - 1}}{\sec\theta}$.

$\qquad$ Alternatively, $\sin\theta = \dfrac{\sin\theta/\cos\theta}{1/\cos\theta} = \dfrac{\tan\theta}{\sec\theta} = \dfrac{\sqrt{\sec^2\theta - 1}}{\sec\theta}$ {$1 + \tan^2\theta = \sec^2\theta$}.

45 $\cos\theta\sec\theta = \cos\theta\,(1/\cos\theta)$ {reciprocal identity} $= 1$

47 $\sin\theta\sec\theta = \sin\theta\,(1/\cos\theta) = \sin\theta/\cos\theta = \tan\theta$ {tangent identity}

49 $\dfrac{\csc\theta}{\sec\theta} = \dfrac{1/\sin\theta}{1/\cos\theta}$ {reciprocal identities} $= \dfrac{\cos\theta}{\sin\theta} = \cot\theta$ {cotangent identity}

51 $(1 + \cos 2\theta)(1 - \cos 2\theta) = 1 - \cos^2 2\theta = \sin^2 2\theta$ {Pythagorean identity}

53 $\cos^2\theta\,(\sec^2\theta - 1) = \cos^2\theta\,(\tan^2\theta)$ {Pythagorean identity}

$\qquad\qquad\qquad = \cos^2\theta \cdot \dfrac{\sin^2\theta}{\cos^2\theta}$ {tangent identity} $= \sin^2\theta$

55 $\dfrac{\sin(\theta/2)}{\csc(\theta/2)} + \dfrac{\cos(\theta/2)}{\sec(\theta/2)} = \dfrac{\sin(\theta/2)}{1/\sin(\theta/2)} + \dfrac{\cos(\theta/2)}{1/\cos(\theta/2)} = \sin^2(\theta/2) + \cos^2(\theta/2) = 1$

57 $(1 + \sin\theta)(1 - \sin\theta) = 1 - \sin^2\theta$ {multiply as a difference of squares} $= \cos^2\theta = \dfrac{1}{\sec^2\theta}$

59 $\sec\theta - \cos\theta = \dfrac{1}{\cos\theta} - \cos\theta = \dfrac{1 - \cos^2\theta}{\cos\theta} = \dfrac{\sin^2\theta}{\cos\theta} = \dfrac{\sin\theta}{\cos\theta} \cdot \sin\theta = \tan\theta\sin\theta$

$\boxed{61}$ $(\cot\theta + \csc\theta)(\tan\theta - \sin\theta)$

$$= \cot\theta\tan\theta - \cot\theta\sin\theta + \csc\theta\tan\theta - \csc\theta\sin\theta \quad \{\text{multiply binomials}\}$$

$$= \frac{1}{\tan\theta}\tan\theta - \frac{\cos\theta}{\sin\theta}\sin\theta + \frac{1}{\sin\theta}\frac{\sin\theta}{\cos\theta} - \frac{1}{\sin\theta}\sin\theta$$

$$\{\text{reciprocal, cotangent, and tangent identities}\}$$

$$= 1 - \cos\theta + \frac{1}{\cos\theta} - 1 \quad \{\text{cancel terms}\}$$

$$= -\cos\theta + \sec\theta = \sec\theta - \cos\theta$$

$\boxed{63}$ $\sec^2 3\theta \csc^2 3\theta = (1 + \tan^2 3\theta)(1 + \cot^2 3\theta) \quad \{\text{Pythagorean identities}\}$

$$= 1 + \tan^2 3\theta + \cot^2 3\theta + 1 \quad \{\text{multiply binomials}\}$$

$$= (1 + \tan^2 3\theta) + (\cot^2 3\theta + 1) \quad \{\text{group terms}\}$$

$$= \sec^2 3\theta + \csc^2 3\theta \quad \{\text{Pythagorean identities}\}$$

$\boxed{65}$ $\log\csc\theta = \log\left(\frac{1}{\sin\theta}\right) \quad \{\text{reciprocal identity}\}$

$$= \log 1 - \log\sin\theta \quad \{\text{property of logarithms}\}$$

$$= 0 - \log\sin\theta \quad \{\log 1 = 0\}$$

$$= -\log\sin\theta$$

$\boxed{67}$ For the point $P(x, y)$, we let r denote the distance from the origin to P. Applying the definition of trigonometric functions of any angle with $x = 4$, $y = -3$, and $r = \sqrt{x^2 + y^2} = \sqrt{4^2 + (-3)^2} = 5$, we obtain the following:

$$\sin\theta = \frac{y}{r} = \frac{-3}{5} = -\frac{3}{5} \qquad\qquad \cos\theta = \frac{x}{r} = \frac{4}{5}$$

$$\tan\theta = \frac{y}{x} = \frac{-3}{4} = -\frac{3}{4} \qquad\qquad \cot\theta = \frac{x}{y} = \frac{4}{-3} = -\frac{4}{3}$$

$$\sec\theta = \frac{r}{x} = \frac{5}{4} \qquad\qquad \csc\theta = \frac{r}{y} = \frac{5}{-3} = -\frac{5}{3}$$

$\boxed{69}$ $x = -2$ and $y = -5 \ \Rightarrow \ r = \sqrt{x^2 + y^2} = \sqrt{(-2)^2 + (-5)^2} = \sqrt{29}$.

$$\sin\theta = \frac{y}{r} = \frac{-5}{\sqrt{29}}, \text{ or } -\frac{5}{29}\sqrt{29} \qquad \cos\theta = \frac{x}{r} = \frac{-2}{\sqrt{29}}, \text{ or } -\frac{2}{29}\sqrt{29}$$

$$\tan\theta = \frac{y}{x} = \frac{-5}{-2} = \frac{5}{2} \qquad\qquad \cot\theta = \frac{x}{y} = \frac{-2}{-5} = \frac{2}{5}$$

$$\sec\theta = \frac{r}{x} = \frac{\sqrt{29}}{-2}, \text{ or } -\frac{1}{2}\sqrt{29} \qquad \csc\theta = \frac{r}{y} = \frac{\sqrt{29}}{-5} = -\frac{1}{5}\sqrt{29}$$

Note: In the following exercises, we will only find the values of x, y, and r. The above definitions $\{$in Exercises 67 and 69$\}$ can then be used to find the values of the trigonometric functions of θ. These values are listed in the usual order in the answer.

[71] Since the terminal side of θ is in QII, choose x to be negative.

If $x = -1$, then $y = 4$ and $(-1, 4)$ is a point on the terminal side of θ.

$x = -1$ and $y = 4 \Rightarrow r = \sqrt{(-1)^2 + 4^2} = \sqrt{17}$.

$$\star \ \frac{4}{\sqrt{17}}, \ -\frac{1}{\sqrt{17}}, \ -4, \ -\frac{1}{4}, \ -\sqrt{17}, \ \frac{\sqrt{17}}{4}$$

[73] Remember that $y = mx$ is an equation of a line that passes through the origin and has slope m. Thus, an equation of the line is $y = \frac{4}{3}x$.

If $x = 3$, then $y = 4$ and $(3, 4)$ is a point on the terminal side of θ.

$x = 3$ and $y = 4 \Rightarrow r = \sqrt{3^2 + 4^2} = 5$. $\star \ \frac{4}{5}, \frac{3}{5}, \frac{4}{3}, \frac{3}{4}, \frac{5}{3}, \frac{5}{4}$

[75] $2y - 7x + 2 = 0 \Leftrightarrow y = \frac{7}{2}x - 1$. Thus, the slope of the given line is $\frac{7}{2}$.

An equation of the line through the origin with that slope is $y = \frac{7}{2}x$.

If $x = -2$, then $y = -7$ and $(-2, -7)$ is a point on the terminal side of θ.

$x = -2$ and $y = -7 \Rightarrow r = \sqrt{(-2)^2 + (-7)^2} = \sqrt{53}$.

$$\star \ -\frac{7}{\sqrt{53}}, \ -\frac{2}{\sqrt{53}}, \ \frac{7}{2}, \frac{2}{7}, \ -\frac{\sqrt{53}}{2}, \ -\frac{\sqrt{53}}{7}$$

[77] *Note:* U denotes *undefined*.

(a) For $\theta = 90°$, choose $x = 0$ and $y = 1$. $r = 1$. $\star$ 1, 0, U, 0, U, 1

(b) For $\theta = 0°$, choose $x = 1$ and $y = 0$. $r = 1$. $\star$ 0, 1, 0, U, 1, U

(c) For $\theta = \frac{7\pi}{2}$, choose $x = 0$ and $y = -1$. $r = 1$. $\star$ -1, 0, U, 0, U, -1

(d) For $\theta = 3\pi$, choose $x = -1$ and $y = 0$. $r = 1$. $\star$ 0, -1, 0, U, -1, U

[79] (a) $\cos\theta > 0$ $\{x > 0\}$ implies that the terminal side of θ is in QI or QIV.

$\sin\theta < 0$ $\{y < 0\}$ implies that the terminal side of θ is in QIII or QIV.

Thus, θ must be in quadrant IV to satisfy both conditions.

(b) $\sin\theta < 0 \Rightarrow \theta$ is in QIII or QIV.

$\cot\theta > 0 \Rightarrow \theta$ is in QI or QIII. $\therefore \theta$ is in QIII.

(c) $\csc\theta > 0 \Rightarrow \theta$ is in QI or QII.

$\sec\theta < 0 \Rightarrow \theta$ is in QII or QIII. $\therefore \theta$ is in QII.

(d) $\sec\theta < 0 \Rightarrow \theta$ is in QII or QIII.

$\tan\theta > 0 \Rightarrow \theta$ is in QI or QIII. $\therefore \theta$ is in QIII.

Note: For Exercises 81–88, steps to determine 2 function values using only the fundamental identities are shown. The other 3 function values are just the reciprocals of those given and are listed in the answer.

$\boxed{81}$ $\tan\theta = -\frac{3}{4}$ and $\sin\theta > 0$ $\Rightarrow$ θ is in QII.

$1 + \tan^2\theta = \sec^2\theta$ $\Rightarrow$ $\sec\theta = \pm\sqrt{1 + \tan^2\theta}$ { Use the "$-$" since the secant is

negative in the second quadrant. } $= -\sqrt{1 + \tan^2\theta} = -\sqrt{1 + \frac{9}{16}} = -\frac{5}{4}$.

$\tan\theta = \frac{\sin\theta}{\cos\theta}$ $\Rightarrow$ $-\frac{3}{4} = \frac{\sin\theta}{-4/5}$ $\Rightarrow$ $\sin\theta = (-\frac{3}{4})(-\frac{4}{5}) = \frac{3}{5}$. $\quad$ ★ $\frac{3}{5}, -\frac{4}{5}, -\frac{3}{4}, -\frac{4}{3}, -\frac{5}{4}, \frac{5}{3}$

$\boxed{83}$ $\sin\theta = -\frac{5}{13}$ and $\sec\theta > 0$ $\Rightarrow$ θ is in QIV.

$\sin^2\theta + \cos^2\theta = 1$ $\Rightarrow$ $\cos^2\theta = 1 - \sin^2\theta$ $\Rightarrow$ $\cos\theta = \pm\sqrt{1 - \sin^2\theta} =$

{ Use the "$+$" since the cosine is positive in the fourth quadrant. } $= \sqrt{1 - \frac{25}{169}} = \frac{12}{13}$.

$\tan\theta = \frac{\sin\theta}{\cos\theta} = \frac{-5/13}{12/13} = -\frac{5}{12}$. $\quad$ ★ $-\frac{5}{13}, \frac{12}{13}, -\frac{5}{12}, -\frac{12}{5}, \frac{13}{12}, -\frac{13}{5}$

$\boxed{85}$ $\cos\theta = -\frac{1}{3}$ and $\sin\theta < 0$ $\Rightarrow$ θ is in QIII. $\sin\theta = -\sqrt{1 - \cos^2\theta} = -\sqrt{1 - \frac{1}{9}} = -\frac{\sqrt{8}}{3}$.

$\tan\theta = \frac{\sin\theta}{\cos\theta} = \frac{-\sqrt{8}/3}{-1/3} = \sqrt{8}$, or $2\sqrt{2}$. $\quad$ ★ $-\frac{\sqrt{8}}{3}, -\frac{1}{3}, \sqrt{8}, \frac{1}{\sqrt{8}}, -3, -\frac{3}{\sqrt{8}}$

$\boxed{87}$ $\sec\theta = -4$ and $\csc\theta > 0$ $\Rightarrow$ θ is in QII. $\tan\theta = -\sqrt{\sec^2\theta - 1} = -\sqrt{16 - 1} = -\sqrt{15}$.

$\tan\theta = \frac{\sin\theta}{\cos\theta}$ $\Rightarrow$ $-\sqrt{15} = \frac{\sin\theta}{-1/4}$ $\Rightarrow$ $\sin\theta = \frac{\sqrt{15}}{4}$.

★ $\frac{\sqrt{15}}{4}, -\frac{1}{4}, -\sqrt{15}, -\frac{1}{\sqrt{15}}, -4, \frac{4}{\sqrt{15}}$

$\boxed{89}$ $\sqrt{\sec^2\theta - 1} = \sqrt{\tan^2\theta}$ { Pythagorean identity }

$= |\tan\theta|$ $\left\{ \sqrt{x^2} = |x| \right\}$ $= -\tan\theta$ since $\tan\theta < 0$ if $\pi/2 < \theta < \pi$.

$\boxed{91}$ $\sqrt{1 + \tan^2\theta} = \sqrt{\sec^2\theta}$ { Pythagorean identity }

$= |\sec\theta|$ $\left\{ \sqrt{x^2} = |x| \right\}$ $= \sec\theta$ since $\sec\theta > 0$ if $3\pi/2 < \theta < 2\pi$.

$\boxed{93}$ $\sqrt{\sin^2(\theta/2)} = |\sin(\theta/2)|$ $\left\{ \sqrt{x^2} = |x| \right\}$

$= -\sin(\theta/2)$ since $\sin(\theta/2) < 0$ if $2\pi < \theta < 4\pi$ { $\pi < \theta/2 < 2\pi$ }.

5.3 Exercises

$\boxed{1}$ Using the definition of the trigonometric functions in terms of a unit circle with

$P(-\frac{15}{17}, \frac{8}{17})$, we have

$$\sin t = y = \frac{8}{17}, \qquad \cos t = x = -\frac{15}{17}, \qquad \tan t = \frac{y}{x} = \frac{8/17}{-15/17} = -\frac{8}{15},$$

$$\csc t = \frac{1}{y} = \frac{1}{8/17} = \frac{17}{8}, \quad \sec t = \frac{1}{x} = \frac{1}{-15/17} = -\frac{17}{15}, \quad \cot t = \frac{x}{y} = \frac{-15/17}{8/17} = -\frac{15}{8}.$$

$\boxed{3}$ As in Exercise 1, $P(\frac{24}{25}, -\frac{7}{25})$ gives us $-\frac{7}{25}, \frac{24}{25}, -\frac{7}{24}, -\frac{24}{7}, \frac{25}{24}, -\frac{25}{7}$.

[5] See *Figure 5.* $P(t) = (\frac{3}{5}, \frac{4}{5})$.

  (a) $(t + \pi)$ will be $\frac{1}{2}$ revolution from (t) in the counterclockwise direction.

                                          Hence, $P(t + \pi) = (-\frac{3}{5}, -\frac{4}{5})$.

  (b) $(t - \pi)$ will be $\frac{1}{2}$ revolution from (t) in the clockwise direction.

        Hence, $P(t - \pi) = (-\frac{3}{5}, -\frac{4}{5})$. Observe that $P(t + \pi) = P(t - \pi)$ for any t.

  (c) $(-t)$ is the angle measure in the opposite direction of (t).

                                          Hence, $P(-t) = (\frac{3}{5}, -\frac{4}{5})$.

  (d) $(-t - \pi)$ will be $\frac{1}{2}$ revolution from $(-t)$ in the clockwise direction.

                                  Hence, $P(-t - \pi) = (-\frac{3}{5}, \frac{4}{5})$.

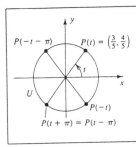

Figure 5

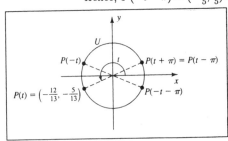

Figure 7

[7] See *Figure 7.*

  (a) As in Exercise 5, $P(t) = (-\frac{12}{13}, -\frac{5}{13}) \Rightarrow P(t + \pi) = (\frac{12}{13}, \frac{5}{13})$.

  (b) $P(t - \pi) = P(t + \pi) = (\frac{12}{13}, \frac{5}{13})$

  (c) $P(-t) = (-\frac{12}{13}, \frac{5}{13})$ (d) $P(-t - \pi) = (\frac{12}{13}, -\frac{5}{13})$

[9] (a) The point P on the unit circle U that corresponds to $t = 2\pi$ has coordinates $(1, 0)$. Thus, we choose $x = 1$ and $y = 0$ and use the definition of the trigonometric functions in terms of a unit circle.

    $\sin t = y \Rightarrow \sin 2\pi = 0.$ $\cos t = x \Rightarrow \cos 2\pi = 1.$

    $\tan t = \frac{y}{x} \Rightarrow \tan 2\pi = \frac{0}{1} = 0.$ $\cot t = \frac{x}{y} \Rightarrow \cot 2\pi = \frac{1}{0}$ is undefined.

    $\sec t = \frac{1}{x} \Rightarrow \sec 2\pi = \frac{1}{1} = 1.$ $\csc t = \frac{1}{y} \Rightarrow \csc 2\pi = \frac{1}{0}$ is undefined.

  (b) $t = -3\pi$ is coterminal with $t = \pi$. The point P on the unit circle U that corresponds to $t = -3\pi$ has coordinates $(-1, 0)$. Thus, we choose $x = -1$ and $y = 0$.

    $\sin t = y \Rightarrow \sin(-3\pi) = 0.$ $\cos t = x \Rightarrow \cos(-3\pi) = -1.$

    $\tan t = \frac{y}{x} \Rightarrow \tan(-3\pi) = \frac{0}{-1} = 0.$ $\cot t = \frac{x}{y} \Rightarrow \cot(-3\pi) = \frac{-1}{0}$ is und.

    $\sec t = \frac{1}{x} \Rightarrow \sec(-3\pi) = \frac{1}{-1} = -1.$ $\csc t = \frac{1}{y} \Rightarrow \csc(-3\pi) = \frac{1}{0}$ is und.

$\boxed{11}$ (a) The point P on the unit circle U that corresponds to $t = \frac{3\pi}{2}$ has coordinates

$(0, -1)$. Thus, we choose $x = 0$ and $y = -1$.

$$\sin t = y \;\Rightarrow\; \sin\frac{3\pi}{2} = -1. \qquad\qquad \cos t = x \;\Rightarrow\; \cos\frac{3\pi}{2} = 0.$$

$$\tan t = \frac{y}{x} \;\Rightarrow\; \tan\frac{3\pi}{2} = \frac{-1}{0} \text{ is und.} \qquad \cot t = \frac{x}{y} \;\Rightarrow\; \cot\frac{3\pi}{2} = \frac{0}{-1} = 0.$$

$$\sec t = \frac{1}{x} \;\Rightarrow\; \sec\frac{3\pi}{2} = \frac{1}{0} \text{ is und.} \qquad \csc t = \frac{1}{y} \;\Rightarrow\; \csc\frac{3\pi}{2} = \frac{1}{-1} = -1.$$

(b) $t = -\frac{7\pi}{2}$ is coterminal with $t = \frac{\pi}{2}$. The point P on the unit circle U that

corresponds to $t = -\frac{7\pi}{2}$ has coordinates $(0, 1)$. Thus, we choose $x = 0$ and $y = 1$.

$$\sin t = y \;\Rightarrow\; \sin\left(-\frac{7\pi}{2}\right) = 1. \qquad\qquad \cos t = x \;\Rightarrow\; \cos\left(-\frac{7\pi}{2}\right) = 0.$$

$$\tan t = \frac{y}{x} \;\Rightarrow\; \tan\left(-\frac{7\pi}{2}\right) = \frac{1}{0} \text{ is und.} \qquad \cot t = \frac{x}{y} \;\Rightarrow\; \cot\left(-\frac{7\pi}{2}\right) = \frac{0}{1} = 0.$$

$$\sec t = \frac{1}{x} \;\Rightarrow\; \sec\left(-\frac{7\pi}{2}\right) = \frac{1}{0} \text{ is und.} \qquad \csc t = \frac{1}{y} \;\Rightarrow\; \csc\left(-\frac{7\pi}{2}\right) = \frac{1}{1} = 1.$$

$\boxed{13}$ (a) $t = \frac{9\pi}{4}$ is coterminal with $t = \frac{\pi}{4}$. The point P on the unit circle U that

corresponds to $t = \frac{9\pi}{4}$ has coordinates $\left(\frac{\sqrt{2}}{2}, \frac{\sqrt{2}}{2}\right)$. Thus, we choose $x = \frac{\sqrt{2}}{2}$ and

$y = \frac{\sqrt{2}}{2}$.

$$\sin t = y \;\Rightarrow\; \sin\frac{9\pi}{4} = \frac{\sqrt{2}}{2}. \qquad\qquad \cos t = x \;\Rightarrow\; \cos\frac{9\pi}{4} = \frac{\sqrt{2}}{2}.$$

$$\tan t = \frac{y}{x} \;\Rightarrow\; \tan\frac{9\pi}{4} = \frac{\sqrt{2}/2}{\sqrt{2}/2} = 1.$$

We now use the reciprocal relationships to find the values of the 3 remaining trigonometric functions.

$$\cot t = \frac{1}{\tan t} = \frac{1}{1} = 1. \qquad\qquad \sec t = \frac{1}{\cos t} = \frac{1}{\sqrt{2}/2} = \frac{2}{\sqrt{2}} = \frac{2^1}{2^{1/2}} = 2^{1/2} = \sqrt{2}.$$

$$\csc t = \frac{1}{\sin t} = \frac{1}{\sqrt{2}/2} = \sqrt{2}.$$

(b) $t = -\frac{5\pi}{4}$ is coterminal with $t = \frac{3\pi}{4}$. The point P on the unit circle U that

corresponds to $t = -\frac{5\pi}{4}$ has coordinates $\left(-\frac{\sqrt{2}}{2}, \frac{\sqrt{2}}{2}\right)$. Thus, we choose $x = -\frac{\sqrt{2}}{2}$

and $y = \frac{\sqrt{2}}{2}$.

$$\sin t = y \;\Rightarrow\; \sin\left(-\frac{5\pi}{4}\right) = \frac{\sqrt{2}}{2}. \qquad\qquad \cos t = x \;\Rightarrow\; \cos\left(-\frac{5\pi}{4}\right) = -\frac{\sqrt{2}}{2}.$$

$$\tan t = \frac{y}{x} \;\Rightarrow\; \tan\left(-\frac{5\pi}{4}\right) = \frac{\sqrt{2}/2}{-\sqrt{2}/2} = -1. \qquad \cot t = \frac{1}{\tan t} = \frac{1}{-1} = -1.$$

$$\sec t = \frac{1}{\cos t} = \frac{1}{-\sqrt{2}/2} = -\sqrt{2}. \qquad\qquad \csc t = \frac{1}{\sin t} = \frac{1}{\sqrt{2}/2} = \sqrt{2}.$$

15 (a) The point P on the unit circle U that corresponds to $t = \frac{5\pi}{4}$ has coordinates $\left(-\frac{\sqrt{2}}{2}, -\frac{\sqrt{2}}{2}\right)$. Thus, we choose $x = -\frac{\sqrt{2}}{2}$ and $y = -\frac{\sqrt{2}}{2}$.

$$\sin t = y \;\Rightarrow\; \sin \tfrac{5\pi}{4} = -\frac{\sqrt{2}}{2}. \qquad\qquad \cos t = x \;\Rightarrow\; \cos \tfrac{5\pi}{4} = -\frac{\sqrt{2}}{2}.$$

$$\tan t = \frac{y}{x} \;\Rightarrow\; \tan \tfrac{5\pi}{4} = \frac{-\sqrt{2}/2}{-\sqrt{2}/2} = 1. \qquad \cot t = \frac{1}{\tan t} = \tfrac{1}{1} = 1.$$

$$\sec t = \frac{1}{\cos t} = \frac{1}{-\sqrt{2}/2} = -\sqrt{2}. \qquad\qquad \csc t = \frac{1}{\sin t} = \frac{1}{-\sqrt{2}/2} = -\sqrt{2}.$$

(b) $t = -\frac{\pi}{4}$ is coterminal with $t = \frac{7\pi}{4}$. The point P on the unit circle U that corresponds to $t = -\frac{\pi}{4}$ has coordinates $\left(\frac{\sqrt{2}}{2}, -\frac{\sqrt{2}}{2}\right)$. Thus, we choose $x = \frac{\sqrt{2}}{2}$ and $y = -\frac{\sqrt{2}}{2}$.

$$\sin t = y \;\Rightarrow\; \sin\left(-\tfrac{\pi}{4}\right) = -\frac{\sqrt{2}}{2}. \qquad\qquad \cos t = x \;\Rightarrow\; \cos\left(-\tfrac{\pi}{4}\right) = \frac{\sqrt{2}}{2}.$$

$$\tan t = \frac{y}{x} \;\Rightarrow\; \tan\left(-\tfrac{\pi}{4}\right) = \frac{-\sqrt{2}/2}{\sqrt{2}/2} = -1. \qquad \cot t = \frac{1}{\tan t} = \tfrac{1}{-1} = -1.$$

$$\sec t = \frac{1}{\cos t} = \frac{1}{\sqrt{2}/2} = \sqrt{2}. \qquad\qquad \csc t = \frac{1}{\sin t} = \frac{1}{-\sqrt{2}/2} = -\sqrt{2}.$$

17 (a) $\sin(-90°) = -\sin 90°$ { since $\sin(-t) = -\sin t$ } $= -1$

(b) $\cos\left(-\tfrac{3\pi}{4}\right) = \cos \tfrac{3\pi}{4}$ { since $\cos(-t) = \cos t$ } $= -\frac{\sqrt{2}}{2}$

(c) $\tan(-45°) = -\tan 45°$ { since $\tan(-t) = -\tan t$ } $= -1$

19 (a) $\cot\left(-\tfrac{3\pi}{4}\right) = -\cot \tfrac{3\pi}{4}$ { si e $\cot(-t) = -\cot t$ } $= -(-1) = 1$

(b) $\sec(-180°) = \sec 180°$ { ice $\sec(-t) = \sec t$ } $= -1$

(c) $\csc\left(-\tfrac{3\pi}{2}\right) = -\csc \tfrac{3\pi}{2}$ { since $\csc(-t) = -\csc t$ } $= -(-1) = 1$

21 $\sin(-x)\sec(-x) = (-\sin x)\sec x$ { formulas for negatives }

$$= (-\sin x)(1/\cos x) \;\{\text{reciprocal identity}\}$$

$$= -\tan x \qquad\qquad \{\text{tangent identity}\}$$

23 $\dfrac{\cot(-x)}{\csc(-x)} = \dfrac{-\cot x}{-\csc x}$ { formulas for negatives }

$$= \frac{\cos x/\sin x}{1/\sin x} \;\{\text{cotangent identity and reciprocal identity}\}$$

$$= \cos x \qquad \{\text{simplify}\}$$

$\boxed{25}$ $\frac{1}{\cos(-x)} - \tan(-x)\sin(-x) = \frac{1}{\cos x} - (-\tan x)(-\sin x)$ { formulas for negatives }

$$= \frac{1}{\cos x} - \frac{\sin x}{\cos x}\sin x \qquad \{\text{tangent identity}\}$$

$$= \frac{1 - \sin^2 x}{\cos x} \qquad \{\text{combine terms}\}$$

$$= \frac{\cos^2 x}{\cos x} \qquad \{\text{Pythagorean identity}\}$$

$$= \cos x \qquad \{\text{cancel } \cos x\}$$

$\boxed{27}$ (a) Using Figure 9 in the text, we see that as x gets close to 0 through numbers greater than 0 (from the *right* of 0), $\sin x$ approaches 0.

(b) As x approaches $-\frac{\pi}{2}$ through numbers less than $-\frac{\pi}{2}$ (from the *left* of $-\frac{\pi}{2}$), $\sin x$ approaches -1.

$\boxed{29}$ (a) Using Figure 10 in the text, we see that as x gets close to $\frac{\pi}{4}$ through numbers greater than $\frac{\pi}{4}$ (from the *right* of $\frac{\pi}{4}$), $\cos x$ approaches $\frac{\sqrt{2}}{2}$. Note that the value $\frac{\sqrt{2}}{2}$ is in the table containing specific values of the cosine function.

(b) As x approaches π through numbers less than π (from the *left* of π), $\cos x$ approaches -1.

$\boxed{31}$ (a) Using Figure 13 in the text, we see that as x gets close to $\frac{\pi}{4}$ through numbers greater than $\frac{\pi}{4}$ (from the *right* of $\frac{\pi}{4}$), $\tan x$ approaches 1.

(b) As x approaches $\frac{\pi}{2}$ through numbers greater than $\frac{\pi}{2}$ (from the *right* of $\frac{\pi}{2}$), $\tan x$ is approaching the vertical asymptote $x = \frac{\pi}{2}$. $\tan x$ is *decreasing* without bound, and we use the notation $\underline{\tan x \to -\infty}$ to denote this.

$\boxed{33}$ (a) Using Figure 17 in the text, we see that as x gets close to $-\frac{\pi}{4}$ through numbers less than $-\frac{\pi}{4}$ (from the *left* of $-\frac{\pi}{4}$), $\cot x$ approaches -1.

(b) As x approaches 0 through numbers greater than 0 (from the *right* of 0), $\cot x$ is approaching the vertical asymptote $x = 0$ { the y-axis }. $\cot x$ is *increasing* without bound, and we use the notation $\underline{\cot x \to \infty}$ to denote this.

$\boxed{35}$ (a) Using Figure 16 in the text, we see that as x gets close to $\frac{\pi}{2}$ through numbers less than $\frac{\pi}{2}$ (from the *left* of $\frac{\pi}{2}$), $\sec x \to \infty$.

(b) As x approaches $\frac{\pi}{4}$ through numbers greater than $\frac{\pi}{4}$ (from the *right* of $\frac{\pi}{4}$), $\sec x$ approaches $\sqrt{2}$. Recall that $\cos\frac{\pi}{4} = \frac{1}{\sqrt{2}}$ and that $\sec x = \frac{1}{\cos x}$.

37 (a) Using Figure 15 in the text, we see that as x gets close to 0 through numbers less than 0 (from the *left* of 0), $\csc x \to -\infty$.

(b) As x approaches $\frac{\pi}{2}$ through numbers greater than $\frac{\pi}{2}$ (from the *right* of $\frac{\pi}{2}$), $\csc x$ approaches 1.

39 Refer to Figure 9 and the accompanying table. We see that $\sin\frac{3\pi}{2} = -1$.

Since the period of the sine is 2π, the second value in $[0, 4\pi]$ is $\frac{3\pi}{2} + 2\pi = \frac{7\pi}{2}$.

41 Recall that $\sin\frac{\pi}{6} = \frac{1}{2}$ and $\sin\frac{5\pi}{6} = \frac{1}{2}$. Since the period of the sine is 2π,

other values in $[0, 4\pi]$ are $\frac{\pi}{6} + 2\pi = \frac{13\pi}{6}$ and $\frac{5\pi}{6} + 2\pi = \frac{17\pi}{6}$.

43 Refer to Figure 11 and the accompanying table. We see that $\cos 0 = \cos 2\pi = 1$.

Since the period of the cosine is 2π, the other value in $[0, 4\pi]$ is $2\pi + 2\pi = 4\pi$.

45 Refer to Figure 11 and the accompanying table. We see that $\cos\frac{\pi}{4} = \cos\frac{7\pi}{4} = \frac{\sqrt{2}}{2}$.

Since the period of the cosine is 2π,

other values in $[0, 4\pi]$ are $\frac{\pi}{4} + 2\pi = \frac{9\pi}{4}$ and $\frac{7\pi}{4} + 2\pi = \frac{15\pi}{4}$.

47 Refer to Figure 14. In the interval $\left(-\frac{\pi}{2}, \frac{\pi}{2}\right)$, $\tan x = 1$ only if $x = \frac{\pi}{4}$. Since the period of the tangent is π, the desired value in the interval $\left(\frac{\pi}{2}, \frac{3\pi}{2}\right)$ is $\frac{\pi}{4} + \pi = \frac{5\pi}{4}$.

49 Refer to Figure 14. In the interval $\left(-\frac{\pi}{2}, \frac{\pi}{2}\right)$, $\tan x = 0$ only if $x = 0$. Since the period of the tangent is π, the desired value in the interval $\left(\frac{\pi}{2}, \frac{3\pi}{2}\right)$ is $0 + \pi = \pi$.

51 $y = \sin x$; $[-2\pi, 2\pi]$; $a = \frac{1}{2}$ • Refer to Figure 9.

$\sin x = \frac{1}{2}$ $\Rightarrow$ $x = \frac{\pi}{6}$ and $\frac{5\pi}{6}$. Also, $\frac{\pi}{6} - 2\pi = -\frac{11\pi}{6}$ and $\frac{5\pi}{6} - 2\pi = -\frac{7\pi}{6}$.

$\sin x > \frac{1}{2}$ when the graph is *above* the horizontal line $y = \frac{1}{2}$.

$\sin x < \frac{1}{2}$ when the graph is *below* the horizontal line $y = \frac{1}{2}$.

★ (a) $-\frac{11\pi}{6}, -\frac{7\pi}{6}, \frac{\pi}{6}, \frac{5\pi}{6}$ (b) $-\frac{11\pi}{6} < x < -\frac{7\pi}{6}$ and $\frac{\pi}{6} < x < \frac{5\pi}{6}$

(c) $-2\pi \le x < -\frac{11\pi}{6}$, $-\frac{7\pi}{6} < x < \frac{\pi}{6}$, and $\frac{5\pi}{6} < x \le 2\pi$

53 $y = \cos x$; $[-2\pi, 2\pi]$; $a = -\frac{1}{2}$ • Refer to Figure 11.

$\cos x = -\frac{1}{2}$ $\Rightarrow$ $x = \frac{2\pi}{3}$ and $\frac{4\pi}{3}$. Also, $\frac{2\pi}{3} - 2\pi = -\frac{4\pi}{3}$ and $\frac{4\pi}{3} - 2\pi = -\frac{2\pi}{3}$.

$\cos x > -\frac{1}{2}$ when the graph is *above* the horizontal line $y = -\frac{1}{2}$.

$\cos x < -\frac{1}{2}$ when the graph is *below* the horizontal line $y = -\frac{1}{2}$.

★ (a) $-\frac{4\pi}{3}, -\frac{2\pi}{3}, \frac{2\pi}{3}, \frac{4\pi}{3}$ (b) $-2\pi \le x < -\frac{4\pi}{3}$, $-\frac{2\pi}{3} < x < \frac{2\pi}{3}$, and $\frac{4\pi}{3} < x \le 2\pi$

(c) $-\frac{4\pi}{3} < x < -\frac{2\pi}{3}$ and $\frac{2\pi}{3} < x < \frac{4\pi}{3}$

55 $y = 2 + \sin x$ • Shift $y = \sin x$ up 2 units.

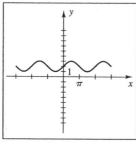

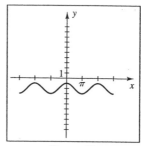

Figure 55   Figure 57

57 $y = \cos x - 2$ • Shift $y = \cos x$ down 2 units.

59 $y = 1 + \tan x$ • Shift $y = \tan x$ up 1 unit.

Since $1 + \tan\left(-\frac{\pi}{4}\right) = 1 + (-1) = 0$, and the period of the tangent is π,

there are x-intercepts at $x = -\frac{\pi}{4} + \pi n$.

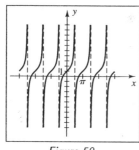

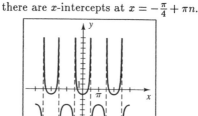

Figure 59   Figure 61

61 $y = \sec x - 2$ •

Shift $y = \sec x$ down 2 units, x-intercepts are at $x = \frac{\pi}{3} + 2\pi n, \frac{5\pi}{3} + 2\pi n$.

63 (a) As we move from left to right, the function increases { goes up } on the

intervals $[-2\pi, -\frac{3\pi}{2}), (-\frac{3\pi}{2}, -\pi], [0, \frac{\pi}{2}), (\frac{\pi}{2}, \pi]$.

(b) As we move from left to right, the function decreases { goes down } on the

intervals $[-\pi, -\frac{\pi}{2}), (-\frac{\pi}{2}, 0], [\pi, \frac{3\pi}{2}), (\frac{3\pi}{2}, 2\pi]$.

65 (a) The tangent function increases on *all* intervals on which it is defined.

Between -2π and 2π,

these intervals are $[-2\pi, -\frac{3\pi}{2}), (-\frac{3\pi}{2}, -\frac{\pi}{2}), (-\frac{\pi}{2}, \frac{\pi}{2}), (\frac{\pi}{2}, \frac{3\pi}{2})$, and $(\frac{3\pi}{2}, 2\pi]$.

(b) The tangent function is *never* decreasing on any interval for which it is defined.

67 This is good advice.

69 (a) From $(1, 0)$, move counterclockwise on the unit circle to the point at the tick

marked 4. The projection of this point on the y-axis,

approximately -0.7 or -0.8, is the value of $\sin 4$.

(b) From $(1, 0)$, move clockwise 1.2 units to about 5.1.

As in part (a), $\sin(-1.2)$ is about -0.9.

(c) Draw the horizontal line $y = 0.5$.

This line intersects the circle at about 0.5 and 2.6.

[71] (a) From $(1, 0)$, move counterclockwise on the unit circle to the point at the tick marked 4. The projection of this point on the x-axis,

approximately -0.6 or -0.7, is the value of $\cos 4$.

(b) Proceeding as in part (a), $\cos(-1.2)$ is about 0.4.

(c) Draw the vertical line $x = -0.6$.

This line intersects the circle at about 2.2 and 4.1.

[73] (a) Note that midnight occurs when $t = -6$.

Time	Temp.	Humidity	Time	Temp.	Humidity
12 A.M.	60	60	12 P.M.	60	60
3 A.M.	52	74	3 P.M.	68	46
6 A.M.	48	80	6 P.M.	72	40
9 A.M.	52	74	9 P.M.	68	46

(b) Since $T(t) = -12\cos\left(\frac{\pi}{12}t\right) + 60$, its maximum is $60 + 12 = 72°$F at $t = 12$ or 6:00 P.M., and its minimum is $60 - 12 = 48°$F at $t = 0$ or 6:00 A.M. Since $H(t) = 20\cos\left(\frac{\pi}{12}t\right) + 60$, its maximum is $60 + 20 = 80\%$ at $t = 0$ or 6:00 A.M., and its minimum is $60 - 20 = 40\%$ at $t = 12$ or 6:00 P.M.

(c) When the temperature increases, the relative humidity decreases and vice versa. As the temperature cools, the air can hold less moisture and the relative humidity increases. Because of this phenomenon, fog often occurs during the evening hours rather than in the middle of the day.

[75] Graph $y = \sin(x^2)$ and $y = 0.5$ on the same coordinate plane. From the graph, we see that $\sin(x^2)$ assumes the value of 0.5 at $x \approx \pm 0.72, \pm 1.62, \pm 2.61, \pm 2.98$.

$[-\pi, \pi, \pi/4]$ by $[-2.09, 2.09]$ $[-2\pi, 2\pi, \pi/2]$ by $[-5.19, 3.19]$

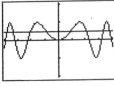   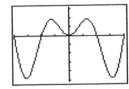

Figure 75   *Figure 77*

[77] We see that the graph of $y = x\sin x$ assumes a maximum value of approximately 1.82 at $x \approx \pm 2.03$, and a minimum value of -4.81 at $x \approx \pm 4.91$.

79 As $x \to 0^+$, $f(x) = \frac{1 - \cos x}{x} \to 0$.

[−1, 1, 0.5] by [−0.67, 0.67, 0.5]

[−2, 2, 0.5] by [−1.33, 1.33, 0.5]

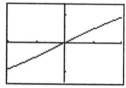

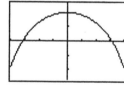

Figure 79

Figure 81

81 As $x \to 0^+$, $f(x) = x \cot x \to 1$.

83 As $x \to 0^+$, $f(x) = \frac{\tan x}{x} \to 1$.

[−1.5, 1.5] by [0, 3]

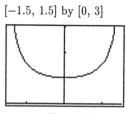

Figure 83

5.4 Exercises

Note: Let θ_C denote the coterminal angle of θ such that $0° \le \theta_C < 360°$ { or $0 \le \theta_C < 2\pi$ }.
The following formulas are then used in the solutions { on text page 398 }.

(1) If θ_C is in QI, then $\theta_R = \theta_C$.

(2) If θ_C is in QII, then $\theta_R = 180° - \theta_C$ { or $\pi - \theta_C$ }.

(3) If θ_C is in QIII, then $\theta_R = \theta_C - 180°$ { or $\theta_C - \pi$ }.

(4) If θ_C is in QIV, then $\theta_R = 360° - \theta_C$ { or $2\pi - \theta_C$ }.

1 (a) Since 240° is in QIII, $\theta_R = 240° - 180° = 60°$.

(b) Since 340° is in QIV, $\theta_R = 360° - 340° = 20°$.

(c) $\theta_C = -202° + 1(360°) = 158° \in$ QII. $\theta_R = 180° - 158° = 22°$.

(d) $\theta_C = -660° + 2(360°) = 60° \in$ QI. $\theta_R = 60°$.

3 (a) Since $\frac{3\pi}{4}$ is in QII, $\theta_R = \pi - \frac{3\pi}{4} = \frac{\pi}{4}$.

(b) Since $\frac{4\pi}{3}$ is in QIII, $\theta_R = \frac{4\pi}{3} - \pi = \frac{\pi}{3}$.

(c) $\theta_C = -\frac{\pi}{6} + 1(2\pi) = \frac{11\pi}{6} \in$ QIV. $\theta_R = 2\pi - \frac{11\pi}{6} = \frac{\pi}{6}$.

(d) $\theta_C = \frac{9\pi}{4} - 1(2\pi) = \frac{\pi}{4} \in$ QI. $\theta_R = \frac{\pi}{4}$.

[5] (a) Since $\frac{\pi}{2} < 3 < \pi$, θ is in QII and $\theta_R = \pi - 3 \approx 0.14$, or $8.1°$.

(b) $\theta_C = -2 + 1(2\pi) = 2\pi - 2 \approx 4.28$.

Since $\pi < 4.28 < \frac{3\pi}{2}$, θ_C is in QIII and $\theta_R = (2\pi - 2) - \pi = \pi - 2 \approx 1.14$, or $65.4°$.

(c) Since $\frac{3\pi}{2} < 5.5 < 2\pi$, θ is in QIV and $\theta_R = 2\pi - 5.5 \approx 0.78$, or $44.9°$.

(d) The number of revolutions formed by θ is $\frac{100}{2\pi} \approx 15.92$, so

$\theta_C = 100 - 15(2\pi) = 100 - 30\pi \approx 5.75$. Since $\frac{3\pi}{2} < 5.75 < 2\pi$,

θ_C is in QIV and $\theta_R = 2\pi - (100 - 30\pi) = 32\pi - 100 \approx 0.53$, or $30.4°$.

Alternatively, if your calculator is capable of computing trigonometric functions

of large values, then computing $\sin^{-1}(\sin 100) \approx -0.53 \Rightarrow \theta_R = 0.53$, or $30.4°$.

Note: For the following problems, we use the theorem on reference angles before

evaluating.

[7] (a) $\sin \frac{2\pi}{3} = \sin \frac{\pi}{3}$ { since the sine is positive in QII

and $\frac{\pi}{3}$ is the reference angle for $\frac{2\pi}{3}$ } $= \frac{\sqrt{3}}{2}$

(b) $\sin\left(-\frac{5\pi}{4}\right) = \sin \frac{3\pi}{4}$ { since $\frac{3\pi}{4}$ is coterminal with $-\frac{5\pi}{4}$ } $=$

$\sin \frac{\pi}{4}$ { since the sine is positive in QII and $\frac{\pi}{4}$ is the reference angle for $\frac{3\pi}{4}$ } $= \frac{\sqrt{2}}{2}$

[9] (a) $\cos 150° = -\cos 30°$ { since the cosine is negative in QII } $= -\frac{\sqrt{3}}{2}$

(b) $\cos(-60°) = \cos 300° = \cos 60°$ { since the cosine is positive in QIV } $= \frac{1}{2}$

[11] (a) $\tan \frac{5\pi}{6} = -\tan \frac{\pi}{6}$ { since the tangent is negative in QII } $= -\frac{\sqrt{3}}{3}$

(b) $\tan\left(-\frac{\pi}{3}\right) = \tan \frac{5\pi}{3} = -\tan \frac{\pi}{3}$ { since the tangent is negative in QIV } $= -\sqrt{3}$

[13] (a) $\cot 120° = -\cot 60°$ { since the cotangent is negative in QII } $= -\frac{\sqrt{3}}{3}$

(b) $\cot(-150°) = \cot 210° = \cot 30°$ { since the cotangent is positive in QIII } $= \sqrt{3}$

[15] (a) $\sec \frac{2\pi}{3} = -\sec \frac{\pi}{3}$ { since the secant is negative in QII } $= -2$

(b) $\sec\left(-\frac{\pi}{6}\right) = \sec \frac{11\pi}{6} = \sec \frac{\pi}{6}$ { since the secant is positive in QIV } $= \frac{2}{\sqrt{3}}$

[17] (a) $\csc 240° = -\csc 60°$ { since the cosecant is negative in QIII } $= -\frac{2}{\sqrt{3}}$

(b) $\csc(-330°) = \csc 30° = 2$

[19] (a) Using the *degree* mode on a calculator, $\sin 73°20' \approx 0.958$.

(b) Using the *radian* mode on a calculator, $\cos 0.68 \approx 0.778$.

[21] (a) Using the *degree* mode on a calculator, $\tan 21°10' \approx 0.387$.

(b) First compute $\tan 1.13$, obtaining 2.11975. Now use the reciprocal key,

usually labeled as either $\boxed{1/x}$ or $\boxed{x^{-1}}$, to obtain $\cot 1.13 \approx 0.472$.

[23] (a) First compute $\cos 67°50'$, obtaining 0.3773.

Now use the reciprocal key to obtain $\sec 67°50' \approx 2.650$.

(b) As in part (a), since $\sin 0.32 \approx 0.3146$, we have $\csc 0.32 \approx 3.179$.

$\boxed{25}$ (a) Using the degree mode, calculate $\cos^{-1}(0.8620)$ to obtain 30.46° to

the nearest one-hundredth of a degree.

(b) Using the answer from part (a), subtract 30 and multiply that result by 60

to obtain 30°27′ to the nearest minute.

$\boxed{27}$ (a) $\tan\theta = 3.7 \Rightarrow \theta = \tan^{-1}(3.7) \approx 74.88°$ (b) $\tan^{-1}(3.7) \approx 74°53′$

$\boxed{29}$ (a) $\sin\theta = 0.4217 \Rightarrow \theta = \sin^{-1}(0.4217) \approx 24.94°$

(b) $\sin^{-1}(0.4217) \approx 24°57′$

$\boxed{31}$ (a) After entering 4.246, use $\boxed{1/x}$ and then $\boxed{\text{INV}}\boxed{\text{COS}}$, or, equivalently, $\boxed{\text{COS}^{-1}}$. Similarly, for the cosecant function use $\boxed{1/x}$ and then $\boxed{\text{INV}}\boxed{\text{SIN}}$ and for the cotangent function use $\boxed{1/x}$ and then $\boxed{\text{INV}}\boxed{\text{TAN}}$. $\sec\theta = 4.246 \Rightarrow \cos\theta = \frac{1}{4.246} \Rightarrow \theta = \cos^{-1}\left(\frac{1}{4.246}\right) \approx 76.38°$.

(b) $\cos^{-1}\left(\frac{1}{4.246}\right) \approx 76°23′$

$\boxed{33}$ (a) $\sin 98°10′ \approx 0.9899$ (b) $\cos 623.7° \approx -0.1097$ (c) $\tan 3 \approx -0.1425$

(d) $\cot 231°40′ = \dfrac{1}{\tan 231°40′} \approx 0.7907$

(e) $\sec 1175.1° = \dfrac{1}{\cos 1175.1°} \approx -11.2493$

(f) $\csc 0.82 = \dfrac{1}{\sin 0.82} \approx 1.3677$

$\boxed{35}$ (a) Use the degree mode. $\sin\theta = -0.5640 \Rightarrow \theta = \sin^{-1}(-0.5640) \approx -34.3° \Rightarrow$ $\theta_R \approx 34.3°$. Since the sine is negative in QIII and QIV, we use θ_R in those quadrants. $180° + 34.3° = \underline{214.3°}$ and $360° - 34.3° = \underline{325.7°}$

(b) $\cos\theta = 0.7490 \Rightarrow \theta = \cos^{-1}(0.7490) \approx 41.5°$. $\theta_R \approx 41.5°$, QI: 41.5°, QIV: 318.5°

(c) $\tan\theta = 2.798 \Rightarrow \theta = \tan^{-1}(2.798) \approx 70.3°$. $\theta_R \approx 70.3°$, QI: 70.3°, QIII: 250.3°

(d) $\cot\theta = -0.9601 \Rightarrow \tan\theta = -\frac{1}{0.9601} \Rightarrow \theta = \tan^{-1}\left(-\frac{1}{0.9601}\right) \approx -46.2°$.

$\theta_R \approx 46.2°$, QII: 133.8°, QIV: 313.8°

(e) $\sec\theta = -1.116 \Rightarrow \cos\theta = -\frac{1}{1.116} \Rightarrow \theta = \cos^{-1}\left(-\frac{1}{1.116}\right) \approx 153.6°$.

$\theta_R \approx 180° - 153.6° = 26.4°$, QII: 153.6°, QIII: 206.4°

(f) $\csc\theta = 1.485 \Rightarrow \sin\theta = \frac{1}{1.485} \Rightarrow \theta = \sin^{-1}\left(\frac{1}{1.485}\right) \approx 42.3°$.

$\theta_R \approx 42.3°$, QI: 42.3°, QII: 137.7°

$\boxed{37}$ (a) Use the radian mode. $\sin\theta = 0.4195 \Rightarrow \theta = \sin^{-1}(0.4195) \approx 0.43$.

$\theta_R \approx 0.43$ is one answer. Since the sine is positive in QI and QII,

we also use the reference angle for θ in quadrant II. QII: $\pi - 0.43 \approx 2.71$

(b) $\cos\theta = -0.1207 \Rightarrow \theta = \cos^{-1}(-0.1207) \approx 1.69$ is one answer.

Since 1.69 is in QII, $\theta_R \approx \pi - 1.69 \approx 1.45$.

The cosine is also negative in QIII. QIII: $\pi + 1.45 \approx 4.59$

(c) $\tan\theta = -3.2504 \Rightarrow \theta = \tan^{-1}(-3.2504) \approx -1.27 \Rightarrow \theta_R \approx 1.27$.

$\qquad\qquad\qquad\qquad\qquad\qquad$ QII: $\pi - 1.27 \approx 1.87$, QIV: $2\pi - 1.27 \approx 5.01$

(d) $\cot\theta = 2.6815 \Rightarrow \tan\theta = \frac{1}{2.6815} \Rightarrow \theta = \tan^{-1}(\frac{1}{2.6815}) \approx 0.36 \Rightarrow$

$\qquad\qquad\qquad\qquad \theta_R \approx 0.36$ is one answer. QIII: $\pi + 0.36 \approx 3.50$

(e) $\sec\theta = 1.7452 \Rightarrow \cos\theta = \frac{1}{1.7452} \Rightarrow \theta = \cos^{-1}(\frac{1}{1.7452}) \approx 0.96 \Rightarrow$

$\qquad\qquad\qquad\qquad \theta_R \approx 0.96$ is one answer. QIV: $2\pi - 0.96 \approx 5.32$

(f) $\csc\theta = -4.8521 \Rightarrow \sin\theta = -\frac{1}{4.8521} \Rightarrow \theta = \sin^{-1}(-\frac{1}{4.8521}) \approx -0.21 \Rightarrow$

$\qquad\qquad\qquad\qquad \theta_R \approx 0.21$. QIII: $\pi + 0.21 \approx 3.35$, QIV: $2\pi - 0.21 \approx 6.07$

$\boxed{39}$ $\ln I_0 - \ln I = kx\sec\theta \Rightarrow \ln\frac{I_0}{I} = kx\sec\theta$ { property of logarithms } $\Rightarrow$

$x = \frac{1}{k\sec\theta}\ln\frac{I_0}{I}$ { solve for x }

$\quad = \frac{1}{1.88\sec 12°}\ln 1.72$ { substitute and approximate } ≈ 0.28 cm.

$\boxed{41}$ (a) Since $\cos\theta$ and $\sin\phi$ are both less than or equal to 1, $R = R_0\cos\theta\sin\phi$, the solar
radiation R will equal its maximum R_0 when $\cos\theta = \sin\phi = 1$. This occurs when
$\theta = 0°$ and $\phi = 90°$, and corresponds to when the sun is just rising in the east.

(b) The sun located in the southeast corresponds to $\phi = 45°$.

$\qquad$ percentage of $R_0 = \dfrac{\text{amount of }R_0}{R_0} = \dfrac{R_0\cos 60°\sin 45°}{R_0} = \dfrac{1}{2}\cdot\dfrac{\sqrt{2}}{2} = \dfrac{\sqrt{2}}{4} \approx 35\%$.

$\boxed{43}$ $\sin\theta = \frac{b}{c} \Rightarrow \sin 60° = \frac{b}{18} \Rightarrow b = 18\sin 60° = 18\cdot\frac{\sqrt{3}}{2} = 9\sqrt{3} \approx 15.6$.

$\cos\theta = \frac{a}{c} \Rightarrow \cos 60° = \frac{a}{18} \Rightarrow a = 18\cos 60° = 18\cdot\frac{1}{2} = 9$.

$\qquad\qquad\qquad\qquad\qquad\qquad$ The hand is located at $(9, 9\sqrt{3})$.

5.5 Exercises

Note: Exercises 1 & 3: We will refer to $y = \sin x$ as just $\sin x$ ($y = \cos x$ as $\cos x$, etc.).
$\qquad$ For the form $y = a\sin bx$, the amplitude is $|a|$ and the period is $\frac{2\pi}{|b|}$.

$\qquad$ These are merely listed in the answer along with the values of the x-intercepts.

$\qquad$ Let n denote any integer.

$\boxed{1}$ (a) $y = 4\sin x$ •   Vertically stretch $\sin x$ by a factor of 4. The x-intercepts are not
$\qquad$ affected by a vertical stretch or compression. $\qquad\qquad$ ★ 4, 2π, x-int. @ πn

(b) $y = \sin 4x$ •   Horizontally compress $\sin x$ by a factor of 4. The x-intercepts
$\qquad$ are affected by a horizontal stretch or compression by the same factor—that is, a
$\qquad$ horizontal compression by a factor of k will move the x-intercepts of $\sin x$ from
$\qquad$ πn to $\frac{\pi}{k}n$, and a horizontal stretch by a factor of k will move the x-intercepts of
$\qquad$ $\sin x$ from πn to $k\pi n$. $\qquad\qquad\qquad\qquad\qquad\qquad$ ★ 1, $\frac{\pi}{2}$, x-int. @ $\frac{\pi}{4}n$

(c) $y = \frac{1}{4}\sin x$ • Vertically compress $\sin x$ by a factor of 4.   ★ $\frac{1}{4}$, 2π, x-int. @ πn

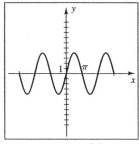

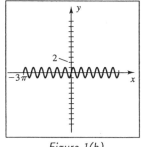

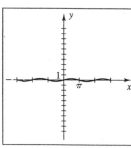

| *Figure 1(a)* | *Figure 1(b)* | *Figure 1(c)* |

(d) $y = \sin\frac{1}{4}x$ • Horizontally stretch $\sin x$ by a factor of 4. ★ 1, 8π, x-int. @ $4\pi n$

(e) $y = 2\sin\frac{1}{4}x$ • Vertically stretch the graph in part (d) by a factor of 2.

★ 2, 8π, x-int. @ $4\pi n$

(f) $y = \frac{1}{2}\sin 4x$ • Vertically compress the graph in part (b) by a factor of 2.

★ $\frac{1}{2}$, $\frac{\pi}{2}$, x-int. @ $\frac{\pi}{4}n$

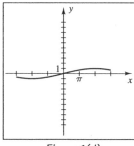

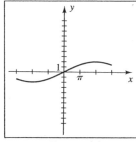

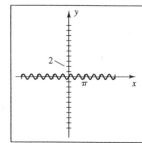

| *Figure 1(d)* | *Figure 1(e)* | *Figure 1(f)* |

(g) $y = -4\sin x$ • Reflect the graph in part (a) through the x-axis.

★ 4, 2π, x-int. @ πn

(h) $y = \sin(-4x) = -\sin 4x$ using a formula for negatives. Reflect the graph in part

(b) through the x-axis.   ★ 1, $\frac{\pi}{2}$, x-int. @ $\frac{\pi}{4}n$

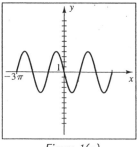

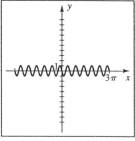

| *Figure 1(g)* | *Figure 1(h)* |

$\boxed{3}$ (a) $y = 3\cos x$ •   Vertically stretch $\cos x$ by a factor of 3.

★ 3, 2π, x-int. @ $\frac{\pi}{2} + \pi n$

(b) $y = \cos 3x$ •   Horizontally compress $\cos x$ by a factor of 3. The x-intercepts are affected by a horizontal stretch or compression by the same factor—that is, horizontal compression by a factor of k will move the x-intercepts of $\cos x$ from $\frac{\pi}{2} + \pi n$ to $\frac{\pi}{2k} + \frac{\pi}{k}n$, and a horizontal stretch by a factor of k will move the x-intercepts of $\cos x$ from $\frac{\pi}{2} + \pi n$ to $\frac{k\pi}{2} + k\pi n$.   ★ 1, $\frac{2\pi}{3}$, x-int. @ $\frac{\pi}{6} + \frac{\pi}{3}n$

(c) $y = \frac{1}{3}\cos x$ •   Vertically compress $\cos x$ by a factor of 3.

★ $\frac{1}{3}$, 2π, x-int. @ $\frac{\pi}{2} + \pi n$

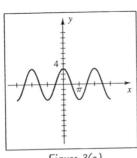 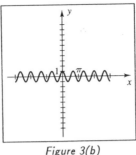 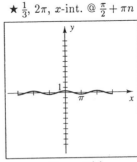

Figure 3(a)   Figure 3(b)   Figure 3(c)

(d) $y = \cos\frac{1}{3}x$ •   Horizontally stretch $\cos x$ by a factor of 3.

★ 1, 6π, x-int. @ $\frac{3\pi}{2} + 3\pi n$

(e) $y = 2\cos\frac{1}{3}x$ •   Vertically stretch the graph in part (d) by a factor of 2.

★ 2, 6π, x-int. @ $\frac{3\pi}{2} + 3\pi n$

(f) $y = \frac{1}{2}\cos 3x$ •   Vertically compress the graph in part (b) by a factor of 2.

★ $\frac{1}{2}$, $\frac{2\pi}{3}$, x-int. @ $\frac{\pi}{6} + \frac{\pi}{3}n$

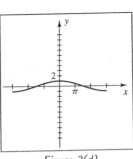 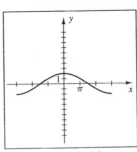 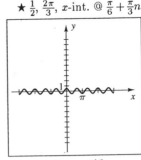

Figure 3(d)   Figure 3(e)   Figure 3(f)

(g) $y = -3\cos x$ •   Reflect the graph in part (a) through the x-axis.

See *Figure 3(g)* on the next page.   ★ 3, 2π, x-int. @ $\frac{\pi}{2} + \pi n$

(h) $y = \cos(-3x) = \cos 3x$ using a formula for negatives. This is the same as the graph in part (b).   ★ 1, $\frac{2\pi}{3}$, x-int. @ $\frac{\pi}{6} + \frac{\pi}{3}n$

See *Figure 3(h)* on the next page.

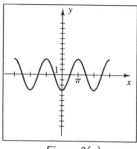

Figure 3(g)

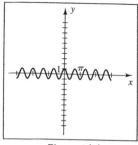

Figure 3(h)

Note: We will write $y = a\sin(bx + c)$ in the form $y = a\sin\left[b\left(x + \frac{c}{b}\right)\right]$. From this form we have the amplitude, $|a|$, the period, $\frac{2\pi}{|b|}$, and the phase shift, $-\frac{c}{b}$. We will also list the interval that corresponds to $[0, 2\pi]$ for the sine functions and to $[-\frac{\pi}{2}, \frac{3\pi}{2}]$ for the cosine functions—this interval gives us one wave (between zeros) of the graph of the function. The work to determine those intervals is shown in each exercise.

5   $y = \sin\left(x - \frac{\pi}{2}\right)$ • $0 \le x - \frac{\pi}{2} \le 2\pi \Rightarrow \frac{\pi}{2} \le x \le \frac{5\pi}{2}$

Phase shift $= -\left(-\frac{\pi}{2}\right) = \frac{\pi}{2}$.

$\bigstar$ 1, 2π, $\frac{\pi}{2}$, $[\frac{\pi}{2}, \frac{5\pi}{2}]$

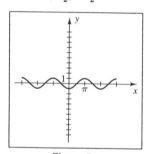

Figure 5

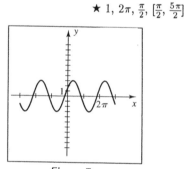

Figure 7

7   $y = 3\sin\left(x + \frac{\pi}{6}\right)$ • $0 \le x + \frac{\pi}{6} \le 2\pi \Rightarrow -\frac{\pi}{6} \le x \le \frac{11\pi}{6}$

Phase shift $= -\left(\frac{\pi}{6}\right) = -\frac{\pi}{6}$.

$\bigstar$ 3, 2π, $-\frac{\pi}{6}$, $[-\frac{\pi}{6}, \frac{11\pi}{6}]$

9   $y = \cos\left(x + \frac{\pi}{2}\right)$ •

$-\frac{\pi}{2} \le x + \frac{\pi}{2} \le \frac{3\pi}{2} \Rightarrow -\pi \le x \le \pi$

Phase shift $= -\left(\frac{\pi}{2}\right) = -\frac{\pi}{2}$.

$\bigstar$ 1, 2π, $-\frac{\pi}{2}$, $[-\pi, \pi]$

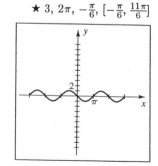

Figure 9

$\boxed{11}$ $y = 4\cos\left(x - \frac{\pi}{4}\right)$ • $-\frac{\pi}{2} \le x - \frac{\pi}{4} \le \frac{3\pi}{2}$ ⇒ $-\frac{\pi}{4} \le x \le \frac{7\pi}{4}$

Phase shift $= -\left(-\frac{\pi}{4}\right) = \frac{\pi}{4}$.

★ $4, 2\pi, \frac{\pi}{4}, \left[-\frac{\pi}{4}, \frac{7\pi}{4}\right]$

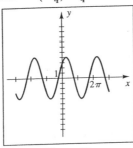

Figure 11

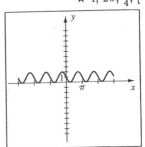

Figure 13

$\boxed{13}$ $y = \sin(2x - \pi) + 1 = \sin\left[2\left(x - \frac{\pi}{2}\right)\right] + 1$. • Period $= \frac{2\pi}{|2|} = \pi$. The normal range

of the sine, -1 to 1, is affected by the "$+1$" at the end of the equation. It shifts

the graph up 1 unit and the resulting range is 0 to 2. It may be easiest to graph

$y = \sin\left[2\left(x - \frac{\pi}{2}\right)\right]$ {1 period of a sine wave with endpoints at $\frac{\pi}{2}$ and $\frac{3\pi}{2}$} and then

make a vertical shift of 1 unit up to complete the graph of $y = \sin\left[2\left(x - \frac{\pi}{2}\right)\right] + 1$.

$0 \le 2x - \pi \le 2\pi$ ⇒ $\pi \le 2x \le 3\pi$ ⇒ $\frac{\pi}{2} \le x \le \frac{3\pi}{2}$ ★ $1, \pi, \frac{\pi}{2}, \left[\frac{\pi}{2}, \frac{3\pi}{2}\right]$

$\boxed{15}$ $y = -\cos(3x + \pi) - 2 = -\cos\left[3\left(x + \frac{\pi}{3}\right)\right] - 2$ • The "$-$" in front of cos has the

effect of reflecting the graph of $y = \cos(3x + \pi)$ through the x-axis. The "-2" at the

end of the equation lowers the range 2 units to -3 to -1. Period $= \frac{2\pi}{|3|} = \frac{2\pi}{3}$, phase

shift $= -\left(\frac{\pi}{3}\right) = -\frac{\pi}{3}$. $-\frac{\pi}{2} \le 3x + \pi \le \frac{3\pi}{2}$ ⇒ $-\frac{3\pi}{2} \le 3x \le \frac{\pi}{2}$ ⇒ $-\frac{\pi}{2} \le x \le \frac{\pi}{6}$

★ $1, \frac{2\pi}{3}, -\frac{\pi}{3}, \left[-\frac{\pi}{2}, \frac{\pi}{6}\right]$

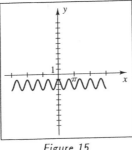

Figure 15

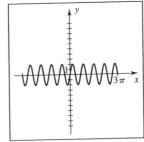

Figure 17

$\boxed{17}$ $y = -2\sin(3x - \pi) = -2\sin\left[3\left(x - \frac{\pi}{3}\right)\right]$. • Amplitude $= |-2| = 2$. The negative

before the "2" has the effect of reflecting the graph of $y = 2\sin(3x - \pi)$ through the

x-axis. Period $= \frac{2\pi}{|3|} = \frac{2\pi}{3}$, phase shift $= -\left(-\frac{\pi}{3}\right) = \frac{\pi}{3}$, $0 \le 3x - \pi \le 2\pi$ ⇒

$\pi \le 3x \le 3\pi$ ⇒ $\frac{\pi}{3} \le x \le \pi$.   ★ $2, \frac{2\pi}{3}, \frac{\pi}{3}, \left[\frac{\pi}{3}, \pi\right]$

19 $y = \sin\left(\frac{1}{2}x - \frac{\pi}{3}\right) = \sin\left[\frac{1}{2}\left(x - \frac{2\pi}{3}\right)\right]$. • Period $= \dfrac{2\pi}{|1/2|} = 4\pi$.

$0 \le \frac{1}{2}x - \frac{\pi}{3} \le 2\pi \;\Rightarrow\; \frac{\pi}{3} \le \frac{1}{2}x \le \frac{7\pi}{3} \;\Rightarrow\; \frac{2\pi}{3} \le x \le \frac{14\pi}{3}$ ★ $1, 4\pi, \frac{2\pi}{3}, [\frac{2\pi}{3}, \frac{14\pi}{3}]$

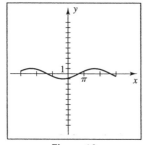

Figure 19

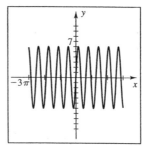

Figure 21

21 $y = 6 \sin \pi x$ • Period $= \dfrac{2\pi}{|\pi|} = 2.$ $0 \le \pi x \le 2\pi \;\Rightarrow\; 0 \le x \le 2$ ★ $6, 2, 0, [0, 2]$

23 $y = 2 \cos \frac{\pi}{2}x$ • Period $= \dfrac{2\pi}{|\pi/2|} = 4.$ $-\frac{\pi}{2} \le \frac{\pi}{2}x \le \frac{3\pi}{2} \;\Rightarrow\; -1 \le x \le 3$

★ $2, 4, 0, [-1, 3]$

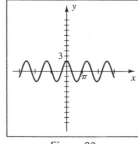

Figure 23

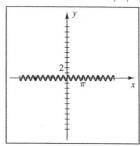

Figure 25

25 $y = \frac{1}{2} \sin 2\pi x$ • $0 \le 2\pi x \le 2\pi \;\Rightarrow\; 0 \le x \le 1$ ★ $\frac{1}{2}, 1, 0, [0, 1]$

27 $y = 5 \sin\left(3x - \frac{\pi}{2}\right) = 5 \sin\left[3\left(x - \frac{\pi}{6}\right)\right]$. •   ★ $5, \frac{2\pi}{3}, \frac{\pi}{6}, [\frac{\pi}{6}, \frac{5\pi}{6}]$

$0 \le 3x - \frac{\pi}{2} \le 2\pi \;\Rightarrow\; \frac{\pi}{2} \le 3x \le \frac{5\pi}{2} \;\Rightarrow\; \frac{\pi}{6} \le x \le \frac{5\pi}{6}.$ To graph this function, draw

one period of a sine wave with endpoints at $\frac{\pi}{6}$ and $\frac{5\pi}{6}$ and an amplitude of 5.

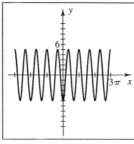

Figure 27

29 $y = 3\cos\left(\frac{1}{2}x - \frac{\pi}{4}\right) = 3\cos\left[\frac{1}{2}\left(x - \frac{\pi}{2}\right)\right].$ •

     $-\frac{\pi}{2} \le \frac{1}{2}x - \frac{\pi}{4} \le \frac{3\pi}{2} \;\Rightarrow\; -\frac{\pi}{4} \le \frac{1}{2}x \le \frac{7\pi}{4} \;\Rightarrow\; -\frac{\pi}{2} \le x \le \frac{7\pi}{2}$ ★ $3,\ 4\pi,\ \frac{\pi}{2},\ [-\frac{\pi}{2}, \frac{7\pi}{2}]$

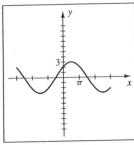

Figure 29

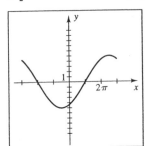

Figure 31

31 $y = -5\cos\left(\frac{1}{3}x + \frac{\pi}{6}\right) = -5\cos\left[\frac{1}{3}\left(x + \frac{\pi}{2}\right)\right].$ •

     $-\frac{\pi}{2} \le \frac{1}{3}x + \frac{\pi}{6} \le \frac{3\pi}{2} \;\Rightarrow\; -\frac{2\pi}{3} \le \frac{1}{3}x \le \frac{4\pi}{3} \;\Rightarrow\; -2\pi \le x \le 4\pi$

                                                  ★ $5,\ 6\pi,\ -\frac{\pi}{2},\ [-2\pi, 4\pi]$

33 $y = 3\cos\left(\pi x + 4\pi\right) = 3\cos\left[\pi\left(x + 4\right)\right].$ •

     $-\frac{\pi}{2} \le \pi x + 4\pi \le \frac{3\pi}{2} \;\Rightarrow\; -\frac{9\pi}{2} \le \pi x \le -\frac{5\pi}{2} \;\Rightarrow\; -\frac{9}{2} \le x \le -\frac{5}{2}$

                                              ★ $3,\ 2,\ -4,\ [-\frac{9}{2}, -\frac{5}{2}]$

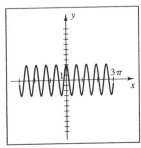

Figure 33

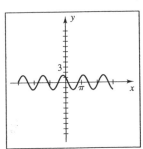

Figure 35

35 $y = -\sqrt{2}\sin\left(\frac{\pi}{2}x - \frac{\pi}{4}\right) = -\sqrt{2}\sin\left[\frac{\pi}{2}\left(x - \frac{1}{2}\right)\right].$ •

     $0 \le \frac{\pi}{2}x - \frac{\pi}{4} \le 2\pi \;\Rightarrow\; \frac{\pi}{4} \le \frac{\pi}{2}x \le \frac{9\pi}{4} \;\Rightarrow\; \frac{1}{2} \le x \le \frac{9}{2}$ ★ $\sqrt{2},\ 4,\ \frac{1}{2},\ [\frac{1}{2}, \frac{9}{2}]$

37 $y = -2\sin\left(2x - \pi\right) + 3 = -2\sin\left[2\left(x - \frac{\pi}{2}\right)\right] + 3.$ •

     $0 \le 2x - \pi \le 2\pi \;\Rightarrow\; \pi \le 2x \le 3\pi \;\Rightarrow\; \frac{\pi}{2} \le x \le \frac{3\pi}{2}.$ The
amplitude of 2 makes the normal sine range of -1 to 1
change to -2 to 2. The "$+3$" at the end of the equation
raises the graph up 3 units and the range is 1 to 5.

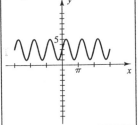

Figure 37

     ★ $2,\ \pi,\ \frac{\pi}{2},\ [\frac{\pi}{2}, \frac{3\pi}{2}]$

39 $y = 5\cos(2x + 2\pi) + 2 = 5\cos[2(x + \pi)] + 2.$ ●

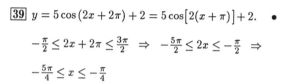

$$-\frac{\pi}{2} \le 2x + 2\pi \le \frac{3\pi}{2} \;\Rightarrow\; -\frac{5\pi}{2} \le 2x \le -\frac{\pi}{2} \;\Rightarrow$$

$$-\frac{5\pi}{4} \le x \le -\frac{\pi}{4}$$

★ $5, \pi, -\pi, [-\frac{5\pi}{4}, -\frac{\pi}{4}]$

Figure 39

41 (a) The amplitude a is 4 and the period { from $-\pi$ to π } is 2π.

The phase shift is the first negative zero that occurs before a maximum, $-\pi$.

(b) Period $= \frac{2\pi}{b} \;\Rightarrow\; 2\pi = \frac{2\pi}{b} \;\Rightarrow\; b = 1.$ Phase shift $= -\frac{c}{b} \;\Rightarrow\; -\pi = -\frac{c}{1} \;\Rightarrow$
$c = \pi.$ Hence, $y = a\sin(bx + c) = 4\sin(x + \pi).$

43 (a) The amplitude a is 2 and the period { from -3 to 1 } is 4.

The phase shift is the first negative zero that occurs before a maximum, -3.

(b) Period $= \frac{2\pi}{b} \;\Rightarrow\; 4 = \frac{2\pi}{b} \;\Rightarrow\; b = \frac{\pi}{2}.$ Phase shift $= -\frac{c}{b} \;\Rightarrow\; -3 = -\frac{c}{\pi/2} \;\Rightarrow$
$c = \frac{3\pi}{2}.$ Hence, $y = a\sin(bx + c) = 2\sin(\frac{\pi}{2}x + \frac{3\pi}{2}).$

45 In the first second, there are 2 complete cycles.

Hence 1 cycle is completed in $\frac{1}{2}$ second and thus, the period is $\frac{1}{2}$.

Also, the period is $\frac{2\pi}{b}$. Equating these expressions yields $\frac{2\pi}{b} = \frac{1}{2} \;\Rightarrow\; b = 4\pi.$

47 We first note that $\frac{1}{2}$ period takes place in $\frac{1}{4}$ second and thus 1 period in $\frac{1}{2}$ second.

As in Exercise 45, $\frac{2\pi}{b} = \frac{1}{2} \;\Rightarrow\; b = 4\pi.$ Since the maximum flow rate is 8
liters/minute, the amplitude is 8. $a = 8$ and $b = 4\pi \;\Rightarrow\; y = 8\sin 4\pi t.$

49 $f(t) = \frac{1}{2}\cos[\frac{\pi}{6}(t - \frac{11}{2})],$ amplitude $= \frac{1}{2},$ period $= \frac{2\pi}{\pi/6} = 12,$ phase shift $= \frac{11}{2}$

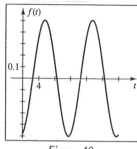

Figure 49

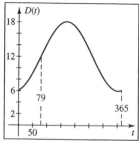

Figure 51

51 $D(t) = 6\sin[\frac{2\pi}{365}(t - 79)] + 12,$ amplitude $= 6,$ period $= \frac{2\pi}{2\pi/365} = 365,$

phase shift $= 79,$ range $= \underline{12 - 6}$ to $\underline{12 + 6}$ or 6 to 18

53 The temperature is 20°F at 9:00 A.M. $(t = 0)$.

It increases to a high of 35°F at 3:00 P.M. $(t = 6)$ and

then decreases to 20°F at 9:00 P.M. $(t = 12)$. It continues

to decrease to a low of 5°F at 3:00 A.M. $(t = 18)$.

It then rises to 20°F at 9:00 A.M. $(t = 24)$.

[0, 24, 4] by [0, 40, 4]

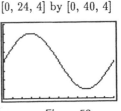

Figure 53

Note: Exer. 55–58: The period is 24 hours. Thus, $24 = \frac{2\pi}{b} \Rightarrow b = \frac{\pi}{12}$.

55 A high of 10°C and a low of −10°C imply that $d = \dfrac{\text{high} + \text{low}}{2} = \dfrac{10 + (-10)}{2} = 0$ and

$a = \text{high} - \text{average} = 10 - 0 = 10$. The average temperature of 0°C will occur 6 hours

{one-half of 12} after the low at 4 A.M., which corresponds to $t = 10$. Letting this

correspond to the first zero of the sine function, we have

$$f(t) = 10\sin\left[\tfrac{\pi}{12}(t - 10)\right] + 0 = 10\sin\left(\tfrac{\pi}{12}t - \tfrac{5\pi}{6}\right) \text{ with } a = 10,\ b = \tfrac{\pi}{12},\ c = -\tfrac{5\pi}{6},\ d = 0.$$

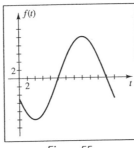

Figure 55

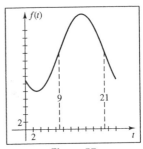

Figure 57

57 A high of 30°C and a low of 10°C imply that $d = \dfrac{30 + 10}{2} = 20$ and $a = 30 - 20 = 10$.

The average temperature of 20°C at 9 A.M. corresponds to $t = 9$.

Letting this correspond to the first zero of the sine function, we have

$$f(t) = 10\sin\left[\tfrac{\pi}{12}(t - 9)\right] + 20 = 10\sin\left(\tfrac{\pi}{12}t - \tfrac{3\pi}{4}\right) + 20 \text{ with}$$

$$a = 10,\ b = \tfrac{\pi}{12},\ c = -\tfrac{3\pi}{4},\ d = 20.$$

59 (b) Since the period is 12 months, $12 = \frac{2\pi}{b} \Rightarrow b = \frac{\pi}{6}$.

The maximum precipitation is 6.1 and the minimum

is 0.2, so the sine wave is centered vertically at

$d = \dfrac{6.1 + 0.2}{2} = 3.15$ and its amplitude is

$a = \dfrac{6.1 - 0.2}{2} = 2.95$. Since the maximum

precipitation occurs at $t = 1$ (January), we must

have $bt + c = \frac{\pi}{2} \Rightarrow \frac{\pi}{6}(1) + c = \frac{\pi}{2} \Rightarrow c = \frac{\pi}{3}$.

Thus, $P(t) = a\sin(bt + c) + d = 2.95\sin\left(\frac{\pi}{6}t + \frac{\pi}{3}\right) + 3.15$.

[0.5, 24.5, 5] by [−1, 8]

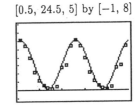

Figure 59

61 (b) Since the period is 12 months, $b = \frac{2\pi}{12} = \frac{\pi}{6}$.

From the table, the maximum number of daylight

hours is 18.72 and the minimum is 5.88.

Thus, the sine wave is centered vertically at

$d = \dfrac{18.72 + 5.88}{2} = 12.3$ and its amplitude is

$a = \dfrac{18.72 - 5.88}{2} = 6.42$. Since the maximum

daylight occurs at $t = 7$ (July), we must have

$bt + c = \frac{\pi}{2} \;\Rightarrow\; \frac{\pi}{6}(7) + c = \frac{\pi}{2} \;\Rightarrow\; c = -\frac{2\pi}{3}$.

Thus, $D(t) = a \sin(bt + c) + d = 6.42 \sin\left(\frac{\pi}{6}t - \frac{2\pi}{3}\right) + 12.3$.

[0.5, 24.5, 5] by [0, 20, 2]

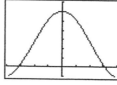

Figure 61

63 As $x \to 0^-$ or as $x \to 0^+$,

y oscillates between -1 and 1 and does not approach a unique value.

[−2, 2, 0.5] by [−1.33, 1.33, 0.5]   [−2, 2, 0.5] by [−0.33, 2.33, 0.5]

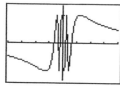   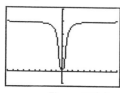

Figure 63   *Figure 65*

65 As $x \to 0^-$ or as $x \to 0^+$, y appears to approach 2.

67 From the first graph, we see that there is a horizontal asymptote of $y = 4$.

The second figure is a graph of the function near the origin.

[−20, 20, 2] by [−1, 5]   [−1, 1, 0.25] by [−0.67, 0.67, 0.25]

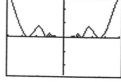   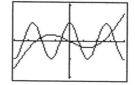

Figure 67(a)   *Figure 67(b)*

69 Graph $Y_1 = \cos 3x$ and $Y_2 = \frac{1}{2}x - \sin x$.

From the graph, Y_1 intersects Y_2 at

$x \approx -1.63, -0.45, 0.61, 1.49, 2.42$.

Thus, $\cos 3x \geq \frac{1}{2}x - \sin x$ on

$[-\pi, -1.63] \cup [-0.45, 0.61] \cup [1.49, 2.42]$.

[−π, π, π/4] by [−2.09, 2.09]

Figure 69

5.6 Exercises

Note: If $y = a\tan(bx + c)$ or $y = a\cot(bx + c)$,

then the periods for the tangent and cotangent graphs are $\pi/|b|$.

If $y = a\sec(bx + c)$ or $y = a\csc(bx + c)$,

then the periods for the secant and cosecant graphs are $2\pi/|b|$.

$\boxed{1}$ $y = 4\tan x$ •   Vertically stretch $\tan x$ by a factor of 4. The *x-intercepts* of $\tan x$ and $\cot x$ are not affected by vertically stretching or compressing their graphs. The *vertical asymptotes* of $\tan x$, $\cot x$, $\sec x$, and $\csc x$ are not affected by vertically stretching or compressing their graphs.   ★ π

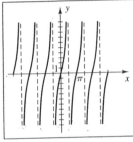

Figure 1

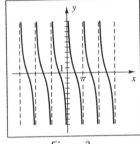

Figure 3

$\boxed{3}$ $y = 3\cot x$ •   Vertically stretch $\cot x$ by a factor of 3.   ★ π

$\boxed{5}$ $y = 2\csc x$ •   Vertically stretch $\csc x$ by a factor of 2.

Note that there is now a minimum value of 2 at $x = \frac{\pi}{2}$.

The range of this function is $(-\infty, -2] \cup [2, \infty)$, or $|y| \geq 2$.   ★ 2π

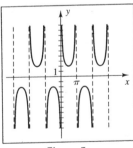

Figure 5

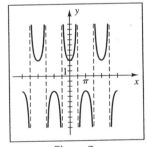

Figure 7

$\boxed{7}$ $y = 3\sec x$ •   Vertically stretch $\sec x$ by a factor of 3.

Note that there is now a minimum value of 3 at $x = 0$.

The range of this function is $(-\infty, -3] \cup [3, \infty)$ or $|y| \geq 3$.   ★ 2π

Note: The vertical asymptotes of each function are denoted by $VA @ x =$. The work to determine two consecutive vertical asymptotes is shown for each exercise. For the tangent and secant functions, the region from $-\frac{\pi}{2}$ to $\frac{\pi}{2}$ is used. For the cotangent and cosecant functions, the region from 0 to π is used.

$\boxed{9}$ $y = \tan\left(x - \frac{\pi}{4}\right)$ • Shift $\tan x$ right $\frac{\pi}{4}$ units, $VA @ x = -\frac{\pi}{4} + \pi n$. Note that the asymptotes remain π units apart. $-\frac{\pi}{2} \le x - \frac{\pi}{4} \le \frac{\pi}{2}$ $\Rightarrow$ $-\frac{\pi}{4} \le x \le \frac{3\pi}{4}$ ★ π

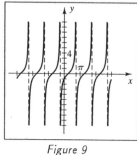

Figure 9

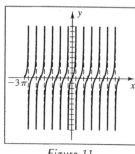

Figure 11

$\boxed{11}$ $y = \tan 2x$ • Horizontally compress $\tan x$ by a factor of 2, $VA @ x = -\frac{\pi}{4} + \frac{\pi}{2} n$. Note that the asymptotes are only $\frac{\pi}{2}$ units apart. $-\frac{\pi}{2} \le 2x \le \frac{\pi}{2}$ $\Rightarrow$ $-\frac{\pi}{4} \le x \le \frac{\pi}{4}$ ★ $\frac{\pi}{2}$

$\boxed{13}$ $y = \tan\frac{1}{4}x$ • Horizontally stretch $\tan x$ by a factor of 4, $VA @ x = -2\pi + 4\pi n$. Note that the asymptotes are 4π units apart. $-\frac{\pi}{2} \le \frac{1}{4}x \le \frac{\pi}{2}$ $\Rightarrow$ $-2\pi \le x \le 2\pi$ ★ 4π

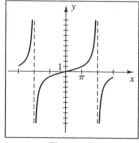

Figure 13

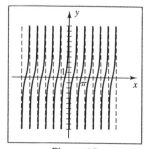

Figure 15

$\boxed{15}$ $y = 2\tan\left(2x + \frac{\pi}{2}\right) = 2\tan\left[2\left(x + \frac{\pi}{4}\right)\right]$. •

The phase shift is $-\frac{\pi}{4}$, the period is $\frac{\pi}{2}$, and we have a vertical stretching factor of 2. $-\frac{\pi}{2} \le 2x + \frac{\pi}{2} \le \frac{\pi}{2}$ $\Rightarrow$ $-\pi \le 2x \le 0$ $\Rightarrow$ $-\frac{\pi}{2} \le x \le 0$, $VA @ x = \frac{\pi}{2} n$ ★ $\frac{\pi}{2}$

$\boxed{17}$ $y = -\frac{1}{4}\tan\left(\frac{1}{2}x + \frac{\pi}{3}\right) = -\frac{1}{4}\tan\left[\frac{1}{2}\left(x + \frac{2\pi}{3}\right)\right]$.   •   Note that the "$-$" in front of the $\frac{1}{4}$

reflects the graph through the x-axis. This changes the appearance of a tangent

graph to that of a cotangent graph { increasing to decreasing }.

$-\frac{\pi}{2} \le \frac{1}{2}x + \frac{\pi}{3} \le \frac{\pi}{2}$ $\Rightarrow$ $-\frac{5\pi}{6} \le \frac{1}{2}x \le \frac{\pi}{6}$ $\Rightarrow$ $-\frac{5\pi}{3} \le x \le \frac{\pi}{3}$, VA @ $x = -\frac{5\pi}{3} + 2\pi n$ ★ 2π

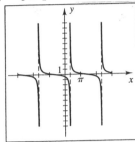

Figure 17

Figure 19

$\boxed{19}$ $y = \cot\left(x - \frac{\pi}{2}\right)$ • $0 \le x - \frac{\pi}{2} \le \pi$ $\Rightarrow$ $\frac{\pi}{2} \le x \le \frac{3\pi}{2}$, VA @ $x = \frac{\pi}{2} + \pi n$ ★ π

$\boxed{21}$ $y = \cot 2x$ • $0 \le 2x \le \pi$ $\Rightarrow$ $0 \le x \le \frac{\pi}{2}$, VA @ $x = \frac{\pi}{2}n$ ★ $\frac{\pi}{2}$

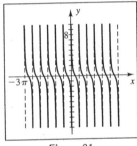

Figure 21

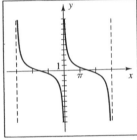

Figure 23

$\boxed{23}$ $y = \cot\frac{1}{3}x$ • $0 \le \frac{1}{3}x \le \pi$ $\Rightarrow$ $0 \le x \le 3\pi$, VA @ $x = 3\pi n$ ★ 3π

$\boxed{25}$ $y = 2\cot\left(2x + \frac{\pi}{2}\right) = 2\cot\left[2\left(x + \frac{\pi}{4}\right)\right]$.   •

$0 \le 2x + \frac{\pi}{2} \le \pi$ $\Rightarrow$ $-\frac{\pi}{2} \le 2x \le \frac{\pi}{2}$ $\Rightarrow$ $-\frac{\pi}{4} \le x \le \frac{\pi}{4}$, VA @ $x = -\frac{\pi}{4} + \frac{\pi}{2}n$ ★ $\frac{\pi}{2}$

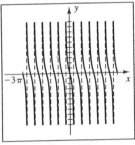

Figure 25

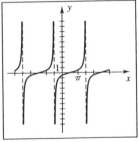

Figure 27

$\boxed{27}$ $y = -\frac{1}{2}\cot\left(\frac{1}{2}x + \frac{\pi}{4}\right) = -\frac{1}{2}\cot\left[\frac{1}{2}\left(x + \frac{\pi}{2}\right)\right]$.   •

$0 \le \frac{1}{2}x + \frac{\pi}{4} \le \pi$ $\Rightarrow$ $-\frac{\pi}{4} \le \frac{1}{2}x \le \frac{3\pi}{4}$ $\Rightarrow$ $-\frac{\pi}{2} \le x \le \frac{3\pi}{2}$, VA @ $x = -\frac{\pi}{2} + 2\pi n$ ★ 2π

29 $y = \sec\left(x - \frac{\pi}{2}\right)$ •   Note that this is the same graph as the graph of $y = \csc x$.

$-\frac{\pi}{2} \le x - \frac{\pi}{2} \le \frac{\pi}{2} \Rightarrow 0 \le x \le \pi$, VA @ $x = \pi n$ ★ 2π

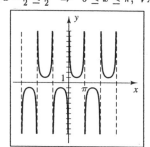

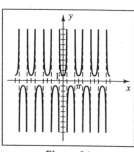

Figure 29   Figure 31

31 $y = \sec 2x$ •   Note that the asymptotes move closer together by a factor of 2.

$-\frac{\pi}{2} \le 2x \le \frac{\pi}{2} \Rightarrow -\frac{\pi}{4} \le x \le \frac{\pi}{4}$, VA @ $x = -\frac{\pi}{4} + \frac{\pi}{2}n$ ★ π

33 $y = \sec\frac{1}{3}x$ •   Note that the asymptotes move farther apart by a factor of 3,

just as they did with the graphs of tan and cot.

$-\frac{\pi}{2} \le \frac{1}{3}x \le \frac{\pi}{2} \Rightarrow -\frac{3\pi}{2} \le x \le \frac{3\pi}{2}$, VA @ $x = -\frac{3\pi}{2} + 3\pi n$ ★ 6π

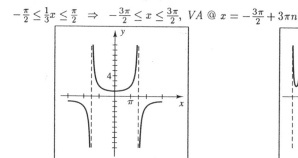

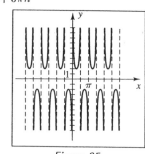

Figure 33   Figure 35

35 $y = 2\sec\left(2x - \frac{\pi}{2}\right) = 2\sec\left[2\left(x - \frac{\pi}{4}\right)\right]$.   •

$-\frac{\pi}{2} \le 2x - \frac{\pi}{2} \le \frac{\pi}{2} \Rightarrow 0 \le 2x \le \pi \Rightarrow 0 \le x \le \frac{\pi}{2}$, VA @ $x = \frac{\pi}{2}n$ ★ π

37 $y = -\frac{1}{3}\sec\left(\frac{1}{2}x + \frac{\pi}{4}\right) = -\frac{1}{3}\sec\left[\frac{1}{2}\left(x + \frac{\pi}{2}\right)\right]$.   •

$-\frac{\pi}{2} \le \frac{1}{2}x + \frac{\pi}{4} \le \frac{\pi}{2} \Rightarrow -\frac{3\pi}{4} \le \frac{1}{2}x \le \frac{\pi}{4} \Rightarrow -\frac{3\pi}{2} \le x \le \frac{\pi}{2}$, VA @ $x = -\frac{3\pi}{2} + 2\pi n$ ★ 4π

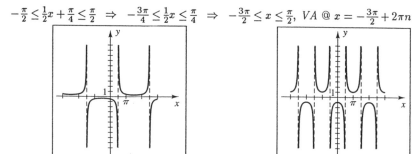

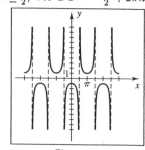

Figure 37   Figure 39

39 $y = \csc\left(x - \frac{\pi}{2}\right)$ • $0 \le x - \frac{\pi}{2} \le \pi \Rightarrow \frac{\pi}{2} \le x \le \frac{3\pi}{2}$, VA @ $x = \frac{\pi}{2} + \pi n$ ★ 2π

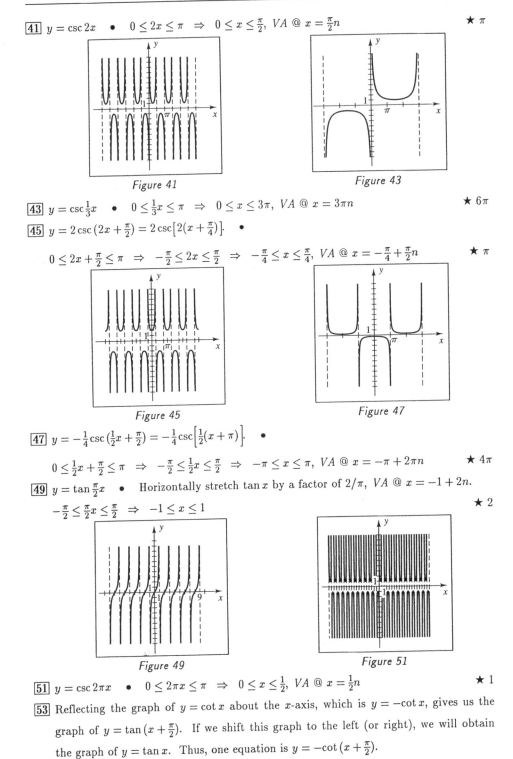

$\boxed{41}$ $y = \csc 2x$ • $0 \le 2x \le \pi \Rightarrow 0 \le x \le \frac{\pi}{2}$, VA @ $x = \frac{\pi}{2}n$ ★ π

Figure 41   *Figure 43*

$\boxed{43}$ $y = \csc \frac{1}{3}x$ • $0 \le \frac{1}{3}x \le \pi \Rightarrow 0 \le x \le 3\pi$, VA @ $x = 3\pi n$ ★ 6π

$\boxed{45}$ $y = 2\csc\left(2x + \frac{\pi}{2}\right) = 2\csc\left[2\left(x + \frac{\pi}{4}\right)\right]$. •

$0 \le 2x + \frac{\pi}{2} \le \pi \Rightarrow -\frac{\pi}{2} \le 2x \le \frac{\pi}{2} \Rightarrow -\frac{\pi}{4} \le x \le \frac{\pi}{4}$, VA @ $x = -\frac{\pi}{4} + \frac{\pi}{2}n$ ★ π

Figure 45   *Figure 47*

$\boxed{47}$ $y = -\frac{1}{4}\csc\left(\frac{1}{2}x + \frac{\pi}{2}\right) = -\frac{1}{4}\csc\left[\frac{1}{2}(x + \pi)\right]$. •

$0 \le \frac{1}{2}x + \frac{\pi}{2} \le \pi \Rightarrow -\frac{\pi}{2} \le \frac{1}{2}x \le \frac{\pi}{2} \Rightarrow -\pi \le x \le \pi$, VA @ $x = -\pi + 2\pi n$ ★ 4π

$\boxed{49}$ $y = \tan \frac{\pi}{2}x$ • Horizontally stretch $\tan x$ by a factor of $2/\pi$, VA @ $x = -1 + 2n$.

$-\frac{\pi}{2} \le \frac{\pi}{2}x \le \frac{\pi}{2} \Rightarrow -1 \le x \le 1$ ★ 2

Figure 49   *Figure 51*

$\boxed{51}$ $y = \csc 2\pi x$ • $0 \le 2\pi x \le \pi \Rightarrow 0 \le x \le \frac{1}{2}$, VA @ $x = \frac{1}{2}n$ ★ 1

$\boxed{53}$ Reflecting the graph of $y = \cot x$ about the x-axis, which is $y = -\cot x$, gives us the graph of $y = \tan\left(x + \frac{\pi}{2}\right)$. If we shift this graph to the left (or right), we will obtain the graph of $y = \tan x$. Thus, one equation is $y = -\cot\left(x + \frac{\pi}{2}\right)$.

55 $y = |\sin x|$ • Reflect the negative values of $y = \sin x$ through the x-axis. In general, when sketching the graph of $y = |f(x)|$, reflect the negative values of $f(x)$ through the x-axis. The absolute value does not affect the nonnegative values.

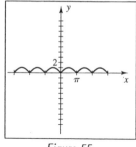

Figure 55

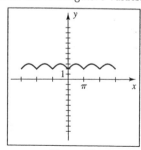

Figure 57

57 $y = |\sin x| + 2$ • Shift $y = |\sin x|$ up 2 units.

59 $y = -|\cos x| + 1$ • Similar to Exercise 55, we first reflect the negative values of $y = \cos x$ through the x-axis. The "$-$" in front of $|\cos x|$ has the effect of reflecting $y = |\cos x|$ through the x-axis. Finally, we shift that graph 1 unit up to obtain the graph of $y = -|\cos x| + 1$.

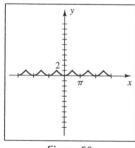

Figure 59

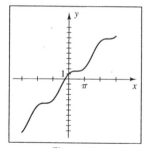

Figure 61

61 $y = x + \cos x$ • The value of $\cos x$ is between -1 and 1—adding this relatively small amount to the value of x has the effect of oscillating the graph about the line $y = x$.

63 $y = 2^{-x} \cos x$ • This graph is similar to the graph in Example 8.

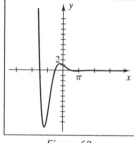

Figure 63

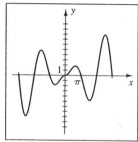

Figure 65

$\boxed{65}$ $y = |x| \sin x$ • See *Figure 65* on the preceding page. The graph will coincide

with the graph of $y = |x|$ if $\sin x = 1$—that is, if $x = \frac{\pi}{2} + 2\pi n$. The graph will

coincide with the graph of $y = -|x|$ if $\sin x = -1$—that is, if $x = \frac{3\pi}{2} + 2\pi n$.

$\boxed{67}$ $f(x) = \tan(0.5x);$ $\qquad$ $g(x) = \tan[0.5(x + \pi/2)]$ • Since $g(x) = f(x + \pi/2),$

the graph of g can be obtained by shifting the graph of f left a distance of $\frac{\pi}{2}$.

$[-2\pi, 2\pi, \pi/2]$ by $[-4, 4]$ $\qquad\qquad$ $[-2\pi, 2\pi, \pi/2]$ by $[-4, 4]$

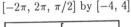

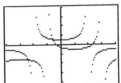

Figure 67 $\qquad\qquad\qquad\qquad\qquad$ Figure 69

$\boxed{69}$ $f(x) = 0.5 \sec 0.5x;$ $\qquad$ $g(x) = 0.5 \sec[0.5(x - \pi/2)] - 1$ •

Since $g(x) = f(x - \pi/2) - 1$, the graph of g can be obtained by shifting the graph of f

horizontally to the right a distance of $\frac{\pi}{2}$ and vertically downward a distance of 1.

$\boxed{71}$ $f(x) = 3 \cos 2x;$ $\qquad$ $g(x) = |3 \cos 2x| - 1$ • Since $g(x) = |f(x)| - 1,$

the graph of g can be obtained from the graph of f by reflecting it through the

x-axis when $f(x) < 0$ and then shifting that graph downward a distance of 1.

$[-2\pi, 2\pi, \pi/2]$ by $[-4, 4]$ $\qquad\qquad$ $[-2\pi, 2\pi, \pi/2]$ by $[-4.19, 4.19]$

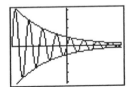

Figure 71 $\qquad\qquad\qquad\qquad\qquad$ Figure 73

$\boxed{73}$ The damping factor of $y = e^{-x/4} \sin 4x$ is $e^{-x/4}$.

$\boxed{75}$ From the graph, we see that the maximum occurs at the approximate coordinates

$(-2.76, 3.09)$, and the minimum occurs at the approximate coordinates $(1.23, -3.68)$.

$[-\pi, \pi, \pi/4]$ by $[-4, 4]$ $\qquad\qquad$ $[-2, 2]$ by $[-1.33, 1.33]$

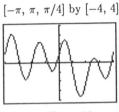

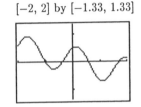

Figure 75 $\qquad\qquad\qquad\qquad\qquad$ Figure 77

$\boxed{77}$ From the graph, we see that f is increasing and one-to-one between

$a \approx -0.70$ and $b \approx 0.12$. Thus, the interval is approximately $[-0.70, 0.12]$.

[79] Graph $Y_1 = \cos(2x - 1) + \sin 3x$ and

$\qquad$ $Y_2 = \sin\frac{1}{3}x + \cos x$. From the graph, Y_1 intersects Y_2

$\qquad$ at $x \approx -1.31,\, 0.11,\, 0.95,\, 2.39$.

$\qquad$ Thus, $\cos(2x - 1) + \sin 3x \geq \sin\frac{1}{3}x + \cos x$ on

$\qquad$ $[-\pi,\, -1.31] \cup [0.11,\, 0.95] \cup [2.39,\, \pi]$.

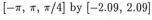

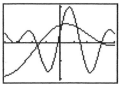

Figure 79

[81] (a) $\theta = 0 \;\Rightarrow\; I = \frac{1}{2}I_0[1 + \cos(\pi \sin 0)] = \frac{1}{2}I_0[1 + \cos(0)] = \frac{1}{2}I_0(2) = I_0$.

$\qquad$ (b) $\theta = \pi/3 \;\Rightarrow\; I = \frac{1}{2}I_0[1 + \cos(\pi \sin(\pi/3))] \approx 0.044I_0$.

$\qquad$ (c) $\theta = \pi/7 \;\Rightarrow\; I = \frac{1}{2}I_0[1 + \cos(\pi \sin(\pi/7))] \approx 0.603I_0$.

[83] (a) The damping factor of $S = A_0 e^{-\alpha z} \sin(kt - \alpha z)$ is $A_0 e^{-\alpha z}$.

$\qquad$ (b) The phase shift at depth z_0 can be found by solving the equation $kt - \alpha z_0 = 0$ for

$\qquad\qquad$ t. Doing so gives us $kt = \alpha z_0$, and hence, $t = \frac{\alpha}{k}z_0$.

$\qquad$ (c) At the surface, $z = 0$. Hence, $S = A_0 \sin kt$ and the amplitude at the surface is

$\qquad\quad$ A_0. Amplitude$_{\text{wave}} = \frac{1}{2}$Amplitude$_{\text{surface}} \;\Rightarrow\;$

$$A_0 e^{-\alpha z} = \tfrac{1}{2}A_0 \;\Rightarrow\; e^{-\alpha z} = \tfrac{1}{2} \;\Rightarrow\; -\alpha z = \ln\tfrac{1}{2} \;\Rightarrow\; z = \frac{-\ln 2}{-\alpha} = \frac{\ln 2}{\alpha}.$$

5.7 Exercises

Note: The missing values are found in terms of the given values.

$\qquad$ We could also use proportions to find the remaining parts.

[1] Since α is given and $\gamma = 90°$, we can easily find β. $\beta = 90° - \alpha = 90° - 30° = 60°$.

$\qquad$ To find a, we will relate it to the given parts, α and b, using the tangent function.

$\qquad\quad$ $\tan\alpha = \frac{a}{b} \;\Rightarrow\; a = b\tan\alpha = 20\tan 30° = 20(\frac{1}{3}\sqrt{3}) = \frac{20}{3}\sqrt{3}$.

$\qquad\quad$ $\sec\alpha = \frac{c}{b} \;\Rightarrow\; c = b\sec\alpha = 20\sec 30° = 20(\frac{2}{3}\sqrt{3}) = \frac{40}{3}\sqrt{3}$.

[3] $\alpha = 90° - \beta = 90° - 45° = 45°$.

$\qquad$ $\cos\beta = \frac{a}{c} \;\Rightarrow\; a = c\cos\beta = 30\cos 45° = 30(\frac{1}{2}\sqrt{2}) = 15\sqrt{2}$.

$\qquad\qquad\qquad\qquad\qquad\qquad\qquad\qquad$ $b = a$ in a $45°$–$45°$–$90°$ $\triangle$.

[5] $\tan\alpha = \frac{a}{b} = \frac{5}{5} = 1 \;\Rightarrow\; \alpha = 45°$. $\beta = 90° - \alpha = 90° - 45° = 45°$. Using the

$\qquad$ Pythagorean theorem, $a^2 + b^2 = c^2 \;\Rightarrow\; c = \sqrt{a^2 + b^2} = \sqrt{25 + 25} = \sqrt{50} = 5\sqrt{2}$.

[7] $\cos\alpha = \frac{b}{c} = \frac{5\sqrt{3}}{10\sqrt{3}} = \frac{1}{2} \;\Rightarrow\; \alpha = 60°$. $\beta = 90° - \alpha = 90° - 60° = 30°$.

$\qquad\qquad\qquad\qquad\qquad\quad$ $a = \sqrt{c^2 - b^2} = \sqrt{300 - 75} = \sqrt{225} = 15$.

[9] $\beta = 90° - \alpha = 90° - 37° = 53°$. $\tan\alpha = \frac{a}{b} \;\Rightarrow\; a = b\tan\alpha = 24\tan 37° \approx 18$.

$\qquad\qquad\qquad\qquad\qquad$ $\sec\alpha = \frac{c}{b} \;\Rightarrow\; c = b\sec\alpha = 24\sec 37° \approx 30$.

$\boxed{11}$ $\alpha = 90° - \beta = 90° - 71°51' = 18°9'.$ $\cot\beta = \frac{a}{b}$ $\Rightarrow$ $a = b\cot\beta = 240.0\cot 71°51' \approx 78.7.$

$\csc\beta = \frac{c}{b}$ $\Rightarrow$ $c = b\csc\beta = 240.0\csc 71°51' \approx 252.6.$

$\boxed{13}$ $\tan\alpha = \frac{a}{b} = \frac{25}{45}$ $\Rightarrow$ $\alpha = \tan^{-1}\frac{25}{45} \approx 29°.$ $\beta = 90° - \alpha \approx 90° - 29° = 61°.$

$c = \sqrt{a^2 + b^2} = \sqrt{25^2 + 45^2} = \sqrt{625 + 2025} = \sqrt{2650} \approx 51.$

$\boxed{15}$ $\cos\alpha = \frac{b}{c} = \frac{2.1}{5.8}$ $\Rightarrow$ $\alpha = \cos^{-1}\frac{21}{58} \approx 69°.$ $\beta = 90° - \alpha \approx 90° - 69° = 21°.$

$a = \sqrt{c^2 - b^2} = \sqrt{(5.8)^2 - (2.1)^2} = \sqrt{33.64 - 4.41} = \sqrt{29.23} \approx 5.4.$

Note: Refer to Figures 1 and 2 in the text for the labeling of the sides and angles.

$\boxed{17}$ We need to find a relationship involving b, c, and α. We want angle α with its adjacent side b and hypotenuse c. The cosine or secant are the functions of α that involve b and c. We choose the cosine since it is easier to solve for b $\{b$ is in the numerator$\}$. $\cos\alpha = \frac{b}{c}$ $\Rightarrow$ $b = c\cos\alpha.$

$\boxed{19}$ We want angle β with its adjacent side a and opposite side b. The tangent or cotangent are the functions of β that involve a and b. $\cot\beta = \frac{a}{b}$ $\Rightarrow$ $a = b\cot\beta.$

$\boxed{21}$ We want angle α with its opposite side a and hypotenuse c. The sine or cosecant are the functions of α that involve a and c. $\csc\alpha = \frac{c}{a}$ $\Rightarrow$ $c = a\csc\alpha.$

$\boxed{23}$ $a^2 + b^2 = c^2$ $\Rightarrow$ $b^2 = c^2 - a^2$ $\Rightarrow$ $b = \sqrt{c^2 - a^2}$

$\boxed{25}$ Let h denote the height of the kite and $x = h - 4.$ $\sin 60° = \frac{x}{500}$ $\Rightarrow$

$x = 500\sin 60° = 500(\frac{1}{2}\sqrt{3}) = 250\sqrt{3}.$ $h = x + 4 = 250\sqrt{3} + 4 \approx 437$ ft.

$\boxed{27}$ $\sin 10° = \frac{5000}{x}$ $\Rightarrow$ $x = \frac{5000}{\sin 10°}$ $\Rightarrow$

$x = 5000\csc 10° \approx 28{,}793.85,$ or $28{,}800$ ft.

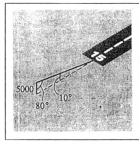

Figure 27

$\boxed{29}$ $\tan\angle PRQ = \frac{d}{50.0}$ $\Rightarrow$ $\tan 72°40' = \frac{d}{50}$ $\Rightarrow$ $d = 50\tan 72°40' \approx 160$ m.

$\boxed{31}$ The 10,000 feet would represent the hypotenuse in a triangle depicting this

information. Let h denote the altitude. $\sin 75° = \frac{h}{10{,}000}$ $\Rightarrow$ $h \approx 9659$ ft.

$\boxed{33}$ (a) The bridge section is 75 feet long. Using the right triangle with the 75 foot section as its hypotenuse and $(d - 15)$ as the side opposite the $35°$ angle, we have

$\sin 35° = \frac{d - 15}{75}$ $\Rightarrow$ $d = 75\sin 35° + 15 \approx 58$ ft.

(b) Let x be the horizontal distance from the end of a bridge section to a point directly underneath the end of the section. $\cos 35° = \frac{x}{75} \Rightarrow x = 75 \cos 35°$.

The distance between the ends of the two sections is (total distance) −

(the 2 horizontal distances under the bridge sections) $= 150 - 2x \approx 27$ ft.

[35] Let α denote the angle of elevation. $\tan \alpha = \frac{5}{4} \Rightarrow \alpha \approx 51°20'$.

[37] Let D denote the position of the duck and t the number of seconds required for a direct hit. The duck will move $(7t)$ cm. and the bullet will travel $(25t)$ cm.

$$\sin \varphi = \frac{\overline{AD}}{\overline{OD}} = \frac{7t}{25t} \Rightarrow \sin \varphi = \frac{7}{25} \Rightarrow \varphi \approx 16.3°.$$

[39] Let h denote the height of the tower.

$$\tan 21°20'24'' = \tan 21.34° = \frac{h}{5280} \Rightarrow h = 5280 \tan 21.34° \approx 2063 \text{ ft.}$$

[41] The central angle of a section of the Pentagon has measure $\frac{360°}{5} = 72°$.

Bisecting that angle, we have an angle of $36°$ whose opposite side is $\frac{921}{2}$.

The height h is given by $\tan 36° = \dfrac{\frac{921}{2}}{h} \Rightarrow h = \dfrac{921}{2 \tan 36°}.$

$$\text{Area} = 5(\tfrac{1}{2}bh) = 5(\tfrac{1}{2})(921)\left(\frac{921}{2 \tan 36°}\right) \approx 1{,}459{,}379 \text{ ft}^2.$$

[43] The diagonal of the base is $\sqrt{8^2 + 6^2} = 10$. $\tan \theta = \frac{4}{10} \Rightarrow \theta \approx 21.8°$

[45] $\cot 53°30' = \frac{x}{h} \Rightarrow x = h \cot 53°30'$. $\cot 26°50' = \frac{x + 25}{h}$

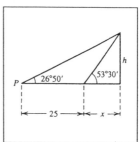

$\Rightarrow x + 25 = h \cot 26°50' \Rightarrow x = h \cot 26°50' - 25.$

We now have two expressions for x.

We will set these equal to each other and solve for h.

Thus, $h \cot 53°30' = h \cot 26°50' - 25 \Rightarrow$

$25 = h \cot 26°50' - h \cot 53°30' \Rightarrow$

$h = \dfrac{25}{\cot 26°50' - \cot 53°30'} \approx 20.2$ m.   *Figure 45*

[47] When the angle of elevation is $19°20'$, $\tan 19°20' = \dfrac{h_1}{110} \Rightarrow h_1 = 110 \tan 19°20'.$

When the angle of elevation is $31°50'$, $\tan 31°50' = \dfrac{h_2}{110} \Rightarrow h_2 = 110 \tan 31°50'.$

The change in elevation is $h_2 - h_1 \approx 68.29 - 38.59 = 29.7$ km.

[49] The distance from the spacelab to the center of the earth is $(380 + r)$ miles.

$\sin 65.8° = \dfrac{r}{r + 380} \Rightarrow r \sin 65.8° + 380 \sin 65.8° = r \Rightarrow$

$r - r \sin 65.8° = 380 \sin 65.8° \Rightarrow r(1 - \sin 65.8°) = 380 \sin 65.8° \Rightarrow$

$$r = \frac{380 \sin 65.8°}{1 - \sin 65.8°} \approx 3944 \text{ mi.}$$

51 Let d be the distance traveled. $\tan 42° = \dfrac{10{,}000}{d} \Rightarrow d = 10{,}000 \cot 42°$.

Converting to mi/hr, we have $\dfrac{10{,}000 \cot 42° \text{ ft}}{1 \text{ minute}} \cdot \dfrac{60 \text{ minutes}}{1 \text{ hour}} \cdot \dfrac{1 \text{ mile}}{5280 \text{ ft}} \approx 126$ mi/hr.

53 (a) As in Exercise 49, there is a right angle formed on the earth's surface.

Bisecting angle θ and forming a right triangle, we have

$\cos\dfrac{\theta}{2} = \dfrac{R}{R+a} = \dfrac{4000}{26{,}300}$. Thus, $\dfrac{\theta}{2} \approx 81.25° \Rightarrow \theta \approx 162.5°$.

The percentage of the equator that is within signal range is $\dfrac{162.5°}{360°} \times 100 \approx 45\%$.

(b) Each satellite has a signal range of more than $120°$,

and thus all 3 will cover all points on the equator.

55 Let $x = h - c$. $\sin\alpha = \dfrac{x}{d} \Rightarrow x = d\sin\alpha$. $h = x + c = d\sin\alpha + c$.

57 Let x denote the distance from the base of the tower to the closer point.

$\cot\beta = \dfrac{x}{h} \Rightarrow x = h\cot\beta$. $\cot\alpha = \dfrac{x+d}{h} \Rightarrow x + d = h\cot\alpha \Rightarrow x = h\cot\alpha - d$.

Thus, $h\cot\beta = h\cot\alpha - d \Rightarrow d = h\cot\alpha - h\cot\beta \Rightarrow d = h(\cot\alpha - \cot\beta) \Rightarrow$

$$h = \dfrac{d}{\cot\alpha - \cot\beta}.$$

59 When the angle of elevation is α, $\tan\alpha = \dfrac{h_1}{d} \Rightarrow h_1 = d\tan\alpha$.

When the angle of elevation is β, $\tan\beta = \dfrac{h_2}{d} \Rightarrow h_2 = d\tan\beta$.

$$h = h_2 - h_1 = d\tan\beta - d\tan\alpha = d(\tan\beta - \tan\alpha).$$

61 The bearing from P to A is $90° - 20° = 70°$ east of north and is denoted by N70°E.

The bearing from P to B is $40°$ west of north and is denoted by N40°W.

The bearing from P to C is $90° - 75° = 15°$ west of south and is denoted by S15°W.

The bearing from P to D is $25°$ east of south and is denoted by S25°E.

63 (a) The first ship travels 2 hours @ 24 mi/hr for a distance of 48 miles. The second ship travels $1\frac{1}{2}$ hours @ 18 mi/hr for a distance of 27 miles.

The paths form a right triangle with legs of 48 miles and 27 miles. The distance between the two ships is

$\sqrt{27^2 + 48^2} \approx 55$ miles.

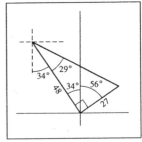

Figure 63

(b) The angle between the side of length 48 and the hypotenuse is found by solving $\tan\alpha = \dfrac{27}{48}$ for α. Now

$\alpha \approx 29°$, so the second ship is approximately $29° + 34° = $ S63°E of the first ship.

$\boxed{65}$ 30 minutes @ 360 mi/hr $= 180$ miles. 45 minutes @ 360 mi/hr $= 270$ miles. $227° - 137° = 90°$, and hence, the plane's flight forms a right triangle with legs 180 miles and 270 miles. The distance from A to the airplane is equal to the hypotenuse of the triangle—that is, $\sqrt{180^2 + 270^2} \approx 324.5$ mi.

$\boxed{67}$ Amplitude, 10 cm; period $= \frac{2\pi}{6\pi} = \frac{1}{3}$ sec; frequency $= \frac{6\pi}{2\pi} = 3$ oscillations/sec. The point is at the origin at $t = 0$. It moves upward with decreasing speed, reaching the point with coordinate 10 when $6\pi t = \frac{\pi}{2}$ or $t = \frac{1}{12}$.

It then reverses direction and moves downward, gaining speed until it reaches the origin when $6\pi t = \pi$ or $t = \frac{1}{6}$. It continues downward with decreasing speed, reaching the point with coordinate -10 when $6\pi t = \frac{3\pi}{2}$ or $t = \frac{1}{4}$.

It then reverses direction and moves upward with increasing speed, returning to the origin when $6\pi t = 2\pi$ or $t = \frac{1}{3}$ to complete one oscillation.

Another approach is to simply model this movement in terms of proportions of the sine curve. For one period, the sine increases for $\frac{1}{4}$ period, decreases for $\frac{1}{2}$ period, and increases for its last $\frac{1}{4}$ period.

$\boxed{69}$ Amplitude, 4 cm; period $= \frac{2\pi}{3\pi/2} = \frac{4}{3}$ sec; frequency $= \frac{3\pi/2}{2\pi} = \frac{3}{4}$ oscillation/sec. The point is at $d = 4$ when $t = 0$. It then decreases in height until $\frac{3\pi}{2}t = \pi$ or $t = \frac{2}{3}$ where it obtains a minimum of $d = -4$. It then reverses direction and increases to a height of $d = 4$ when $\frac{3\pi}{2}t = 2\pi$ or $t = \frac{4}{3}$ to complete one oscillation.

$\boxed{71}$ Period $= 3$ $\Rightarrow$ $\frac{2\pi}{\omega} = 3$ $\Rightarrow$ $\omega = \frac{2\pi}{3}$. Amplitude $= 5$ $\Rightarrow$ $a = 5$. $d = 5\cos\frac{2\pi}{3}t$

$\boxed{73}$ (a) period $= 30$ $\Rightarrow$ $\frac{2\pi}{\omega} = 30$ $\Rightarrow$ $\omega = \frac{\pi}{15}$. When $t = 0$, the wave is at its highest

point, thus, we use the cosine function. $y = 25\cos\frac{\pi}{15}t$, where t is in minutes.

(b) 180 ft/sec $= 10{,}800$ ft/min.

10,800 ft/min for 30 minutes is a distance of 324,000 ft, or $61\frac{4}{11}$ miles.

Chapter 5 Review Exercises

$\boxed{1}$ $330° \cdot \frac{\pi}{180} = \frac{11 \cdot 30\pi}{6 \cdot 30} = \frac{11\pi}{6}$; $405° \cdot \frac{\pi}{180} = \frac{9 \cdot 45\pi}{4 \cdot 45} = \frac{9\pi}{4}$

$-150° \cdot \frac{\pi}{180} = -\frac{5 \cdot 30\pi}{6 \cdot 30} = -\frac{5\pi}{6}$; $240° \cdot \frac{\pi}{180} = \frac{4 \cdot 60\pi}{3 \cdot 60} = \frac{4\pi}{3}$;

$36° \cdot \frac{\pi}{180} = \frac{36\pi}{5 \cdot 36} = \frac{\pi}{5}$

$\boxed{2}$ $\frac{9\pi}{2} \cdot \left(\frac{180}{\pi}\right)° = \left(\frac{9 \cdot 90 \cdot 2\pi}{2\pi}\right)° = 810°$; $-\frac{2\pi}{3} \cdot \left(\frac{180}{\pi}\right)° = -\left(\frac{2 \cdot 60 \cdot 3\pi}{3\pi}\right)° = -120°$

$\boxed{3}$ (a) $\theta = \frac{s}{r} = \frac{20\,\text{cm}}{2\,\text{m}} = \frac{20\,\text{cm}}{2\,(100)\,\text{cm}} = 0.1\,\text{radian}$ (b) $A = \frac{1}{2}r^2\theta = \frac{1}{2}(2)^2(0.1) = 0.2$ m^2

$\boxed{4}$ (a) $s = r\theta = (15 \cdot \frac{1}{2})(70 \cdot \frac{\pi}{180}) = \frac{35\pi}{12} \approx 9.16$ cm

     (b) $A = \frac{1}{2}r^2\theta = \frac{1}{2}(15 \cdot \frac{1}{2})^2(70 \cdot \frac{\pi}{180}) = \frac{175\pi}{16} \approx 34.4$ cm^2

$\boxed{5}$ As in Exercise 43 in Section 5.1, we take the number of revolutions per minute times the number of radians per revolution and obtain $(33\frac{1}{3})(2\pi) = \frac{200\pi}{3}$ and $45(2\pi) = 90\pi$.

$\boxed{7}$ $\sin 60° = \frac{9}{x} \Rightarrow \frac{\sqrt{3}}{2} = \frac{9}{x} \Rightarrow x = 6\sqrt{3};\ \tan 60° = \frac{9}{y} \Rightarrow \sqrt{3} = \frac{9}{y} \Rightarrow y = 3\sqrt{3}$

$\boxed{9}$ $1 + \tan^2\theta = \sec^2\theta \Rightarrow \tan^2\theta = \sec^2\theta - 1 \Rightarrow \tan\theta = \sqrt{\sec^2\theta - 1}$

$\boxed{11}$ $\sin\theta\,(\csc\theta - \sin\theta) = \sin\theta\,\csc\theta - \sin^2\theta$ { multiply terms }

                 $= \sin\theta \cdot \dfrac{1}{\sin\theta} - \sin^2\theta$ { reciprocal identity }

                 $= 1 - \sin^2\theta$ { simplify }

                 $= \cos^2\theta$ { Pythagorean identity }

$\boxed{13}$ $(\cos^2\theta - 1)(\tan^2\theta + 1) = (\cos^2\theta - 1)(\sec^2\theta)$ { Pythagorean identity }

                 $= \cos^2\theta\,\sec^2\theta - \sec^2\theta$ { multiply terms }

                 $= 1 - \sec^2\theta$ { reciprocal identity }

$\boxed{15}$ $\dfrac{1 + \tan^2\theta}{\tan^2\theta} = \dfrac{1}{\tan^2\theta} + \dfrac{\tan^2\theta}{\tan^2\theta}$ { split up the fraction }

                 $= \cot^2\theta + 1$ { reciprocal identity, simplify }

                 $= \csc^2\theta$ { Pythagorean identity }

$\boxed{17}$ $\dfrac{\cot\theta - 1}{1 - \tan\theta} = \dfrac{\dfrac{\cos\theta}{\sin\theta} - 1}{1 - \dfrac{\sin\theta}{\cos\theta}}$ { put in terms of sines and cosines }

           $= \dfrac{\dfrac{\cos\theta - \sin\theta}{\sin\theta}}{\dfrac{\cos\theta - \sin\theta}{\cos\theta}}$ $\left\{ \begin{array}{l} \text{make the numerator and the} \\ \text{denominator each a single fraction} \end{array} \right\}$

           $= \dfrac{(\cos\theta - \sin\theta)\cos\theta}{(\cos\theta - \sin\theta)\sin\theta}$ { simplify a complex fraction }

           $= \dfrac{\cos\theta}{\sin\theta}$ { cancel like term }

           $= \cot\theta$ { cotangent identity }

$\boxed{19}$ $\dfrac{\tan(-\theta) + \cot(-\theta)}{\tan\theta} = \dfrac{-\tan\theta - \cot\theta}{\tan\theta}$ { formulas for negatives }

           $= -\dfrac{\tan\theta}{\tan\theta} - \dfrac{\cot\theta}{\tan\theta}$ { split up fraction }

           $= -1 - \cot^2\theta$ { simplify, reciprocal identity }

           $= -(1 + \cot^2\theta)$ { factor out -1 }

           $= -\csc^2\theta$ { Pythagorean identity }

$\boxed{21}$ $\text{opp} = \sqrt{\text{hyp}^2 - \text{adj}^2} = \sqrt{7^2 - 4^2} = \sqrt{33}.$ $\qquad \star \dfrac{\sqrt{33}}{7}, \dfrac{4}{7}, \dfrac{\sqrt{33}}{4}, \dfrac{4}{\sqrt{33}}, \dfrac{7}{4}, \dfrac{7}{\sqrt{33}}$

$\boxed{22}$ (a) $x = 30$ and $y = -40 \;\Rightarrow\; r = \sqrt{30^2 + (-40)^2} = 50.$ $\quad \star$ (a) $-\frac{4}{5}, \frac{3}{5}, -\frac{4}{3}, -\frac{3}{4}, \frac{5}{3}, -\frac{5}{4}$

(b) $2x + 3y + 6 = 0 \;\Leftrightarrow\; y = -\frac{2}{3}x - 2$, so the slope of the given line is $-\frac{2}{3}$.

The line through the origin with that slope is $y = -\frac{2}{3}x$.

If $x = -3$, then $y = 2$ and $(-3, 2)$ is a point on the terminal side of θ.

$$x = -3 \text{ and } y = 2 \;\Rightarrow\; r = \sqrt{(-3)^2 + 2^2} = \sqrt{13}.$$

$$\star \text{ (b) } \dfrac{2}{\sqrt{13}}, -\dfrac{3}{\sqrt{13}}, -\dfrac{2}{3}, -\dfrac{3}{2}, -\dfrac{\sqrt{13}}{3}, \dfrac{\sqrt{13}}{2}$$

(c) For $\theta = -90°$, choose $x = 0$ and $y = -1$. r is 1. $\qquad \star$ (c) $-1, 0, U, 0, U, -1$

$\boxed{23}$ (a) $\sec\theta < 0 \;\Rightarrow\;$ terminal side of θ is in QII or QIII.

$\qquad\qquad\quad \sin\theta > 0 \;\Rightarrow\;$ terminal side of θ is in QI or QII. Hence, θ is in QII.

(b) $\cot\theta > 0 \;\Rightarrow\;$ terminal side of θ is in QI or QIII.

$\qquad\qquad\quad \csc\theta < 0 \;\Rightarrow\;$ terminal side of θ is in QIII or QIV. Hence, θ is in QIII.

(c) $\cos\theta > 0 \;\Rightarrow\;$ terminal side of θ is in QI or QIV.

$\qquad\qquad\quad \tan\theta < 0 \;\Rightarrow\;$ terminal side of θ is in QII or QIV. Hence, θ is in QIV.

$\boxed{24}$ (b) $\cot\theta = \dfrac{\cos\theta}{\sin\theta} = \dfrac{\csc\theta}{\sec\theta} \;\Rightarrow\; -\dfrac{3}{2} = \dfrac{\sqrt{13}/2}{\sec\theta} \;\Rightarrow\; \sec\theta = -\dfrac{\sqrt{13}}{3};$

$\qquad\qquad\qquad\qquad\qquad\qquad\qquad\qquad$ the other values are just the reciprocals.

$\boxed{25}$ 7π is coterminal with π. $P(7\pi) = P(\pi) = (-1, 0)$.

$-\frac{5\pi}{2}$ is coterminal with $-\frac{\pi}{2}$. $P(-\frac{5\pi}{2}) = P(-\frac{\pi}{2}) = (0, -1)$. $P(\frac{9\pi}{2}) = P(\frac{\pi}{2}) = (0, 1)$.

$P(-\frac{3\pi}{4}) = (-\frac{\sqrt{2}}{2}, -\frac{\sqrt{2}}{2})$. $P(18\pi) = P(0) = (1, 0)$. $P(\frac{\pi}{6}) = (\frac{\sqrt{3}}{2}, \frac{1}{2})$.

$\boxed{26}$ $P(t) = (-\frac{3}{5}, -\frac{4}{5})$ is in QIII. $P(t + \pi)$ is in QI and will have the same coordinates as

$P(t)$, but with opposite (positive) signs. $P(t + 3\pi) = P(t + \pi) = P(t - \pi) = (\frac{3}{5}, \frac{4}{5})$.

$\boxed{27}$ (a) $\theta = \frac{5\pi}{4} \;\Rightarrow\; \theta_R = \frac{5\pi}{4} - \pi = \frac{\pi}{4}$.

$\qquad \theta = -\frac{5\pi}{6} \;\Rightarrow\; \theta_C = \frac{7\pi}{6}$ and $\theta_R = \frac{7\pi}{6} - \pi = \frac{\pi}{6}$.

$\qquad \theta = -\frac{9\pi}{8} \;\Rightarrow\; \theta_C = \frac{7\pi}{8}$ and $\theta_R = \pi - \frac{7\pi}{8} = \frac{\pi}{8}$.

(b) $\theta = 245° \;\Rightarrow\; \theta_R = 245° - 180° = 65°$.

$\qquad \theta = 137° \;\Rightarrow\; \theta_R = 180° - 137° = 43°$.

$\qquad \theta = 892° \;\Rightarrow\; \theta_C = 172°$ and $\theta_R = 180° - 172° = 8°$.

$\boxed{28}$ (b) For $\theta = -\frac{5\pi}{4}$, choose $x = -1$ and $y = 1$. $r = \sqrt{2}$.

$$\star \text{ (b) } \dfrac{\sqrt{2}}{2}, -\dfrac{\sqrt{2}}{2}, -1, -1, -\sqrt{2}, \sqrt{2}$$

(d) For $\theta = \frac{11\pi}{6}$, choose $x = \sqrt{3}$ and $y = -1$. $r = 2$.

$$\star \text{ (d) } -\dfrac{1}{2}, \dfrac{\sqrt{3}}{2}, -\dfrac{\sqrt{3}}{3}, -\sqrt{3}, \dfrac{2}{\sqrt{3}}, -2$$

29 (a) $\cos 225° = -\cos 45° = -\frac{\sqrt{2}}{2}$

(b) $\tan 150° = -\tan 30° = -\frac{\sqrt{3}}{3}$

(c) $\sin\left(-\frac{\pi}{6}\right) = -\sin\frac{\pi}{6} = -\frac{1}{2}$

(d) $\sec\frac{4\pi}{3} = -\sec\frac{\pi}{3} = -2$

(e) $\cot\frac{7\pi}{4} = -\cot\frac{\pi}{4} = -1$

(f) $\csc 300° = -\csc 60° = -\frac{2}{\sqrt{3}}$

30 $\sin\theta = -0.7604 \;\Rightarrow\; \theta = \sin^{-1}(-0.7604) \approx -49.5° \;\Rightarrow\; \theta_R \approx 49.5°$.

Since the sine is negative in QIII and QIV, and the secant is positive in QIV,

we want the fourth-quadrant angle having $\theta_R = 49.5°$. $360° - 49.5° = 310.5°$

31 $\tan\theta = 2.7381 \;\Rightarrow\; \theta = \tan^{-1}(2.7381) \approx 1.2206 \;\Rightarrow\; \theta_R \approx 1.2206$. Since the tangent

is positive in QI and QIII, we need to add π to θ_R to find the value in QIII.

$\pi + \theta_R \approx 4.3622$

33 $y = 5\cos x$ • Vertically stretch $\cos x$ by a factor of 5. ★ 5, 2π, x-int. @ $\frac{\pi}{2} + \pi n$

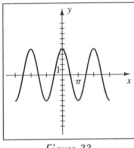

Figure 33

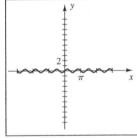

Figure 35

35 $y = \frac{1}{3}\sin 3x$ • Horizontally compress $\sin x$ by a factor of 3 and vertically compress

that graph by a factor of 3. ★ $\frac{1}{3}$, $\frac{2\pi}{3}$, x-int. @ $\frac{\pi}{3}n$

37 $y = -3\cos\frac{1}{2}x$ • Horizontally stretch $\cos x$ by a factor of 2, vertically stretch that

graph by a factor of 3, and reflect that graph through the x-axis.

★ 3, 4π, x-int. @ $\pi + 2\pi n$

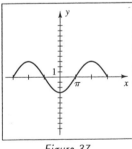

Figure 37

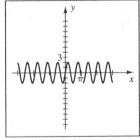

Figure 39

39 $y = 2\sin\pi x$ • Horizontally compress $\sin x$ by a factor of π, and then vertically

stretch that graph by a factor of 2. ★ 2, 2, x-int. @ n

Note: Let a denote the amplitude and p the period.

$\boxed{41}$ (a) $a = |-1.43| = 1.43$. We have $\frac{3}{4}$ of a period from $(0, 0)$ to $(1.5, -1.43)$.

$$\tfrac{3}{4}p = 1.5 \;\Rightarrow\; p = 2.$$

(b) Since the period p is given by $\frac{2\pi}{b}$, we can solve for b, obtaining $b = \frac{2\pi}{p}$.

$$\text{Hence, } b = \frac{2\pi}{p} = \frac{2\pi}{2} = \pi \text{ and consequently, } y = 1.43 \sin \pi x.$$

$\boxed{43}$ (a) Since the y-intercept is -3, $a = |-3| = 3$. The second positive x-intercept is π,

so $\frac{3}{4}$ of a period occurs from $(0, -3)$ to $(\pi, 0)$. Thus, $\frac{3}{4}p = \pi \;\Rightarrow\; p = \frac{4\pi}{3}$.

(b) $b = \frac{2\pi}{p} = \frac{2\pi}{4\pi/3} = \frac{3}{2}$, $y = -3 \cos \frac{3}{2}x$.

$\boxed{45}$ $y = 2 \sin \left(x - \frac{2\pi}{3}\right)$ • $0 \le x - \frac{2\pi}{3} \le 2\pi \;\Rightarrow\; \frac{2\pi}{3} \le x \le \frac{8\pi}{3}$.

There are x-intercepts at $x = \frac{2\pi}{3} + \pi n$.

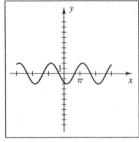

Figure 45

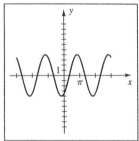

Figure 47

$\boxed{47}$ $y = -4 \cos \left(x + \frac{\pi}{6}\right)$ • $-\frac{\pi}{2} \le x + \frac{\pi}{6} \le \frac{3\pi}{2} \;\Rightarrow\; -\frac{2\pi}{3} \le x \le \frac{4\pi}{3}$.

There are x-intercepts at $x = -\frac{2\pi}{3} + \pi n$.

$\boxed{49}$ $y = 2 \tan \left(\frac{1}{2}x - \pi\right) = 2 \tan \left[\frac{1}{2}(x - 2\pi)\right]$. •

$-\frac{\pi}{2} \le \frac{1}{2}x - \pi \le \frac{\pi}{2} \;\Rightarrow\; \frac{\pi}{2} \le \frac{1}{2}x \le \frac{3\pi}{2} \;\Rightarrow\; \pi \le x \le 3\pi$, VA @ $x = \pi + 2\pi n$

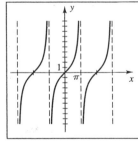

Figure 49

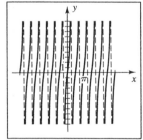

Figure 51

$\boxed{51}$ $y = -4 \cot \left(2x - \frac{\pi}{2}\right) = -4 \cot \left[2\left(x - \frac{\pi}{4}\right)\right]$. •

$0 \le 2x - \frac{\pi}{2} \le \pi \;\Rightarrow\; \frac{\pi}{2} \le 2x \le \frac{3\pi}{2} \;\Rightarrow\; \frac{\pi}{4} \le x \le \frac{3\pi}{4}$, VA @ $x = \frac{\pi}{4} + \frac{\pi}{2}n$

[53] $y = \sec\left(\frac{1}{2}x + \pi\right) = \sec\left[\frac{1}{2}(x + 2\pi)\right].$ •

$-\frac{\pi}{2} \le \frac{1}{2}x + \pi \le \frac{\pi}{2} \Rightarrow -\frac{3\pi}{2} \le \frac{1}{2}x \le -\frac{\pi}{2} \Rightarrow -3\pi \le x \le -\pi,$ VA @ $x = -3\pi + 2\pi n$

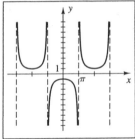

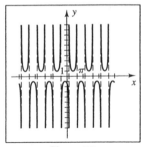

Figure 53   Figure 55

[55] $y = \csc\left(2x - \frac{\pi}{4}\right) = \csc\left[2\left(x - \frac{\pi}{8}\right)\right].$ •

$0 \le 2x - \frac{\pi}{4} \le \pi \Rightarrow \frac{\pi}{4} \le 2x \le \frac{5\pi}{4} \Rightarrow \frac{\pi}{8} \le x \le \frac{5\pi}{8},$ VA @ $x = \frac{\pi}{8} + \frac{\pi}{2}n$

[57] α and β are complementary, so $\alpha = 90° - \beta = 90° - 60° = 30°.$

$\cot\beta = \frac{a}{b} \Rightarrow a = b\cot\beta = 40\cot 60° = 40(\frac{1}{3}\sqrt{3}) \approx 23.$

$\csc\beta = \frac{c}{b} \Rightarrow c = b\csc\beta = 40\csc 60° = 40(\frac{2}{3}\sqrt{3}) \approx 46.$

[59] $\tan\alpha = \frac{a}{b} = \frac{62}{25} \Rightarrow \alpha \approx 68°.$ $\beta = 90° - \alpha \approx 90° - 68° = 22°.$

$c = \sqrt{a^2 + b^2} = \sqrt{62^2 + 25^2} = \sqrt{3844 + 625} = \sqrt{4469} \approx 67.$

[61] (a) $\left(\frac{545\text{ rev}}{1\text{ min}}\right)\left(\frac{2\pi\text{ rad}}{1\text{ rev}}\right)\left(\frac{1\text{ min}}{60\text{ sec}}\right) = \frac{109\pi}{6}$ rad/sec ≈ 57 rad/sec

(b) $d = 22.625$ ft $\Rightarrow C = \pi d = 22.625\pi$ ft.

$\left(\frac{22.625\pi\text{ ft}}{1\text{ rev}}\right)\left(\frac{545\text{ rev}}{1\text{ min}}\right)\left(\frac{1\text{ mile}}{5280\text{ ft}}\right)\left(\frac{60\text{ min}}{1\text{ hour}}\right) \approx 440.2$ mi/hr

[63] $\Delta f = \frac{2fv}{c} \Rightarrow v = \frac{c(\Delta f)}{2f} = \frac{186,000 \times 10^8}{2 \times 10^{14}} = 0.093$ mi/sec

[65] Let d denote the distance from Venus to the sun in the third figure.

$\sin 47° = \frac{d}{92,900,000} \Rightarrow d \approx 67,942,759$ mi., or approximately $67,900,000$ mi.

[67] Let A denote the point at the $36°$ angle. Let Q denote the highest point of the mountain and P denote Q's projection such that $\angle APQ = 90°.$ Let x denote the distance from the left end of the mountain to P and let y denote the distance from the right end of the mountain to P.

$\cot 36° = \frac{x + 200}{260} \Rightarrow 260\cot 36° = x + 200 \Rightarrow x = 260\cot 36° - 200.$

$\cot 47° = \frac{y + 150}{260} \Rightarrow 260\cot 47° = y + 150 \Rightarrow y = 260\cot 47° - 150.$ $x + y \approx 250$ ft.

69 (a) Let $x = \overline{QR}$. $\cot \beta = \frac{x}{h} \Rightarrow x = h \cot \beta$. $\cot \alpha = \frac{d+x}{h} \Rightarrow x = h \cot \alpha - d$.

Thus, $h \cot \beta = h \cot \alpha - d \Rightarrow d = h(\cot \alpha - \cot \beta) \Rightarrow h = \frac{d}{\cot \alpha - \cot \beta}$.

(b) $d = 2$ miles, $\alpha = 15°$, and $\beta = 20° \Rightarrow h = \frac{2}{\cot 15° - \cot 20°} \approx 2.03$, or 2 miles.

71 (a) $\cos \theta = \frac{15}{s} \Rightarrow s = \frac{15}{\cos \theta}$. $E = \frac{5000 \cos \theta}{s^2} = \frac{5000 \cos \theta}{225/\cos^2 \theta} = \frac{200}{9} \cos^3 \theta$.

$\theta = 30° \Rightarrow E = \frac{200}{9}(\frac{1}{2}\sqrt{3})^3 = \frac{200}{9}(\frac{3}{8}\sqrt{3}) = \frac{25}{3}\sqrt{3} \approx 14.43$ ft-candles

(b) $E = \frac{1}{2}E_{\max} \Rightarrow \frac{200}{9}\cos^3 \theta = \frac{1}{2}(\frac{200}{9}\cos^3 0°) \Rightarrow \cos^3 \theta = \frac{1}{2} \Rightarrow \cos \theta = \sqrt[3]{\frac{1}{2}} \Rightarrow$

$$\theta = \cos^{-1}\sqrt[3]{\frac{1}{2}} \Rightarrow \theta \approx 37.47°$$

73 (a) Let d denote the distance from the end of the bracket to the wall.

$$\cos 30° = \frac{d}{85.5} \Rightarrow d = (85.5)(\frac{1}{2}\sqrt{3}) \approx 74.05 \text{ in.}$$

(b) Let y denote the side opposite 30°. $\sin 30° = \frac{y}{85.5} \Rightarrow y = 42.75$ in.

Thus, the distance from the ceiling to the top of the screen is the

(length of the bracket $+ y -$ one-half the height of the screen) =

$$(18'' + 42.75'' - 36'') = 24.75''.$$

75 (a) Let $\theta = \angle PCQ$. Since $\angle BPC = 90°$, $\cos \theta = \frac{R}{R+h} \Rightarrow$

$$R + h = R \sec \theta \Rightarrow h = R \sec \theta - R. \text{ Since } \theta = \frac{s}{R}, h = R \sec \frac{s}{R} - R.$$

(b) $h = R\left(\sec \frac{s}{R} - 1\right) = 4000\left(\sec \frac{50}{4000} - 1\right) \approx 0.31252 \text{ mi} \approx 1650 \text{ ft.}$

77 The range of temperatures is 0.6°F, so $a = 0.3$. $\frac{2\pi}{b} = 24 \Rightarrow b = \frac{\pi}{12}$. The average

temperature occurs at 11 A.M., 6 hours before the high at 5 P.M., which corresponds

to $t = 11$. The argument of the sine is then $\frac{\pi}{12}(t - 11)$, or $\frac{\pi}{12}t - \frac{11\pi}{12}$. Thus,

$$y = 98.6 + (0.3)\sin\left(\frac{\pi}{12}t - \frac{11\pi}{12}\right) \text{ \{ or equivalently, } y = 98.6 + (0.3)\sin\left(\frac{\pi}{12}t + \frac{13\pi}{12}\right) \}.$$

79 (a) For $D(t) = 2000 \sin \frac{\pi}{90}t + 4000$, the period $p = \frac{2\pi}{\pi/90} = 180 \text{ days.}$

(b) The greatest demand will occur when the argument of the sine is $\frac{\pi}{2}$.

$\frac{\pi}{90}t = \frac{\pi}{2} \Rightarrow t = 45$ days into summer.

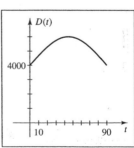

Figure 79

> ## Chapter 5 Discussion Exercises

1. On the TI-82/83 with $a = 15$, there is an indication that there are 15 sine waves on each side of the y-axis, but the minimums and maximums do not get to -1 and 1, respectively. With $a = 30$, the number of sine waves is undetectable—there simply aren't enough pixels for any degree of clarity. With $a = 45$, there are 2 sine waves on each side of the y-axis—there should be 45!

3. The sum on the left-side of the equation can never be greater than 3—no solutions.

5. The graph of $y_1 = x$, $y_2 = \sin x$, and $y_3 = \tan x$ in the suggested viewing rectangle, $[-0.1, 0.1]$ by $[-0.1, 0.1]$, indicates that their values are very close to each other near $x = 0$—in fact, the graphs of the functions are indistinguishable. Creating a table of values on the order of 10^{-10} also shows that all three functions are nearly equal.

7. (a) S is at $(0, -1)$ on the rectangular coordinate system. Starting at S, subtract 1 from the 2 km to get to the circular portion of the track. Now consider $t = (1 + \frac{3\pi}{2})$ as the radian measurement (starting from S) of the circular track in Discussion Exercise 6. $\{\frac{3\pi}{2}$ to get to the bottom of the circle and 1 to make the second km.$\}$ $x = \cos t + 1 \approx 1.8415$ and $y = \sin t \approx -0.5403$.

   (b) The perimeter of the track is $(4 + 2\pi)$ km. $\frac{500}{4 + 2\pi} \approx 48.623066$ laps. Subtracting the 48 whole laps, we have a portion of a lap left, 0.623066 lap. Multiply by $4 + 2\pi$ to see how many km we are from S. $(4 + 2\pi)(0.623066) \approx 6.4071052$. To determine where this places us on the track, start at S and subtract 1 {to get to $(1, -1)$}, subtract π {to get to $(1, 1)$}, and subtract 2 {to get to $(-1, 1)$}. $6.4071052 - (3 + \pi) \approx 0.2655126$. Now consider $t = (0.2655126 + \frac{\pi}{2})$ as the radian measurement of the circular track in Discussion Exercise 6 (but with S at the bottom). $x \approx \cos t - 1 \approx -1.2624$ and $y \approx \sin t \approx 0.9650$.

Chapter 6: Analytic Trigonometry

1 $\csc\theta - \sin\theta = \dfrac{1}{\sin\theta} - \sin\theta = \dfrac{1-\sin^2\theta}{\sin\theta} = \dfrac{\cos^2\theta}{\sin\theta} = \dfrac{\cos\theta}{\sin\theta}\cdot\cos\theta = \cot\theta\,\cos\theta$

3 $\dfrac{\sec^2 2u - 1}{\sec^2 2u} = 1 - \dfrac{1}{\sec^2 2u}\ \{\,\text{split up fraction}\,\} = 1 - \cos^2 2u = \sin^2 2u$

5 $\dfrac{\csc^2\theta}{1+\tan^2\theta} = \dfrac{\csc^2\theta}{\sec^2\theta} = \dfrac{1/\sin^2\theta}{1/\cos^2\theta} = \dfrac{\cos^2\theta}{\sin^2\theta} = \left(\dfrac{\cos\theta}{\sin\theta}\right)^2 = \cot^2\theta$

7 $\dfrac{1+\cos 3t}{\sin 3t} + \dfrac{\sin 3t}{1+\cos 3t} = \dfrac{(1+\cos 3t)^2 + \sin^2 3t}{\sin 3t\,(1+\cos 3t)}$ $\{\,\text{combine fractions}\,\}$

$\qquad\qquad\qquad = \dfrac{1 + 2\cos 3t + \cos^2 3t + \sin^2 3t}{\sin 3t\,(1+\cos 3t)}$ $\{\,\text{expand}\,\}$

$\qquad\qquad\qquad = \dfrac{2 + 2\cos 3t}{\sin 3t\,(1+\cos 3t)}$ $\{\,\cos^2 3t + \sin^2 3t = 1\,\}$

$\qquad\qquad\qquad = \dfrac{2(1+\cos 3t)}{\sin 3t\,(1+\cos 3t)}$ $\{\,\text{factor out } 2\,\}$

$\qquad\qquad\qquad = \dfrac{2}{\sin 3t}$ $\{\,\text{cancel like term}\,\}$

$\qquad\qquad\qquad = 2\csc 3t$ $\{\,\text{reciprocal identity}\,\}$

9 $\dfrac{1}{1-\cos\gamma} + \dfrac{1}{1+\cos\gamma} = \dfrac{1+\cos\gamma+1-\cos\gamma}{1-\cos^2\gamma} = \dfrac{2}{\sin^2\gamma} = 2\csc^2\gamma$

11 $(\sec u - \tan u)(\csc u + 1) = \left(\dfrac{1}{\cos u} - \dfrac{\sin u}{\cos u}\right)\left(\dfrac{1}{\sin u} + 1\right)$ $\{\,\text{change to sines and cosines}\,\}$

$\qquad\qquad\qquad = \left(\dfrac{1-\sin u}{\cos u}\right)\left(\dfrac{1+\sin u}{\sin u}\right)$ $\{\,\text{combine fractions}\,\}$

$\qquad\qquad\qquad = \dfrac{1-\sin^2 u}{\cos u\,\sin u}$ $\{\,\text{multiply terms}\,\}$

$\qquad\qquad\qquad = \dfrac{\cos^2 u}{\cos u\,\sin u} = \dfrac{\cos u}{\sin u} = \cot u$

13 $\csc^4 t - \cot^4 t = (\csc^2 t)^2 - (\cot^2 t)^2$ $\{\,\text{recognize as the diff. of 2 squares}\,\}$

$\qquad\qquad = (\csc^2 t + \cot^2 t)(\csc^2 t - \cot^2 t)$ $\{\,\text{factor}\,\}$

$\qquad\qquad = (\csc^2 t + \cot^2 t)(1)$ $\{\,\text{Pythagorean id., } 1 + \cot^2 t = \csc^2 t\,\}$

$\qquad\qquad = \csc^2 t + \cot^2 t$

$\boxed{15}$ The first step in the following verification is to multiply the denominator, $1 - \sin \beta$, by its conjugate, $1 + \sin \beta$. This procedure will give us the difference of two squares, $1 - \sin^2\beta$, which is equal to $\cos^2\beta$. This step is often helpful when simplifying trigonometric expressions because manipulation of the Pythagorean identities often allows us to reduce the resulting expression to a single term.

$$\frac{\cos \beta}{1 - \sin \beta} = \frac{\cos \beta}{1 - \sin \beta} \cdot \frac{1 + \sin \beta}{1 + \sin \beta} = \frac{\cos \beta \, (1 + \sin \beta)}{1 - \sin^2\beta} = \frac{\cos \beta \, (1 + \sin \beta)}{\cos^2\beta} = \frac{1 + \sin \beta}{\cos \beta} =$$

$$\frac{1}{\cos \beta} + \frac{\sin \beta}{\cos \beta} = \sec \beta + \tan \beta$$

$\boxed{17}$ $\dfrac{\tan^2x}{\sec x + 1} = \dfrac{\sec^2x - 1}{\sec x + 1} = \dfrac{(\sec x + 1)(\sec x - 1)}{\sec x + 1} = \sec x - 1 = \dfrac{1}{\cos x} - 1 = \dfrac{1 - \cos x}{\cos x}$

$\boxed{19}$ $\dfrac{\cot 4u - 1}{\cot 4u + 1} = \dfrac{\dfrac{1}{\tan 4u} - 1}{\dfrac{1}{\tan 4u} + 1} = \dfrac{\dfrac{1}{\tan 4u} - \dfrac{\tan 4u}{\tan 4u}}{\dfrac{1}{\tan 4u} + \dfrac{\tan 4u}{\tan 4u}} = \dfrac{\dfrac{1 - \tan 4u}{\tan 4u}}{\dfrac{1 + \tan 4u}{\tan 4u}} = \dfrac{1 - \tan 4u}{1 + \tan 4u}$

$\boxed{21}$ $\sin^4r - \cos^4r = (\sin^2r - \cos^2r)(\sin^2r + \cos^2r) = (\sin^2r - \cos^2r)(1) = \sin^2r - \cos^2r$

$\boxed{23}$ $\tan^4k - \sec^4k = (\tan^2k - \sec^2k)(\tan^2k + \sec^2k)$ { factor as the diff. of 2 squares }

$\qquad\qquad = (-1)(\sec^2k - 1 + \sec^2k)$ { Pythagorean identity }

$\qquad\qquad = (-1)(2\sec^2k - 1) = 1 - 2\sec^2k$

$\boxed{25}$ $(\sec t + \tan t)^2 = \left(\dfrac{1}{\cos t} + \dfrac{\sin t}{\cos t}\right)^2 = \left(\dfrac{1 + \sin t}{\cos t}\right)^2 = \dfrac{(1 + \sin t)^2}{\cos^2t} =$

$$\frac{(1 + \sin t)^2}{1 - \sin^2t} = \frac{(1 + \sin t)^2}{(1 + \sin t)(1 - \sin t)} = \frac{1 + \sin t}{1 - \sin t}$$

$\boxed{27}$ $(\sin^2\theta + \cos^2\theta)^3 = (1)^3 = 1$

$\boxed{29}$ $\dfrac{1 + \csc \beta}{\cot \beta + \cos \beta} = \dfrac{1 + \dfrac{1}{\sin \beta}}{\dfrac{\cos \beta}{\sin \beta} + \cos \beta} = \dfrac{\dfrac{\sin \beta + 1}{\sin \beta}}{\dfrac{\cos \beta + \cos \beta \sin \beta}{\sin \beta}} = \dfrac{\sin \beta + 1}{\cos \beta \, (1 + \sin \beta)} = \dfrac{1}{\cos \beta} = \sec \beta$

$\boxed{31}$ As demonstrated in the following verification, it may be beneficial to try to combine expressions rather than expand them. $(\csc t - \cot t)^4(\csc t + \cot t)^4 =$

$$[(\csc t - \cot t)(\csc t + \cot t)]^4 = (\csc^2t - \cot^2t)^4 = (1)^4 = 1$$

$\boxed{33}$ $\text{RS} = \dfrac{\tan \alpha + \tan \beta}{1 - \tan \alpha \tan \beta} = \dfrac{\dfrac{\sin \alpha}{\cos \alpha} + \dfrac{\sin \beta}{\cos \beta}}{1 - \dfrac{\sin \alpha}{\cos \alpha} \cdot \dfrac{\sin \beta}{\cos \beta}} = \dfrac{\dfrac{\sin \alpha \cos \beta + \cos \alpha \sin \beta}{\cos \alpha \cos \beta}}{\dfrac{\cos \alpha \cos \beta - \sin \alpha \sin \beta}{\cos \alpha \cos \beta}} =$

$$\frac{\sin \alpha \cos \beta + \cos \alpha \sin \beta}{\cos \alpha \cos \beta - \sin \alpha \sin \beta} = \text{LS}$$

Note: We could obtain the RS by dividing

the numerator and denominator of the LS by $(\cos \alpha \, \cos \beta)$.

$\boxed{35}$ $\dfrac{\tan\alpha}{1+\sec\alpha} + \dfrac{1+\sec\alpha}{\tan\alpha} = \dfrac{\tan^2\alpha + (1+\sec\alpha)^2}{(1+\sec\alpha)\tan\alpha} = \dfrac{(\sec^2\alpha - 1) + (1 + 2\sec\alpha + \sec^2\alpha)}{(1+\sec\alpha)\tan\alpha} =$

$\dfrac{2\sec^2\alpha + 2\sec\alpha}{(1+\sec\alpha)\tan\alpha} = \dfrac{2\sec\alpha(\sec\alpha + 1)\cot\alpha}{1+\sec\alpha} = \dfrac{2}{\cos\alpha}\cdot\dfrac{\cos\alpha}{\sin\alpha} = \dfrac{2}{\sin\alpha} = 2\csc\alpha$

$\boxed{37}$ $\dfrac{1}{\tan\beta + \cot\beta} = \dfrac{1}{\dfrac{\sin\beta}{\cos\beta} + \dfrac{\cos\beta}{\sin\beta}} = \dfrac{1}{\dfrac{\sin^2\beta + \cos^2\beta}{\cos\beta\,\sin\beta}} = \dfrac{1}{\dfrac{1}{\cos\beta\,\sin\beta}} = \sin\beta\,\cos\beta$

$\boxed{39}$ $\sec\theta + \csc\theta - \cos\theta - \sin\theta = \dfrac{1}{\cos\theta} + \dfrac{1}{\sin\theta} - \cos\theta - \sin\theta$

$$= \left(\dfrac{1}{\cos\theta} - \cos\theta\right) + \left(\dfrac{1}{\sin\theta} - \sin\theta\right)$$

$$= \dfrac{1 - \cos^2\theta}{\cos\theta} + \dfrac{1 - \sin^2\theta}{\sin\theta}$$

$$= \dfrac{\sin^2\theta}{\cos\theta} + \dfrac{\cos^2\theta}{\sin\theta}$$

$$= \left(\sin\theta \cdot \dfrac{\sin\theta}{\cos\theta}\right) + \left(\cos\theta \cdot \dfrac{\cos\theta}{\sin\theta}\right) = \sin\theta\,\tan\theta + \cos\theta\,\cot\theta$$

$\boxed{41}$ $\mathrm{RS} = \sec^4\phi - 4\tan^2\phi = (\sec^2\phi)^2 - 4\tan^2\phi = (1 + \tan^2\phi)^2 - 4\tan^2\phi =$

$(1 + 2\tan^2\phi + \tan^4\phi) - 4\tan^2\phi = 1 - 2\tan^2\phi + \tan^4\phi = (1 - \tan^2\phi)^2 = \mathrm{LS}$

$\boxed{43}$ $\dfrac{\cot(-t) + \tan(-t)}{\cot t} = \dfrac{-\cot t - \tan t}{\cot t} = -\dfrac{\cot t}{\cot t} - \dfrac{\tan t}{\cot t} = -(1 + \tan^2 t) = -\sec^2 t$

$\boxed{45}$ $\log 10^{\tan t} = \log_{10} 10^{\tan t} = \tan t$, since $\log_a a^x = x$

$\boxed{47}$ $\ln\cot x = \ln(\cot x) = \ln(\tan x)^{-1} = -\ln(\tan x)$ $\{$since $\ln a^x = x\ln a\} = -\ln\tan x$

$\boxed{49}$ $\ln|\sec\theta + \tan\theta| = \ln\left|\dfrac{(\sec\theta + \tan\theta)(\sec\theta - \tan\theta)}{\sec\theta - \tan\theta}\right|$ $\{$multiply by the conjugate$\}$

$$= \ln\left|\dfrac{\sec^2\theta - \tan^2\theta}{\sec\theta - \tan\theta}\right| \qquad \{\text{simplify}\}$$

$$= \ln\left|\dfrac{1}{\sec\theta - \tan\theta}\right| \qquad \{\text{Pythagorean identity}\}$$

$$= \ln\dfrac{|1|}{|\sec\theta - \tan\theta|} \qquad \left\{\left|\dfrac{a}{b}\right| = \dfrac{|a|}{|b|}\right\}$$

$$= \ln|1| - \ln|\sec\theta - \tan\theta| \qquad \left\{\ln\dfrac{a}{b} = \ln a - \ln b\right\}$$

$$= -\ln|\sec\theta - \tan\theta| \qquad \{\ln 1 = 0\}$$

$\boxed{51}$ $\cos^2 t = 1 - \sin^2 t \Rightarrow \cos t = \pm\sqrt{1 - \sin^2 t}$. Since the given equation is

$\cos t = +\sqrt{1 - \sin^2 t}$, we may choose any t such that $\cos t < 0$.

Using $t = \pi$, $\mathrm{LS} = \cos\pi = -1$. $\mathrm{RS} = \sqrt{1 - \sin^2\pi} = 1$. Since $-1 \neq 1$, $\mathrm{LS} \neq \mathrm{RS}$.

$\boxed{53}$ $\sqrt{\sin^2 t} = |\sin t| = \pm \sin t$. Hence, choose any t such that $\sin t < 0$.

   Using $t = \frac{3\pi}{2}$, LS $= \sqrt{(-1)^2} = 1$. RS $= \sin \frac{3\pi}{2} = -1$. Since $1 \neq -1$, LS $\neq$ RS.

$\boxed{55}$ $(\sin \theta + \cos \theta)^2 = \sin^2\theta + 2\sin \theta \cos \theta + \cos^2\theta$. Since the right side of the given

   equation is only $\sin^2\theta + \cos^2\theta$, we may choose any θ such that $2\sin \theta \cos \theta \neq 0$.

   Using $\theta = \frac{\pi}{4}$, LS $= (\frac{1}{2}\sqrt{2} + \frac{1}{2}\sqrt{2})^2 = (\sqrt{2})^2 = 2$.

   RS $= (\frac{1}{2}\sqrt{2})^2 + (\frac{1}{2}\sqrt{2})^2 = \frac{1}{2} + \frac{1}{2} = 1$. Since $2 \neq 1$, LS $\neq$ RS.

$\boxed{57}$ $\cos(-t) = -\cos t$ •   Since $\cos(-t) = \cos t$, we may choose any t such that

   $\cos t \neq -\cos t$—that is, any t such that $\cos t \neq 0$. Using $t = \pi$, LS $= \cos(-\pi) = -1$.

   RS $= -\cos \pi = -(-1) = 1$. Since $-1 \neq 1$, LS $\neq$ RS.

$\boxed{59}$ Don't confuse $\cos(\sec t) = 1$ with $\cos t \cdot \sec t = 1$. The former is true if $\sec t = 2\pi$ or

   an integer multiple of 2π. The latter is true for any value of t as long as $\sec t$ is

   defined. Choose any t such that $\sec t \neq 2\pi n$.

   Using $t = \frac{\pi}{4}$, LS $= \cos(\sec \frac{\pi}{4}) = \cos \sqrt{2} \neq 1 = $ RS.

$\boxed{61}$ $\sin^2 t - 4\sin t - 5 = 0 \Rightarrow (\sin t - 5)(\sin t + 1) = 0 \Rightarrow \sin t = 5$ or $\sin t = -1$.

   Since $\sin t$ cannot equal 5, we may choose any t such that $\sin t \neq -1$.

   Using $t = \pi$, LS $= -5 \neq 0 = $ RS.

Note: Exer. 63–66: Use $\sqrt{a^2 - x^2} = a \cos \theta$ because

   $$\sqrt{a^2 - x^2} = \sqrt{a^2 - a^2 \sin^2\theta} = \sqrt{a^2(1 - \sin^2\theta)} = \sqrt{a^2 \cos^2\theta} = |a| \, |\cos \theta| = a \cos \theta$$

   since $\cos \theta > 0$ if $-\frac{\pi}{2} < \theta < \frac{\pi}{2}$ and $a > 0$.

$\boxed{63}$ $(a^2 - x^2)^{3/2} = (\sqrt{a^2 - x^2})^3 = (a \cos \theta)^3$ {see above note} $= a^3 \cos^3\theta$

$\boxed{65}$ $\dfrac{x^2}{\sqrt{a^2 - x^2}} = \dfrac{a^2 \sin^2\theta}{a \cos \theta} = a \cdot \dfrac{\sin \theta}{\cos \theta} \cdot \sin \theta = a \tan \theta \sin \theta$

Note: Exer. 67–70: Use $\sqrt{a^2 + x^2} = a \sec \theta$ because

   $$\sqrt{a^2 + x^2} = \sqrt{a^2 + a^2 \tan^2\theta} = \sqrt{a^2(1 + \tan^2\theta)} = \sqrt{a^2 \sec^2\theta} = |a| \, |\sec \theta| =$$

   $a \sec \theta$ since $\sec \theta > 0$ if $-\frac{\pi}{2} < \theta < \frac{\pi}{2}$ and $a > 0$.

$\boxed{67}$ $\sqrt{a^2 + x^2} = a \sec \theta$ {see above note}

$\boxed{69}$ $\dfrac{1}{x^2 + a^2} = \dfrac{1}{(\sqrt{a^2 + x^2})^2} = \dfrac{1}{(a \sec \theta)^2} = \dfrac{1}{a^2 \sec^2\theta} = \dfrac{1}{a^2} \cos^2\theta$

Note: Exer. 71–74: Use $\sqrt{x^2 - a^2} = a \tan \theta$ because

   $$\sqrt{x^2 - a^2} = \sqrt{a^2 \sec^2\theta - a^2} = \sqrt{a^2(\sec^2\theta - 1)} = \sqrt{a^2 \tan^2\theta} = |a| \, |\tan \theta| =$$

   $a \tan \theta$ since $\tan \theta > 0$ if $0 < \theta < \frac{\pi}{2}$ and $a > 0$.

$\boxed{71}$ $\sqrt{x^2 - a^2} = a \tan \theta$ {see above note}

$\boxed{73}$ $x^3 \sqrt{x^2 - a^2} = (a^3 \sec^3\theta)(a \tan \theta) = a^4 \sec^3\theta \tan \theta$

$\boxed{75}$ The graph of $f(x) = \dfrac{\sin^2 x - \sin^4 x}{(1 - \sec^2 x)\cos^4 x}$ appears to be that of the horizontal line

$y = g(x) = -1$. Verifying this identity, we have

$$\frac{\sin^2 x - \sin^4 x}{(1 - \sec^2 x)\cos^4 x} = \frac{\sin^2 x(1 - \sin^2 x)}{-\tan^2 x \, \cos^4 x} = \frac{\sin^2 x \, \cos^2 x}{-(\sin^2 x/\cos^2 x)\cos^4 x} = \frac{\sin^2 x \, \cos^2 x}{-\sin^2 x \, \cos^2 x} = -1.$$

$\boxed{77}$ The graph of $f(x) = \sec x \, (\sin x \, \cos x + \cos^2 x) - \sin x$ appears to be that of

$y = g(x) = \cos x$. Verifying this identity, we have

$$\sec x \, (\sin x \, \cos x + \cos^2 x) - \sin x = \sec x \, \cos x \, (\sin x + \cos x) - \sin x$$

$$= (\sin x + \cos x) - \sin x = \cos x.$$

6.2 Exercises

$\boxed{1}$ In $[0, 2\pi)$, $\sin x = -\dfrac{\sqrt{2}}{2}$ only if $x = \frac{5\pi}{4}, \frac{7\pi}{4}$. <u>All solutions</u> would include these angles

plus all angles coterminal with them. Hence, $x = \frac{5\pi}{4} + 2\pi n, \frac{7\pi}{4} + 2\pi n$.

$\boxed{3}$ $\tan\theta = \sqrt{3} \implies \theta = \frac{\pi}{3} + \pi n$. *Note:* This solution could be written as $\frac{\pi}{3} + 2\pi n$ and

$\frac{4\pi}{3} + 2\pi n$. Since the period of the tangent (and the cotangent) is π, we use the

abbreviated form $\frac{\pi}{3} + \pi n$ to describe all solutions. We will use "πn" for solutions of

exercises involving the tangent and cotangent functions.

$\boxed{5}$ $\sec\beta = 2 \implies \cos\beta = \frac{1}{2} \implies \beta = \frac{\pi}{3} + 2\pi n, \frac{5\pi}{3} + 2\pi n$.

$\boxed{7}$ $\sin x = \frac{\pi}{2}$ has no solution since $\frac{\pi}{2} > 1$, which is not in the range $[-1, 1]$. Your

calculator will give some kind of error message if you attempt to find a solution to

this problem or any similar type problem.

$\boxed{9}$ $\cos\theta = \dfrac{1}{\sec\theta}$ is true for all values *for which the equation is defined.*

$\bigstar$ All θ except $\theta = \frac{\pi}{2} + \pi n$

$\boxed{11}$ $2\cos 2\theta - \sqrt{3} = 0 \implies \cos 2\theta = \dfrac{\sqrt{3}}{2}$ { 2θ is just an angle — so we solve this equation

for 2θ and then divide those solutions by 2 } $\implies$

$$2\theta = \frac{\pi}{6} + 2\pi n, \frac{11\pi}{6} + 2\pi n \implies \theta = \frac{\pi}{12} + \pi n, \frac{11\pi}{12} + \pi n$$

$\boxed{13}$ $\sqrt{3}\tan\frac{1}{3}t = 1 \implies \tan\frac{1}{3}t = \dfrac{1}{\sqrt{3}} \implies \frac{1}{3}t = \frac{\pi}{6} + \pi n \implies t = \frac{\pi}{2} + 3\pi n$

$\boxed{15}$ $\sin\left(\theta + \frac{\pi}{4}\right) = \frac{1}{2} \implies \theta + \frac{\pi}{4} = \frac{\pi}{6} + 2\pi n, \frac{5\pi}{6} + 2\pi n \implies \theta = -\frac{\pi}{12} + 2\pi n, \frac{7\pi}{12} + 2\pi n$

$\boxed{17}$ $\sin\left(2x - \frac{\pi}{3}\right) = \frac{1}{2} \implies 2x - \frac{\pi}{3} = \frac{\pi}{6} + 2\pi n, \frac{5\pi}{6} + 2\pi n \implies 2x = \frac{\pi}{2} + 2\pi n, \frac{7\pi}{6} + 2\pi n \implies$

$$x = \frac{\pi}{4} + \pi n, \frac{7\pi}{12} + \pi n$$

$\boxed{19}$ $2\cos t + 1 = 0 \implies 2\cos t = -1 \implies \cos t = -\frac{1}{2} \implies t = \frac{2\pi}{3} + 2\pi n, \frac{4\pi}{3} + 2\pi n$

$\boxed{21}$ $\tan^2 x = 1 \implies \tan x = \pm 1 \implies x = \frac{\pi}{4} + \pi n, \frac{3\pi}{4} + \pi n,$ or simply $\frac{\pi}{4} + \frac{\pi}{2}n$

23 $(\cos\theta - 1)(\sin\theta + 1) = 0 \;\Rightarrow\; (\cos\theta - 1) = 0$ or $(\sin\theta + 1) = 0 \;\Rightarrow$

$$\cos\theta = 1 \text{ or } \sin\theta = -1 \;\Rightarrow\; \theta = 2\pi n \text{ or } \theta = \tfrac{3\pi}{2} + 2\pi n$$

25 $\sec^2\alpha - 4 = 0 \;\Rightarrow\; \sec^2\alpha = 4 \;\Rightarrow\; \sec\alpha = \pm 2 \;\Rightarrow$

$$\alpha = \tfrac{\pi}{3} + 2\pi n, \tfrac{5\pi}{3} + 2\pi n, \tfrac{2\pi}{3} + 2\pi n, \tfrac{4\pi}{3} + 2\pi n, \text{ or simply } \tfrac{\pi}{3} + \pi n, \tfrac{2\pi}{3} + \pi n$$

27 $\sqrt{3} + 2\sin\beta = 0 \;\Rightarrow\; 2\sin\beta = -\sqrt{3} \;\Rightarrow\; \sin\beta = -\dfrac{\sqrt{3}}{2} \;\Rightarrow\; \beta = \tfrac{4\pi}{3} + 2\pi n, \tfrac{5\pi}{3} + 2\pi n$

29 $\cot^2 x - 3 = 0 \;\Rightarrow\; \cot^2 x = 3 \;\Rightarrow\; \cot x = \pm\sqrt{3} \;\Rightarrow\; x = \tfrac{\pi}{6} + \pi n, \tfrac{5\pi}{6} + \pi n$

31 $(2\sin\theta + 1)(2\cos\theta + 3) = 0 \;\Rightarrow\; \sin\theta = -\tfrac{1}{2} \text{ or } \sin\theta = -\tfrac{3}{2} \;\Rightarrow$

$$\theta = \tfrac{7\pi}{6} + 2\pi n, \tfrac{11\pi}{6} + 2\pi n \;\{\sin\theta = -\tfrac{3}{2} \text{ has no solutions}\}$$

33 $\sin 2x\,(\csc 2x - 2) = 0 \;\Rightarrow\; 1 - 2\sin 2x = 0 \;\Rightarrow\; \sin 2x = \tfrac{1}{2} \;\Rightarrow$

$$2x = \tfrac{\pi}{6} + 2\pi n, \tfrac{5\pi}{6} + 2\pi n \;\Rightarrow\; x = \tfrac{\pi}{12} + \pi n, \tfrac{5\pi}{12} + \pi n$$

35 $\cos(\ln x) = 0 \;\Rightarrow\; \ln x = \tfrac{\pi}{2} + \pi n \;\Rightarrow\; x = e^{(\pi/2) + \pi n} \;\{\text{since } \ln x = y \;\Leftrightarrow\; x = e^y\}$

37 $\cos\left(2x - \tfrac{\pi}{4}\right) = 0 \;\Rightarrow\; 2x - \tfrac{\pi}{4} = \tfrac{\pi}{2} + \pi n \;\Rightarrow\; 2x = \tfrac{3\pi}{4} + \pi n \;\Rightarrow\; x = \tfrac{3\pi}{8} + \tfrac{\pi}{2}n.$

x will be in the interval $[0, 2\pi)$ if $n = 0$, 1, 2, or 3. Thus, $x = \tfrac{3\pi}{8}, \tfrac{7\pi}{8}, \tfrac{11\pi}{8}, \tfrac{15\pi}{8}.$

39 $2 - 8\cos^2 t = 0 \;\Rightarrow\; \cos^2 t = \tfrac{1}{4} \;\Rightarrow\; \cos t = \pm\tfrac{1}{2} \;\Rightarrow\; t = \tfrac{\pi}{3}, \tfrac{2\pi}{3}, \tfrac{4\pi}{3}, \tfrac{5\pi}{3}$

41 $2\sin^2 u = 1 - \sin u \;\Rightarrow\; 2\sin^2 u + \sin u - 1 = 0 \;\Rightarrow\; (2\sin u - 1)(\sin u + 1) = 0 \;\Rightarrow$

$$\sin u = \tfrac{1}{2}, -1 \;\Rightarrow\; u = \tfrac{\pi}{6}, \tfrac{5\pi}{6}, \tfrac{3\pi}{2}$$

43 $\tan^2 x \sin x = \sin x \;\Rightarrow\; \tan^2 x$

$\sin x - \sin x = 0 \;\Rightarrow\; \sin x\,(\tan^2 x - 1) = 0 \;\Rightarrow\; \sin x = 0 \;$ or $\; \tan x = \pm 1 \;\Rightarrow\; x = 0,$

$\pi, \tfrac{\pi}{4}, \tfrac{3\pi}{4}, \tfrac{5\pi}{4}, \tfrac{7\pi}{4}.$ *Note:* A common mistake is to divide both sides of the given

equation by $\sin x$—doing so results in losing the solutions for $\sin x = 0$.

45 $2\cos^2\gamma + \cos\gamma = 0 \;\Rightarrow\; \cos\gamma\,(2\cos\gamma + 1) = 0 \;\Rightarrow\; \cos\gamma = 0, -\tfrac{1}{2} \;\Rightarrow$

$$\gamma = \tfrac{\pi}{2}, \tfrac{3\pi}{2}, \tfrac{2\pi}{3}, \tfrac{4\pi}{3}$$

47 $\sin^2\theta + \sin\theta - 6 = 0 \;\Rightarrow\; (\sin\theta + 3)(\sin\theta - 2) = 0 \;\Rightarrow\; \sin\theta = -3, 2.$

There are *no solutions* for either equation.

49 $1 - \sin t = \sqrt{3}\cos t$ • Square both sides to obtain an equation in either sin or cos.

$(1 - \sin t)^2 = (\sqrt{3}\cos t)^2 \;\Rightarrow\; 1 - 2\sin t + \sin^2 t = 3\cos^2 t \;\Rightarrow$

$\sin^2 t - 2\sin t + 1 = 3(1 - \sin^2 t) \;\Rightarrow\; 4\sin^2 t - 2\sin t - 2 = 0 \;\Rightarrow$

$2\sin^2 t - \sin t - 1 = 0 \;\Rightarrow\; (2\sin t + 1)(\sin t - 1) = 0 \;\Rightarrow\; \sin t = -\tfrac{1}{2}, 1 \;\Rightarrow$

$t = \tfrac{7\pi}{6}, \tfrac{11\pi}{6}, \tfrac{\pi}{2}.$ Since each side of the equation was squared, the solutions must be

checked in the original equation.

Check $\tfrac{7\pi}{6}$: LS $= 1 - \sin\tfrac{7\pi}{6} = 1 - (-\tfrac{1}{2}) = \tfrac{3}{2}$ and RS $= \sqrt{3}\cos\tfrac{7\pi}{6} = \sqrt{3}\left(-\dfrac{\sqrt{3}}{2}\right) = -\tfrac{3}{2}.$

Since LS $\ne$ RS, $\tfrac{7\pi}{6}$ is an extraneous solution.

Check $\tfrac{11\pi}{6}$: LS $= 1 - \sin\tfrac{11\pi}{6} = 1 - (-\tfrac{1}{2}) = \tfrac{3}{2}$ and RS $= \sqrt{3}\cos\tfrac{11\pi}{6} = \sqrt{3}\left(\dfrac{\sqrt{3}}{2}\right) = \tfrac{3}{2}.$

Since LS $=$ RS, $\tfrac{11\pi}{6}$ is a valid solution. (continued)

Check $\frac{\pi}{2}$: LS $= 1 - 1 = 0$ and RS $= \sqrt{3} \cdot 0 = 0$. Since LS $=$ RS, $\frac{\pi}{2}$ is a valid solution.

Thus, $t = \frac{\pi}{2}$ and $t = \frac{11\pi}{6}$ are the only solutions of the equation.

$\boxed{51}$ $\cos\alpha + \sin\alpha = 1$ $\Rightarrow$ $\cos\alpha = 1 - \sin\alpha$ { square both sides } $\Rightarrow$

$\cos^2\alpha = 1 - 2\sin\alpha + \sin^2\alpha$

$\quad$ { change $\cos^2\alpha$ to $1 - \sin^2\alpha$ to obtain an equation involving only $\sin\alpha$ } $\Rightarrow$

$1 - \sin^2\alpha = 1 - 2\sin\alpha + \sin^2\alpha$ $\Rightarrow$ $2\sin^2\alpha - 2\sin\alpha = 0$ $\Rightarrow$

$2\sin\alpha(\sin\alpha - 1) = 0$ $\Rightarrow$ $\sin\alpha = 0, 1$ $\Rightarrow$ $\alpha = 0, \pi, \frac{\pi}{2}$. π is an extraneous solution.

$\boxed{53}$ $2\tan t - \sec^2 t = 0$ $\Rightarrow$ $2\tan t - (1 + \tan^2 t) = 0$ $\Rightarrow$ $\tan^2 t - 2\tan t + 1 = 0$ $\Rightarrow$

$$(\tan t - 1)^2 = 0 \Rightarrow \tan t = 1 \Rightarrow t = \tfrac{\pi}{4}, \tfrac{5\pi}{4}$$

$\boxed{55}$ $\cot\alpha + \tan\alpha = \csc\alpha\sec\alpha$ $\Rightarrow$ $\dfrac{\cos\alpha}{\sin\alpha} + \dfrac{\sin\alpha}{\cos\alpha} = \dfrac{1}{\sin\alpha\cos\alpha}$ $\Rightarrow$

$\dfrac{\cos^2\alpha + \sin^2\alpha}{\sin\alpha\cos\alpha} = \dfrac{1}{\sin\alpha\cos\alpha}$. This is an identity and is true for *all numbers in* $[0, 2\pi)$

$\quad$ *except* $0, \frac{\pi}{2}, \pi$, and $\frac{3\pi}{2}$ since these values make the original equation undefined.

$\boxed{57}$ $2\sin^3 x + \sin^2 x - 2\sin x - 1 = 0$ { factor by grouping since there are four terms } $\Rightarrow$

$\sin^2 x(2\sin x + 1) - 1(2\sin x + 1) = 0$ $\Rightarrow$

$$(\sin^2 x - 1)(2\sin x + 1) = 0 \Rightarrow \sin x = \pm 1, -\tfrac{1}{2} \Rightarrow x = \tfrac{\pi}{2}, \tfrac{3\pi}{2}, \tfrac{7\pi}{6}, \tfrac{11\pi}{6}$$

$\boxed{59}$ $2\tan t\csc t + 2\csc t + \tan t + 1 = 0$ $\Rightarrow$ $2\csc t(\tan t + 1) + 1(\tan t + 1)$ $\Rightarrow$

$(2\csc t + 1)(\tan t + 1) = 0$ $\Rightarrow$ $\csc t = -\frac{1}{2}$ or $\tan t = -1$ $\Rightarrow$

$$t = \tfrac{3\pi}{4}, \tfrac{7\pi}{4} \text{ \{ since } \csc t \neq -\tfrac{1}{2}\}$$

$\boxed{61}$ $\sin^2 t - 4\sin t + 1 = 0$ $\Rightarrow$ $\sin t = \dfrac{4 \pm \sqrt{12}}{2} = 2 \pm \sqrt{3}$.

$\quad (2 + \sqrt{3}) > 1$ is not in the range of the sine, so $\sin t = 2 - \sqrt{3}$ $\Rightarrow$

$$t = 15°30' \text{ or } 164°30' \text{ \{ to the nearest ten minutes \}}$$

$\boxed{63}$ $\tan^2\theta + 3\tan\theta + 2 = 0$ $\Rightarrow$ $(\tan\theta + 1)(\tan\theta + 2) = 0$ $\Rightarrow$

$$\tan\theta = -1, -2 \Rightarrow \theta = 135°, 315°, 116°30', 296°30'$$

$\boxed{65}$ $12\sin^2 u - 5\sin u - 2 = 0$ $\Rightarrow$ $(3\sin u - 2)(4\sin u + 1) = 0$ $\Rightarrow$ $\sin u = \frac{2}{3}, -\frac{1}{4}$ $\Rightarrow$

$$u = 41°50', 138°10', 194°30', 345°30'$$

$\boxed{67}$ The top of the wave will be above the sea wall when its height is greater than 12.5.

$y > 12.5$ $\Rightarrow$ $25\cos\frac{\pi}{15}t > 12.5$ $\Rightarrow$ $\cos\frac{\pi}{15}t > \frac{1}{2}$ $\Rightarrow$

$\quad$ { To visualize this step, it may help to look at a unit circle and draw a vertical line

$\quad\quad$ through 0.5 on the x-axis — $x > 0.5$ is the same as $\cos\frac{\pi}{15}t > 0.5$. }

$-\frac{\pi}{3} < \frac{\pi}{15}t < \frac{\pi}{3}$ $\Rightarrow$ $-5 < t < 5$ { multiply by $\frac{15}{\pi}$ } $\Rightarrow$

$\quad\quad y > 12.5$ for about $5 - (-5) = 10$ minutes of each 30-minute period.

69 (a)   [1, 25, 5] by [0, 100, 10]

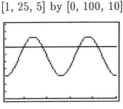

Figure 69

(b) July: $T(7) = 83°\text{F}$; October: $T(10) = 56.5°\text{F}$.

(c) Graph $Y_1 = 26.5 \sin\left(\frac{\pi}{6}x - \frac{2\pi}{3}\right) + 56.5$ and $Y_2 = 69$. Their graphs intersect at $t \approx 4.94$, 9.06 on [1, 13]. The average high temperature is above 69°F approximately May through September.

(d) A sine function is periodic and varies between a maximum and minimum value. Average monthly high temperatures are also seasonal with a 12-month period. Therefore, a sine function is a reasonable function to model these temperatures.

71 $I = \frac{1}{2}I_\text{M}$ and $D = 12$ $\Rightarrow$ $\frac{1}{2}I_\text{M} = I_\text{M}\sin^3\frac{\pi}{12}t$ $\Rightarrow$ $\sin^3\frac{\pi}{12}t = \frac{1}{2}$ $\Rightarrow$

$\sin\frac{\pi}{12}t = \sqrt[3]{\frac{1}{2}}$ $\Rightarrow$ $\frac{\pi}{12}t \approx 0.9169$ and 2.2247 { $\pi - 0.9169 \approx 2.2247$ is the reference angle for 0.9169 in QII. } $\Rightarrow$ $t \approx 3.50$ and $t \approx 8.50$

73 75% of the maximum intensity $= 0.75\,I_M$

(a) $I > 0.75\,I_M$ $\Rightarrow$ $I_M\sin^3\frac{\pi}{12}t > 0.75\,I_M$ $\Rightarrow$ $\sin^3\frac{\pi}{12}t > \frac{3}{4}$ $\Rightarrow$ $\sin\frac{\pi}{12}t > \sqrt[3]{\frac{3}{4}}$ $\Rightarrow$

$1.1398 < \frac{\pi}{12}t < 2.0018$ $\Rightarrow$ $4.3538 < t < 7.6462$, or approximately 3.29 hours.

(b) $I > 0.75\,I_M$ $\Rightarrow$ $I_M\sin^2\frac{\pi}{12}t > 0.75\,I_M$ $\Rightarrow$ $\sin^2\frac{\pi}{12}t > \frac{3}{4}$ $\Rightarrow$ $\sin\frac{\pi}{12}t > \frac{1}{2}\sqrt{3}$ $\Rightarrow$

$\frac{\pi}{3} < \frac{\pi}{12}t < \frac{2\pi}{3}$ $\Rightarrow$ $4 < t < 8$, or 4 hours.

75 (a) $N(t) = 1000\cos\frac{\pi}{5}t + 4000$, amplitude $= 1000$,

period $= \frac{2\pi}{\pi/5} = 10$ years

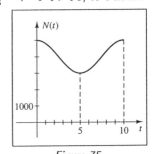

(b) $N > 4500$ $\Rightarrow$ $1000\cos\frac{\pi}{5}t + 4000 > 4500$ $\Rightarrow$

$\cos\frac{\pi}{5}t > \frac{1}{2}$ $\Rightarrow$ { The cosine function is greater

than $\frac{1}{2}$ on $[0, \frac{\pi}{3})$ and $(\frac{5\pi}{3}, 2\pi]$. }

$0 \le \frac{\pi}{5}t < \frac{\pi}{3}$ and $\frac{5\pi}{3} < \frac{\pi}{5}t \le 2\pi$ $\Rightarrow$

$0 \le t < \frac{5}{3}$ and $\frac{25}{3} < t \le 10$

Figure 75

77 $\frac{1}{2} + \cos x = 0$ $\Rightarrow$ $\cos x = -\frac{1}{2}$ $\Rightarrow$ $x = -\frac{4\pi}{3}, -\frac{2\pi}{3}, \frac{2\pi}{3}$, and $\frac{4\pi}{3}$ { for x in $[-2\pi, 2\pi]$ }

for A, B, C, and D, respectively. The corresponding y values are found by using

$y = \frac{1}{2}x + \sin x$ with each of the above values. The points are:

$A\left(-\frac{4\pi}{3}, -\frac{2\pi}{3} + \frac{1}{2}\sqrt{3}\right)$, $B\left(-\frac{2\pi}{3}, -\frac{\pi}{3} - \frac{1}{2}\sqrt{3}\right)$, $C\left(\frac{2\pi}{3}, \frac{\pi}{3} + \frac{1}{2}\sqrt{3}\right)$, and $D\left(\frac{4\pi}{3}, \frac{2\pi}{3} - \frac{1}{2}\sqrt{3}\right)$

79 $I(t) = k \Rightarrow 20\sin(60\pi t - 6\pi) = -10 \Rightarrow \sin(60\pi t - 6\pi) = -\frac{1}{2} \Rightarrow$

$60\pi t_1 - 6\pi = \frac{7\pi}{6} + 2\pi n$ or $60\pi t_2 - 6\pi = \frac{11\pi}{6} + 2\pi n$ {We use t_1 and t_2 to distinguish

between angles coterminal with $\frac{7\pi}{6}$ and those coterminal with $\frac{11\pi}{6}$.} $\Rightarrow$

$60t_1 = \frac{43}{6} + 2n$ or $60t_2 = \frac{47}{6} + 2n \Rightarrow t_1 = \frac{43}{360} + \frac{1}{30}n$ or $t_2 = \frac{47}{360} + \frac{1}{30}n$. We must now

find the smallest positive value of t_1 or t_2. $t_1 > 0 \Rightarrow \frac{1}{30}n > -\frac{43}{360} \Rightarrow$

$n > -\frac{43}{12} \approx -3.58$. The last result indicates that n must be an integer in the set

$\{-3, -2, -1, \ldots\}$. If $n = -3$, then $t_1 = \frac{7}{360}$. Greater values of n yield greater values

of t_1. Similarly, for t_2, $t_2 > 0 \Rightarrow \frac{1}{30}n > -\frac{47}{360} \Rightarrow n > -\frac{47}{12}$. If $n = -3$, then

$t_1 = \frac{11}{360}$. Thus, the smallest exact value of t for which $I(t) = -10$ is $t = \frac{7}{360}$ sec.

81 Graph $y = \cos x$ and $y = 0.3$ on the same coordinate plane. See *Figure 81* on the next

page. The points of intersection are located at $x \approx 1.27$, 5.02, and $\cos x$ is less than

0.3 between these values. Therefore, $\cos x \geq 0.3$ on $[0, 1.27] \cup [5.02, 2\pi]$.

$[0, 2\pi, \pi/4]$ by $[-2.09, 2.09]$ $[0, 2\pi, \pi/4]$ by $[-2.09, 2.09]$

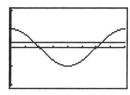   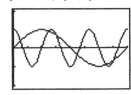

Figure 81   *Figure 83*

83 Graph $y = \cos 3x$ and $y = \sin x$ on the same coordinate plane.

The points of intersection are located at $x \approx 0.39$, 1.96, 2.36, 3.53, 5.11, 5.50.

From the graph, we see that $\cos 3x$ is less than $\sin x$ on

$$(0.39, 1.96) \cup (2.36, 3.53) \cup (5.11, 5.50).$$

85 (a) The largest zero occurs when $x \approx 0.6366$.

(b) As x becomes large, the graph of $f(x) = \cos(1/x)$ approaches the horizontal

asymptote $y = 1$.

(c) There appears to be an infinite number of zeros on $[0, c]$ for any $c > 0$.

$[0, 3]$ by $[-1.5, 1.5]$

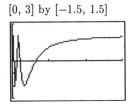

Figure 85

Note: Exer. 87–90: Graph $Y_1 = M$ and $Y_2 = \theta + e\sin\theta$ and approximate the value of θ

such that $Y_1 = Y_2$.

87 Mercury: $Y_1 = 5.241$ and $Y_2 = \theta + 0.206\sin\theta$ intersect when $\theta \approx 5.400$ (radians).

[0, 12] by [0, 8]   [0, 12] by [0, 8]

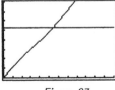   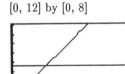

      Figure 87   *Figure 89*

89 Earth: $Y_1 = 3.611$ and $Y_2 = \theta + 0.0167\sin\theta$ intersect when $\theta \approx 3.619$.

91 Graph $y = \sin 2x$ and $y = 2 - x^2$. From the graph, we see that there are two points

of intersection. The x-coordinates of these points are $x \approx -1.48, \, 1.08$.

$[-\pi, \pi, \pi/4]$ by $[-2.09, 2.09]$ $[-\pi, \pi, \pi/4]$ by $[-2.09, 2.09]$

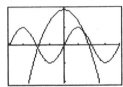   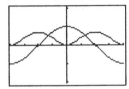

      Figure 91   *Figure 93*

93 Graph $y = \ln(1 + \sin^2 x)$ and $y = \cos x$. From the graph, we see that there are two

points of intersection. The x-coordinates of these points are $x \approx \pm 1.00$.

95 Graph $y = 3\cos^4 x - 2\cos^3 x + \cos x - 1$.

The graph has x-intercepts at $x \approx \pm 0.64, \, \pm 2.42$.

$[-\pi, \pi, \pi/4]$ by $[-2.09, 2.09]$

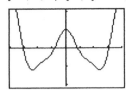

Figure 95

97 (a) $g = 9.8 \;\Rightarrow\; 9.8 = 9.8066(1 - 0.00264\cos 2\theta) \;\Rightarrow$

$\dfrac{9.8}{9.8066} = 1 - 0.00264\cos 2\theta \;\Rightarrow\; 0.00264\cos 2\theta = 1 - \dfrac{9.8}{9.8066} \;\Rightarrow$

$0.00264\cos 2\theta = \dfrac{0.0066}{9.8066} \;\Rightarrow\; \cos 2\theta = \dfrac{0.0066}{(9.8066)(0.00264)} \;\Rightarrow$

$2\theta = \cos^{-1}\left(\dfrac{0.0066}{0.025889424}\right) \approx 75.2° \;\Rightarrow\; \theta \approx 37.6°$

(b) At the equator, $g_0 = 9.8066(1 - 0.00264\cos 0°) = 9.8066(0.99736)$. Since the weight W of a person on the earth's surface is directly proportional to the force of gravity, we have $W = kg$.

At $\theta = 0°$, $W = kg \Rightarrow 150 = kg_0 \Rightarrow k = \dfrac{150}{g_0} \Rightarrow W = \dfrac{150}{g_0}g$.

$W = 150.5 \Rightarrow 150.5 = \dfrac{150}{g_0}g \Rightarrow 150.5 = \dfrac{150 \cdot 9.8066(1 - 0.00264\cos 2\theta)}{9.8066(0.99736)} \Rightarrow$

$150.5 = \dfrac{150}{0.99736}(1 - 0.00264\cos 2\theta) \Rightarrow 0.00264\cos 2\theta = 1 - \dfrac{150.5(0.99736)}{150} \Rightarrow$

$\cos 2\theta \approx -0.2593 \Rightarrow 2\theta \approx 105.0° \Rightarrow \theta \approx 52.5°$.

6.3 Exercises

$\boxed{1}$ *Note:* Use the cofunction formulas with $\left(\frac{\pi}{2} - u\right)$ for the argument if you're working with radian measure, $(90° - u)$ if you're working with degree measure.

(a) $\sin 46°37' = \cos(90° - 46°37') = \cos 43°23'$

(b) $\cos 73°12' = \sin(90° - 73°12') = \sin 16°48'$

(c) $\tan\frac{\pi}{6} = \cot\left(\frac{\pi}{2} - \frac{\pi}{6}\right) = \cot\left(\frac{3\pi}{6} - \frac{\pi}{6}\right) = \cot\frac{2\pi}{6} = \cot\frac{\pi}{3}$

(d) $\sec 17.28° = \csc(90° - 17.28°) = \csc 72.72°$

$\boxed{3}$ (a) $\cos\frac{7\pi}{20} = \sin\left(\frac{\pi}{2} - \frac{7\pi}{20}\right) = \sin\frac{3\pi}{20}$ (b) $\sin\frac{1}{4} = \cos\left(\frac{\pi}{2} - \frac{1}{4}\right) = \cos\left(\frac{2\pi-1}{4}\right)$

(c) $\tan 1 = \cot\left(\frac{\pi}{2} - 1\right) = \cot\left(\frac{\pi-2}{2}\right)$ (d) $\csc 0.53 = \sec\left(\frac{\pi}{2} - 0.53\right)$

$\boxed{5}$ (a) $\cos\frac{\pi}{4} + \cos\frac{\pi}{6} = \dfrac{\sqrt{2}}{2} + \dfrac{\sqrt{3}}{2} = \dfrac{\sqrt{2}+\sqrt{3}}{2}$

(b) $\cos\frac{5\pi}{12} = \cos\left(\frac{\pi}{4} + \frac{\pi}{6}\right) = \cos\frac{\pi}{4}\cos\frac{\pi}{6} - \sin\frac{\pi}{4}\sin\frac{\pi}{6} = \dfrac{\sqrt{2}}{2} \cdot \dfrac{\sqrt{3}}{2} - \dfrac{\sqrt{2}}{2} \cdot \dfrac{1}{2} = \dfrac{\sqrt{6}-\sqrt{2}}{4}$

$\boxed{7}$ (a) $\tan 60° + \tan 225° = \sqrt{3} + 1$

(b) $\tan 285° = \tan(60° + 225°) =$

$$\frac{\tan 60° + \tan 225°}{1 - \tan 60°\tan 225°} = \frac{\sqrt{3}+1}{1 - (\sqrt{3})(1)} \cdot \frac{1+\sqrt{3}}{1+\sqrt{3}} = \frac{4+2\sqrt{3}}{-2} = -2 - \sqrt{3}$$

$\boxed{9}$ (a) $\sin\frac{3\pi}{4} - \sin\frac{\pi}{6} = \dfrac{\sqrt{2}}{2} - \dfrac{1}{2} = \dfrac{\sqrt{2}-1}{2}$

(b) $\sin\frac{7\pi}{12} = \sin\left(\frac{3\pi}{4} - \frac{\pi}{6}\right) = \sin\frac{3\pi}{4}\cos\frac{\pi}{6} - \cos\frac{3\pi}{4}\sin\frac{\pi}{6} = \dfrac{\sqrt{2}}{2} \cdot \dfrac{\sqrt{3}}{2} - \left(-\dfrac{\sqrt{2}}{2}\right) \cdot \dfrac{1}{2} = \dfrac{\sqrt{6}+\sqrt{2}}{4}$

$\boxed{11}$ Since the expression is of the form "cos cos plus sin sin", we recognize it as the subtraction formula for the cosine.

$$\cos 48° \cos 23° + \sin 48° \sin 23° = \cos(48° - 23°) = \cos 25°$$

$\boxed{13}$ $\cos 10° \sin 5° - \sin 10° \cos 5° = \sin(5° - 10°) = \sin(-5°)$

15 Since we have angle arguments of 2, 3, and -2, we want to change one of them so that we can apply one of the formulas. We recognize that $\cos 2 = \cos(-2)$ and this is probably the simplest change. $\cos 3 \sin(-2) - \cos 2 \sin 3 =$

$$\sin(-2) \cos 3 - \cos(-2) \sin 3 = \sin(-2 - 3) = \sin(-5)$$

17 See *Figure 17* for a drawing of angles α and β.

   (a) $\sin(\alpha + \beta) = \sin \alpha \cos \beta + \cos \alpha \sin \beta = \frac{3}{5} \cdot \frac{15}{17} + \frac{4}{5} \cdot \frac{8}{17} = \frac{77}{85}$

   (b) $\cos(\alpha + \beta) = \cos \alpha \cos \beta - \sin \alpha \sin \beta = \frac{4}{5} \cdot \frac{15}{17} - \frac{3}{5} \cdot \frac{8}{17} = \frac{36}{85}$

   (c) Since the sine and cosine of $(\alpha + \beta)$ are positive, $(\alpha + \beta)$ is in QI.

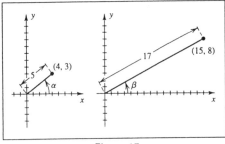 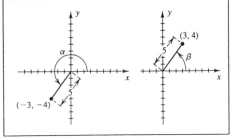

          Figure 17   *Figure 19*

19 See *Figure 19* for a drawing of angles α and β.

   (a) $\sin(\alpha + \beta) = \sin \alpha \cos \beta + \cos \alpha \sin \beta = \left(-\frac{4}{5}\right) \cdot \frac{3}{5} + \left(-\frac{3}{5}\right) \cdot \frac{4}{5} = -\frac{24}{25}$

   (b) $\tan(\alpha + \beta) = \dfrac{\tan \alpha + \tan \beta}{1 - \tan \alpha \tan \beta} = \dfrac{\frac{4}{3} + \frac{4}{3}}{1 - \frac{4}{3} \cdot \frac{4}{3}} \cdot \dfrac{9}{9} = \dfrac{12 + 12}{9 - 16} = -\dfrac{24}{7}$

   (c) Since the sine and tangent of $(\alpha + \beta)$ are negative, $(\alpha + \beta)$ is in QIV.

21 (a) $\sin(\alpha - \beta) = \sin \alpha \cos \beta - \cos \alpha \sin \beta$

$$= \left(-\frac{\sqrt{21}}{5}\right)\left(-\frac{3}{5}\right) - \left(-\frac{2}{5}\right) \cdot \left(-\frac{4}{5}\right) = \frac{3\sqrt{21} - 8}{25} \approx 0.23$$

   (b) $\cos(\alpha - \beta) = \cos \alpha \cos \beta + \sin \alpha \sin \beta$

$$= \left(-\frac{2}{5}\right) \cdot \left(-\frac{3}{5}\right) + \left(-\frac{\sqrt{21}}{5}\right) \cdot \left(-\frac{4}{5}\right) = \frac{4\sqrt{21} + 6}{25} \approx 0.97$$

   (c) Since the sine and cosine of $(\alpha - \beta)$ are positive, $(\alpha - \beta)$ is in QI.

23 $\sin(\theta + \pi) = \sin \theta \cos \pi + \cos \theta \sin \pi = \sin \theta (-1) + \cos \theta (0) = -\sin \theta$

25 $\sin\left(x - \frac{5\pi}{2}\right) = \sin x \cos \frac{5\pi}{2} - \cos x \sin \frac{5\pi}{2} = \sin x (0) - \cos x (1) = -\cos x$

27 $\cos(\theta - \pi) = \cos \theta \cos \pi + \sin \theta \sin \pi = \cos \theta (-1) + \sin \theta (0) = -\cos \theta$

29 $\cos\left(x + \frac{3\pi}{2}\right) = \cos x \cos \frac{3\pi}{2} - \sin x \sin \frac{3\pi}{2} = \cos x (0) - \sin x (-1) = \sin x$

31 $\tan\left(x - \frac{\pi}{2}\right) = \dfrac{\sin\left(x - \frac{\pi}{2}\right)}{\cos\left(x - \frac{\pi}{2}\right)} = \dfrac{\sin x \cos \frac{\pi}{2} - \cos x \sin \frac{\pi}{2}}{\cos x \cos \frac{\pi}{2} + \sin x \sin \frac{\pi}{2}} = \dfrac{-\cos x}{\sin x} = -\cot x$

[33] The tangent of a sum formula won't work here since $\tan \frac{\pi}{2}$ is undefined.

We will use a cofunction identity to verify the identity.

$$\tan\left(\theta + \tfrac{\pi}{2}\right) = \cot\left[\tfrac{\pi}{2} - \left(\theta + \tfrac{\pi}{2}\right)\right] = \cot\left(-\theta\right) = -\cot\theta$$

Alternatively, we could also write $\tan\left(\theta + \tfrac{\pi}{2}\right)$ as $\dfrac{\sin\left(\theta + \tfrac{\pi}{2}\right)}{\cos\left(\theta + \tfrac{\pi}{2}\right)}$ and then simplify.

[35] $\sin\left(\theta + \tfrac{\pi}{4}\right) = \sin\theta \cos\tfrac{\pi}{4} + \cos\theta \sin\tfrac{\pi}{4} = \dfrac{\sqrt{2}}{2}\sin\theta + \dfrac{\sqrt{2}}{2}\cos\theta = \dfrac{\sqrt{2}}{2}\left(\sin\theta + \cos\theta\right)$

[37] $\tan\left(u + \tfrac{\pi}{4}\right) = \dfrac{\tan u + \tan\tfrac{\pi}{4}}{1 - \tan u \tan\tfrac{\pi}{4}} = \dfrac{\tan u + 1}{1 - \tan u (1)} = \dfrac{1 + \tan u}{1 - \tan u}$

[39] $\cos\left(u + v\right) + \cos\left(u - v\right) = \left(\cos u \cos v - \sin u \sin v\right) + \left(\cos u \cos v + \sin u \sin v\right) =$

$$2\cos u \cos v$$

[41] $\sin\left(u + v\right) \cdot \sin\left(u - v\right) = \left(\sin u \cos v + \cos u \sin v\right) \cdot \left(\sin u \cos v - \cos u \sin v\right)$

$$\{\text{ addition and subtraction formulas for the sine }\}$$

$$= \sin^2 u \cos^2 v - \cos^2 u \sin^2 v$$

$$\{\text{ recognize as the difference of two squares }\}$$

$$= \sin^2 u \left(1 - \sin^2 v\right) - \left(1 - \sin^2 u\right)\sin^2 v$$

$$\{\text{ change to terms only involving sine }\}$$

$$= \sin^2 u - \sin^2 u \sin^2 v - \sin^2 v + \sin^2 u \sin^2 v = \sin^2 u - \sin^2 v$$

[43] $\dfrac{1}{\cot\alpha - \cot\beta} = \dfrac{1}{\dfrac{\cos\alpha}{\sin\alpha} - \dfrac{\cos\beta}{\sin\beta}} = \dfrac{1}{\dfrac{\cos\alpha \sin\beta - \cos\beta \sin\alpha}{\sin\alpha \sin\beta}} = \dfrac{\sin\alpha \sin\beta}{\sin\left(\beta - \alpha\right)}$

[45] $\sin\left(u + v + w\right) = \sin\left[\left(u + v\right) + w\right]$

$$= \sin\left(u + v\right)\cos w + \cos\left(u + v\right)\sin w$$

$$= \left(\sin u \cos v + \cos u \sin v\right)\cos w + \left(\cos u \cos v - \sin u \sin v\right)\sin w$$

$$= \sin u \cos v \cos w + \cos u \sin v \cos w + \cos u \cos v \sin w - \sin u \sin v \sin w$$

[47] The question usually asked here is "Why divide by $\sin u \sin v$?" Since we know the

form we want to end up with, we need to "force" the term "$-\sin u \sin v$" to equal

"-1", hence divide all terms by "$\sin u \sin v$."

$$\cot\left(u + v\right) = \dfrac{\cos\left(u + v\right)}{\sin\left(u + v\right)} = \dfrac{\left(\cos u \cos v - \sin u \sin v\right)(1/\sin u \sin v)}{\left(\sin u \cos v + \cos u \sin v\right)(1/\sin u \sin v)} = \dfrac{\cot u \cot v - 1}{\cot v + \cot u}$$

[49] $\sin\left(u - v\right) = \sin\left[u + \left(-v\right)\right] = \sin u \cos\left(-v\right) + \cos u \sin\left(-v\right) = \sin u \cos v - \cos u \sin v$

[51] $\dfrac{f(x + h) - f(x)}{h} = \dfrac{\cos\left(x + h\right) - \cos x}{h} = \dfrac{\cos x \cos h - \sin x \sin h - \cos x}{h} =$

$$\dfrac{\cos x \cos h - \cos x}{h} - \dfrac{\sin x \sin h}{h} = \cos x \left(\dfrac{\cos h - 1}{h}\right) - \sin x \left(\dfrac{\sin h}{h}\right)$$

$\boxed{53}$ $\sin 4t \cos t = \sin t \cos 4t$ $\Rightarrow$ $\sin 4t \cos t - \sin t \cos 4t = 0$ $\Rightarrow$ $\sin(4t - t) = 0$ $\Rightarrow$

$\qquad$ $\sin 3t = 0$ $\Rightarrow$ $3t = \pi n$ $\Rightarrow$ $t = \frac{\pi}{3}n$. In $[0, \pi)$, $t = 0, \frac{\pi}{3}, \frac{2\pi}{3}$.

$\boxed{55}$ $\cos 5t \cos 2t = -\sin 5t \sin 2t$ $\Rightarrow$ $\cos 5t \cos 2t + \sin 5t \sin 2t = 0$ $\Rightarrow$

$\qquad$ $\cos(5t - 2t) = 0$ $\Rightarrow$ $\cos 3t = 0$ $\Rightarrow$ $3t = \frac{\pi}{2} + \pi n$ $\Rightarrow$ $t = \frac{\pi}{6} + \frac{\pi}{3}n$.

$\qquad\qquad\qquad\qquad\qquad\qquad\qquad\qquad\qquad$ In $[0, \pi)$, $t = \frac{\pi}{6}, \frac{\pi}{2}, \frac{5\pi}{6}$.

$\boxed{57}$ $\tan 2t + \tan t = 1 - \tan 2t \tan t$ $\Rightarrow$ $\dfrac{\tan 2t + \tan t}{1 - \tan 2t \tan t} = 1$ $\Rightarrow$ $\tan(2t + t) = 1$ $\Rightarrow$

$\qquad$ $\tan 3t = 1$ $\Rightarrow$ $3t = \frac{\pi}{4} + \pi n$ $\Rightarrow$ $t = \frac{\pi}{12} + \frac{\pi}{3}n$. In $[0, \pi)$, $t = \frac{\pi}{12}, \frac{5\pi}{12}, \frac{3\pi}{4}$.

$\qquad$ However, $\tan 2t$ is undefined if $t = \frac{3\pi}{4}$, so exclude this value of t.

$\boxed{59}$ (a) $f(x) = \sqrt{3} \cos 2x + \sin 2x$ $\quad\bullet\quad$ $A = \sqrt{(\sqrt{3})^2 + 1^2} = 2$. $\tan C = \frac{1}{\sqrt{3}}$ $\Rightarrow$ $C = \frac{\pi}{6}$.

$\qquad\qquad\qquad\qquad\qquad\qquad$ $f(x) = 2 \cos(2x - \frac{\pi}{6}) = 2 \cos[2(x - \frac{\pi}{12})]$

$\qquad$ (b) amplitude $= 2$, period $= \frac{2\pi}{2} = \pi$, phase shift $= \frac{\pi}{12}$

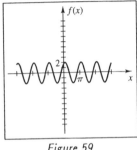
Figure 59

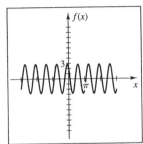
Figure 61

$\boxed{61}$ (a) $f(x) = 2 \cos 3x - 2 \sin 3x$ $\quad\bullet\quad$ $A = \sqrt{2^2 + 2^2} = 2\sqrt{2}$. $\tan C = \frac{-2}{2} = -1$ $\Rightarrow$

$\qquad\qquad\qquad\qquad\qquad\qquad$ $C = -\frac{\pi}{4}$. $f(x) = 2\sqrt{2} \cos(3x + \frac{\pi}{4}) = 2\sqrt{2} \cos[3(x + \frac{\pi}{12})]$

$\qquad$ (b) amplitude $= 2\sqrt{2}$, period $= \frac{2\pi}{3}$, phase shift $= -\frac{\pi}{12}$

$\boxed{63}$ $y = 50 \sin 60\pi t + 40 \cos 60\pi t$ $\quad\bullet\quad$ $A = \sqrt{50^2 + 40^2} = 10\sqrt{41}$. $\tan C = \frac{50}{40}$ $\Rightarrow$

$\qquad$ $C = \tan^{-1} \frac{5}{4} \approx 0.8961$. $y = 10\sqrt{41} \cos(60\pi t - \tan^{-1} \frac{5}{4}) \approx 10\sqrt{41} \cos(60\pi t - 0.8961)$.

$\boxed{65}$ (a) $y = 2 \cos t + 3 \sin t$; $A = \sqrt{2^2 + 3^2} = \sqrt{13}$; $\tan C = \frac{3}{2}$ $\Rightarrow$ $C \approx 0.98$;

$\qquad\qquad\qquad$ $y = \sqrt{13} \cos(t - C)$; amplitude $= \sqrt{13}$, period $= \frac{2\pi}{1} = 2\pi$

$\qquad$ (b) $y = 0$ $\Rightarrow$ $\sqrt{13} \cos(t - C) = 0$ $\Rightarrow$ $t - C = \frac{\pi}{2} + \pi n$ $\Rightarrow$

$\qquad\qquad\qquad$ $t = C + \frac{\pi}{2} + \pi n \approx 2.5536 + \pi n$ for every nonnegative integer n.

$\boxed{67}$ (a) $p(t) = A \sin \omega t + B \sin(\omega t + \tau)$

$\qquad\qquad$ $= A \sin \omega t + B(\sin \omega t \cos \tau + \cos \omega t \sin \tau)$

$\qquad\qquad$ $= A \underline{\sin \omega t} + B \cos \tau \underline{\sin \omega t} + B \sin \tau \underline{\cos \omega t}$

$\qquad\qquad$ $= (B \sin \tau) \cos \omega t + (A + B \cos \tau) \sin \omega t$

$\qquad\qquad$ $= a \cos \omega t + b \sin \omega t$ with $a = B \sin \tau$ and $b = A + B \cos \tau$

(b) $C^2 = (B \sin \tau)^2 + (A + B \cos \tau)^2$

$\qquad = B^2 \sin^2 \tau + A^2 + 2AB \cos \tau + B^2 \cos^2 \tau$

$\qquad = A^2 + B^2(\sin^2 \tau + \cos^2 \tau) + 2AB \cos \tau$

$\qquad = A^2 + B^2 + 2AB \cos \tau$

69 (a) $C^2 = A^2 + B^2 + 2AB \cos \tau \le A^2 + B^2 + 2AB$, since $\cos \tau \le 1$ and

$\qquad A > 0$, $B > 0$. Thus, $C^2 \le (A + B)^2$, and hence $C \le A + B$.

(b) $C = A + B$ if $\cos \tau = 1$, or $\tau = 0, 2\pi$.

(c) Constructive interference will occur if $C > A$. $C > A \Rightarrow C^2 > A^2 \Rightarrow$

$\qquad A^2 + B^2 + 2AB \cos \tau > A^2 \Rightarrow B^2 + 2AB \cos \tau > 0 \Rightarrow B(B + 2A \cos \tau) > 0$.

$\qquad$ Since $B > 0$, the product will be positive if $B + 2A \cos \tau > 0$, i.e., $\cos \tau > -\dfrac{B}{2A}$.

71 Graph $y = 3 \sin 2t + 2 \sin (4t + 1)$. Constructive

interference will occur when $y > 3$ or $y < -3$.

From the graph, we see that this occurs on the intervals

$(-2.97, -2.69)$, $(-1.00, -0.37)$, $(0.17, 0.46)$,

and $(2.14, 2.77)$.

$[-\pi, \pi, \pi/4]$ by $[-5, 5]$

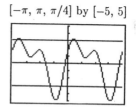

Figure 71

6.4 Exercises

1 From *Figure 1*, $\sin \theta = \frac{4}{5}$ and $\cos \theta = \frac{3}{5}$. Thus, $\sin 2\theta = 2 \sin \theta \cos \theta = 2(\frac{4}{5})(\frac{3}{5}) = \frac{24}{25}$.

$\qquad \cos 2\theta = \cos^2 \theta - \sin^2 \theta = (\frac{3}{5})^2 - (\frac{4}{5})^2 = -\frac{7}{25}$. $\tan 2\theta = \dfrac{\sin 2\theta}{\cos 2\theta} = \dfrac{24/25}{-7/25} = -\dfrac{24}{7}$.

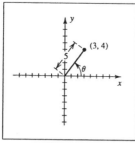

Figure 1

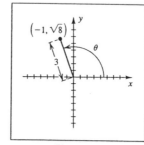

Figure 3

3 From *Figure 3*, $\sin \theta = \sqrt{8}/3 = \frac{2}{3}\sqrt{2}$ and $\cos \theta = -\frac{1}{3}$.

Thus, $\sin 2\theta = 2 \sin \theta \cos \theta = 2(\frac{2}{3}\sqrt{2})(-\frac{1}{3}) = -\frac{4}{9}\sqrt{2}$.

$\cos 2\theta = \cos^2 \theta - \sin^2 \theta = (-\frac{1}{3})^2 - (\frac{2}{3}\sqrt{2})^2 = \frac{1}{9} - \frac{8}{9} = -\frac{7}{9}$.

$\qquad \qquad \qquad \qquad \tan 2\theta = \dfrac{\sin 2\theta}{\cos 2\theta} = \dfrac{-4\sqrt{2}/9}{-7/9} = \dfrac{4\sqrt{2}}{7}$.

[5] $\sec\theta = \frac{5}{4} \implies \cos\theta = \frac{4}{5}$. θ acute implies that $\frac{\theta}{2}$ is acute, so all 6 trigonometric

functions of $\frac{\theta}{2}$ are positive and we use the "$+$" sign for the sine and cosine.

$$\sin\frac{\theta}{2} = \sqrt{\frac{1-\cos\theta}{2}} = \sqrt{\frac{1-\frac{4}{5}}{2}} = \sqrt{\frac{\frac{1}{5}}{2}} = \sqrt{\frac{1}{10}\cdot\frac{10}{10}} = \frac{\sqrt{10}}{10}.$$

$$\cos\frac{\theta}{2} = \sqrt{\frac{1+\cos\theta}{2}} = \sqrt{\frac{1+\frac{4}{5}}{2}} = \sqrt{\frac{\frac{9}{5}}{2}} = \sqrt{\frac{9}{10}\cdot\frac{10}{10}} = \frac{3\sqrt{10}}{10}.$$

$$\tan\frac{\theta}{2} = \frac{\sin\frac{\theta}{2}}{\cos\frac{\theta}{2}} = \frac{\sqrt{10}/10}{3\sqrt{10}/10} = \frac{1}{3}.$$

[7] From *Figure 7* $(a > 0)$,

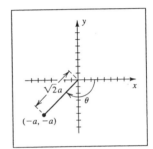

$$\cos\theta = -\frac{a}{\sqrt{2}\,a} = -\frac{\sqrt{2}}{2} \text{ and } \frac{\theta}{2} \text{ is in QIV.}$$

$$\sin\frac{\theta}{2} = -\sqrt{\frac{1-\cos\theta}{2}}$$

$$= -\sqrt{\frac{1+\sqrt{2}/2}{2}} = -\sqrt{\frac{2+\sqrt{2}}{4}} = -\frac{1}{2}\sqrt{2+\sqrt{2}}.$$

Figure 7

$$\cos\frac{\theta}{2} = \sqrt{\frac{1+\cos\theta}{2}}$$

$$= \sqrt{\frac{1-\sqrt{2}/2}{2}} = \sqrt{\frac{2-\sqrt{2}}{4}} = \frac{1}{2}\sqrt{2-\sqrt{2}}.$$

$$\tan\frac{\theta}{2} = \frac{1-\cos\theta}{\sin\theta} = \frac{1+\sqrt{2}/2}{-\sqrt{2}/2}\cdot\frac{2}{2} = \frac{2+\sqrt{2}}{-\sqrt{2}} = -\sqrt{2}-1.$$

[9] (a) $\cos 67°30' = \sqrt{\frac{1+\cos 135°}{2}} = \sqrt{\frac{1-\sqrt{2}/2}{2}} = \sqrt{\frac{2-\sqrt{2}}{4}} = \frac{1}{2}\sqrt{2-\sqrt{2}}.$

(b) $\sin 15° = \sqrt{\frac{1-\cos 30°}{2}} = \sqrt{\frac{1-\sqrt{3}/2}{2}} = \sqrt{\frac{2-\sqrt{3}}{4}} = \frac{1}{2}\sqrt{2-\sqrt{3}}.$

(c) $\tan\frac{3\pi}{8} = \frac{1-\cos\frac{3\pi}{4}}{\sin\frac{3\pi}{4}} = \frac{1+\sqrt{2}/2}{\sqrt{2}/2}\cdot\frac{2}{2} = \frac{2+\sqrt{2}}{\sqrt{2}} = \sqrt{2}+1.$

[11] Recognize that $10\theta = 2\cdot 5\theta$ and use the double-angle formula for the sine.

$$\sin 10\theta = \sin(2\cdot 5\theta) = 2\sin 5\theta\,\cos 5\theta$$

[13] We recognize the product of the sine and the cosine as being of the form of the right

side of the double-angle formula for the sine, and apply that formula "in reverse."

$$4\sin\frac{x}{2}\cos\frac{x}{2} = 2\cdot 2\sin\frac{x}{2}\cos\frac{x}{2} = 2\sin\left(2\cdot\frac{x}{2}\right) = 2\sin x$$

[15] $(\sin t + \cos t)^2 = \sin^2 t + 2\sin t\cos t + \cos^2 t = (\sin^2 t + \cos^2 t) + (2\sin t\cos t) =$

$$1 + \sin 2t$$

$\boxed{17}$ $\sin 3u = \sin (2u + u)$

$\qquad = \sin 2u \, \cos u + \cos 2u \, \sin u$ $\qquad$ { addition formula for sine }

$\qquad = (2 \sin u \, \cos u) \cos u + (1 - 2 \sin^2 u) \sin u$ $\quad$ { double-angle formulas for sin & cos }

$\qquad = 2 \sin u \, \cos^2 u + \sin u - 2 \sin^3 u$

$\qquad = 2 \sin u (1 - \sin^2 u) + \sin u - 2 \sin^3 u$ $\qquad$ { $\sin^2 u + \cos^2 u = 1$ }

$\qquad = 2 \sin u - 2 \sin^3 u + \sin u - 2 \sin^3 u$

$\qquad = 3 \sin u - 4 \sin^3 u$

$\qquad = \sin u (3 - 4 \sin^2 u)$

$\boxed{19}$ $\cos 4\theta = \cos (2 \cdot 2\theta) = 2 \cos^2 2\theta - 1$ $\qquad$ { double-angle formula for cos }

$\qquad = 2(2 \cos^2 \theta - 1)^2 - 1$ $\qquad$ { double-angle formula for cos }

$\qquad = 2(4 \cos^4 \theta - 4 \cos^2 \theta + 1) - 1$ $\quad$ { square binomial }

$\qquad = (8 \cos^4 \theta - 8 \cos^2 \theta + 2) - 1$

$\qquad = 8 \cos^4 \theta - 8 \cos^2 \theta + 1$

$\boxed{21}$ $\sin^4 t = (\sin^2 t)^2 \; = \left(\dfrac{1 - \cos 2t}{2} \right)^2$

$\qquad = \tfrac{1}{4}(1 - 2 \cos 2t + \cos^2 2t)$

$\qquad = \tfrac{1}{4} - \tfrac{1}{2} \cos 2t + \tfrac{1}{4}\left(\dfrac{1 + \cos 4t}{2} \right)$ $\qquad$ { half-angle identity (2) }

$\qquad = \tfrac{1}{4} - \tfrac{1}{2} \cos 2t + \tfrac{1}{8} + \tfrac{1}{8} \cos 4t$

$\qquad = \tfrac{3}{8} - \tfrac{1}{2} \cos 2t + \tfrac{1}{8} \cos 4t$

$\boxed{23}$ We do not have a formula for $\sec 2\theta$, so we will write $\sec 2\theta$ in terms of $\cos 2\theta$ in order to apply the double-angle formula for the cosine.

$$\sec 2\theta = \frac{1}{\cos 2\theta} = \frac{1}{2 \cos^2 \theta - 1} = \frac{1}{2\left(\dfrac{1}{\sec^2 \theta}\right) - 1} = \frac{1}{\dfrac{2 - \sec^2 \theta}{\sec^2 \theta}} = \frac{\sec^2 \theta}{2 - \sec^2 \theta}$$

$\boxed{25}$ We need to match the arguments of the trigonometric functions involved—that is, either write both of them in terms of $2t$ or in terms of $4t$. Converting $\cos 4t$ to an expression with $2t$ as the angle argument gives us

$$2 \sin^2 2t + \cos 4t = 2 \sin^2 2t + \cos (2 \cdot 2t) = 2 \sin^2 2t + (1 - 2 \sin^2 2t) = 1.$$

Alternatively, if we write both arguments in terms of $4t$, we have

$$2 \sin^2 2t + \cos 4t = 2 \cdot \frac{1 - \cos 4t}{2} + \cos 4t = 1 - \cos 4t + \cos 4t = 1.$$

$\boxed{27}$ $\tan 3u = \tan(2u + u) = \dfrac{\tan 2u + \tan u}{1 - \tan 2u \tan u}$

$$= \dfrac{\dfrac{2\tan u}{1 - \tan^2 u} + \tan u}{1 - \dfrac{2\tan u}{1 - \tan^2 u} \cdot \tan u}$$

$$= \dfrac{\dfrac{2\tan u + \tan u - \tan^3 u}{1 - \tan^2 u}}{\dfrac{1 - \tan^2 u - 2\tan^2 u}{1 - \tan^2 u}}$$

$$= \dfrac{3\tan u - \tan^3 u}{1 - 3\tan^2 u}$$

$$= \dfrac{\tan u\,(3 - \tan^2 u)}{1 - 3\tan^2 u}$$

$\boxed{29}$ $\cos^4 \dfrac{\theta}{2} = \left(\cos^2 \dfrac{\theta}{2}\right)^2 = \left(\dfrac{1 + \cos\theta}{2}\right)^2 = \dfrac{1 + 2\cos\theta + \cos^2\theta}{4}$

$$= \tfrac{1}{4} + \tfrac{1}{2}\cos\theta + \tfrac{1}{4}\left(\dfrac{1 + \cos 2\theta}{2}\right)$$

$$= \tfrac{1}{4} + \tfrac{1}{2}\cos\theta + \tfrac{1}{8} + \tfrac{1}{8}\cos 2\theta = \tfrac{3}{8} + \tfrac{1}{2}\cos\theta + \tfrac{1}{8}\cos 2\theta$$

$\boxed{31}$ $\sin^4 2x = (\sin^2 2x)^2 = \left(\dfrac{1 - \cos 4x}{2}\right)^2 = \dfrac{1 - 2\cos 4x + \cos^2 4x}{4}$

$$= \tfrac{1}{4} - \tfrac{2}{4}\cos 4x + \tfrac{1}{4}\cos^2 4x = \tfrac{1}{4} - \tfrac{1}{2}\cos 4x + \tfrac{1}{4}\left(\dfrac{1 + \cos 8x}{2}\right)$$

$$= \tfrac{1}{4} - \tfrac{1}{2}\cos 4x + \tfrac{1}{8} + \tfrac{1}{8}\cos 8x = \tfrac{3}{8} - \tfrac{1}{2}\cos 4x + \tfrac{1}{8}\cos 8x$$

$\boxed{33}$ $\sin 2t + \sin t = 0 \;\Rightarrow\; 2\sin t\cos t + \sin t = 0 \;\Rightarrow\; \sin t\,(2\cos t + 1) = 0 \;\Rightarrow$

$\qquad\qquad\qquad\qquad\qquad \sin t = 0 \text{ or } \cos t = -\tfrac{1}{2} \;\Rightarrow\; t = 0,\ \pi \text{ or } \tfrac{2\pi}{3},\ \tfrac{4\pi}{3}$

$\boxed{35}$ $\cos u + \cos 2u = 0 \;\Rightarrow\; \cos u + 2\cos^2 u - 1 = 0 \;\Rightarrow\; 2\cos^2 u + \cos u - 1 = 0 \;\Rightarrow$

$\qquad\qquad (2\cos u - 1)(\cos u + 1) = 0 \;\Rightarrow\; \cos u = \tfrac{1}{2},\ -1 \;\Rightarrow\; u = \tfrac{\pi}{3},\ \tfrac{5\pi}{3},\ \pi$

$\boxed{37}$ A first approach uses the concept that if $\tan\alpha = \tan\beta$, then $\alpha = \beta + \pi n$.

$\qquad\qquad \tan 2x = \tan x \;\Rightarrow\; 2x = x + \pi n \;\Rightarrow\; x = \pi n \;\Rightarrow\; x = 0,\ \pi.$

$\qquad$ Another approach is:

$\qquad\qquad \tan 2x = \tan x \;\Rightarrow\; \dfrac{\sin 2x}{\cos 2x} = \dfrac{\sin x}{\cos x} \;\Rightarrow\; \sin 2x \cos x = \sin x \cos 2x \;\Rightarrow$

$\qquad\qquad \sin 2x \cos x - \sin x \cos 2x = 0 \;\Rightarrow\; \sin(2x - x) = 0 \;\Rightarrow\; \sin x = 0 \;\Rightarrow\; x = 0,\ \pi.$

$\boxed{39}$ $\sin \tfrac{1}{2}u + \cos u = 1 \;\Rightarrow\; \sin\tfrac{1}{2}u + \cos\left[2\cdot(\tfrac{1}{2}u)\right] = 1 \;\Rightarrow\; \sin\tfrac{1}{2}u + (1 - 2\sin^2\tfrac{1}{2}u) = 1 \;\Rightarrow$

$\qquad \sin\tfrac{1}{2}u - 2\sin^2\tfrac{1}{2}u = 0 \;\Rightarrow\; \sin\tfrac{1}{2}u\,(1 - 2\sin\tfrac{1}{2}u) = 0 \;\Rightarrow\; \sin\tfrac{1}{2}u = 0,\ \tfrac{1}{2} \;\Rightarrow$

$\qquad\qquad\qquad\qquad\qquad \tfrac{1}{2}u = 0,\ \tfrac{\pi}{6},\ \tfrac{5\pi}{6} \;\Rightarrow\; u = 0,\ \tfrac{\pi}{3},\ \tfrac{5\pi}{3}$

41 $RS = \sqrt{a^2 + b^2} \sin(u + v) = \sqrt{a^2 + b^2} (\sin u \cos v + \cos u \sin v)$

$$= \sqrt{a^2 + b^2} \sin u \cos v + \sqrt{a^2 + b^2} \cos u \sin v.$$

$$= (\sqrt{a^2 + b^2} \cos v) \sin u + (\sqrt{a^2 + b^2} \sin v) \cos u.$$

$LS = a \sin u + b \cos u.$ Equating coefficients of $\sin u$ and $\cos u$ gives us

$a = \sqrt{a^2 + b^2} \cos v$ and $b = \sqrt{a^2 + b^2} \sin v \Rightarrow$

$\cos v = \dfrac{a}{\sqrt{a^2 + b^2}}$ and $\sin v = \dfrac{b}{\sqrt{a^2 + b^2}}.$ Since $0 < u < \frac{\pi}{2}$, $\sin u > 0$ and $\cos v > 0.$

Now $a > 0$ and $b > 0$ combine with the above to imply that $\cos v > 0$ and $\sin v > 0.$

Thus, $0 < v < \frac{\pi}{2}.$

43 (a) $\cos 2x + 2 \cos x = 0 \Rightarrow 2 \cos^2 x + 2 \cos x - 1 = 0 \Rightarrow$

$$\cos x = \frac{-2 \pm \sqrt{12}}{4} = \frac{-1 \pm \sqrt{3}}{2} \approx 0.366 \left\{ \cos x \neq \frac{-1 - \sqrt{3}}{2} < -1 \right\}.$$

Thus, $x \approx 1.20$ and $5.09.$

(b) $\sin 2x + \sin x = 0 \Rightarrow 2 \sin x \cos x + \sin x = 0 \Rightarrow \sin x (2 \cos x + 1) = 0 \Rightarrow$

$\sin x = 0$ or $\cos x = -\frac{1}{2} \Rightarrow x = 0, \pi, 2\pi$ or $\frac{2\pi}{3}, \frac{4\pi}{3}.$

$$P(\tfrac{2\pi}{3}, -1.5), \ Q(\pi, -1), \ R(\tfrac{4\pi}{3}, -1.5)$$

45 (a) $\cos 3x - 3 \cos x = 0 \Rightarrow 4 \cos^3 x - 3 \cos x - 3 \cos x = 0 \Rightarrow$

$4 \cos^3 x - 6 \cos x = 0 \Rightarrow 2 \cos x (2 \cos^2 x - 3) = 0 \Rightarrow \cos x = 0, \pm \sqrt{3/2} \Rightarrow$

$$x = -\tfrac{3\pi}{2}, -\tfrac{\pi}{2}, \tfrac{\pi}{2}, \tfrac{3\pi}{2} \left\{ \cos x \neq \pm \sqrt{3/2} \text{ since } \sqrt{3/2} > 1 \right\}$$

(b) $\sin 3x - \sin x = 0 \Rightarrow 3 \sin x - 4 \sin^3 x - \sin x = 0 \Rightarrow 4 \sin^3 x - 2 \sin x = 0 \Rightarrow$

$2 \sin x (2 \sin^2 x - 1) = 0 \Rightarrow \sin x = 0, \pm 1/\sqrt{2} \Rightarrow$

$$x = 0, \pm \pi, \pm 2\pi, \pm \tfrac{\pi}{4}, \pm \tfrac{3\pi}{4}, \pm \tfrac{5\pi}{4}, \pm \tfrac{7\pi}{4}$$

47 (a) Let $y = \overline{BC}.$ Form a right triangle with hypotenuse y, side opposite θ, 20, and

side adjacent θ, $x.$ $\sin \theta = \dfrac{20}{y} \Rightarrow y = \dfrac{20}{\sin \theta}.$

$\cos \theta = \frac{x}{y} \Rightarrow x = y \cos \theta = \dfrac{20}{\sin \theta} \cdot \cos \theta = \dfrac{20 \cos \theta}{\sin \theta}.$

Now $d = (40 - x) + y = 40 - \dfrac{20 \cos \theta}{\sin \theta} + \dfrac{20}{\sin \theta} = 20 \left(\dfrac{1 - \cos \theta}{\sin \theta} \right) + 40 = 20 \tan \dfrac{\theta}{2} + 40.$

(b) $50 = 20 \tan \frac{\theta}{2} + 40 \Rightarrow \tan \frac{\theta}{2} = \frac{1}{2} \Rightarrow \dfrac{1 - \cos \theta}{\sin \theta} = \frac{1}{2} \Rightarrow 2 - 2 \cos \theta = \sin \theta \Rightarrow$

$4 - 8 \cos \theta + 4 \cos^2 \theta = \sin^2 \theta = 1 - \cos^2 \theta \Rightarrow 5 \cos^2 \theta - 8 \cos \theta + 3 = 0 \Rightarrow$

$(5 \cos \theta - 3)(\cos \theta - 1) = 0 \Rightarrow \cos \theta = \frac{3}{5}, 1. \ \{ \cos \theta = 1 \Rightarrow$

$\theta = 0$ and 0 is extraneous $\}.$ $\cos \theta = \frac{3}{5} \Rightarrow \sin \theta = \frac{4}{5}$ and $y = \frac{20}{4/5} = 25.$

$\cos \theta = \frac{x}{y}$ and $\cos \theta = \frac{3}{5} \Rightarrow \frac{x}{25} = \frac{3}{5} \Rightarrow$

$x = 15$, which means that B would be 25 miles from $A.$

49 (a) From Example 8, the area A of a cross section is

$$A = \tfrac{1}{2}(\text{side})^2(\text{sine of included angle}) = \tfrac{1}{2}\left(\tfrac{1}{2}\right)^2\sin\theta = \tfrac{1}{8}\sin\theta.$$

The volume $V = (\text{length of gutter})(\text{area of cross section}) = 20\left(\tfrac{1}{8}\sin\theta\right) = \tfrac{5}{2}\sin\theta.$

(b) $V = 2 \;\Rightarrow\; \tfrac{5}{2}\sin\theta = 2 \;\Rightarrow\; \sin\theta = \tfrac{4}{5} \;\Rightarrow\; \theta \approx 53.13°.$

51 (a) Let $y = \overline{DB}$ and x denote the distance from D to the midpoint of $\overline{BC}$.

$$\sin\tfrac{\theta}{2} = \frac{b/2}{y} \;\Rightarrow\; y = \frac{b}{2}\cdot\frac{1}{\sin(\theta/2)} \quad\text{and}\quad \tan\tfrac{\theta}{2} = \frac{b/2}{x} \;\Rightarrow\; x = \frac{b}{2}\cdot\frac{\cos(\theta/2)}{\sin(\theta/2)}.$$

$$l = (a-x) + y = a - \frac{b}{2}\cdot\frac{\cos(\theta/2)}{\sin(\theta/2)} + \frac{b}{2}\cdot\frac{1}{\sin(\theta/2)} = a + \frac{b}{2}\cdot\frac{1-\cos(\theta/2)}{\sin(\theta/2)} =$$

$$a + \frac{b}{2}\tan\left(\frac{\theta/2}{2}\right) = a + \frac{b}{2}\tan\frac{\theta}{4}.$$

(b) $a = 10$ mm, $b = 6$ mm, and $\theta = 156° \;\Rightarrow\; l = 10 + 3\tan 39° \approx 12.43$ mm.

53 The graph of f appears to be that of $y = g(x) = \tan x$.

$$\frac{\sin 2x + \sin x}{\cos 2x + \cos x + 1} = \frac{2\sin x\cos x + \sin x}{(2\cos^2 x - 1) + \cos x + 1} = \frac{\sin x(2\cos x + 1)}{\cos x(2\cos x + 1)} = \frac{\sin x}{\cos x} = \tan x$$

55 Graph $Y_1 = \tan(0.5x + 1)$ and $Y_2 = \sin 0.5x$ on $[-2\pi,\, 2\pi]$ in Dot mode. There are two points of intersection. They occur at $x \approx -3.55,\, 5.22.$

$[-2\pi,\, 2\pi,\, \pi/2]$ by $[-4,\, 4]$ $[-\pi,\, \pi,\, \pi/4]$ by $[-4,\, 4]$ $[-2\pi,\, 2\pi,\, \pi/2]$ by $[-4,\, 4]$

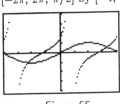
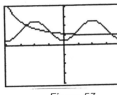
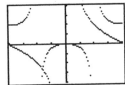

Figure 55   Figure 57   Figure 59

57 Graph $Y_1 = 1/\sin(0.25x + 1)$ and $Y_2 = 1.5 - \cos 2x$ on $[-\pi,\, \pi]$. There are four points of intersection. They occur at $x \approx -2.03,\, -0.72,\, 0.58,\, 2.62.$

59 Graph $Y_1 = 2/\tan(.25x)$ and $Y_2 = 1 - 1/\cos(.5x)$ on $[-2\pi,\, 2\pi]$ in Dot mode. There is one point of intersection. It occurs at $x \approx -2.59.$

6.5 Exercises

Note: We will reference the product-to-sum formulas as [P1]–[P4] and the sum-to-product formulas as [S1]–[S4] in the order they appear in the text. The formulas $\cos(-kx) = \cos kx$ and $\sin(-kx) = -\sin kx$ will be used without mention.

1 $\sin 7t\,\sin 3t = [\text{P4}]\; \tfrac{1}{2}\left[\cos(7t - 3t) - \cos(7t + 3t)\right] = \tfrac{1}{2}\cos 4t - \tfrac{1}{2}\cos 10t$

3 $\cos 6u\,\cos(-4u) = [\text{P3}]\; \tfrac{1}{2}\left\{\cos[6u + (-4u)] + \cos[6u - (-4u)]\right\} = \tfrac{1}{2}\cos 2u + \tfrac{1}{2}\cos 10u$

5 $2\sin 9\theta\,\cos 3\theta = [\text{P1}]\; 2\cdot\tfrac{1}{2}\left[\sin(9\theta + 3\theta) + \sin(9\theta - 3\theta)\right] = \sin 12\theta + \sin 6\theta$

$\boxed{7}$ $3\cos x \sin 2x = $ [P2] $3 \cdot \frac{1}{2}[\sin(x+2x) - \sin(x-2x)] = \frac{3}{2}\sin 3x - \frac{3}{2}\sin(-x) = $

$$\frac{3}{2}\sin 3x + \frac{3}{2}\sin x$$

$\boxed{9}$ $\sin 6\theta + \sin 2\theta = $ [S1] $2\sin\dfrac{6\theta + 2\theta}{2}\cos\dfrac{6\theta - 2\theta}{2} = 2\sin 4\theta \cos 2\theta$

$\boxed{11}$ $\cos 5x - \cos 3x = $ [S4] $-2\sin\dfrac{5x + 3x}{2}\sin\dfrac{5x - 3x}{2} = -2\sin 4x \sin x$

$\boxed{13}$ $\sin 3t - \sin 7t = $ [S2] $2\cos\dfrac{3t + 7t}{2}\sin\dfrac{3t - 7t}{2} = 2\cos 5t \sin(-2t) = -2\cos 5t \sin 2t$

$\boxed{15}$ $\cos x + \cos 2x = $ [S3] $2\cos\dfrac{x + 2x}{2}\cos\dfrac{x - 2x}{2} = 2\cos\frac{3}{2}x \cos(-\frac{1}{2}x) = 2\cos\frac{3}{2}x \cos\frac{1}{2}x$

$\boxed{17}$ $\dfrac{\sin 4t + \sin 6t}{\cos 4t - \cos 6t} = \dfrac{[S1]\ 2\sin 5t \cos(-t)}{[S4]\ -2\sin 5t \sin(-t)} = \dfrac{\cos t}{\sin t} = \cot t$

$\boxed{19}$ $\dfrac{\sin u + \sin v}{\cos u + \cos v} = \dfrac{[S1]\ 2\sin\frac{1}{2}(u+v)\cos\frac{1}{2}(u-v)}{[S3]\ 2\cos\frac{1}{2}(u+v)\cos\frac{1}{2}(u-v)} = \dfrac{\sin\frac{1}{2}(u+v)}{\cos\frac{1}{2}(u+v)} = \tan\frac{1}{2}(u+v)$

$\boxed{21}$ $\dfrac{\sin u - \sin v}{\sin u + \sin v} = \dfrac{[S2]\ 2\cos\frac{1}{2}(u+v)\sin\frac{1}{2}(u-v)}{[S1]\ 2\sin\frac{1}{2}(u+v)\cos\frac{1}{2}(u-v)} = \cot\frac{1}{2}(u+v)\tan\frac{1}{2}(u-v) = \dfrac{\tan\frac{1}{2}(u-v)}{\tan\frac{1}{2}(u+v)}$

$\boxed{23}$ Since the arguments on the right side are all even multiples of x ($2x$, $4x$, and $6x$), we begin by grouping the terms with the odd multiples of x together, and operate on them with a product-to-sum formula, which will convert these expressions to expressions with even multiples of x.

$4\cos x \cos 2x \sin 3x = 2\cos 2x (2\sin 3x \cos x)$

$\qquad\qquad\qquad = 2\cos 2x ([P1]\ \sin 4x + \sin 2x)$

$\qquad\qquad\qquad = (2\cos 2x \sin 4x) + (2\cos 2x \sin 2x)$

$\qquad\qquad\qquad = [[P2]\ \sin 6x - \sin(-2x)] + ([P2]\ \sin 4x - \sin 0)$

$\qquad\qquad\qquad = \sin 2x + \sin 4x + \sin 6x$

$\boxed{25}$ $(\sin ax)(\cos bx) = $ [P1] $\frac{1}{2}[\sin(ax + bx) + \sin(ax - bx)] = \frac{1}{2}\sin[(a+b)x] + \frac{1}{2}\sin[(a-b)x]$

$\boxed{27}$ $\sin 5t + \sin 3t = 0 \ \Rightarrow \ $ [S1] $2\sin 4t \cos t = 0 \ \Rightarrow \ \sin 4t = 0$ or $\cos t = 0 \ \Rightarrow$

$\qquad\qquad 4t = \pi n$ or $t = \frac{\pi}{2} + \pi n \ \Rightarrow \ t = \frac{\pi}{4}n$ {which includes $t = \frac{\pi}{2} + \pi n$}

$\boxed{29}$ $\cos x = \cos 3x \ \Rightarrow \ \cos x - \cos 3x = 0 \ \Rightarrow \ $ [S4] $-2\sin 2x \sin(-x) = 0 \ \Rightarrow$

$\qquad \sin 2x = 0$ or $\sin x = 0 \ \Rightarrow \ 2x = \pi n$ or $x = \pi n \ \Rightarrow \ x = \frac{\pi}{2}n$ {which includes $x = \pi n$}

$\boxed{31}$ $\cos 3x + \cos 5x = \cos x \ \Rightarrow \ $ [S3] $2\cos 4x \cos(-x) - \cos x = 0 \ \Rightarrow$

$\qquad \cos x (2\cos 4x - 1) = 0 \ \Rightarrow \ \cos x = 0$ or $\cos 4x = \frac{1}{2} \ \Rightarrow$

$\qquad\qquad x = \frac{\pi}{2} + \pi n$ or $4x = \frac{\pi}{3} + 2\pi n, \frac{5\pi}{3} + 2\pi n \ \Rightarrow \ x = \frac{\pi}{2} + \pi n, \frac{\pi}{12} + \frac{\pi}{2}n, \frac{5\pi}{12} + \frac{\pi}{2}n$

$\boxed{33}$ $\sin 2x - \sin 5x = 0 \ \Rightarrow \ $ [S2] $2\cos\frac{7}{2}x \sin(-\frac{3}{2}x) = 0 \ \Rightarrow \ \frac{7}{2}x = \frac{\pi}{2} + \pi n$ or $\frac{3}{2}x = \pi n \ \Rightarrow$

$\qquad\qquad\qquad\qquad x = \frac{\pi}{7} + \frac{2\pi}{7}n$ or $x = \frac{2\pi}{3}n$

35 $\cos x + \cos 3x = 0 \Rightarrow$ [S3] $2\cos 2x \cos(-x) = 0 \Rightarrow \cos 2x = 0$ or $\cos x = 0 \Rightarrow$

$2x = \frac{\pi}{2} + \pi n$ or $x = \frac{\pi}{2} + \pi n \Rightarrow x = \frac{\pi}{4} + \frac{\pi}{2}n$ or $x = \frac{\pi}{2} + \pi n \Rightarrow$

$x = \frac{\pi}{4}, \frac{3\pi}{4}, \frac{5\pi}{4}, \frac{7\pi}{4}, \frac{\pi}{2}, \frac{3\pi}{2}$ for $0 \le x \le 2\pi$

37 $\sin 3x - \sin x = 0 \Rightarrow$ [S2] $2\cos 2x \sin x = 0 \Rightarrow \cos 2x = 0$ or $\sin x = 0 \Rightarrow$

$2x = \frac{\pi}{2} + \pi n$ or $x = \pi n \Rightarrow x = \frac{\pi}{4} + \frac{\pi}{2}n$ or $x = \pi n \Rightarrow$

$x = 0, \pm \pi, \pm 2\pi, \pm \frac{\pi}{4}, \pm \frac{3\pi}{4}, \pm \frac{5\pi}{4}, \pm \frac{7\pi}{4}$ for $-2\pi \le x \le 2\pi$.

39 $f(x) = \sin\left(\frac{\pi n}{l}x\right)\cos\left(\frac{k\pi n}{l}t\right)$

$= $ [P1] $\frac{1}{2}\left[\sin\left(\frac{\pi n}{l}x + \frac{k\pi n}{l}t\right) + \sin\left(\frac{\pi n}{l}x - \frac{k\pi n}{l}t\right)\right]$

$= \frac{1}{2}\left[\sin\frac{\pi n}{l}(x + kt) + \sin\frac{\pi n}{l}(x - kt)\right]$

$= \frac{1}{2}\sin\frac{\pi n}{l}(x + kt) + \frac{1}{2}\sin\frac{\pi n}{l}(x - kt)$

41 (a) Estimating the x-intercepts, we have $x \approx 0, \pm 1.05, \pm 1.57, \pm 2.09, \pm 3.14$.

(b) $\sin 4x + \sin 2x = 2\sin 3x \cos x = 0 \Rightarrow \sin 3x = 0$ or $\cos x = 0$.

$\sin 3x = 0 \Rightarrow 3x = \pi n \Rightarrow x = \frac{\pi}{3}n \Rightarrow x = 0, \pm\frac{\pi}{3}, \pm\frac{2\pi}{3}, \pm\pi$.

$\cos x = 0 \Rightarrow x = \pm\frac{\pi}{2}$. The x-intercepts are $0, \pm\frac{\pi}{3}, \pm\frac{\pi}{2}, \pm\frac{2\pi}{3}, \pm\pi$.

$[-\pi, \pi, \pi/4]$ by $[-2.09, 2.09]$

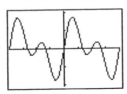

Figure 41

43 Graphing on an interval of $[-\pi, \pi]$ gives us a figure that resembles a tangent function. It appears that there is a vertical asymptote at about 0.78. Recognizing that this value is about $\frac{\pi}{4}$, we might make the conjecture that we have "halved" the period of the tangent and that the graph of $f(x) = \dfrac{\sin x + \sin 2x + \sin 3x}{\cos x + \cos 2x + \cos 3x}$ appears to be that of $y = g(x) = \tan 2x$. Verifying this identity, we have

$\dfrac{\sin x + \sin 2x + \sin 3x}{\cos x + \cos 2x + \cos 3x} = \dfrac{\sin 2x + (\sin 3x + \sin x)}{\cos 2x + (\cos 3x + \cos x)} = \dfrac{\sin 2x + 2\sin 2x \cos x}{\cos 2x + 2\cos 2x \cos x} = $

$\dfrac{\sin 2x(1 + 2\cos x)}{\cos 2x(1 + 2\cos x)} = \dfrac{\sin 2x}{\cos 2x} = \tan 2x.$

6.6 Exercises

$\boxed{1}$ (a) $\sin^{-1}\left(-\frac{\sqrt{2}}{2}\right) = -\frac{\pi}{4}$ since $\sin\left(-\frac{\pi}{4}\right) = -\frac{\sqrt{2}}{2}$ and $-\frac{\pi}{2} \leq -\frac{\pi}{4} \leq \frac{\pi}{2}$

(b) $\cos^{-1}\left(-\frac{1}{2}\right) = \frac{2\pi}{3}$ since $\cos\frac{2\pi}{3} = -\frac{1}{2}$ and $0 \leq \frac{2\pi}{3} \leq \pi$

(c) $\tan^{-1}(-\sqrt{3}) = -\frac{\pi}{3}$ since $\tan\left(-\frac{\pi}{3}\right) = -\sqrt{3}$ and $-\frac{\pi}{2} < -\frac{\pi}{3} < \frac{\pi}{2}$

$\boxed{3}$ (a) $\arcsin\frac{\sqrt{3}}{2} = \frac{\pi}{3}$ since $\sin\frac{\pi}{3} = \frac{\sqrt{3}}{2}$ and $-\frac{\pi}{2} \leq \frac{\pi}{3} \leq \frac{\pi}{2}$

(b) $\arccos\frac{\sqrt{2}}{2} = \frac{\pi}{4}$ since $\cos\frac{\pi}{4} = \frac{\sqrt{2}}{2}$ and $0 \leq \frac{\pi}{4} \leq \pi$

(c) $\arctan\frac{1}{\sqrt{3}} = \frac{\pi}{6}$ since $\tan\frac{\pi}{6} = \frac{1}{\sqrt{3}}$ and $-\frac{\pi}{2} < \frac{\pi}{6} < \frac{\pi}{2}$

$\boxed{5}$ (a) $\sin^{-1}\frac{\pi}{3}$ is <u>not defined</u> since $\frac{\pi}{3} > 1$, that is, $\frac{\pi}{3} \notin [-1, 1]$

(b) $\cos^{-1}\frac{\pi}{2}$ is <u>not defined</u> since $\frac{\pi}{2} > 1$, that is, $\frac{\pi}{2} \notin [-1, 1]$

(c) $\tan^{-1}1 = \frac{\pi}{4}$ since $\tan\frac{\pi}{4} = 1$ and $-\frac{\pi}{2} < \frac{\pi}{4} < \frac{\pi}{2}$

Note: Exercises 7–10 refer to the boxed properties of $\sin^{-1}$, $\cos^{-1}$, and $\tan^{-1}$.

$\boxed{7}$ (a) $\sin\left[\arcsin\left(-\frac{3}{10}\right)\right] = -\frac{3}{10}$ since $-1 \leq -\frac{3}{10} \leq 1$

(b) $\cos\left(\arccos\frac{1}{2}\right) = \frac{1}{2}$ since $-1 \leq \frac{1}{2} \leq 1$

(c) $\tan\left(\arctan 14\right) = 14$ since $\tan\left(\arctan x\right) = x$ for every x

$\boxed{9}$ (a) $\sin^{-1}\left(\sin\frac{\pi}{3}\right) = \frac{\pi}{3}$ since $-\frac{\pi}{2} \leq \frac{\pi}{3} \leq \frac{\pi}{2}$

(b) $\cos^{-1}\left[\cos\left(\frac{5\pi}{6}\right)\right] = \frac{5\pi}{6}$ since $0 \leq \frac{5\pi}{6} \leq \pi$

(c) $\tan^{-1}\left[\tan\left(-\frac{\pi}{6}\right)\right] = -\frac{\pi}{6}$ since $-\frac{\pi}{2} < -\frac{\pi}{6} < \frac{\pi}{2}$

$\boxed{11}$ (a) $\arcsin\left(\sin\frac{5\pi}{4}\right) = \arcsin\left(-\frac{\sqrt{2}}{2}\right) = -\frac{\pi}{4}$

(b) $\arccos\left(\cos\frac{5\pi}{4}\right) = \arccos\left(-\frac{\sqrt{2}}{2}\right) = \frac{3\pi}{4}$

(c) $\arctan\left(\tan\frac{7\pi}{4}\right) = \arctan(-1) = -\frac{\pi}{4}$

$\boxed{13}$ (a) $\sin\left[\cos^{-1}\left(-\frac{1}{2}\right)\right] = \sin\frac{2\pi}{3} = \frac{\sqrt{3}}{2}$

(b) $\cos\left(\tan^{-1}1\right) = \cos\frac{\pi}{4} = \frac{\sqrt{2}}{2}$

(c) $\tan\left[\sin^{-1}(-1)\right] = \tan\left(-\frac{\pi}{2}\right)$, which is <u>not defined</u>.

15 (a) Let $\theta = \sin^{-1}\frac{2}{3}$. From *Figure 15(a)*, $\cot\left(\sin^{-1}\frac{2}{3}\right) = \cot\theta = \frac{x}{y} = \frac{\sqrt{5}}{2}$.

(b) Let $\theta = \tan^{-1}\left(-\frac{3}{5}\right)$. From *Figure 15(b)*, $\sec\left[\tan^{-1}\left(-\frac{3}{5}\right)\right] = \sec\theta = \frac{r}{x} = \frac{\sqrt{34}}{5}$.

(c) Let $\theta = \cos^{-1}\left(-\frac{1}{4}\right)$. From *Figure 15(c)*, $\csc\left[\cos^{-1}\left(-\frac{1}{4}\right)\right] = \csc\theta = \frac{r}{y} = \frac{4}{\sqrt{15}}$.

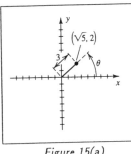 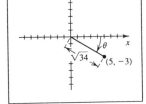 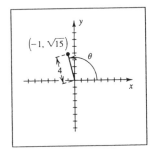

Figure 15(a)   Figure 15(b)   Figure 15(c)

17 (a) $\sin\left(\arcsin\frac{1}{2} + \arccos 0\right) = \sin\left(\frac{\pi}{6} + \frac{\pi}{2}\right) = \sin\frac{2\pi}{3} = \frac{\sqrt{3}}{2}$.

(b) Remember, $\arctan\left(-\frac{3}{4}\right)$ and $\arcsin\frac{4}{5}$ are <u>angles</u>. To abbreviate this solution, we let $\alpha = \arctan\left(-\frac{3}{4}\right)$ and $\beta = \arcsin\frac{4}{5}$. Using the difference identity for the cosine and figures as in Exercises 15 and 16, we have $\cos\left[\arctan\left(-\frac{3}{4}\right) - \arcsin\frac{4}{5}\right]$

$$= \cos(\alpha - \beta) = \cos\alpha\cos\beta + \sin\alpha\sin\beta = \frac{4}{5}\cdot\frac{3}{5} + \left(-\frac{3}{5}\right)\cdot\frac{4}{5} = 0.$$

(c) Let $\alpha = \arctan\frac{4}{3}$ and $\beta = \arccos\frac{8}{17}$. $\tan\left(\arctan\frac{4}{3} + \arccos\frac{8}{17}\right) =$

$$\tan(\alpha + \beta) = \frac{\tan\alpha + \tan\beta}{1 - \tan\alpha\tan\beta} = \frac{\frac{4}{3} + \frac{15}{8}}{1 - \frac{4}{3}\cdot\frac{15}{8}}\cdot\frac{24}{24} = \frac{32 + 45}{24 - 60} = -\frac{77}{36}.$$

19 (a) We first recognize this expression as being of the form $\sin(\textit{twice an angle})$, where $\arccos\left(-\frac{3}{5}\right)$ is the angle. Let $\alpha = \arccos\left(-\frac{3}{5}\right)$. It may help to draw a figure as in Exercise 15 to determine that α is a second quadrant angle and that $\sin\alpha = \frac{4}{5}$. Applying the double-angle formula for the sine gives us

$$\sin\left[2\arccos\left(-\frac{3}{5}\right)\right] = \sin 2\alpha = 2\sin\alpha\cos\alpha = 2\left(\frac{4}{5}\right)\left(-\frac{3}{5}\right) = -\frac{24}{25}.$$

(b) Let $\alpha = \sin^{-1}\frac{15}{17}$. Thus, $\cos\alpha = \frac{8}{17}$ and we apply the double-angle formula for the cosine. $\cos\left(2\sin^{-1}\frac{15}{17}\right) = \cos 2\alpha = \cos^2\alpha - \sin^2\alpha = \left(\frac{8}{17}\right)^2 - \left(\frac{15}{17}\right)^2 = -\frac{161}{289}$.

(c) Let $\alpha = \tan^{-1}\frac{3}{4}$. Thus, $\tan\alpha = \frac{3}{4}$ and we apply the double-angle formula for the tangent. $\tan\left(2\tan^{-1}\frac{3}{4}\right) = \tan 2\alpha = \frac{2\tan\alpha}{1 - \tan^2\alpha} = \frac{2\cdot\frac{3}{4}}{1 - \left(\frac{3}{4}\right)^2}\cdot\frac{16}{16} = \frac{24}{16 - 9} = \frac{24}{7}$.

21 (a) We first recognize this expression as being of the form $\sin(\text{one-half an angle})$, where $\sin^{-1}(-\frac{7}{25})$ is the angle. Let $\alpha = \sin^{-1}(-\frac{7}{25})$. Hence, α is a fourth quadrant angle and $\cos\alpha = \frac{24}{25}$. $-\frac{\pi}{2} < \alpha < 0 \Rightarrow -\frac{\pi}{4} < \frac{1}{2}\alpha < 0$. We need to know the sign of $\sin\frac{1}{2}\alpha$ in order to correctly apply the half-angle formula for the sine.

Since $-\frac{\pi}{4} < \frac{1}{2}\alpha < 0$ { QIV }, $\sin\frac{1}{2}\alpha < 0$.

$$\sin\left[\tfrac{1}{2}\sin^{-1}\left(-\tfrac{7}{25}\right)\right] = \sin\tfrac{1}{2}\alpha = -\sqrt{\frac{1-\cos\alpha}{2}} = -\sqrt{\frac{1-\frac{24}{25}}{2}} = -\sqrt{\frac{1}{50}\cdot\frac{2}{2}} = -\frac{1}{10}\sqrt{2}.$$

(b) Let $\alpha = \tan^{-1}\frac{8}{15}$. $0 < \alpha < \frac{\pi}{2} \Rightarrow 0 < \frac{1}{2}\alpha < \frac{\pi}{4}$ and $\cos\frac{1}{2}\alpha > 0$.

Applying the half-angle formula for the cosine,

$$\cos\left(\tfrac{1}{2}\tan^{-1}\tfrac{8}{15}\right) = \cos\tfrac{1}{2}\alpha = \sqrt{\frac{1+\cos\alpha}{2}} = \sqrt{\frac{1+\frac{15}{17}}{2}} = \sqrt{\frac{16}{17}\cdot\frac{17}{17}} = \frac{4}{17}\sqrt{17}.$$

(c) Let $\alpha = \cos^{-1}\frac{3}{5}$. Applying the half-angle formula for the tangent,

$$\tan\left(\tfrac{1}{2}\cos^{-1}\tfrac{3}{5}\right) = \tan\tfrac{1}{2}\alpha = \frac{1-\cos\alpha}{\sin\alpha} = \frac{1-\frac{3}{5}}{\frac{4}{5}} = \frac{1}{2}.$$

23 We recognize this expression as being of the form $\sin(\text{some angle})$. The angle is *the angle whose tangent is* α. The easiest way to picture this angle is to consider it as being the angle whose ratio of opposite side to adjacent side is x to 1. Let $\alpha = \tan^{-1}x$. From *Figure 23*, $\sin(\tan^{-1}x) = \sin\alpha = \dfrac{x}{\sqrt{x^2+1}}$.

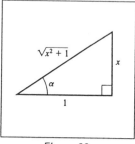

Figure 23

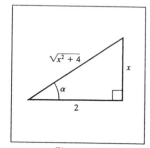

Figure 25

25 Let $\alpha = \sin^{-1}\dfrac{x}{\sqrt{x^2+4}}$. From *Figure 25*, $\sec\left(\sin^{-1}\dfrac{x}{\sqrt{x^2+4}}\right) = \sec\alpha = \dfrac{\sqrt{x^2+4}}{2}$.

27 Let $\alpha = \sin^{-1} x$. From *Figure 27*,

$$\sin\left(2\sin^{-1} x\right) = \sin 2\alpha = 2\sin\alpha \cos\alpha = 2 \cdot \frac{x}{1} \cdot \frac{\sqrt{1-x^2}}{1} = 2x\sqrt{1-x^2}.$$

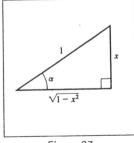

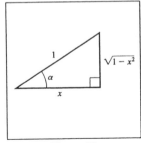

Figure 27   Figure 29

29 Let $\alpha = \arccos x$. $0 \le \alpha \le \pi \;\Rightarrow\; 0 \le \frac{1}{2}\alpha \le \frac{\pi}{2}$.

Thus $\cos\frac{1}{2}\alpha > 0$ and we use the "+" in the half-angle formula for the cosine.

$$\cos\left(\tfrac{1}{2}\arccos x\right) = \cos\tfrac{1}{2}\alpha = \sqrt{\frac{1+\cos\alpha}{2}} = \sqrt{\frac{1+x}{2}}.$$

31 (a) See text Figure 2. As $x \to -1^+$, $\sin^{-1} x \to$ $\;-\frac{\pi}{2}\;$.

(b) See text Figure 5. As $x \to 1^-$, $\cos^{-1} x \to$ $\;0\;$.

(c) See text Figure 8. As $x \to \infty$, $\tan^{-1} x \to$ $\;\frac{\pi}{2}\;$.

33 $y = \sin^{-1} 2x$ •   Horizontally compress $y = \sin^{-1} x$ by a factor of 2.

Note that the domain changes from $[-1,\,1]$ to $[-\frac{1}{2},\,\frac{1}{2}]$.

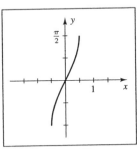

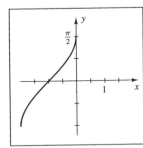

Figure 33   Figure 35

35 $y = \sin^{-1}(x+1)$ •   Shift $y = \sin^{-1} x$ left 1 unit.

The domain changes from $[-1,\,1]$ to $[-2,\,0]$.

37 $y = \cos^{-1}\frac{1}{2}x$ • Horizontally stretch $y = \cos^{-1}x$ by a factor of 2.

The domain changes from $[-1, 1]$ to $[-2, 2]$.

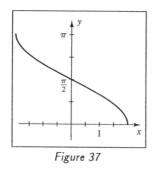

Figure 37

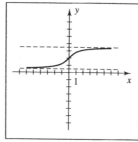

Figure 39

39 $y = 2 + \tan^{-1}x$ • Shift $y = \tan^{-1}x$ up 2 units. The range changes from

$(-\frac{\pi}{2}, \frac{\pi}{2})$ to $(2 - \frac{\pi}{2}, 2 + \frac{\pi}{2})$, which is approximately $(0.43, 3.57)$.

41 If $\alpha = \arccos x$, then $\cos\alpha = x$, where $0 \le \alpha \le \pi$.

Hence, $y = \sin(\arccos x)$

$$= \sin\alpha = \sqrt{1 - \cos^2\alpha} = \sqrt{1 - x^2}.$$

Thus, we have the graph of the semicircle

$y = \sqrt{1 - x^2}$ on the interval $[-1, 1]$.

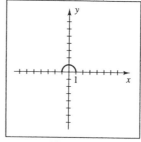

Figure 41

43 (a) Since the domain of the arcsine function is $[-1, 1]$, we know that $x - 3$ must be

in that interval. Thus, $-1 \le x - 3 \le 1 \ \Rightarrow \ 2 \le x \le 4$.

(b) The range of the arcsine function is $[-\frac{\pi}{2}, \frac{\pi}{2}]$.

Thus, $-\frac{\pi}{2} \le \sin^{-1}(x - 3) \le \frac{\pi}{2} \ \Rightarrow \ -\frac{\pi}{4} \le \frac{1}{2}\sin^{-1}(x - 3) \le \frac{\pi}{4} \ \Rightarrow \ -\frac{\pi}{4} \le y \le \frac{\pi}{4}.$

(c) $y = \frac{1}{2}\sin^{-1}(x - 3) \ \Rightarrow \ 2y = \sin^{-1}(x - 3) \ \Rightarrow \ \sin 2y = x - 3 \ \Rightarrow \ x = \sin 2y + 3$

45 (a) $-1 \le \frac{2}{3}x \le 1 \ \Rightarrow \ -\frac{3}{2} \le x \le \frac{3}{2}$

(b) $0 \le \cos^{-1}\frac{2}{3}x \le \pi \ \Rightarrow \ 0 \le 4\cos^{-1}\frac{2}{3}x \le 4\pi \ \Rightarrow \ 0 \le y \le 4\pi$

(c) $y = 4\cos^{-1}\frac{2}{3}x \ \Rightarrow \ \frac{1}{4}y = \cos^{-1}\frac{2}{3}x \ \Rightarrow \ \cos\frac{1}{4}y = \frac{2}{3}x \ \Rightarrow \ x = \frac{3}{2}\cos\frac{1}{4}y$

47 $y = -3 - \sin x \ \Rightarrow \ y + 3 = -\sin x \ \Rightarrow \ -(y + 3) = \sin x \ \Rightarrow \ x = \sin^{-1}(-y - 3)$

49 $y = 15 - 2\cos x \ \Rightarrow \ 2\cos x = 15 - y \ \Rightarrow \ \cos x = \frac{1}{2}(15 - y) \ \Rightarrow \ x = \cos^{-1}\left[\frac{1}{2}(15 - y)\right]$

51 $\frac{\sin x}{3} = \frac{\sin y}{4}$ $\Rightarrow$ $\sin x = \frac{3}{4}\sin y$. Since $0 < y < \pi$, we know that $\sin y > 0$. Thus,

solving $\sin x = \left(\frac{3}{4}\sin y\right)$ is similar to solving $\sin x = a$, where $0 < a < 1$. Remember

that when solving equations of this form, there are two solutions, a first quadrant

angle and a second quadrant angle. The reference angle for x is $x_R = \sin^{-1}\left(\frac{3}{4}\sin y\right)$,

where $0 < \frac{3}{4}\sin y \leq \frac{3}{4} < 1$. If $0 < x < \frac{\pi}{2}$, then $x = x_R$. If $\frac{\pi}{2} < x < \pi$, then $x = \pi - x_R$.

53 $\cos^2 x + 2\cos x - 1 = 0$ $\Rightarrow$ $\cos x = -1 \pm \sqrt{2} \approx 0.4142,\ -2.4142$.

Since $-2.4142 < -1$, $x = \cos^{-1}(-1 + \sqrt{2}) \approx 1.1437$ is one answer.

$$x = 2\pi - \cos^{-1}(-1 + \sqrt{2}) \approx 2\pi - 1.1437 \approx 5.1395 \text{ is the other.}$$

55 $2\tan^2 t + 9\tan t + 3 = 0$ $\Rightarrow$ $\tan t = \frac{-9 \pm \sqrt{81 - 24}}{4}$ $\Rightarrow$ $t = \tan^{-1}\frac{1}{4}(-9 \pm \sqrt{57})$

$$\tan^{-1}\tfrac{1}{4}(-9 + \sqrt{57}) \approx -0.3478,\ \tan^{-1}\tfrac{1}{4}(-9 - \sqrt{57}) \approx -1.3337$$

57 $15\cos^4 x - 14\cos^2 x + 3 = 0$ $\Rightarrow$ $(5\cos^2 x - 3)(3\cos^2 x - 1) = 0$ $\Rightarrow$ $\cos^2 x = \frac{3}{5}, \frac{1}{3}$ $\Rightarrow$

$\cos x = \pm\frac{1}{5}\sqrt{15},\ \pm\frac{1}{3}\sqrt{3}$ $\Rightarrow$ $x = \cos^{-1}\left(\pm\frac{1}{5}\sqrt{15}\right),\ \cos^{-1}\left(\pm\frac{1}{3}\sqrt{3}\right)$.

$\cos^{-1}\frac{1}{5}\sqrt{15} \approx 0.6847,\ \cos^{-1}\left(-\frac{1}{5}\sqrt{15}\right) \approx 2.4569,$

$$\cos^{-1}\tfrac{1}{3}\sqrt{3} \approx 0.9553,\ \cos^{-1}\left(-\tfrac{1}{3}\sqrt{3}\right) \approx 2.1863$$

59 $6\sin^3\theta + 18\sin^2\theta - 5\sin\theta - 15 = 0$ $\Rightarrow$ $6\sin^2\theta(\sin\theta + 3) - 5(\sin\theta + 3) = 0$ $\Rightarrow$

$(6\sin^2\theta - 5)(\sin\theta + 3) = 0$ $\Rightarrow$ $\sin\theta = \pm\frac{1}{6}\sqrt{30}$ $\Rightarrow$ $\theta = \sin^{-1}\left(\pm\frac{1}{6}\sqrt{30}\right) \approx \pm 1.1503$

61 $(\cos x)(15\cos x + 4) = 3$ $\Rightarrow$ $15\cos^2 x + 4\cos x - 3 = 0$ $\Rightarrow$

$(5\cos x + 3)(3\cos x - 1) = 0$ $\Rightarrow$ $\cos x = -\frac{3}{5}, \frac{1}{3}$ $\Rightarrow$

$x = \cos^{-1}\left(-\frac{3}{5}\right) \approx 2.2143,\ \cos^{-1}\frac{1}{3} \approx 1.2310$. These angles are in the second quadrant

and the first quadrant. In $[0, 2\pi)$, we must also have a third quadrant angle and a

fourth quadrant angle that satisfy the original equation. These angles are

$2\pi - \cos^{-1}\left(-\frac{3}{5}\right) \approx 4.0689$ and $2\pi - \cos^{-1}\frac{1}{3} \approx 5.0522$.

63 $3\cos 2x - 7\cos x + 5 = 0$ $\Rightarrow$ $3(2\cos^2 x - 1) - 7\cos x + 5 = 0$ $\Rightarrow$

$6\cos^2 x - 7\cos x + 2 = 0$ $\Rightarrow$ $(3\cos x - 2)(2\cos x - 1) = 0$ $\Rightarrow$ $\cos x = \frac{2}{3}, \frac{1}{2}$ $\Rightarrow$

$$x = \cos^{-1}\tfrac{2}{3} \approx 0.8411,\ 2\pi - \cos^{-1}\tfrac{2}{3} \approx 5.4421,\ \tfrac{\pi}{3} \approx 1.0472,\ \tfrac{5\pi}{3} \approx 5.2360.$$

65 (a) $S = 4$, $D = 3.5$, $d = 1$ $\Rightarrow$ $M = \frac{S}{2}\left(1 - \frac{2}{\pi}\tan^{-1}\frac{d}{D}\right) = \frac{4}{2}\left(1 - \frac{2}{\pi}\tan^{-1}\frac{1}{3.5}\right) \approx 1.65$ m

(b) $d = 4$ $\Rightarrow$ $M = \frac{S}{2}\left(1 - \frac{2}{\pi}\tan^{-1}\frac{d}{D}\right) = \frac{4}{2}\left(1 - \frac{2}{\pi}\tan^{-1}\frac{4}{3.5}\right) \approx 0.92$ m

(c) $d = 10$ $\Rightarrow$ $M = \frac{S}{2}\left(1 - \frac{2}{\pi}\tan^{-1}\frac{d}{D}\right) = \frac{4}{2}\left(1 - \frac{2}{\pi}\tan^{-1}\frac{10}{3.5}\right) \approx 0.43$ m

67 Form a right triangle with the center line of the fairway and half the width of the

fairway as the legs of the triangle.

$$\text{opp} = \tfrac{1}{2}(30) = 15 \text{ and hyp} = 280 \Rightarrow \sin\theta = \tfrac{15}{280} \Rightarrow \theta = \sin^{-1}\tfrac{15}{280} \approx 3.07°$$

[69] (a) Let β denote the angle by the sailboat with opposite side d and hypotenuse k.

Now $\sin \beta = \frac{d}{k} \;\Rightarrow\; \beta = \sin^{-1}\frac{d}{k}$. Using alternate interior angles,

we see that $\alpha + \beta = \theta$. Thus, $\alpha = \theta - \beta = \theta - \sin^{-1}\frac{d}{k}$.

(b) $d = 50$, $k = 210$, and $\theta = 53.4° \;\Rightarrow\; \alpha = 53.4° - \sin^{-1}\frac{50}{210} \approx 39.63°$, or $40°$.

Note: The following is a general outline that can be used for verifying trigonometric

identities involving inverse trigonometric functions.

(1) Define angles and their ranges—make sure the range of values for one side of

the equation is equal to the range of values for the other side.

(2) Choose a trigonometric function T that is one-to-one on the range of values

listed in part (1).

(3) Show that $T(\text{LS}) = T(\text{RS})$. Note that $T(\text{LS}) = T(\text{RS}) \not\Rightarrow \text{LS} = \text{RS}$.

(4) Conclude that since T is one-to-one on the range of values, $\text{LS} = \text{RS}$.

[71] Let $\alpha = \sin^{-1}x$ and $\beta = \tan^{-1}\dfrac{x}{\sqrt{1-x^2}}$ with $-\frac{\pi}{2} < \alpha < \frac{\pi}{2}$ and $-\frac{\pi}{2} < \beta < \frac{\pi}{2}$.

Thus, $\sin \alpha = x$ and $\sin \beta = x$. Since the sine function is one-to-one on $(-\frac{\pi}{2}, \frac{\pi}{2})$,

we have $\alpha = \beta$—that is, $\sin^{-1}x = \tan^{-1}\dfrac{x}{\sqrt{1-x^2}}$.

[73] Let $\alpha = \arcsin(-x)$ and $\beta = \arcsin x$ with $-\frac{\pi}{2} \le \alpha \le \frac{\pi}{2}$ and $-\frac{\pi}{2} \le \beta \le \frac{\pi}{2}$.

Thus, $\sin \alpha = -x$ and $\sin \beta = x$. Consequently, $\sin \alpha = -\sin \beta = \sin(-\beta)$.

Since the sine function is one-to-one on $[-\frac{\pi}{2}, \frac{\pi}{2}]$,

we have $\alpha = -\beta$—that is, $\arcsin(-x) = -\arcsin x$.

[75] Let $\alpha = \arctan x$ and $\beta = \arctan(1/x)$.

Since $x > 0$, we have $0 < \alpha < \frac{\pi}{2}$ and $0 < \beta < \frac{\pi}{2}$, and hence $0 < \alpha + \beta < \pi$.

Thus, $\tan(\alpha + \beta) = \dfrac{\tan \alpha + \tan \beta}{1 - \tan \alpha \tan \beta} = \dfrac{x + (1/x)}{1 - x \cdot (1/x)} = \dfrac{x + (1/x)}{0}$.

Since the denominator is 0, $\tan(\alpha + \beta)$ is undefined and hence $\alpha + \beta = \frac{\pi}{2}$ since $\frac{\pi}{2}$ is

the only value between 0 and π for which the tangent is undefined.

[77] The domain of $\sin^{-1}(x - 1)$ is $[0, 2]$ and the

domain of $\cos^{-1}\frac{1}{2}x$ is $[-2, 2]$. The domain of

f is the intersection of $[0, 2]$ and $[-2, 2]$, i.e., $[0, 2]$.

From the graph, we see that the function is

increasing and its range is $[-\frac{\pi}{2}, \pi]$.

$[-3, 6]$ by $[-2, 4]$

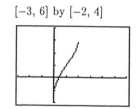

Figure 77

[79] Graph $y = \sin^{-1} 2x$ and $y = \tan^{-1}(1 - x)$.

From the graph, we see that there is one solution at $x \approx 0.29$.

$[-3, 3]$ by $[-2, 2]$ $[0, \pi/2, 0.2]$ by $[0, 1.05, 0.2]$

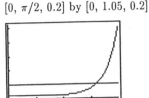

Figure 79   *Figure 81*

[81] Make the assignments $Y_1 = \sin^{-1}(\sin x/1.52)$, $Y_2 = x - Y_1$, $Y_3 = x + Y_1$, and $Y_4 = .5((\sin Y_2)^2/(\sin Y_3)^2 + (\tan Y_2)^2/(\tan Y_3)^2)$. Now turn *off* Y_1, Y_2, and Y_3—leaving only Y_4 *on* to graph. From the graph, we see that when $f(\theta) = 0.2$, $\theta \approx 1.25$, or approximately $72°$.

[83] Actual distance between x-ticks is equal to $x_A = \frac{3 \text{ units}}{3 \text{ ticks}} = 1$ unit between ticks.

Actual distance between y-ticks is equal to $y_A = \frac{2 \text{ units}}{2 \text{ ticks}} = 1$ unit between ticks.

The ratio is $m_A = \frac{y_A}{x_A} = \frac{1}{1} = 1$. The graph will make an angle of $\theta = \tan^{-1} 1 = 45°$.

$[0, 3]$ by $[0, 2]$ $[0, 3]$ by $[0, 4]$

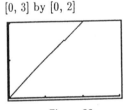 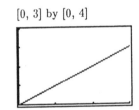

Figure 83   *Figure 85*

[85] $x_A = \frac{3 \text{ units}}{3 \text{ ticks}} = 1$, $y_A = \frac{2 \text{ units}}{4 \text{ ticks}} = \frac{1}{2}$ $\Rightarrow$ $m_A = \frac{1/2}{1} = \frac{1}{2}$ $\Rightarrow$ $\theta = \tan^{-1} \frac{1}{2} \approx 26.6°$.

Chapter 6 Review Exercises

[1] $(\cot^2 x + 1)(1 - \cos^2 x) = (\csc^2 x)(\sin^2 x) = (1/\sin^2 x)(\sin^2 x) = 1$

[3] $\dfrac{(\sec^2 \theta - 1)\cot \theta}{\tan \theta \sin \theta + \cos \theta} = \dfrac{(\tan^2 \theta)\cot \theta}{\dfrac{\sin \theta}{\cos \theta} \cdot \sin \theta + \cos \theta}$ { Pythagorean and tangent identities }

$= \dfrac{\tan \theta (\tan \theta \cot \theta)}{\dfrac{\sin^2 \theta}{\cos \theta} + \cos \theta}$ { combine terms }

$= \dfrac{\tan \theta}{\dfrac{\sin^2 \theta + \cos^2 \theta}{\cos \theta}}$ { reciprocal identity, common denominator }

(continued)

$$= \frac{\sin\theta / \cos\theta}{1/\cos\theta} \qquad \{\text{Pythagorean and tangent identities}\}$$

$$= \sin\theta \qquad \{\text{simplify}\}$$

$\boxed{5}$ $\dfrac{1}{1+\sin t} = \dfrac{1}{1+\sin t}\cdot\dfrac{1-\sin t}{1-\sin t}$ $\left\{\begin{array}{l}\text{multiply the numerator and the denominator}\\ \text{by conjugate of the denominator}\end{array}\right\}$

$$= \frac{1-\sin t}{1-\sin^2 t} \qquad \{\text{the difference of two squares}\}$$

$$= \frac{1-\sin t}{\cos^2 t} \qquad \{\text{Pythagorean identity}\}$$

$$= \frac{1-\sin t}{\cos t}\cdot\frac{1}{\cos t} \qquad \{\text{break up since we want } \sec t \text{ on the right side}\}$$

$$= \left(\frac{1}{\cos t}-\frac{\sin t}{\cos t}\right)\cdot\sec t \qquad \{\text{split up fraction}\}$$

$$= (\sec t - \tan t)\sec t \qquad \{\text{reciprocal and tangent identities}\}$$

$\boxed{6}$ $\dfrac{\sin(\alpha-\beta)}{\cos(\alpha+\beta)} = \dfrac{\sin\alpha\cos\beta - \cos\alpha\sin\beta}{\cos\alpha\cos\beta - \sin\alpha\sin\beta}$ $\left\{\begin{array}{l}\text{subtraction formula for the sine}\\ \text{addition formula for the cosine}\end{array}\right\}$

The first term in the denominator is $\cos\alpha\cos\beta$, but looking ahead, we see that the first term in the denominator of the expression we want to obtain is 1. Hence, we will divide both the numerator and the denominator by $\cos\alpha\cos\beta$.

$$= \frac{(\sin\alpha\cos\beta - \cos\alpha\sin\beta)\,/\,\cos\alpha\cos\beta}{(\cos\alpha\cos\beta - \sin\alpha\sin\beta)\,/\,\cos\alpha\cos\beta}$$

$$= \frac{\dfrac{\sin\alpha\cos\beta}{\cos\alpha\cos\beta} - \dfrac{\cos\alpha\sin\beta}{\cos\alpha\cos\beta}}{\dfrac{\cos\alpha\cos\beta}{\cos\alpha\cos\beta} - \dfrac{\sin\alpha\sin\beta}{\cos\alpha\cos\beta}} = \frac{\dfrac{\sin\alpha}{\cos\alpha} - \dfrac{\sin\beta}{\cos\beta}}{1 - \dfrac{\sin\alpha}{\cos\alpha}\cdot\dfrac{\sin\beta}{\cos\beta}} = \frac{\tan\alpha - \tan\beta}{1-\tan\alpha\tan\beta}$$

$\boxed{7}$ $\tan 2u = \dfrac{2\tan u}{1-\tan^2 u} \qquad \{\text{apply the double-angle formula for the tangent}\}$

$$= \frac{2\cdot\dfrac{1}{\cot u}}{1-\dfrac{1}{\cot^2 u}} \qquad \{\text{put in terms of cot since it appears on the right side}\}$$

$$= \frac{\dfrac{2}{\cot u}}{\dfrac{\cot^2 u - 1}{\cot^2 u}} \qquad \{\text{combine into one fraction}\}$$

$$= \frac{2\cot u}{\cot^2 u - 1} \qquad \{\text{simplify complex fraction}\}$$

$$= \frac{2\cot u}{(\csc^2 u - 1) - 1} \qquad \{\text{Pythagorean identity}\}$$

$$= \frac{2\cot u}{\csc^2 u - 2} \qquad \{\text{simplify}\}$$

$\boxed{9}$　Factor the numerator using the difference of two cubes formula.

$$\frac{\tan^3\phi - \cot^3\phi}{\tan^2\phi + \csc^2\phi} = \frac{(\tan\phi - \cot\phi)(\tan^2\phi + \tan\phi\,\cot\phi + \cot^2\phi)}{\tan^2\phi + (1 + \cot^2\phi)}$$

$$= \frac{(\tan\phi - \cot\phi)(\tan^2\phi + 1 + \cot^2\phi)}{\tan^2\phi + 1 + \cot^2\phi} = \tan\phi - \cot\phi$$

$\boxed{10}$　$\mathrm{LS} = \dfrac{\sin u + \sin v}{\csc u + \csc v} = \dfrac{\sin u + \sin v}{\dfrac{1}{\sin u} + \dfrac{1}{\sin v}} = \dfrac{\sin u + \sin v}{\dfrac{\sin v + \sin u}{\sin u \sin v}} = \sin u \sin v$

At this point, there is no apparent "next step." Thus, we will stop working with the

left side and try to simplify the right side to the same expression, $\sin u \sin v$.

$\mathrm{RS} = \dfrac{1 - \sin u \sin v}{-1 + \csc u \csc v} = \dfrac{1 - \sin u \sin v}{-1 + \dfrac{1}{\sin u \sin v}} = \dfrac{1 - \sin u \sin v}{\dfrac{1 - \sin u \sin v}{\sin u \sin v}} = \sin u \sin v$

Since the LS and RS equal the same expression and the steps are reversible,

the identity is verified.

$\boxed{11}$　$\left(\dfrac{\sin^2 x}{\tan^4 x}\right)^3 \left(\dfrac{\csc^3 x}{\cot^6 x}\right)^2 = \left(\dfrac{\sin^6 x}{\tan^{12} x}\right)\left(\dfrac{\csc^6 x}{\cot^{12} x}\right) = \dfrac{(\sin x \csc x)^6}{(\tan x \cot x)^{12}} = \dfrac{(1)^6}{(1)^{12}} = \dfrac{1}{1} = 1$

$\boxed{12}$　$\dfrac{\cos\gamma}{1 - \tan\gamma} + \dfrac{\sin\gamma}{1 - \cot\gamma} = \dfrac{\cos\gamma}{1 - \dfrac{\sin\gamma}{\cos\gamma}} + \dfrac{\sin\gamma}{1 - \dfrac{\cos\gamma}{\sin\gamma}} = \dfrac{\cos\gamma}{\dfrac{\cos\gamma - \sin\gamma}{\cos\gamma}} + \dfrac{\sin\gamma}{\dfrac{\sin\gamma - \cos\gamma}{\sin\gamma}} =$

$\dfrac{\cos^2\gamma}{\cos\gamma - \sin\gamma} + \dfrac{\sin^2\gamma}{\sin\gamma - \cos\gamma} = \dfrac{\cos^2\gamma}{\cos\gamma - \sin\gamma} - \dfrac{\sin^2\gamma}{-(\sin\gamma - \cos\gamma)} =$

$\dfrac{\cos^2\gamma}{\cos\gamma - \sin\gamma} - \dfrac{\sin^2\gamma}{\cos\gamma - \sin\gamma} = \dfrac{\cos^2\gamma - \sin^2\gamma}{\cos\gamma - \sin\gamma} = \dfrac{(\cos\gamma + \sin\gamma)(\cos\gamma - \sin\gamma)}{\cos\gamma - \sin\gamma} =$

$\cos\gamma + \sin\gamma$

$\boxed{13}$　$\dfrac{\cos(-t)}{\sec(-t) + \tan(-t)} = \dfrac{\cos t}{\sec t + (-\tan t)} = \dfrac{\cos t}{\sec t - \tan t} = \dfrac{\cos t}{\dfrac{1}{\cos t} - \dfrac{\sin t}{\cos t}} = \dfrac{\cos t}{\dfrac{1 - \sin t}{\cos t}} =$

$\dfrac{\cos^2 t}{1 - \sin t} = \dfrac{1 - \sin^2 t}{1 - \sin t} = \dfrac{(1 - \sin t)(1 + \sin t)}{1 - \sin t} = 1 + \sin t$

$\boxed{15}$　In the following solution, we could multiply both the numerator and the denominator

by *either* the conjugate of the numerator *or* the conjugate of the denominator. Since

the numerator on the right side looks like the numerator on the left side, we'll change

the denominator.

$$\sqrt{\dfrac{1 - \cos t}{1 + \cos t}} = \sqrt{\dfrac{(1 - \cos t)}{(1 + \cos t)} \cdot \dfrac{(1 - \cos t)}{(1 - \cos t)}} = \sqrt{\dfrac{(1 - \cos t)^2}{1 - \cos^2 t}} = \sqrt{\dfrac{(1 - \cos t)^2}{\sin^2 t}} =$$

$$\dfrac{\sqrt{(1 - \cos t)^2}}{\sqrt{\sin^2 t}} = \dfrac{|1 - \cos t|}{|\sin t|} = \dfrac{1 - \cos t}{|\sin t|}, \text{ since } (1 - \cos t) \geq 0.$$

$\boxed{17}$ $\cos\left(x - \frac{5\pi}{2}\right) = \cos x \cos\frac{5\pi}{2} + \sin x \sin\frac{5\pi}{2} = (\cos x)(0) + (\sin x)(1) = \sin x$

$\boxed{19}$ We need to break down the angle argument of 4β into terms with only β as their argument. We can do this by using either the double-angle formula for the sine or the addition formula for the sine.

$\frac{1}{4}\sin 4\beta = \frac{1}{4}\sin(2 \cdot 2\beta) = \frac{1}{4}(2\sin 2\beta \, \cos 2\beta) = \frac{1}{2}(2\sin\beta \, \cos\beta)(\cos^2\beta - \sin^2\beta) =$

$$\sin\beta \, \cos^3\beta - \cos\beta \, \sin^3\beta$$

$\boxed{20}$ $\tan\frac{1}{2}\theta = \frac{1 - \cos\theta}{\sin\theta}$ { apply the half-angle formula for the tangent }

$= \frac{1}{\sin\theta} - \frac{\cos\theta}{\sin\theta}$ { split up the fraction }

$= \csc\theta - \cot\theta$ { reciprocal and cotangent identities }

$\boxed{21}$ $\sin 8\theta = \sin(2 \cdot 4\theta)$

$= 2(\sin 4\theta)(\cos 4\theta)$ { double angle formula for sine }

$= 2(2\underline{\sin 2\theta} \, \cos 2\theta)(1 - 2\sin^2 2\theta)$ $\left\{\begin{array}{l}\text{double angle formulas}\\ \text{for sine and cosine}\end{array}\right\}$

$= 2\big[2 \cdot 2\underline{\sin\theta \, \cos\theta} \, (1 - 2\sin^2\theta)\big]\big[1 - 2(\sin 2\theta)^2\big]$ $\left\{\begin{array}{l}\text{double angle formulas}\\ \text{for sine and cosine}\end{array}\right\}$

$= 8\sin\theta \, \cos\theta \, (1 - 2\sin^2\theta)[1 - 2(2\sin\theta \, \cos\theta)^2]$ { double angle formula for sine }

$= 8\sin\theta \, \cos\theta \, (1 - 2\sin^2\theta)[1 - 2(4\sin^2\theta \, \cos^2\theta)]$

$= 8\sin\theta \, \cos\theta \, (1 - 2\sin^2\theta)(1 - 8\sin^2\theta \, \cos^2\theta)$

$\boxed{22}$ Let $\alpha = \arctan x$ and $\beta = \arctan\dfrac{2x}{1 - x^2}$.

Because $-1 < x < 1$, $\arctan(-1) < \arctan(x) < \arctan(1)$, and hence $-\frac{\pi}{4} < \alpha < \frac{\pi}{4}$.

Thus, $\tan\alpha = x$ and $\tan\beta = \dfrac{2x}{1 - x^2} = \dfrac{2\tan\alpha}{1 - \tan^2\alpha} = \tan 2\alpha$.

Since the tangent function is one-to-one on $\left(-\frac{\pi}{2}, \frac{\pi}{2}\right)$, we have $\beta = 2\alpha$ or, equivalently,

$$\alpha = \frac{1}{2}\beta. \text{ In terms of } x, \text{ we have } \arctan x = \frac{1}{2}\arctan\frac{2x}{1 - x^2}.$$

$\boxed{23}$ $2\cos^3\theta - \cos\theta = 0 \ \Rightarrow \ \cos\theta(2\cos^2\theta - 1) = 0 \ \Rightarrow \ \cos\theta = 0, \pm\dfrac{\sqrt{2}}{2} \ \Rightarrow$

$$\theta = \frac{\pi}{2}, \frac{3\pi}{2}, \frac{\pi}{4}, \frac{7\pi}{4}, \frac{3\pi}{4}, \frac{5\pi}{4}$$

$\boxed{24}$ $2\cos\alpha + \tan\alpha = \sec\alpha \ \Rightarrow \ 2\cos\alpha + \frac{\sin\alpha}{\cos\alpha} = \frac{1}{\cos\alpha} \ \Rightarrow \ \{\text{multiply by the lcd, } \cos\alpha\}$

$2\cos^2\alpha + \sin\alpha = 1 \ \Rightarrow \ 2(1 - \sin^2\alpha) + \sin\alpha = 1 \ \Rightarrow \ 2 - 2\sin^2\alpha + \sin\alpha = 1 \ \Rightarrow$

$2\sin^2\alpha - \sin\alpha - 1 = 0 \ \Rightarrow \ (2\sin\alpha + 1)(\sin\alpha - 1) = 0 \ \Rightarrow$

$\sin\alpha = -\frac{1}{2}, 1 \ \Rightarrow \ \alpha = \frac{7\pi}{6}, \frac{11\pi}{6}, \frac{\pi}{2}$.

Checking these values in the original equation, we see that $\tan\frac{\pi}{2}$ is undefined so

exclude $\frac{\pi}{2}$ and our solution is $\alpha = \frac{7\pi}{6}, \frac{11\pi}{6}$.

25 $\sin\theta = \tan\theta \;\Rightarrow\; \sin\theta - \dfrac{\sin\theta}{\cos\theta} = 0 \;\Rightarrow\; \sin\theta\left(1 - \dfrac{1}{\cos\theta}\right) = 0 \;\Rightarrow$

     $\sin\theta = 0$ or $1 = \dfrac{1}{\cos\theta} \;\Rightarrow\; \sin\theta = 0$ or $\cos\theta = 1 \;\Rightarrow\; \theta = 0,\,\pi$ or $\theta = 0 \;\Rightarrow\; \theta = 0,\,\pi$

27 $2\cos^3 t + \cos^2 t - 2\cos t - 1 = 0 \;\Rightarrow\; \cos^2 t\,(2\cos t + 1) - 1(2\cos t + 1) = 0 \;\Rightarrow$

     $(\cos^2 t - 1)(2\cos t + 1) = 0 \;\Rightarrow\; \cos t = \pm 1,\,-\tfrac{1}{2} \;\Rightarrow\; t = 0,\,\pi,\,\tfrac{2\pi}{3},\,\tfrac{4\pi}{3}$

29 $\sin\beta + 2\cos^2\beta = 1 \;\Rightarrow\; \sin\beta + 2(1 - \sin^2\beta) = 1 \;\Rightarrow\; 2\sin^2\beta - \sin\beta - 1 = 0 \;\Rightarrow$

     $(2\sin\beta + 1)(\sin\beta - 1) = 0 \;\Rightarrow\; \sin\beta = -\tfrac{1}{2},\,1 \;\Rightarrow\; \beta = \tfrac{7\pi}{6},\,\tfrac{11\pi}{6},\,\tfrac{\pi}{2}$

31 $2\sec u\,\sin u + 2 = 4\sin u + \sec u \;\Rightarrow\; 2\sec u\,\sin u - 4\sin u - \sec u + 2 = 0 \;\Rightarrow$

     $2\sin u\,(\sec u - 2) - 1(\sec u - 2) = 0 \;\Rightarrow\; (2\sin u - 1)(\sec u - 2) = 0 \;\Rightarrow$

     $\sin u = \tfrac{1}{2}$ or $\sec u = 2\;\{\cos u = \tfrac{1}{2}\} \;\Rightarrow\; u = \tfrac{\pi}{6},\,\tfrac{5\pi}{6},\,\tfrac{\pi}{3},\,\tfrac{5\pi}{3}$

32 $\tan 2x\,\cos 2x = \sin 2x \;\Rightarrow\; \sin 2x = \sin 2x$. This is an identity and is true for all

values of x in $[0,\,2\pi)$ except those that make $\tan 2x$ undefined, or, equivalently, those

that make $\cos 2x$ equal to 0. $\cos 2x = 0 \;\Rightarrow\; 2x = \tfrac{\pi}{2} + \pi n \;\Rightarrow\; x = \tfrac{\pi}{4} + \tfrac{\pi}{2}n$.

     Hence, the solutions are all x in $[0,\,2\pi)$ except $\tfrac{\pi}{4},\,\tfrac{3\pi}{4},\,\tfrac{5\pi}{4},\,\tfrac{7\pi}{4}$.

33 $2\cos 3x\,\cos 2x = 1 - 2\sin 3x\,\sin 2x \;\Rightarrow\; 2\cos 3x\,\cos 2x + 2\sin 3x\,\sin 2x = 1 \;\Rightarrow$

     $2(\cos 3x\,\cos 2x + \sin 3x\,\sin 2x) = 1 \;\Rightarrow\; \cos(3x - 2x) = \tfrac{1}{2} \;\Rightarrow$

                                      $\cos x = \tfrac{1}{2} \;\Rightarrow\; x = \tfrac{\pi}{3},\,\tfrac{5\pi}{3}$

35 $\cos\pi x + \sin\pi x = 0 \;\Rightarrow\; \sin\pi x = -\cos\pi x \;\Rightarrow$

     $\dfrac{\sin\pi x}{\cos\pi x} = \dfrac{-\cos\pi x}{\cos\pi x}\;\{\text{divide by } \cos\pi x\} \;\Rightarrow\; \tan\pi x = -1 \;\Rightarrow\; \pi x = \tfrac{3\pi}{4} + \pi n \;\Rightarrow$

                 $x = \tfrac{3}{4} + n \;\Rightarrow\; x = \tfrac{3}{4},\,\tfrac{7}{4},\,\tfrac{11}{4},\,\tfrac{15}{4},\,\tfrac{19}{4},\,\tfrac{23}{4}$

37 $2\cos^2\tfrac{1}{2}\theta - 3\cos\theta = 0 \;\Rightarrow\; 2\left(\dfrac{1 + \cos\theta}{2}\right) - 3\cos\theta = 0 \;\Rightarrow$

         $(1 + \cos\theta) - 3\cos\theta = 0 \;\Rightarrow\; 1 - 2\cos\theta = 0 \;\Rightarrow\; \cos\theta = \tfrac{1}{2} \;\Rightarrow\; \theta = \tfrac{\pi}{3},\,\tfrac{5\pi}{3}$

39 $\sin 5x = \sin 3x \;\Rightarrow\; \sin 5x - \sin 3x = 0 \;\Rightarrow\; [\text{S2}]\; 2\cos\dfrac{5x + 3x}{2}\,\sin\dfrac{5x - 3x}{2} = 0 \;\Rightarrow$

     $\cos 4x\,\sin x = 0 \;\Rightarrow\; 4x = \tfrac{\pi}{2} + \pi n$ or $x = \pi n \;\Rightarrow\; x = \tfrac{\pi}{8} + \tfrac{\pi}{4}n$ or $x = 0,\,\pi \;\Rightarrow$

            $x = 0,\,\tfrac{\pi}{8},\,\tfrac{3\pi}{8},\,\tfrac{5\pi}{8},\,\tfrac{7\pi}{8},\,\pi,\,\tfrac{9\pi}{8},\,\tfrac{11\pi}{8},\,\tfrac{13\pi}{8},\,\tfrac{15\pi}{8}$

41 $\cos 75° = \cos(45° + 30°) = \cos 45°\cos 30° - \sin 45°\sin 30° = \dfrac{\sqrt{2}}{2}\cdot\dfrac{\sqrt{3}}{2} - \dfrac{\sqrt{2}}{2}\cdot\dfrac{1}{2} = \dfrac{\sqrt{6} - \sqrt{2}}{4}$

43 $\sin 195° = \sin(135° + 60°)$

     $= \sin 135°\cos 60° + \cos 135°\sin 60° = \dfrac{\sqrt{2}}{2}\cdot\dfrac{1}{2} + \left(-\dfrac{\sqrt{2}}{2}\right)\cdot\dfrac{\sqrt{3}}{2} = \dfrac{\sqrt{2} - \sqrt{6}}{4}$

44 $\csc\dfrac{\pi}{8} = \dfrac{1}{\sin\frac{\pi}{8}} = \dfrac{1}{\sin\left(\frac{1}{2}\cdot\frac{\pi}{4}\right)} = \dfrac{1}{\sqrt{\dfrac{1 - \cos\frac{\pi}{4}}{2}}} = \dfrac{1}{\sqrt{\dfrac{1 - \sqrt{2}/2}{2}}} = \dfrac{1}{\sqrt{\dfrac{2 - \sqrt{2}}{4}}} = \dfrac{2}{\sqrt{2 - \sqrt{2}}}$

45 $\csc\theta = \tfrac{5}{3}$ and $\cos\phi = \tfrac{8}{17} \;\Rightarrow\; \sin\theta = \tfrac{3}{5},\,\cos\theta = \tfrac{4}{5},\,\tan\theta = \tfrac{3}{4}$ and $\sin\phi = \tfrac{15}{17},\,\tan\phi = \tfrac{15}{8}$.

     $\sin(\theta + \phi) = \sin\theta\,\cos\phi + \cos\theta\,\sin\phi = \tfrac{3}{5}\cdot\tfrac{8}{17} + \tfrac{4}{5}\cdot\tfrac{15}{17} = \tfrac{84}{85}$

$\boxed{47}$ $\tan(\phi+\theta) = \tan(\theta+\phi) = \dfrac{\sin(\theta+\phi)}{\cos(\theta+\phi)} = \dfrac{84/85 \ \{\text{from Exercise 45}\}}{-13/85 \ \{\text{from Exercise 46}\}} = -\dfrac{84}{13}$

$\boxed{49}$ $\sin(\phi-\theta) = \sin\phi\,\cos\theta - \cos\phi\,\sin\theta = \dfrac{15}{17}\cdot\dfrac{4}{5} - \dfrac{8}{17}\cdot\dfrac{3}{5} = \dfrac{36}{85}$

$\boxed{51}$ $\sin 2\phi = 2\sin\phi\,\cos\phi = 2\cdot\dfrac{15}{17}\cdot\dfrac{8}{17} \ \{\text{from Exercise 45}\} = \dfrac{240}{289}$

$\boxed{52}$ $\cos 2\phi = \cos^2\phi - \sin^2\phi = \left(\dfrac{8}{17}\right)^2 - \left(\dfrac{15}{17}\right)^2 = -\dfrac{161}{289}$

$\boxed{53}$ $\tan 2\theta = \dfrac{2\tan\theta}{1-\tan^2\theta} = \dfrac{2\cdot\frac{3}{4}}{1-\left(\frac{3}{4}\right)^2} = \dfrac{\frac{3}{2}}{1-\frac{9}{16}}\cdot\dfrac{16}{16} = \dfrac{24}{16-9} = \dfrac{24}{7}$

$\boxed{55}$ $\tan\frac{1}{2}\theta = \dfrac{1-\cos\theta}{\sin\theta} = \dfrac{1-\frac{4}{5}}{\frac{3}{5}} = \dfrac{\frac{1}{5}}{\frac{3}{5}} = \dfrac{1}{3}$

$\boxed{57}$ (a) $\sin 7t \sin 4t = \text{[P4]} \ \frac{1}{2}[\cos(7t-4t) - \cos(7t+4t)] = \frac{1}{2}\cos 3t - \frac{1}{2}\cos 11t$

(b) $\cos\frac{1}{4}u \ \cos(-\frac{1}{6}u) = \text{[P3]} \ \frac{1}{2}\left\{\cos\left[\frac{1}{4}u + (-\frac{1}{6}u)\right] + \cos\left[\frac{1}{4}u - (-\frac{1}{6}u)\right]\right\} =$

$\frac{1}{2}(\cos\frac{2}{24}u + \cos\frac{10}{24}u) = \frac{1}{2}\cos\frac{1}{12}u + \frac{1}{2}\cos\frac{5}{12}u$

(c) $6\cos 5x \sin 3x = \text{[P2]} \ 6\cdot\frac{1}{2}[\sin(5x+3x) - \sin(5x-3x)] = 3\sin 8x - 3\sin 2x$

(d) $4\sin 3\theta \cos 7\theta = \text{[P1]} \ 4\cdot\frac{1}{2}[\sin(3\theta+7\theta) + \sin(3\theta-7\theta)] = 2\sin 10\theta - 2\sin 4\theta$

$\boxed{58}$ (b) $\cos 3\theta - \cos 8\theta = \text{[S4]} \ -2\sin\dfrac{3\theta+8\theta}{2} \sin\dfrac{3\theta-8\theta}{2} = -2\sin\frac{11}{2}\theta \sin(-\frac{5}{2}\theta) =$

$2\sin\frac{11}{2}\theta \sin\frac{5}{2}\theta$

(c) $\sin\frac{1}{4}t - \sin\frac{1}{5}t = \text{[S2]} \ 2\cos\dfrac{\frac{1}{4}t+\frac{1}{5}t}{2} \sin\dfrac{\frac{1}{4}t-\frac{1}{5}t}{2} = 2\cos\dfrac{\frac{5}{20}t+\frac{4}{20}t}{2} \sin\dfrac{\frac{5}{20}t-\frac{4}{20}t}{2} =$

$2\cos\frac{9}{40}t \sin\frac{1}{40}t$

$\boxed{59}$ $\cos^{-1}\left(\dfrac{\sqrt{3}}{2}\right) = \frac{\pi}{6}$ since $\cos\frac{\pi}{6} = \dfrac{\sqrt{3}}{2}$ and $0 \le \frac{\pi}{6} \le \pi$

$\boxed{61}$ $\arctan\sqrt{3} = \frac{\pi}{3}$ since $\tan\frac{\pi}{3} = \sqrt{3}$ and $-\frac{\pi}{2} < \frac{\pi}{3} < \frac{\pi}{2}$

$\boxed{63}$ $\arcsin\left(\sin\dfrac{5\pi}{4}\right) = \arcsin\left(-\dfrac{\sqrt{2}}{2}\right) = -\dfrac{\pi}{4}$

$\boxed{65}$ $\sin\left[\arccos\left(-\dfrac{\sqrt{3}}{2}\right)\right] = \sin\frac{5\pi}{6} = \frac{1}{2}$

$\boxed{67}$ $\sec(\sin^{-1}\frac{3}{2})$ is not defined since $\frac{3}{2} > 1$ and the domain of $\sin^{-1}$ is -1 to $+1$.

$\boxed{69}$ Let $\alpha = \sin^{-1}\frac{15}{17}$ and $\beta = \sin^{-1}\frac{8}{17}$.

$\cos(\sin^{-1}\frac{15}{17} - \sin^{-1}\frac{8}{17}) = \cos(\alpha-\beta) = \cos\alpha\,\cos\beta + \sin\alpha\,\sin\beta = \frac{8}{17}\cdot\frac{15}{17} + \frac{15}{17}\cdot\frac{8}{17} = \frac{240}{289}$.

$\boxed{70}$ Let $\alpha = \sin^{-1}\frac{4}{5}$. $\cos(2\sin^{-1}\frac{4}{5}) = \cos(2\alpha) = \cos^2\alpha - \sin^2\alpha = \left(\frac{3}{5}\right)^2 - \left(\frac{4}{5}\right)^2 = -\frac{7}{25}$.

[71] $y = \cos^{-1} 3x$ • horizontally compress $y = \cos^{-1} x$ by a factor of 3

[73] $y = 1 - \sin^{-1} x = -\sin^{-1} x + 1$ •

Reflect $y = \sin^{-1} x$ through the x-axis and shift it up 1 unit.

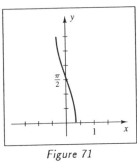

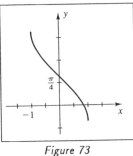

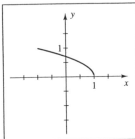

Figure 71   *Figure 73*   *Figure 74*

[74] If $\alpha = \cos^{-1} x$, then $\cos \alpha = x$, where $0 \le \alpha \le \pi$.

Hence, $y = \sin\left(\frac{1}{2}\cos^{-1} x\right) = \sin \frac{1}{2}\alpha = \sqrt{\frac{1 - \cos \alpha}{2}} = \sqrt{\frac{1 - x}{2}}$.

Thus, we have the graph of the half-parabola $y = \sqrt{\frac{1}{2}(1 - x)}$ on the interval $[-1, 1]$.

[75] $\cos(\alpha + \beta + \gamma)$

$$= \cos\left[(\alpha + \beta) + \gamma\right]$$
$$= \cos(\alpha + \beta)\cos \gamma - \sin(\alpha + \beta)\sin \gamma$$
$$= (\cos \alpha \cos \beta - \sin \alpha \sin \beta)\cos \gamma - (\sin \alpha \cos \beta + \cos \alpha \sin \beta)\sin \gamma$$
$$= \cos \alpha \cos \beta \cos \gamma - \sin \alpha \sin \beta \cos \gamma - \sin \alpha \cos \beta \sin \gamma - \cos \alpha \sin \beta \sin \gamma$$

[77] $\cos x - \cos 2x + \cos 3x = 0 \;\Rightarrow\; (\cos x + \cos 3x) - \cos 2x \;\Rightarrow$

[S3] $2\cos\dfrac{x + 3x}{2}\cos\dfrac{x - 3x}{2} - \cos 2x = 0 \;\Rightarrow\; 2\cos 2x \cos x - \cos 2x = 0 \;\Rightarrow$

$\cos 2x(2\cos x - 1) = 0 \;\Rightarrow\; \cos 2x = 0 \text{ or } \cos x = \frac{1}{2} \;\Rightarrow$

$2x = \frac{\pi}{2} + \pi n \text{ (or } x = \frac{\pi}{4} + \frac{\pi}{2}n\text{) or } x = \frac{\pi}{3} + 2\pi n, \frac{5\pi}{3} + 2\pi n.$

In the figure on $[-2\pi, 2\pi]$, $x = \pm\frac{\pi}{4}, \pm\frac{3\pi}{4}, \pm\frac{5\pi}{4}, \pm\frac{7\pi}{4}, \pm\frac{\pi}{3}, \pm\frac{5\pi}{3}$.

[79] (a) Bisect θ to form two right triangles.

$$\cos\tfrac{1}{2}\theta = \frac{r}{d + r} \;\Rightarrow\; d + r = \frac{r}{\cos\frac{1}{2}\theta} \;\Rightarrow\; d = r\sec\tfrac{1}{2}\theta - r = r\left(\sec\tfrac{1}{2}\theta - 1\right).$$

(b) $d = 300$ and $r = 4000 \;\Rightarrow\; \cos\tfrac{1}{2}\theta = \frac{r}{d + r} = \frac{4000}{4300} \;\Rightarrow\; \tfrac{1}{2}\theta \approx 21.5° \;\Rightarrow\; \theta \approx 43°.$

Chapter 6 Discussion Exercises

$\boxed{1}$ $\dfrac{\tan x}{1-\cot x}+\dfrac{\cot x}{1-\tan x}=\dfrac{\frac{\sin x}{\cos x}}{1-\frac{\cos x}{\sin x}}+\dfrac{\frac{\cos x}{\sin x}}{1-\frac{\sin x}{\cos x}}=$

$$\dfrac{\sin^2 x}{\cos x\,(\sin x-\cos x)}+\dfrac{\cos^2 x}{\sin x\,(\cos x-\sin x)}=\dfrac{\sin^2 x}{\cos x\,(\sin x-\cos x)}-\dfrac{\cos^2 x}{\sin x\,(\sin x-\cos x)}=$$

$$\dfrac{\sin^3 x-\cos^3 x}{\cos x\,\sin x\,(\sin x-\cos x)}=\dfrac{(\sin x-\cos x)(\sin^2 x+\sin x\,\cos x+\cos^2 x)}{\cos x\,\sin x\,(\sin x-\cos x)}=$$

$$\dfrac{1+\sin x\,\cos x}{\cos x\,\sin x}=\dfrac{1}{\cos x\,\sin x}+1=1+\sec x\,\csc x$$

$\boxed{3}$ *Note:* Graphing on a TI-82/83 doesn't really help to solve this problem.

$3\cos 45x+4\sin 45x=5 \;\Rightarrow\; \{\text{by Example 6 in Section 6.3}\}$

$5\cos\left(45x-\tan^{-1}\tfrac{4}{3}\right)=5 \;\Rightarrow\; \cos\left(45x-\tan^{-1}\tfrac{4}{3}\right)=1 \;\Rightarrow\; 45x-\tan^{-1}\tfrac{4}{3}=2\pi n \;\Rightarrow\;$

$45x=2\pi n+\tan^{-1}\tfrac{4}{3} \;\Rightarrow\; x=\dfrac{2\pi n+\tan^{-1}\frac{4}{3}}{45}.$ $n=0,\,1,\,\ldots,\,44$ will yield x values in

$[0,\,2\pi)$. *Note:* After using Example 6, you might notice that this is a function with period $2\pi/45$, and it will obtain 45 maximums on an interval of length 2π. The largest value of x is approximately 6.164 {when $n=44$}.

$\boxed{5}$ Let $\alpha=\tan^{-1}(\tfrac{1}{239})$ and $\theta=\tan^{-1}(\tfrac{1}{5})$. $\tfrac{\pi}{4}=4\theta-\alpha \;\Rightarrow\; \tfrac{\pi}{4}+\alpha=4\theta$. Both sides are acute angles, and we will show that the tangent of each side is equal to the same value, hence proving the identity.

$\text{LS}=\tan\left(\tfrac{\pi}{4}+\alpha\right)=\dfrac{\tan\frac{\pi}{4}+\tan\alpha}{1-\tan\frac{\pi}{4}\tan\alpha}=\dfrac{1+\frac{1}{239}}{1-1\cdot\frac{1}{239}}=\dfrac{\frac{240}{239}}{\frac{238}{239}}=\dfrac{240}{238}=\dfrac{120}{119}.$

$\text{RS}=\tan 4\theta=$

$$\dfrac{2\tan 2\theta}{1-\tan^2(2\theta)}=\dfrac{2\cdot\frac{2\tan\theta}{1-\tan^2\theta}}{1-\left(\frac{2\tan\theta}{1-\tan^2\theta}\right)^2}=\dfrac{2\cdot\frac{\frac{2}{5}}{1-\frac{1}{25}}}{1-\left(\frac{\frac{2}{5}}{1-\frac{1}{25}}\right)^2}=\dfrac{\frac{\frac{4}{5}}{\frac{24}{25}}}{1-\frac{\frac{4}{25}}{\frac{24\cdot24}{25\cdot25}}}=\dfrac{\frac{5}{6}}{\frac{119}{144}}=\dfrac{120}{119}.$$

Similarly, for the second relationship, we could write $\tfrac{\pi}{4}=\alpha+\beta+\gamma$ and show that $\tan\left(\tfrac{\pi}{4}-\alpha\right)=\tan(\beta+\gamma)=\tfrac{1}{3}$.

Chapter 7: Applications of Trigonometry

1 $\beta = 180° - \alpha - \gamma = 180° - 41° - 77° = 62°.$

$$\frac{b}{\sin \beta} = \frac{a}{\sin \alpha} \;\Rightarrow\; b = \frac{a \sin \beta}{\sin \alpha} = \frac{10.5 \sin 62°}{\sin 41°} \approx 14.1.$$

$$\frac{c}{\sin \gamma} = \frac{a}{\sin \alpha} \;\Rightarrow\; c = \frac{a \sin \gamma}{\sin \alpha} = \frac{10.5 \sin 77°}{\sin 41°} \approx 15.6.$$

3 $\gamma = 180° - \alpha - \beta = 180° - 27°40' - 52°10' = 100°10'.$

$$\frac{b}{\sin \beta} = \frac{a}{\sin \alpha} \;\Rightarrow\; b = \frac{a \sin \beta}{\sin \alpha} = \frac{32.4 \sin 52°10'}{\sin 27°40'} \approx 55.1.$$

$$\frac{c}{\sin \gamma} = \frac{a}{\sin \alpha} \;\Rightarrow\; c = \frac{a \sin \gamma}{\sin \alpha} = \frac{32.4 \sin 100°10'}{\sin 27°40'} \approx 68.7.$$

5 $\beta = 180° - \alpha - \gamma = 180° - 42°10' - 61°20' = 76°30'.$

$$\frac{a}{\sin \alpha} = \frac{b}{\sin \beta} \;\Rightarrow\; a = \frac{b \sin \alpha}{\sin \beta} = \frac{19.7 \sin 42°10'}{\sin 76°30'} \approx 13.6.$$

$$\frac{c}{\sin \gamma} = \frac{b}{\sin \beta} \;\Rightarrow\; c = \frac{b \sin \gamma}{\sin \beta} = \frac{19.7 \sin 61°20'}{\sin 76°30'} \approx 17.8.$$

7 $\dfrac{\sin \beta}{b} = \dfrac{\sin \gamma}{c} \;\Rightarrow\; \beta = \sin^{-1}\left(\dfrac{b \sin \gamma}{c}\right) = \sin^{-1}\left(\dfrac{12 \sin 81°}{11}\right) \approx \sin^{-1}(1.0775).$ Since 1.0775

is not in the domain of the inverse sine function, which is $[-1, 1]$, *no triangle exists.*

9 $\dfrac{\sin \alpha}{a} = \dfrac{\sin \gamma}{c} \;\Rightarrow\; \alpha = \sin^{-1}\left(\dfrac{a \sin \gamma}{c}\right) = \sin^{-1}\left(\dfrac{140 \sin 53°20'}{115}\right) \approx \sin^{-1}(0.9765) \approx$

$77°30'$ or $102°30'$ { rounded to the nearest 10 minutes }. When using the inverse sine function to solve for an angle, remember that there are always *two* values if $\theta \neq 90°$. The first angle is the one you obtain using your calculator, and the second is the reference angle for the first angle in the second quadrant. After obtaining these values, we need to check the sum of the angles. If this sum is less than 180°, a triangle is formed. If this sum is greater than or equal to 180°, no triangle is formed. For this exercise, there are two triangles possible since in either case $\alpha + \gamma < 180°$.

$\beta = (180° - \gamma) - \alpha \approx (180° - 53°20') - (77°30' \text{ or } 102°30') = 49°10' \text{ or } 24°10'.$

$$\frac{b}{\sin \beta} = \frac{c}{\sin \gamma} \;\Rightarrow\; b = \frac{c \sin \beta}{\sin \gamma} \approx \frac{115 \sin (49°10' \text{ or } 24°10')}{\sin 53°20'} \approx 108 \text{ or } 58.7.$$

11 $\dfrac{\sin \alpha}{a} = \dfrac{\sin \gamma}{c} \;\Rightarrow\; \alpha = \sin^{-1}\left(\dfrac{a \sin \gamma}{c}\right) = \sin^{-1}\left(\dfrac{131.08 \sin 47.74°}{97.84}\right) \approx \sin^{-1}(0.9915) \approx$

$82.54°$ or $97.46°$. There are two triangles possible since in either case $\alpha + \gamma < 180°$.

$\beta = (180° - \gamma) - \alpha \approx (180° - 47.74°) - (82.54° \text{ or } 97.46°) = 49.72° \text{ or } 34.80°.$

$$\frac{b}{\sin \beta} = \frac{c}{\sin \gamma} \;\Rightarrow\; b = \frac{c \sin \beta}{\sin \gamma} \approx \frac{97.84 \sin (49.72° \text{ or } 34.80°)}{\sin 47.74°} \approx 100.85 \text{ or } 75.45.$$

$\boxed{13}$ $\frac{\sin\beta}{b} = \frac{\sin\alpha}{a}$ $\Rightarrow$ $\beta = \sin^{-1}\left(\frac{b\sin\alpha}{a}\right) = \sin^{-1}\left(\frac{18.9\sin 65°10'}{21.3}\right) \approx \sin^{-1}(0.8053) \approx$

53°40′ or 126°20′ {rounded to the nearest 10 minutes}. Reject 126°20′ because then

$\alpha + \beta \geq 180°$. $\gamma = 180° - \alpha - \beta \approx 180° - 65°10' - 53°40' = 61°10'$.

$$\frac{c}{\sin\gamma} = \frac{a}{\sin\alpha} \Rightarrow c = \frac{a\sin\gamma}{\sin\alpha} \approx \frac{21.3\sin 61°10'}{\sin 65°10'} \approx 20.6.$$

$\boxed{15}$ $\frac{\sin\gamma}{c} = \frac{\sin\beta}{b}$ $\Rightarrow$ $\gamma = \sin^{-1}\left(\frac{c\sin\beta}{b}\right) = \sin^{-1}\left(\frac{0.178\sin 121.624°}{0.283}\right) \approx \sin^{-1}(0.5356) \approx$

32.383° or 147.617°. Reject 147.617° because then $\beta + \gamma \geq 180°$.

$\alpha = 180° - \beta - \gamma \approx 180° - 121.624° - 32.383° = 25.993°$.

$$\frac{a}{\sin\alpha} = \frac{b}{\sin\beta} \Rightarrow a = \frac{b\sin\alpha}{\sin\beta} \approx \frac{0.283\sin 25.993°}{\sin 121.624°} \approx 0.146.$$

$\boxed{17}$ $\angle ABC = 180° - 54°10' - 63°20' = 62°30'$. $\frac{\overline{AB}}{\sin 54°10'} = \frac{240}{\sin 62°30'}$ $\Rightarrow$ $\overline{AB} \approx 219.36$ yd

$\boxed{19}$ (a) $\angle ABP = 180° - 65° = 115°$. $\angle APB = 180° - 21° - 115° = 44°$.

$$\frac{\overline{AP}}{\sin 115°} = \frac{1.2}{\sin 44°} \Rightarrow \overline{AP} \approx 1.57, \text{ or } 1.6 \text{ mi.}$$

(b) $\sin 21° = \frac{\text{height of } P}{\overline{AP}}$ $\Rightarrow$ height of $P = \frac{1.2\sin 115°\sin 21°}{\sin 44°}$

{from part (a)} ≈ 0.56, or 0.6 mi.

$\boxed{21}$ Let C denote the base of the balloon and P its projection on the ground.

$\angle ACB = 180° - 24°10' - 47°40' = 108°10'$. $\frac{\overline{AC}}{\sin 47°40'} = \frac{8.4}{\sin 108°10'}$ $\Rightarrow$ $\overline{AC} \approx 6.5$ mi.

$$\sin 24°10' = \frac{\overline{PC}}{AC} \Rightarrow \overline{PC} = \frac{8.4\sin 47°40'\sin 24°10'}{\sin 108°10'} \approx 2.7 \text{ mi.}$$

$\boxed{23}$ $\angle APQ = 57° - 22° = 35°$. $\angle AQP = 180° - (63° - 22°) = 139°$.

$$\angle PAQ = 180° - 139° - 35° = 6°. \quad \frac{\overline{AP}}{\sin 139°} = \frac{100}{\sin 6°} \Rightarrow \overline{AP} = \frac{100\sin 139°}{\sin 6°} \approx 628 \text{ m.}$$

$\boxed{25}$ $\angle FAB = 90° - 27°10' = 62°50'$.

$\angle FBA = 90° - 52°40' = 37°20'$.

$\angle AFB = 180° - 62°50' - 37°20' = 79°50'$.

$\frac{\overline{AF}}{\sin 37°20'} = \frac{6}{\sin 79°50'}$ $\Rightarrow$ $\overline{AF} \approx 3.70$ mi.

$\frac{\overline{BF}}{\sin 62°50'} = \frac{6}{\sin 79°50'}$ $\Rightarrow$ $\overline{BF} \approx 5.42$ mi.

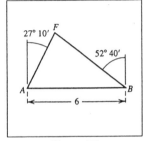

Figure 25

27 Let A denote the base of the hill, B the base of the cathedral, and C the top of the spire. The angle at the base of the hill is $180° - 48° = 132°$. The angle at the top

of the spire is $180° - 132° - 41° = 7°$. $\dfrac{\overline{AC}}{\sin 41°} = \dfrac{200}{\sin 7°} \Rightarrow \overline{AC} = \dfrac{200 \sin 41°}{\sin 7°} \approx 1077$ ft.

$\angle BAC = 48° - 32° = 16°$. $\angle ACB = 90° - 48° = 42°$. $\angle ABC = 180° - 42° - 16° = 122°$.

$$\frac{\overline{BC}}{\sin 16°} = \frac{\overline{AC}}{\sin 122°} \Rightarrow \overline{BC} = \frac{200 \sin 41° \sin 16°}{\sin 7° \sin 122°} \approx 350 \text{ ft.}$$

29 (a) In the triangle that forms the base, the third angle is $180° - 103° - 52° = 25°$.

Let l denote the length of the dashed line. $\dfrac{l}{\sin 103°} = \dfrac{12.0}{\sin 25°} \Rightarrow l \approx 27.7$ units.

Now $\tan 34° = \dfrac{h}{l} \Rightarrow h \approx 18.7$ units.

(b) Draw a line from the $103°$ angle that is perpendicular to l and call it d.

$\sin 52° = \dfrac{d}{12} \Rightarrow d \approx 9.5$ units. The area of the triangular base is $B = \frac{1}{2}ld$.

The volume V is $\frac{1}{3}(\frac{1}{2}ld)h = 288 \sin 52° \sin^2 103° \tan 34° \csc^2 25° \approx 814$ cubic units.

31 Draw a line through P perpendicular to the x-axis. Locate points A and B on this line so that $\angle PAQ = \angle PBR = 90°$. $\overline{AP} = 5127.5 - 3452.8 = 1674.7$,

$\overline{AQ} = 3145.8 - 1487.7 = 1658.1$, and $\tan \angle APQ = \frac{1658.1}{1674.7} \Rightarrow \angle APQ \approx 44°43'$.

Thus, $\angle BPR \approx 180° - 55°50' - 44°43' = 79°27'$.

By the distance formula, $\overline{PQ} \approx \sqrt{(1674.7)^2 + (1658.1)^2} \approx 2356.7$.

Now $\dfrac{\overline{PR}}{\sin 65°22'} = \dfrac{\overline{PQ}}{\sin(180° - 55°50' - 65°22')} \Rightarrow \overline{PR} = \dfrac{2356.7 \sin 65°22'}{\sin 58°48'} \approx 2504.5$.

$\sin \angle BPR = \dfrac{\overline{BR}}{\overline{PR}} \Rightarrow \overline{BR} \approx (2504.5)(\sin 79°27') \approx 2462.2$. $\cos \angle BPR = \dfrac{\overline{BP}}{\overline{PR}} \Rightarrow$

$\overline{BP} \approx (2504.5)(\cos 79°27') \approx 458.6$. Using the coordinates of P, we see that

$$R(x, y) \approx (1487.7 + 2462.2, 3452.8 - 458.6) = (3949.9, 2994.2).$$

7.2 Exercises

Note: These formulas will be used to solve problems that involve the law of cosines.

(1) $a^2 = b^2 + c^2 - 2bc \cos \alpha \Rightarrow a = \sqrt{b^2 + c^2 - 2bc \cos \alpha}$

(Similar formulas are used for b and c.)

(2) $a^2 = b^2 + c^2 - 2bc \cos \alpha \Rightarrow 2bc \cos \alpha = b^2 + c^2 - a^2 \Rightarrow$

$\cos \alpha = \left(\dfrac{b^2 + c^2 - a^2}{2bc}\right) \Rightarrow \alpha = \cos^{-1}\left(\dfrac{b^2 + c^2 - a^2}{2bc}\right)$

(Similar formulas are used for β and γ.)

$\boxed{1}$ See the note on the previous page.

$$a = \sqrt{b^2 + c^2 - 2bc\,\cos\alpha} = \sqrt{20^2 + 30^2 - 2(20)(30)\,\cos 60^\circ} = \sqrt{700} \approx 26.$$

$$\beta = \cos^{-1}\!\left(\frac{a^2 + c^2 - b^2}{2ac}\right) = \cos^{-1}\!\left(\frac{700 + 30^2 - 20^2}{2(\sqrt{700})(30)}\right) \approx \cos^{-1}(0.7559) \approx 41^\circ.$$

$$\gamma = 180^\circ - \alpha - \beta \approx 180^\circ - 60^\circ - 41^\circ = 79^\circ.$$

$\boxed{3}$ $b = \sqrt{a^2 + c^2 - 2ac\,\cos\beta} = \sqrt{23{,}400 + 4500\sqrt{3}} \approx 177$, or 180.

$$\alpha = \cos^{-1}\!\left(\frac{b^2 + c^2 - a^2}{2bc}\right) \approx \cos^{-1}(0.9054) \approx 25^\circ 10', \text{ or } 25^\circ.$$

Note: We do not have to worry about an "ambiguous" case of the law of cosines as we did with the law of sines. This is true because the inverse cosine function gives us a unique angle from 0° to 180° for all values in its domain.

$$\gamma = 180^\circ - \alpha - \beta \approx 180^\circ - 25^\circ 10' - 150^\circ = 4^\circ 50', \text{ or } 5^\circ.$$

$\boxed{5}$ $c = \sqrt{a^2 + b^2 - 2ab\,\cos\gamma} \approx \sqrt{7.58} \approx 2.75.$

$$\alpha = \cos^{-1}\!\left(\frac{b^2 + c^2 - a^2}{2bc}\right) \approx \cos^{-1}(0.9324) \approx 21^\circ 10'.$$

$$\beta = 180^\circ - \alpha - \gamma \approx 180^\circ - 21^\circ 10' - 115^\circ 10' = 43^\circ 40'.$$

$\boxed{7}$ $\alpha = \cos^{-1}\!\left(\dfrac{b^2 + c^2 - a^2}{2bc}\right) = \cos^{-1}(0.875) \approx 29^\circ.$

$$\beta = \cos^{-1}\!\left(\frac{a^2 + c^2 - b^2}{2ac}\right) = \cos^{-1}(0.6875) \approx 47^\circ.$$

$$\gamma = 180^\circ - \alpha - \beta \approx 180^\circ - 29^\circ - 47^\circ = 104^\circ.$$

$\boxed{9}$ $\alpha = \cos^{-1}\!\left(\dfrac{b^2 + c^2 - a^2}{2bc}\right) \approx \cos^{-1}(0.9766) \approx 12^\circ 30'.$

$$\beta = \cos^{-1}\!\left(\frac{a^2 + c^2 - b^2}{2ac}\right) = \cos^{-1}(-0.725) \approx 136^\circ 30'.$$

$$\gamma = 180^\circ - \alpha - \beta \approx 180^\circ - 12^\circ 30' - 136^\circ 30' = 31^\circ 00'.$$

Note: When deciding whether to use the law of cosines or the law of sines, it may be helpful to remember that the law of cosines must be used when you have:

(1) 3 sides, or

(2) 2 sides and the angle between them.

$\boxed{11}$ We have two sides and the angle between them. Hence, we apply the law of cosines.

$$\text{Third side} = \sqrt{175^2 + 150^2 - 2(175)(150)\cos 73^\circ 40'} \approx 196 \text{ feet.}$$

$\boxed{13}$ 20 minutes $= \frac{1}{3}$ hour $\Rightarrow$

the cars have traveled $60(\frac{1}{3}) = 20$ miles and $45(\frac{1}{3}) = 15$ miles, respectively.

$$\text{The distance } d \text{ apart is } d = \sqrt{20^2 + 15^2 - 2(20)(15)\cos 84^\circ} \approx 24 \text{ miles.}$$

15 The first ship travels $(24)(2) = 48$ miles in two hours. The second ship travels $(18)(1\frac{1}{2}) = 27$ miles in $1\frac{1}{2}$ hours. The angle between the paths is $20° + 35° = 55°$.

$$\overline{AB} = \sqrt{27^2 + 48^2 - 2(27)(48)\cos 55°} \approx 39 \text{ miles.}$$

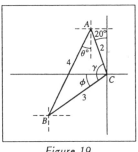

Figure 15

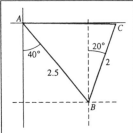

Figure 17

17 $\angle ABC = 40° + 20° = 60°$.

$$\overline{AB} = \left(\frac{1 \text{ mile}}{8 \text{ min}} \cdot 20 \text{ min}\right) = 2.5 \text{ miles and } \overline{BC} = \left(\frac{1 \text{ mile}}{8 \text{ min}} \cdot 16 \text{ min}\right) = 2 \text{ miles.}$$

$$\overline{AC} = \sqrt{2.5^2 + 2^2 - 2(2)(2.5)\cos 60°} = \sqrt{5.25} \approx 2.3 \text{ miles.}$$

19 $\gamma = \cos^{-1}\left(\frac{2^2 + 3^2 - 4^2}{2 \cdot 2 \cdot 3}\right) = \cos^{-1}(-0.25) \approx 104°29'$. $\phi \approx 104°29' - 70° = 34°29'$.

The direction that the third side was traversed is approximately

$$\text{N}(90° - 34°29')\text{E} = \text{N}55°31'\text{ E.}$$

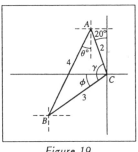

Figure 19

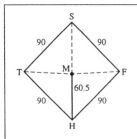

Figure 21

21 Let H denote home plate, M the mound, F first base, S second base, and T third base. $\overline{HS} = \sqrt{90^2 + 90^2} = 90\sqrt{2} \approx 127.3$ ft. $\overline{MS} = 90\sqrt{2} - 60.5 \approx 66.8$ ft.

$\angle MHF = 45°$ so $\overline{MF} = \sqrt{60.5^2 + 90^2 - 2(60.5)(90)\cos 45°} \approx 63.7$ ft.

$$\overline{MT} = \overline{MF} \text{ by the symmetry of the field.}$$

23 $\angle RTP = 21°$ and $\angle RSP = 37°$. $\sin \angle RSP = \frac{10,000}{\overline{SP}} \Rightarrow \overline{SP} = 10,000 \csc 37° \approx$

$16,616$ ft. $\sin \angle RTP = \frac{10,000}{\overline{TP}} \Rightarrow \overline{TP} = 10,000 \csc 21° \approx 27,904$ ft.

$$\overline{ST} = \sqrt{\overline{SP}^2 + \overline{TP}^2 - 2(\overline{SP})(\overline{TP})\cos 110°} \approx 37,039 \text{ ft} \approx 7 \text{ miles.}$$

[25] Let $d = \overline{ES}$. $d^2 = R^2 + R^2 - 2RR\cos\theta \Rightarrow d^2 = 2R^2(1 - \cos\theta) \Rightarrow$

$d^2 = 4R^2\left(\dfrac{1 - \cos\theta}{2}\right) \Rightarrow d = 2R\sqrt{\dfrac{1 - \cos\theta}{2}} \Rightarrow d = 2R\sin\dfrac{\theta}{2}.$

Since $d = vt$, $t = \dfrac{d}{v} = \dfrac{2R}{v}\sin\dfrac{\theta}{2}.$

[27] (a) $\angle BCP = \frac{1}{2}(\angle BCD) = \frac{1}{2}(72°) = 36°$. $\triangle BPC$ is isosceles so

$\angle BPC = \angle PBC$ and $2\angle BPC = 180° - 36° \Rightarrow \underline{\angle BPC = 72°}.$

$\angle APB = 180° - \angle BPC = 180° - 72° = \underline{108°}.$

$\angle ABP = 180° - \angle APB - \angle BAP = 180° - 108° - 36° = \underline{36°}.$

(b) $\overline{BP} = \sqrt{\overline{BC}^2 + \overline{PC}^2 - 2(\overline{BC})(\overline{PC})\cos 36°} = \sqrt{1^2 + 1^2 - 2(1)(1)\cos 36°} \approx 0.62.$

(c) $\text{Area}_{\text{kite}} = 2(\text{Area of } \triangle BPC) = 2 \cdot \frac{1}{2}(\overline{CB})(\overline{CP})\sin\angle BCP = \sin 36° \approx 0.59.$

$\text{Area}_{\text{dart}} = 2(\text{Area of } \triangle ABP) = 2 \cdot \frac{1}{2}(\overline{AB})(\overline{BP})\sin\angle ABP = \overline{BP}\sin 36° \approx 0.36.$

$\{\overline{BP} \text{ was found in part (b)}\}$

Note: Exer. 29–36: $\mathcal{A}$ (the area) is measured in square units.

[29] Since α is the angle between sides b and c, we may apply the area of a triangle

formula listed in this section. $\mathcal{A} = \frac{1}{2}bc\sin\alpha = \frac{1}{2}(20)(30)\sin 60° = 300(\sqrt{3}/2) \approx 260.$

[31] $\gamma = 180° - \alpha - \beta = 180° - 40.3° - 62.9° = 76.8°.$

$\dfrac{a}{\sin\alpha} = \dfrac{b}{\sin\beta} \Rightarrow a = \dfrac{b\sin\alpha}{\sin\beta} = \dfrac{5.63\sin 40.3°}{\sin 62.9°}. \quad \mathcal{A} = \frac{1}{2}ab\sin\gamma \approx 11.21.$

[33] $\dfrac{\sin\beta}{b} = \dfrac{\sin\alpha}{a} \Rightarrow \sin\beta = \dfrac{b\sin\alpha}{a} = \dfrac{3.4\sin 80.1°}{8.0} \approx 0.4187 \Rightarrow \beta \approx 24.8°$ or $155.2°$.

Reject $155.2°$ because then $\alpha + \beta = 235.3° \geq 180°$. $\gamma \approx 180° - 80.1° - 24.8° = 75.1°.$

$\mathcal{A} = \frac{1}{2}ab\sin\gamma = \frac{1}{2}(8.0)(3.4)\sin 75.1° \approx 13.1.$

[35] Given the lengths of the 3 sides of a triangle, we compute the semiperimeter and then

use Heron's formula to find the area of the triangle.

$s = \frac{1}{2}(a + b + c) = \frac{1}{2}(25.0 + 80.0 + 60.0) = 82.5.$

$\mathcal{A} = \sqrt{s(s-a)(s-b)(s-c)} = \sqrt{(82.5)(57.5)(2.5)(22.5)} \approx 516.56$, or $517.0.$

[37] $s = \frac{1}{2}(a + b + c) = \frac{1}{2}(115 + 140 + 200) = 227.5.$ $\mathcal{A} = \sqrt{s(s-a)(s-b)(s-c)} =$

$\sqrt{(227.5)(112.5)(87.5)(27.5)} \approx 7847.6 \text{ yd}^2$, or $\mathcal{A}/4840 \approx 1.62$ acres.

[39] The area of the parallelogram is twice the area of the triangle formed by the two

sides and the included angle. $\mathcal{A} = 2(\frac{1}{2})(12.0)(16.0)\sin 40° \approx 123.4 \text{ ft}^2.$

7.3 Exercises

[1] $\mathbf{a} + \mathbf{b} = \langle 2, -3\rangle + \langle 1, 4\rangle = \langle 2 + 1, -3 + 4\rangle = \langle 3, 1\rangle.$

$\mathbf{a} - \mathbf{b} = \langle 2, -3\rangle - \langle 1, 4\rangle = \langle 2 - 1, -3 - 4\rangle = \langle 1, -7\rangle.$

$$4\mathbf{a} + 5\mathbf{b} = 4\langle 2, -3\rangle + 5\langle 1, 4\rangle = \langle 4(2), 4(-3)\rangle + \langle 5(1), 5(4)\rangle$$
$$= \langle 8, -12\rangle + \langle 5, 20\rangle = \langle 8 + 5, -12 + 20\rangle = \langle 13, 8\rangle.$$

From above, $4\mathbf{a} - 5\mathbf{b} = \langle 8, -12\rangle - \langle 5, 20\rangle = \langle 8 - 5, -12 - 20\rangle = \langle 3, -32\rangle.$

3. Simplifying $\mathbf{a}$ and $\mathbf{b}$ first, we have $\mathbf{a} = -\langle 7, -2\rangle = \langle -(7), -(-2)\rangle = \langle -7, 2\rangle$ and
$$\mathbf{b} = 4\langle -2, 1\rangle = \langle 4(-2), 4(1)\rangle = \langle -8, 4\rangle.$$

$\mathbf{a} + \mathbf{b} = \langle -7, 2\rangle + \langle -8, 4\rangle = \langle -7 + (-8), 2 + 4\rangle = \langle -15, 6\rangle.$

$\mathbf{a} - \mathbf{b} = \langle -7, 2\rangle - \langle -8, 4\rangle = \langle -7 - (-8), 2 - 4\rangle = \langle 1, -2\rangle.$

$4\mathbf{a} + 5\mathbf{b} = 4\langle -7, 2\rangle + 5\langle -8, 4\rangle = \langle -28, 8\rangle + \langle -40, 20\rangle = \langle -68, 28\rangle.$

From above, $4\mathbf{a} - 5\mathbf{b} = \langle -28, 8\rangle - \langle -40, 20\rangle = \langle 12, -12\rangle.$

5. $\mathbf{a} + \mathbf{b} = (\mathbf{i} + 2\mathbf{j}) + (3\mathbf{i} - 5\mathbf{j}) = (1 + 3)\mathbf{i} + (2 - 5)\mathbf{j} = 4\mathbf{i} - 3\mathbf{j}.$

$\mathbf{a} - \mathbf{b} = (\mathbf{i} + 2\mathbf{j}) - (3\mathbf{i} - 5\mathbf{j}) = (1 - 3)\mathbf{i} + (2 - (-5))\mathbf{j} = -2\mathbf{i} + 7\mathbf{j}.$

$4\mathbf{a} + 5\mathbf{b} = 4(\mathbf{i} + 2\mathbf{j}) + 5(3\mathbf{i} - 5\mathbf{j}) = (4\mathbf{i} + 8\mathbf{j}) + (15\mathbf{i} - 25\mathbf{j}) = 19\mathbf{i} - 17\mathbf{j}.$

From above, $4\mathbf{a} - 5\mathbf{b} = (4\mathbf{i} + 8\mathbf{j}) - (15\mathbf{i} - 25\mathbf{j}) = -11\mathbf{i} + 33\mathbf{j}.$

7. $\mathbf{a} = 3\mathbf{i} + 2\mathbf{j}$ and $\mathbf{b} = -\mathbf{i} + 5\mathbf{j} \Rightarrow \mathbf{a} + \mathbf{b} = 2\mathbf{i} + 7\mathbf{j}, \ 2\mathbf{a} = 6\mathbf{i} + 4\mathbf{j},$ and $-3\mathbf{b} = 3\mathbf{i} - 15\mathbf{j}.$

Terminal points of the vectors are $(3, 2), (-1, 5), (2, 7), (6, 4),$ and $(3, -15).$

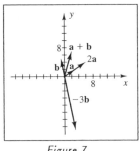

Figure 7

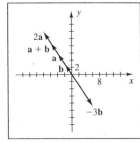

Figure 9

9. $\mathbf{a} = \langle -4, 6\rangle$ and $\mathbf{b} = \langle -2, 3\rangle \Rightarrow \mathbf{a} + \mathbf{b} = \langle -6, 9\rangle, \ 2\mathbf{a} = \langle -8, 12\rangle,$ and $-3\mathbf{b} = \langle 6, -9\rangle.$

Terminal points of the vectors are $(-4, 6), (-2, 3), (-6, 9), (-8, 12),$ and $(6, -9).$

11. $\mathbf{a} + \mathbf{b} = \langle 2, 0\rangle + \langle -1, 0\rangle = \langle 1, 0\rangle = -\langle -1, 0\rangle = -\mathbf{b}$

13. $\mathbf{b} + \mathbf{e} = \langle -1, 0\rangle + \langle 2, 2\rangle = \langle 1, 2\rangle = \mathbf{f}$

15. $\mathbf{b} + \mathbf{d} = \langle -1, 0\rangle + \langle 0, -1\rangle = \langle -1, -1\rangle = -\frac{1}{2}\langle 2, 2\rangle = -\frac{1}{2}\mathbf{e}$

17. $\mathbf{a} + (\mathbf{b} + \mathbf{c}) = \langle a_1, a_2\rangle + (\langle b_1, b_2\rangle + \langle c_1, c_2\rangle)$
$$= \langle a_1, a_2\rangle + \langle b_1 + c_1, b_2 + c_2\rangle$$
$$= \langle a_1 + b_1 + c_1, a_2 + b_2 + c_2\rangle$$
$$= \langle a_1 + b_1, a_2 + b_2\rangle + \langle c_1, c_2\rangle$$
$$= (\langle a_1, a_2\rangle + \langle b_1, b_2\rangle) + \langle c_1, c_2\rangle$$
$$= (\mathbf{a} + \mathbf{b}) + \mathbf{c}$$

$\boxed{19}$ $\mathbf{a} + (-\mathbf{a}) = \langle a_1, a_2 \rangle + (-\langle a_1, a_2 \rangle)$
$= \langle a_1, a_2 \rangle + \langle -a_1, -a_2 \rangle$
$= \langle a_1 - a_1, a_2 - a_2 \rangle$
$= \langle 0, 0 \rangle = \mathbf{0}$

$\boxed{21}$ $(mn)\mathbf{a} = (mn)\langle a_1, a_2 \rangle$
$= \langle (mn)a_1, (mn)a_2 \rangle$
$= \langle mna_1, mna_2 \rangle$
$= m\langle na_1, na_2 \rangle \quad$ or $n\langle ma_1, ma_2 \rangle$
$= m(n\langle a_1, a_2 \rangle) \quad$ or $n(m\langle a_1, a_2 \rangle)$
$= m(n\mathbf{a}) \quad\quad$ or $n(m\mathbf{a})$

$\boxed{23}$ $0\mathbf{a} = 0\langle a_1, a_2 \rangle = \langle 0a_1, 0a_2 \rangle = \langle 0, 0 \rangle = \mathbf{0}.$

Also, $m\mathbf{0} = m\langle 0, 0 \rangle = \langle m0, m0 \rangle = \langle 0, 0 \rangle = \mathbf{0}.$

$\boxed{25}$ $-(\mathbf{a} + \mathbf{b}) = -(\langle a_1, a_2 \rangle + \langle b_1, b_2 \rangle)$
$= -(\langle a_1 + b_1, a_2 + b_2 \rangle)$
$= \langle -(a_1 + b_1), -(a_2 + b_2) \rangle$
$= \langle -a_1 - b_1, -a_2 - b_2 \rangle$
$= \langle -a_1, -a_2 \rangle + \langle -b_1, -b_2 \rangle$
$= -\mathbf{a} + (-\mathbf{b}) = -\mathbf{a} - \mathbf{b}$

$\boxed{27}$ $\| 2\mathbf{v} \| = \| 2\langle a, b \rangle \| = \| \langle 2a, 2b \rangle \| = \sqrt{(2a)^2 + (2b)^2} = \sqrt{4a^2 + 4b^2} = \sqrt{4(a^2 + b^2)} =$
$2\sqrt{a^2 + b^2} = 2\| \langle a, b \rangle \| = 2\| \mathbf{v} \|$

$\boxed{29}$ $\| \mathbf{a} \| = \sqrt{3^2 + (-3)^2} = \sqrt{18} = 3\sqrt{2}.$ $\tan\theta = \frac{-3}{3} = -1$ and θ in QIV $\Rightarrow$ $\theta = \frac{7\pi}{4}.$

$\boxed{31}$ $\| \mathbf{a} \| = 5.$ The terminal side of θ is on the negative x-axis $\Rightarrow$ $\theta = \pi.$

$\boxed{33}$ $\| \mathbf{a} \| = \sqrt{41}.$ $\tan\theta = \frac{5}{-4}$ and θ in QII $\Rightarrow$ $\theta = \tan^{-1}\left(-\frac{5}{4}\right) + \pi.$

$\boxed{35}$ $\| \mathbf{a} \| = 18.$ The terminal side of θ is on the negative y-axis $\Rightarrow$ $\theta = \frac{3\pi}{2}.$

Note: Exercises 37–42: Each resultant force is found by completing the parallelogram and then applying the law of cosines.

$\boxed{37}$ $\| \mathbf{r} \| = \sqrt{40^2 + 70^2 - 2(40)(70)\cos 135°} = \sqrt{6500 + 2800\sqrt{2}} \approx 102.3,$ or 102 lb.

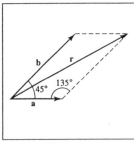

Figure 37

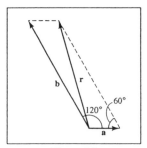

Figure 39

$\boxed{39}$ $\| \mathbf{r} \| = \sqrt{2^2 + 8^2 - 2(2)(8)\cos 60°} = \sqrt{68 - 16} = \sqrt{52} \approx 7.2$ lb.

41 $\|\mathbf{r}\| = \sqrt{90^2 + 60^2 - 2(90)(60)\cos 70°} \approx 89.48$, or 89 lb.

Using the law of cosines, $\alpha = \cos^{-1}\left(\dfrac{90^2 + \|\mathbf{r}\|^2 - 60^2}{2(90)(\|\mathbf{r}\|)}\right) \approx$

$\cos^{-1}(0.7765) \approx 39°$, which is 24° under the negative x-axis.

This angle is 204°, or S66°W.

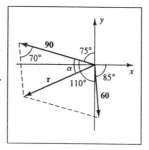

Figure 41

43 We will use a component approach for this exercise.

(a) $= \langle 6\cos 110°,\ 6\sin 110°\rangle \approx \langle -2.05,\ 5.64\rangle$.

(b) $= \langle 2\cos 215°,\ 2\sin 215°\rangle \approx \langle -1.64,\ -1.15\rangle$.

$\mathbf{a} + \mathbf{b} \approx \langle -3.69,\ 4.49\rangle$ and $\|\mathbf{a} + \mathbf{b}\| \approx 5.8$ lb.

$\tan\theta \approx \dfrac{4.49}{-3.69} \Rightarrow \theta \approx 129°$ since θ is in QII.

45 Horizontal $= 50\cos 35° \approx 40.96$. Vertical $= 50\sin 35° \approx 28.68$.

47 Horizontal $= 20\cos 108° \approx -6.18$. Vertical $= 20\sin 108° \approx 19.02$.

Note: In Exercises 49–52, let $\mathbf{u}$ denote the unit vector in the direction of $\mathbf{a}$.

49 (a) The unit vector in the direction of $\mathbf{a}$ is $\dfrac{1}{\|\mathbf{a}\|}\mathbf{a}$.

$\mathbf{a} = -8\mathbf{i} + 15\mathbf{j} \Rightarrow \|\mathbf{a}\| = \sqrt{(-8)^2 + 15^2} = \sqrt{64 + 225} = \sqrt{289} = 17$.

$\mathbf{u} = \dfrac{1}{\|\mathbf{a}\|}\mathbf{a} = \frac{1}{17}(-8\mathbf{i} + 15\mathbf{j}) = -\frac{8}{17}\mathbf{i} + \frac{15}{17}\mathbf{j}$.

(b) The unit vector $-\mathbf{u}$ has the opposite direction of $\mathbf{u}$ and hence,

the opposite direction of $\mathbf{a}$. $-\mathbf{u} = -(-\frac{8}{17}\mathbf{i} + \frac{15}{17}\mathbf{j}) = \frac{8}{17}\mathbf{i} - \frac{15}{17}\mathbf{j}$.

51 (a) As in Exercise 49, $\mathbf{a} = \langle 2, -5\rangle \Rightarrow$

$\|\mathbf{a}\| = \sqrt{2^2 + (-5)^2} = \sqrt{29}$ and $\mathbf{u} = \dfrac{\mathbf{a}}{\|\mathbf{a}\|} = \left\langle \dfrac{2}{\sqrt{29}},\ -\dfrac{5}{\sqrt{29}}\right\rangle$.

(b) $-\mathbf{u} = -\left\langle \dfrac{2}{\sqrt{29}},\ -\dfrac{5}{\sqrt{29}}\right\rangle = \left\langle -\dfrac{2}{\sqrt{29}},\ \dfrac{5}{\sqrt{29}}\right\rangle$

53 (a) $2\mathbf{a}$ has twice the magnitude of $\mathbf{a}$ and the same direction as $\mathbf{a}$.

Hence, $2\langle -6, 3\rangle = \langle -12, 6\rangle$.

(b) As in part (a), $\frac{1}{2}\langle -6, 3\rangle = \langle -3, \frac{3}{2}\rangle$.

Note: In Exercises 55–56, let $\mathbf{v}$ denote the desired vector.

55 The unit vector $\dfrac{\mathbf{a}}{\|\mathbf{a}\|}$ has the same direction as $\mathbf{a}$. The vector $-6\left(\dfrac{\mathbf{a}}{\|\mathbf{a}\|}\right)$ will have a

magnitude of 6 and the opposite direction of $\mathbf{a}$. $\mathbf{a} = 4\mathbf{i} - 7\mathbf{j} \Rightarrow$

$\|\mathbf{a}\| = \sqrt{16 + 49} = \sqrt{65}$. Thus, $-6\left(\dfrac{\mathbf{a}}{\|\mathbf{a}\|}\right) = -6\left(\dfrac{4}{\sqrt{65}}\mathbf{i} - \dfrac{7}{\sqrt{65}}\mathbf{j}\right) = -\dfrac{24}{\sqrt{65}}\mathbf{i} + \dfrac{42}{\sqrt{65}}\mathbf{j}$.

57 (a) $\mathbf{F} = \mathbf{F}_1 + \mathbf{F}_2 + \mathbf{F}_3 = \langle 4, 3\rangle + \langle -2, -3\rangle + \langle 5, 2\rangle = \langle 7, 2\rangle$.

(b) $\mathbf{F} + \mathbf{G} = 0 \Rightarrow \mathbf{G} = -\mathbf{F} = \langle -7, -2\rangle$.

59 (a) $\mathbf{F} = \mathbf{F_1} + \mathbf{F_2} = \langle 6\cos 130°,\ 6\sin 130°\rangle + \langle 4\cos(-120°),\ 4\sin(-120°)\rangle \approx \langle -5.86,\ 1.13\rangle$.

(b) $\mathbf{F} + \mathbf{G} = \mathbf{0} \ \Rightarrow\ \mathbf{G} = -\mathbf{F} \approx \langle 5.86,\ -1.13\rangle$.

61 The vertical components of the forces must add up to zero for the large ship to move along the line segment AB. The vertical component of the smaller tug is $3200\sin(-30°) = -1600$. The vertical component of the larger tug is $4000\sin\theta$.

$$4000\sin\theta = 1600 \ \Rightarrow\ \theta = \sin^{-1}\left(\tfrac{1600}{4000}\right) = \sin^{-1}(0.4) \approx 23.6°.$$

Note: Exercises 63–68: Measure angles from the positive x-axis.

63 $\mathbf{p} = \langle 200\cos 40°,\ 200\sin 40°\rangle \approx \langle 153.21,\ 128.56\rangle$. $\ \mathbf{w} = \langle 40\cos 0°,\ 40\sin 0°\rangle = \langle 40,\ 0\rangle$.

$\mathbf{p} + \mathbf{w} \approx \langle 193.21,\ 128.56\rangle$ and $\|\mathbf{p} + \mathbf{w}\| \approx 232.07$, or 232 mi/hr.

$$\tan\theta \approx \tfrac{128.56}{193.21} \ \Rightarrow\ \theta \approx 34°. \text{ The true course is then N}(90° - 34°)\text{E, or N56°E.}$$

65 $\mathbf{w} = \langle 50\cos 90°,\ 50\sin 90°\rangle = \langle 0,\ 50\rangle$.

$\mathbf{r} = \langle 400\cos 200°,\ 400\sin 200°\rangle \approx \langle -375.88,\ -136.81\rangle$, where $\mathbf{r}$ is the desired resultant of $\mathbf{p} + \mathbf{w}$. Since $\mathbf{r} = \mathbf{p} + \mathbf{w}$, $\mathbf{p} = \mathbf{r} - \mathbf{w} \approx \langle -375.88,\ -186.81\rangle$. $\|\mathbf{p}\| \approx 419.74$, or 420 mi/hr.

$\tan\theta \approx \tfrac{-186.81}{-375.88}$ and θ is in QIII $\ \Rightarrow\ \theta \approx 206°$ from the positive x-axis,

or 244° using the directional form.

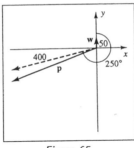

Figure 65

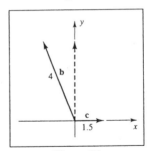

Figure 67

67 Let the vectors $\mathbf{c}$, $\mathbf{b}$, and $\mathbf{r}$ denote the current, the boat, and the resultant, respectively. $\mathbf{c} = \langle 1.5\cos 0°,\ 1.5\sin 0°\rangle = \langle 1.5,\ 0\rangle$. $\ \mathbf{r} = \langle s\cos 90°,\ s\sin 90°\rangle = \langle 0,\ s\rangle$, where s is the resulting speed. $\mathbf{b} = \langle 4\cos\theta,\ 4\sin\theta\rangle$. Also, $\mathbf{b} = \mathbf{r} - \mathbf{c} = \langle -1.5,\ s\rangle$.

$$4\cos\theta = -1.5 \ \Rightarrow\ \theta \approx 112°, \text{ or N22°W.}$$

69 In the figure in the text, suppose $\mathbf{v_1}$ was pointing upwards instead of downwards. If we then write $\mathbf{v_1}$ in terms of its vertical and horizontal components, we have $\mathbf{v_1} = \|\mathbf{v_1}\|\cos\theta_1\,\mathbf{j} - \|\mathbf{v_1}\|\sin\theta_1\,\mathbf{i}$. We want the negative of this vector since $\mathbf{v_1}$ is actually pointing downward, so using $\|\mathbf{v_1}\| = 8.2$ and $\theta_1 = 30°$, we obtain

$$\mathbf{v_1} = \|\mathbf{v_1}\|\sin\theta_1\,\mathbf{i} - \|\mathbf{v_1}\|\cos\theta_1\,\mathbf{j} = 8.2(\tfrac{1}{2})\,\mathbf{i} - 8.2(\sqrt{3}/2)\,\mathbf{j} = 4.1\,\mathbf{i} - 4.1\sqrt{3}\,\mathbf{j} \approx 4.1\,\mathbf{i} - 7.10\,\mathbf{j}.$$

The angle θ_2 can now be computed using the given relationship in the text.

(continued)

$$\frac{\|\mathbf{v}_1\|}{\|\mathbf{v}_2\|} = \frac{\tan\theta_1}{\tan\theta_2} \;\Rightarrow\; \tan\theta_2 = \frac{\|\mathbf{v}_2\|}{\|\mathbf{v}_1\|}\tan\theta_1 = \frac{3.8}{8.2}\times\frac{1}{\sqrt{3}} \;\Rightarrow\; \tan\theta_2 \approx 0.2676 \;\Rightarrow\;$$

$\theta_2 \approx 14.98°$. We now write $\mathbf{v}_2$ in terms of its horizontal and vertical components, and

using $\|\mathbf{v}_2\| = 3.8$, it follows that $\mathbf{v}_2 = \|\mathbf{v}_2\|\sin\theta_2\,\mathbf{i} - \|\mathbf{v}_2\|\cos\theta_2\,\mathbf{j} \approx 0.98\,\mathbf{i} - 3.67\,\mathbf{j}$.

[71] (a) $\mathbf{a} = 15\cos 40°\mathbf{i} + 15\sin 40°\mathbf{j} \approx 11.49\,\mathbf{i} + 9.64\,\mathbf{j}$.

      $\mathbf{b} = 17\cos 40°\mathbf{i} + 17\sin 40°\mathbf{j} \approx 13.02\,\mathbf{i} + 10.93\,\mathbf{j}$.

$$\overrightarrow{PR} = \mathbf{a} + \mathbf{b} \approx 24.51\,\mathbf{i} + 20.57\,\mathbf{j} \;\Rightarrow\; R \approx (24.51,\ 20.57).$$

   (b) $\mathbf{c} = 15\cos(40° + 85°)\mathbf{i} + 15\sin 125°\mathbf{j} \approx -8.60\,\mathbf{i} + 12.29\,\mathbf{j}$.

      $\mathbf{d} = 17\cos(40° + 85° + 35°)\mathbf{i} + 17\sin 160°\mathbf{j} \approx -15.97\,\mathbf{i} + 5.81\,\mathbf{j}$.

$$\overrightarrow{PR} = \mathbf{c} + \mathbf{d} \approx -24.57\,\mathbf{i} + 18.10\,\mathbf{j} \;\Rightarrow\; R \approx (-24.57,\ 18.10).$$

[73] Break the force into a horizontal and a vertical component. The group of 550 people

had to contribute a force equal to the vertical component up the ramp. The vertical

component is $99,000\sin 9° \approx 15,487$ lb. $\dfrac{15,487}{550} \approx 28.2$ lb/person. (Since friction was

ignored, the actual force would have been greater.)

7.4 Exercises

[1] (a) The dot product of the two vectors is

$$\langle -2,\ 5\rangle \cdot \langle 3,\ 6\rangle = (-2)(3) + (5)(6) = -6 + 30 = 24.$$

   (b) The angle between the two vectors is

$$\theta = \cos^{-1}\left(\frac{\langle -2,\ 5\rangle \cdot \langle 3,\ 6\rangle}{\|\langle -2,\ 5\rangle\|\,\|\langle 3,\ 6\rangle\|}\right) = \cos^{-1}\left(\frac{24}{\sqrt{29}\,\sqrt{45}}\right) \approx 48°22'.$$

[3] (a) $(4\mathbf{i} - \mathbf{j}) \cdot (-3\mathbf{i} + 2\mathbf{j}) = (4)(-3) + (-1)(2) = -12 - 2 = -14$

   (b) $\theta = \cos^{-1}\left(\dfrac{(4\mathbf{i} - \mathbf{j}) \cdot (-3\mathbf{i} + 2\mathbf{j})}{\|4\mathbf{i} - \mathbf{j}\|\,\|-3\mathbf{i} + 2\mathbf{j}\|}\right) = \cos^{-1}\left(\dfrac{-14}{\sqrt{17}\,\sqrt{13}}\right) \approx 160°21'$

[5] (a) $(9\mathbf{i}) \cdot (5\mathbf{i} + 4\mathbf{j}) = (9)(5) + (0)(4) = 45 + 0 = 45$

   (b) $\theta = \cos^{-1}\left(\dfrac{(9\mathbf{i}) \cdot (5\mathbf{i} + 4\mathbf{j})}{\|9\mathbf{i}\|\,\|5\mathbf{i} + 4\mathbf{j}\|}\right) = \cos^{-1}\left(\dfrac{45}{\sqrt{81}\,\sqrt{41}}\right) \approx 38°40'$

[7] (a) $\langle 10,\ 7\rangle \cdot \langle -2,\ -\frac{7}{5}\rangle = (10)(-2) + (7)(-\frac{7}{5}) = -\frac{149}{5}$

   (b) $\theta = \cos^{-1}\left(\dfrac{\langle 10,\ 7\rangle \cdot \langle -2,\ -\frac{7}{5}\rangle}{\|\langle 10,\ 7\rangle\|\,\|\langle -2,\ -\frac{7}{5}\rangle\|}\right) = \cos^{-1}\left(\dfrac{-149/5}{\sqrt{149}\,\sqrt{149/25}}\right) = \cos^{-1}(-1) = 180°$

     This result $\{\theta = 180°\}$ indicates that the vectors have the opposite direction.

[9] $\langle 4,\ -1\rangle \cdot \langle 2,\ 8\rangle = 8 - 8 = 0 \;\Rightarrow\;$ vectors are orthogonal since their dot product is zero.

[11] $(-4\mathbf{j}) \cdot (-7\mathbf{i}) = 0 + 0 = 0 \;\Rightarrow\;$ vectors are orthogonal.

13 We first find the angle between the two vectors.

If this angle is 0 or π, then the vectors are parallel.

$$\cos\theta = \frac{\mathbf{a}\cdot\mathbf{b}}{\|\mathbf{a}\|\|\mathbf{b}\|} = \frac{(3)(-\frac{12}{7}) + (-5)(\frac{20}{7})}{\sqrt{9+25}\,\sqrt{\frac{144}{49} + \frac{400}{49}}} = \frac{-\frac{136}{7}}{\sqrt{\frac{18,496}{49}}} = \frac{-\frac{136}{7}}{\frac{136}{7}} = -1 \;\Rightarrow$$

$\theta = \cos^{-1}(-1) = \pi.$ $\mathbf{b} = m\mathbf{a} \;\Rightarrow\; -\frac{12}{7}\mathbf{i} + \frac{20}{7}\mathbf{j} = 3m\mathbf{i} - 5m\mathbf{j} \;\Rightarrow$

$3m = -\frac{12}{7}$ and $-5m = \frac{20}{7} \;\Rightarrow\; m = -\frac{4}{7} < 0 \;\Rightarrow\;$ **a** and **b** have the opposite direction.

15 $\cos\theta = \dfrac{\mathbf{a}\cdot\mathbf{b}}{\|\mathbf{a}\|\|\mathbf{b}\|} = \dfrac{(\frac{2}{3})(8) + (\frac{1}{2})(6)}{\sqrt{\frac{4}{9} + \frac{1}{4}}\,\sqrt{64+36}} = \dfrac{\frac{25}{3}}{\sqrt{\frac{25}{36}}\cdot 10} = 1 \;\Rightarrow\; \theta = \cos^{-1}1 = 0.$

$\mathbf{b} = m\mathbf{a} \;\Rightarrow\; 8\mathbf{i} + 6\mathbf{j} = \frac{2}{3}m\mathbf{i} + \frac{1}{2}m\mathbf{j} \;\Rightarrow\; 8 = \frac{2}{3}m$ and $6 = \frac{1}{2}m \;\Rightarrow\; m = 12 > 0 \;\Rightarrow$

a and **b** have the same direction.

17 We need to have the dot product of the two vectors equal 0.

$$(3\mathbf{i} - 2\mathbf{j})\cdot(4\mathbf{i} + 5m\mathbf{j}) = 0 \;\Rightarrow\; 12 - 10m = 0 \;\Rightarrow\; m = \tfrac{6}{5}.$$

19 $(9\mathbf{i} - 16m\mathbf{j})\cdot(\mathbf{i} + 4m\mathbf{j}) = 0 \;\Rightarrow\; 9 - 64m^2 = 0 \;\Rightarrow\; m^2 = \frac{9}{64} \;\Rightarrow\; m = \pm\frac{3}{8}.$

21 (a) $\mathbf{a}\cdot(\mathbf{b}+\mathbf{c}) = \langle 2, -3\rangle\cdot(\langle 3, 4\rangle + \langle -1, 5\rangle) = \langle 2, -3\rangle\cdot\langle 2, 9\rangle = 4 - 27 = -23$

(b) $\mathbf{a}\cdot\mathbf{b} + \mathbf{a}\cdot\mathbf{c} = \langle 2, -3\rangle\cdot\langle 3, 4\rangle + \langle 2, -3\rangle\cdot\langle -1, 5\rangle = (6-12) + (-2-15) = -23$

23 $(2\mathbf{a}+\mathbf{b})\cdot(3\mathbf{c}) = (2\langle 2, -3\rangle + \langle 3, 4\rangle)\cdot(3\langle -1, 5\rangle)$

$= (\langle 4, -6\rangle + \langle 3, 4\rangle)\cdot\langle -3, 15\rangle$

$= \langle 7, -2\rangle\cdot\langle -3, 15\rangle = -21 - 30 = -51$

25 $\operatorname{comp}_{\mathbf{c}}\mathbf{b} = \dfrac{\mathbf{b}\cdot\mathbf{c}}{\|\mathbf{c}\|} = \dfrac{\langle 3, 4\rangle\cdot\langle -1, 5\rangle}{\|\langle -1, 5\rangle\|} = \dfrac{17}{\sqrt{26}} \approx 3.33$

27 $\operatorname{comp}_{\mathbf{b}}(\mathbf{a}+\mathbf{c}) = \dfrac{(\mathbf{a}+\mathbf{c})\cdot\mathbf{b}}{\|\mathbf{b}\|} = \dfrac{(\langle 2, -3\rangle + \langle -1, 5\rangle)\cdot\langle 3, 4\rangle}{\|\langle 3, 4\rangle\|} = \dfrac{\langle 1, 2\rangle\cdot\langle 3, 4\rangle}{5} = \dfrac{11}{5} = 2.2$

29 $\mathbf{c}\cdot\overrightarrow{PQ} = \langle 3, 4\rangle\cdot\langle 5, -2\rangle = 15 - 8 = 7.$

31 We want a vector with initial point at the origin and terminal point located so that this vector has the same magnitude and direction as $\overrightarrow{PQ}$. Following the hint in the text, $\mathbf{b} = \overrightarrow{PQ} \;\Rightarrow\; \langle b_1, b_2\rangle = \langle 4-2, 3-(-1)\rangle \;\Rightarrow\; \langle b_1, b_2\rangle = \langle 2, 4\rangle.$

$$\mathbf{c}\cdot\mathbf{b} = \langle 6, 4\rangle\cdot\langle 2, 4\rangle = 12 + 16 = 28.$$

33 The force is described by the vector $\langle 0, 4\rangle$.

The work done is $\langle 0, 4\rangle\cdot\langle 8, 3\rangle = 0 + 12 = 12.$

35 $\mathbf{a}\cdot\mathbf{a} = \langle a_1, a_2\rangle\cdot\langle a_1, a_2\rangle = a_1^2 + a_2^2 = (\sqrt{a_1^2 + a_2^2})^2 = \|\mathbf{a}\|^2$

37 $(m\mathbf{a})\cdot\mathbf{b} = (m\langle a_1, a_2\rangle)\cdot\langle b_1, b_2\rangle$

$= \langle ma_1, ma_2\rangle\cdot\langle b_1, b_2\rangle$

$= ma_1b_1 + ma_2b_2$

$= m(a_1b_1 + a_2b_2) = m(\mathbf{a}\cdot\mathbf{b})$

$\boxed{39}$ $\mathbf{0} \cdot \mathbf{a} = \langle 0, 0 \rangle \cdot \langle a_1, a_2 \rangle = 0(a_1) + 0(a_2) = 0 + 0 = 0$

$\boxed{41}$ Using the horizontal and vertical components of a vector from Section 7.3,

we have the force vector as $\langle 20 \cos 30°, 20 \sin 30° \rangle = \langle 10\sqrt{3}, 10 \rangle$.

The distance (direction vector) can be described by the vector $\langle 100, 0 \rangle$.

The work done is $\langle 10\sqrt{3}, 10 \rangle \cdot \langle 100, 0 \rangle = 1000\sqrt{3} \approx 1732$ ft-lb.

$\boxed{43}$ (a) The horizontal component has magnitude 93×10^6 and

the vertical component has magnitude 0.432×10^6.

Thus, $\mathbf{v} = (93 \times 10^6)\mathbf{i} + (0.432 \times 10^6)\mathbf{j}$ and $\mathbf{w} = (93 \times 10^6)\mathbf{i} - (0.432 \times 10^6)\mathbf{j}$.

(b) $\cos\theta = \dfrac{\mathbf{v} \cdot \mathbf{w}}{\|\mathbf{v}\| \|\mathbf{w}\|} = \dfrac{(93 \times 10^6)^2 - (0.432 \times 10^6)^2}{\sqrt{(93 \times 10^6)^2 + (0.432 \times 10^6)^2}\ \sqrt{(93 \times 10^6)^2 + (0.432 \times 10^6)^2}} \approx$

$0.99995685 \ \Rightarrow \ \theta \approx 0.53°$.

$\boxed{45}$ $\mathbf{R} = 2(\mathbf{N} \cdot \mathbf{L})\mathbf{N} - \mathbf{L} = 2(\langle 0, 1 \rangle \cdot \langle -\frac{4}{5}, \frac{3}{5} \rangle) \langle 0, 1 \rangle - \langle -\frac{4}{5}, \frac{3}{5} \rangle = 2(\frac{3}{5})\langle 0, 1 \rangle - \langle -\frac{4}{5}, \frac{3}{5} \rangle =$

$\langle 0, \frac{6}{5} \rangle - \langle -\frac{4}{5}, \frac{3}{5} \rangle = \langle \frac{4}{5}, \frac{3}{5} \rangle$

$\boxed{47}$ Let horizontal ground be represented by $\mathbf{b} = \langle 1, 0 \rangle$ (it could be any $\langle a, 0 \rangle$).

$\text{comp}_{\mathbf{b}}\,\mathbf{a} = \dfrac{\mathbf{a} \cdot \mathbf{b}}{\|\mathbf{b}\|} = \dfrac{\langle 2.6, 4.5 \rangle \cdot \langle 1, 0 \rangle}{\|\langle 1, 0 \rangle\|} = \dfrac{2.6}{1} = 2.6 \ \Rightarrow \ |\,\text{comp}_{\mathbf{b}}\,\mathbf{a}\,| = 2.6$

$\boxed{49}$ Let the direction of the ground be represented by $\mathbf{b} = \langle \cos\theta, \sin\theta \rangle = \langle \cos 12°, \sin 12° \rangle$.

$\text{comp}_{\mathbf{b}}\,\mathbf{a} = \dfrac{\mathbf{a} \cdot \mathbf{b}}{\|\mathbf{b}\|} = \dfrac{\langle 25.7, -3.9 \rangle \cdot \langle \cos 12°, \sin 12° \rangle}{\|\langle \cos 12°, \sin 12° \rangle\|} \approx \dfrac{24.33}{1} = 24.33 \ \Rightarrow$

$|\,\text{comp}_{\mathbf{b}}\,\mathbf{a}\,| = 24.33$

$\boxed{51}$ From the "Theorem on the Dot Product," we know that $\mathbf{F} \cdot \mathbf{v} = \|\mathbf{F}\| \|\mathbf{v}\| \cos\theta$, so

$P = \frac{1}{550}(\mathbf{F} \cdot \mathbf{v}) = \frac{1}{550}\|\mathbf{F}\| \|\mathbf{v}\| \cos\theta = \frac{1}{550}(2200)(8) \cos 30° = 16\sqrt{3} \approx 27.7$ horsepower.

7.5 Exercises

$\boxed{1}$ $|3 - 4i| = \sqrt{3^2 + (-4)^2} = \sqrt{9 + 16} = \sqrt{25} = 5$

$\boxed{3}$ $|-6 - 7i| = \sqrt{(-6)^2 + (-7)^2} = \sqrt{36 + 49} = \sqrt{85}$

$\boxed{5}$ $|8i| = |0 + 8i| = \sqrt{0^2 + 8^2} = \sqrt{64} = 8$

$\boxed{7}$ From Section 1.5, $i^m = i, -1, -i,$ or 1. Since all of these are 1 unit from the origin,

$|i^m| = 1$ for any integer m. As an alternate solution,

$|i^{500}| = |(i^4)^{125}| = |(1)^{125}| = |1| = |1 + 0i| = \sqrt{1^2 + 0^2} = \sqrt{1} = 1.$

$\boxed{9}$ $|0| = |0 + 0i| = \sqrt{0^2 + 0^2} = \sqrt{0} = 0$

11 $4 + 2i$ 13 $3 - 5i$ 15 $-(3 - 6i) = -3 + 6i$

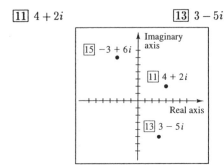   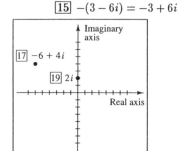

Figure for Exercises 11, 13, 15   *Figure for Exercises 17, 19*

17 $2i(2 + 3i) = 4i + 6i^2 = 4i - 6 = -6 + 4i$

19 $(1 + i)^2 = 1 + 2(1)(i) + i^2 = 1 + 2i - 1 = 2i$

Note: For each of the following exercises, we need to find r and θ.

If $z = a + bi$, then $r = \sqrt{a^2 + b^2}$. To find θ, we will use the fact that

$\tan\theta = \frac{b}{a}$ and our knowledge of what quadrant the terminal side of θ is in.

21 $z = 1 - i \Rightarrow r = \sqrt{1 + (-1)^2} = \sqrt{2}$. $\tan\theta = \frac{-1}{1} = -1$ and θ in QIV $\Rightarrow \theta = \frac{7\pi}{4}$.

Thus, $z = 1 - i = \sqrt{2}\left(\cos\frac{7\pi}{4} + i\sin\frac{7\pi}{4}\right)$, or simply $\sqrt{2}\operatorname{cis}\frac{7\pi}{4}$.

23 $z = -4\sqrt{3} + 4i \Rightarrow r = \sqrt{(-4\sqrt{3})^2 + 4^2} = \sqrt{64} = 8$.

$\tan\theta = \frac{4}{-4\sqrt{3}} = -\frac{1}{\sqrt{3}}$ and θ in QII $\Rightarrow \theta = \frac{5\pi}{6}$. $z = 8\operatorname{cis}\frac{5\pi}{6}$.

25 $z = 2\sqrt{3} + 2i \Rightarrow r = \sqrt{(2\sqrt{3})^2 + 2^2} = \sqrt{16} = 4$.

$\tan\theta = \frac{2}{2\sqrt{3}} = \frac{1}{\sqrt{3}}$ and θ in QI $\Rightarrow \theta = \frac{\pi}{6}$. $z = 4\operatorname{cis}\frac{\pi}{6}$.

27 $z = -4 - 4i \Rightarrow r = \sqrt{(-4)^2 + (-4)^2} = \sqrt{32} = 4\sqrt{2}$.

$\tan\theta = \frac{-4}{-4} = 1$ and θ in QIII $\Rightarrow \theta = \frac{5\pi}{4}$. $z = 4\sqrt{2}\operatorname{cis}\frac{5\pi}{4}$.

29 $z = -20i \Rightarrow r = 20$. θ on the negative y-axis $\Rightarrow \theta = \frac{3\pi}{2}$. $z = 20\operatorname{cis}\frac{3\pi}{2}$.

31 $z = 12 \Rightarrow r = 12$. θ on the positive x-axis $\Rightarrow \theta = 0$. $z = 12\operatorname{cis}0$.

33 $z = -7 \Rightarrow r = 7$. θ on the negative x-axis $\Rightarrow \theta = \pi$. $z = 7\operatorname{cis}\pi$.

35 $z = 6i \Rightarrow r = 6$. θ on the positive y-axis $\Rightarrow \theta = \frac{\pi}{2}$. $z = 6\operatorname{cis}\frac{\pi}{2}$.

37 $z = -5 - 5\sqrt{3}\,i \Rightarrow r = \sqrt{(-5)^2 + (-5\sqrt{3})^2} = \sqrt{100} = 10$.

$\tan\theta = \frac{-5\sqrt{3}}{-5} = \sqrt{3}$ and θ in QIII $\Rightarrow \theta = \frac{4\pi}{3}$. $z = 10\operatorname{cis}\frac{4\pi}{3}$.

39 $z = 2 + i \Rightarrow r = \sqrt{2^2 + 1^2} = \sqrt{5}$. $\tan\theta = \frac{1}{2}$ and θ in QI $\Rightarrow \theta = \tan^{-1}\frac{1}{2}$.

$z = \sqrt{5}\operatorname{cis}\left(\tan^{-1}\frac{1}{2}\right)$.

41 $z = -3 + i \Rightarrow r = \sqrt{(-3)^2 + 1^2} = \sqrt{10}$.

$\tan\theta = \frac{1}{-3}$ and θ in QII $\Rightarrow \theta = \tan^{-1}\left(-\frac{1}{3}\right) + \pi$. We must add π to $\tan^{-1}\left(-\frac{1}{3}\right)$

because $-\frac{\pi}{2} < \tan^{-1}\left(-\frac{1}{3}\right) < 0$ and we want θ to be in the interval $(\frac{\pi}{2}, \pi)$.

$$z = \sqrt{10}\,\mathrm{cis}\left[\tan^{-1}\left(-\frac{1}{3}\right) + \pi\right].$$

43 $z = -5 - 3i \Rightarrow r = \sqrt{(-5)^2 + (-3)^2} = \sqrt{34}$. $\tan\theta = \frac{-3}{-5} = \frac{3}{5}$ and

θ in QIII $\Rightarrow \theta = \tan^{-1}\frac{3}{5} + \pi$. We must add π to $\tan^{-1}\frac{3}{5}$ because $0 < \tan^{-1}\frac{3}{5} < \frac{\pi}{2}$

and we want θ to be in the interval $(\pi, \frac{3\pi}{2})$. $z = \sqrt{34}\,\mathrm{cis}\left(\tan^{-1}\frac{3}{5} + \pi\right)$.

45 $z = 4 - 3i \Rightarrow r = \sqrt{4^2 + (-3)^2} = \sqrt{25} = 5$.

$\tan\theta = \frac{-3}{4}$ and θ in QIV $\Rightarrow \theta = \tan^{-1}\left(-\frac{3}{4}\right) + 2\pi$. We must add 2π to $\tan^{-1}\left(-\frac{3}{4}\right)$

because $-\frac{\pi}{2} < \tan^{-1}\left(-\frac{3}{4}\right) < 0$ and we want θ to be in the interval $(\frac{3\pi}{2}, 2\pi)$.

$$z = 5\,\mathrm{cis}\left[\tan^{-1}\left(-\frac{3}{4}\right) + 2\pi\right].$$

47 $4\left(\cos\frac{\pi}{4} + i\sin\frac{\pi}{4}\right) = 4\left(\frac{\sqrt{2}}{2} + \frac{\sqrt{2}}{2}i\right) = 2\sqrt{2} + 2\sqrt{2}\,i$

49 $6\left(\cos\frac{2\pi}{3} + i\sin\frac{2\pi}{3}\right) = 6\left(-\frac{1}{2} + \frac{\sqrt{3}}{2}i\right) = -3 + 3\sqrt{3}\,i$

51 $5\left(\cos\pi + i\sin\pi\right) = 5\left(-1 + 0i\right) = -5$

53 For any given angle θ, where $\theta = \tan^{-1}\frac{y}{x}$, we have $r = \sqrt{x^2 + y^2}$. In this case,

$\theta = \tan^{-1}\frac{3}{5}$, so $r = \sqrt{3^2 + 5^2} = \sqrt{34}$. You may want to draw a figure to represent x,

y, and r, as we did in previous chapters. Also, note that $\cos\theta = \frac{x}{r}$ and $\sin\theta = \frac{y}{r}$.

$\sqrt{34}\,\mathrm{cis}\left(\tan^{-1}\frac{3}{5}\right) = \sqrt{34}\left[\cos\left(\tan^{-1}\frac{3}{5}\right) + i\sin\left(\tan^{-1}\frac{3}{5}\right)\right] = \sqrt{34}\left(\frac{5}{\sqrt{34}} + \frac{3}{\sqrt{34}}i\right) = 5 + 3i$

55 $\sqrt{5}\,\mathrm{cis}\left[\tan^{-1}\left(-\frac{1}{2}\right)\right] = \sqrt{5}\left\{\cos\left[\tan^{-1}\left(-\frac{1}{2}\right)\right] + i\sin\left[\tan^{-1}\left(-\frac{1}{2}\right)\right]\right\} =$

$$\sqrt{5}\left(\frac{2}{\sqrt{5}} - \frac{1}{\sqrt{5}}i\right) = 2 - i$$

Note: For Exercises 57–64, the trigonometric forms for z_1 and z_2 are listed and then used

in the theorem in this section. The trigonometric forms can be found as in

Exercises 21–46.

57 $z_1 = \sqrt{2}\,\mathrm{cis}\frac{3\pi}{4}$ and $z_2 = \sqrt{2}\,\mathrm{cis}\frac{\pi}{4}$. $z_1 z_2 = \sqrt{2}\cdot\sqrt{2}\,\mathrm{cis}\left(\frac{3\pi}{4} + \frac{\pi}{4}\right) = 2\,\mathrm{cis}\,\pi = -2 + 0i$.

$$\frac{z_1}{z_2} = \frac{\sqrt{2}}{\sqrt{2}}\,\mathrm{cis}\left(\frac{3\pi}{4} - \frac{\pi}{4}\right) = 1\,\mathrm{cis}\frac{\pi}{2} = 0 + i$$

59 $z_1 = 4\,\mathrm{cis}\frac{4\pi}{3}$ and $z_2 = 5\,\mathrm{cis}\frac{\pi}{2}$. $z_1 z_2 = 4\cdot 5\,\mathrm{cis}\left(\frac{4\pi}{3} + \frac{\pi}{2}\right) = 20\,\mathrm{cis}\frac{11\pi}{6} = 10\sqrt{3} - 10i$.

$$\frac{z_1}{z_2} = \frac{4}{5}\,\mathrm{cis}\left(\frac{4\pi}{3} - \frac{\pi}{2}\right) = \frac{4}{5}\,\mathrm{cis}\frac{5\pi}{6} = -\frac{2}{5}\sqrt{3} + \frac{2}{5}i.$$

61 $z_1 = 10 \operatorname{cis} \pi$ and $z_2 = 4 \operatorname{cis} \pi$. $z_1 z_2 = 10 \cdot 4 \operatorname{cis}(\pi + \pi) = 40 \operatorname{cis} 2\pi = 40 + 0i$.

$$\frac{z_1}{z_2} = \frac{10}{4} \operatorname{cis}(\pi - \pi) = \frac{5}{2} \operatorname{cis} 0 = \frac{5}{2} + 0i.$$

63 $z_1 = 4 \operatorname{cis} 0$ and $z_2 = \sqrt{5} \operatorname{cis}\left[\tan^{-1}\left(-\frac{1}{2}\right)\right]$. Let $\theta = \tan^{-1}\left(-\frac{1}{2}\right)$.

Thus $\cos\theta = \dfrac{2}{\sqrt{5}}$ and $\sin\theta = -\dfrac{1}{\sqrt{5}}$.

$$z_1 z_2 = 4 \cdot \sqrt{5} \operatorname{cis}(0 + \theta) = 4\sqrt{5}(\cos\theta + i\sin\theta) = 4\sqrt{5}\left(\frac{2}{\sqrt{5}} + \frac{-1}{\sqrt{5}}i\right) = 8 - 4i.$$

$$\frac{z_1}{z_2} = \frac{4}{\sqrt{5}} \operatorname{cis}(0 - \theta) = \frac{4}{\sqrt{5}}[\cos(-\theta) + i\sin(-\theta)]$$

$$= \frac{4}{\sqrt{5}}(\cos\theta - i\sin\theta) = \frac{4}{\sqrt{5}}\left(\frac{2}{\sqrt{5}} + \frac{1}{\sqrt{5}}i\right) = \frac{8}{5} + \frac{4}{5}i.$$

65 Let $z_1 = r_1 \operatorname{cis}\theta_1$ and $z_2 = r_2 \operatorname{cis}\theta_2$.

$$\frac{z_1}{z_2} = \frac{r_1 \operatorname{cis}\theta_1}{r_2 \operatorname{cis}\theta_2} = \frac{r_1(\cos\theta_1 + i\sin\theta_1)(\cos\theta_2 - i\sin\theta_2)}{r_2(\cos\theta_2 + i\sin\theta_2)(\cos\theta_2 - i\sin\theta_2)}$$

{ multiplying by the conjugate of the denominator }

$$= \frac{r_1[(\cos\theta_1 \cos\theta_2 + \sin\theta_1 \sin\theta_2) + i(\sin\theta_1 \cos\theta_2 - \sin\theta_2 \cos\theta_1)]}{r_2[(\cos^2\theta_2 + \sin^2\theta_2) + i(\sin\theta_2 \cos\theta_2 - \cos\theta_2 \sin\theta_2)]}$$

$$= \frac{r_1[\cos(\theta_1 - \theta_2) + i\sin(\theta_1 - \theta_2)]}{r_2(1 + 0i)} = \frac{r_1}{r_2}\operatorname{cis}(\theta_1 - \theta_2).$$

67 The unknown quantity is V: $I = V/Z \Rightarrow$

$V = IZ = (10 \operatorname{cis} 35°)(3 \operatorname{cis} 20°) = (10 \times 3) \operatorname{cis}(35° + 20°) = 30 \operatorname{cis} 55° \approx 17.21 + 24.57i$.

69 The unknown quantity is Z: $I = V/Z \Rightarrow$

$$Z = \frac{V}{I} = \frac{115 \operatorname{cis} 45°}{8 \operatorname{cis} 5°} = (115 \div 8) \operatorname{cis}(45° - 5°) = 14.375 \operatorname{cis} 40° \approx 11.01 + 9.24i$$

71 $Z = 14 - 13i \Rightarrow |Z| = \sqrt{14^2 + (-13)^2} = \sqrt{365} \approx 19.1$ ohms

73 $I = \dfrac{V}{Z} \Rightarrow V = IZ = (4 \operatorname{cis} 90°)[18 \operatorname{cis}(-78°)] = 72 \operatorname{cis} 12° \approx 70.43 + 14.97i$.

7.6 Exercises

Note: In this section, it is assumed that the reader can transform complex numbers to their trigonometric from. If this is not true, see Exercises 7.5.

1 $(3 + 3i)^5 = \left(3\sqrt{2} \operatorname{cis}\frac{\pi}{4}\right)^5 = (3\sqrt{2})^5 \operatorname{cis}\left(5 \cdot \frac{\pi}{4}\right)$

$$= (3\sqrt{2})^5 \operatorname{cis}\frac{5\pi}{4} = 972\sqrt{2}\left(-\frac{\sqrt{2}}{2} - \frac{\sqrt{2}}{2}i\right) = -972 - 972i$$

3 $(1-i)^{10} = (\sqrt{2}\operatorname{cis}\frac{7\pi}{4})^{10} = (\sqrt{2})^{10}\operatorname{cis}(10\cdot\frac{7\pi}{4}) = (\sqrt{2})^{10}\operatorname{cis}\frac{35\pi}{2} =$

$$2^5\operatorname{cis}\left(16\pi+\tfrac{3\pi}{2}\right) = 32\operatorname{cis}\tfrac{3\pi}{2} = 32(0-i) = -32i$$

5 $(1-\sqrt{3}\,i)^3 = (2\operatorname{cis}\frac{5\pi}{3})^3 = 2^3\operatorname{cis}5\pi = 8\operatorname{cis}\pi = 8(-1+0i) = -8$

7 $\left(-\dfrac{\sqrt{2}}{2}+\dfrac{\sqrt{2}}{2}i\right)^{15} = (1\operatorname{cis}\frac{3\pi}{4})^{15} = 1^{15}\operatorname{cis}\frac{45\pi}{4} = \operatorname{cis}\frac{5\pi}{4} = -\dfrac{\sqrt{2}}{2}-\dfrac{\sqrt{2}}{2}i$

9 $\left(-\dfrac{\sqrt{3}}{2}-\dfrac{1}{2}i\right)^{20} = (1\operatorname{cis}\frac{7\pi}{6})^{20} = 1^{20}\operatorname{cis}\frac{70\pi}{3} = \operatorname{cis}\frac{4\pi}{3} = -\dfrac{1}{2}-\dfrac{\sqrt{3}}{2}i$

11 $(\sqrt{3}+i)^7 = (2\operatorname{cis}\frac{\pi}{6})^7 = 2^7\operatorname{cis}\frac{7\pi}{6} = 128\left(-\dfrac{\sqrt{3}}{2}-\dfrac{1}{2}i\right) = -64\sqrt{3}-64i$

13 $1+\sqrt{3}\,i = 2\operatorname{cis}60°.$ $w_k = \sqrt{2}\operatorname{cis}\left(\dfrac{60°+360°k}{2}\right)$ for $k=0,\,1$.

$$w_0 = \sqrt{2}\operatorname{cis}30° = \sqrt{2}\left(\dfrac{\sqrt{3}}{2}+\dfrac{1}{2}i\right) = \dfrac{\sqrt{6}}{2}+\dfrac{\sqrt{2}}{2}i.$$

$$w_1 = \sqrt{2}\operatorname{cis}210° = \sqrt{2}\left(-\dfrac{\sqrt{3}}{2}-\dfrac{1}{2}i\right) = -\dfrac{\sqrt{6}}{2}-\dfrac{\sqrt{2}}{2}i.$$

15 $-1-\sqrt{3}\,i = 2\operatorname{cis}240°.$ $w_k = \sqrt[4]{2}\operatorname{cis}\left(\dfrac{240°+360°k}{4}\right)$ for $k=0,\,1,\,2,\,3$.

$$w_0 = \sqrt[4]{2}\operatorname{cis}60° = \sqrt[4]{2}\left(\dfrac{1}{2}+\dfrac{\sqrt{3}}{2}i\right) = \dfrac{\sqrt[4]{2}}{2}+\dfrac{\sqrt[4]{18}}{2}i.$$

$$\{\text{since } \sqrt[4]{2}\cdot\sqrt{3} = \sqrt[4]{2}\cdot\sqrt[4]{9} = \sqrt[4]{18}\,\}$$

$$w_1 = \sqrt[4]{2}\operatorname{cis}150° = \sqrt[4]{2}\left(-\dfrac{\sqrt{3}}{2}+\dfrac{1}{2}i\right) = -\dfrac{\sqrt[4]{18}}{2}+\dfrac{\sqrt[4]{2}}{2}i.$$

$$w_2 = \sqrt[4]{2}\operatorname{cis}240° = \sqrt[4]{2}\left(-\dfrac{1}{2}-\dfrac{\sqrt{3}}{2}i\right) = -\dfrac{\sqrt[4]{2}}{2}-\dfrac{\sqrt[4]{18}}{2}i.$$

$$w_3 = \sqrt[4]{2}\operatorname{cis}330° = \sqrt[4]{2}\left(\dfrac{\sqrt{3}}{2}-\dfrac{1}{2}i\right) = \dfrac{\sqrt[4]{18}}{2}-\dfrac{\sqrt[4]{2}}{2}i.$$

17 $-27i = 27\operatorname{cis}270°.$ $w_k = \sqrt[3]{27}\operatorname{cis}\left(\dfrac{270°+360°k}{3}\right)$ for $k=0,\,1,\,2$.

$$w_0 = 3\operatorname{cis}90° = 3(0+i) = 3i.$$

$$w_1 = 3\operatorname{cis}210° = 3\left(-\dfrac{\sqrt{3}}{2}-\dfrac{1}{2}i\right) = -\dfrac{3\sqrt{3}}{2}-\dfrac{3}{2}i.$$

$$w_2 = 3\operatorname{cis}330° = 3\left(\dfrac{\sqrt{3}}{2}-\dfrac{1}{2}i\right) = \dfrac{3\sqrt{3}}{2}-\dfrac{3}{2}i.$$

19 $1 = 1 \operatorname{cis} 0°.$ $w_k = \sqrt[6]{1} \operatorname{cis}\left(\frac{0° + 360°k}{6}\right)$ for $k = 0, 1, 2, 3, 4, 5.$

$w_0 = 1 \operatorname{cis} 0° = 1 + 0i.$ $w_1 = 1 \operatorname{cis} 60° = \frac{1}{2} + \frac{\sqrt{3}}{2}i.$

$w_2 = 1 \operatorname{cis} 120° = -\frac{1}{2} + \frac{\sqrt{3}}{2}i.$ $w_3 = 1 \operatorname{cis} 180° = -1 + 0i.$

$w_4 = 1 \operatorname{cis} 240° = -\frac{1}{2} - \frac{\sqrt{3}}{2}i.$ $w_5 = 1 \operatorname{cis} 300° = \frac{1}{2} - \frac{\sqrt{3}}{2}i.$

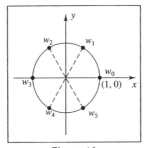   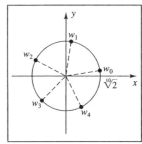

Figure 19   Figure 21

21 $1 + i = \sqrt{2} \operatorname{cis} 45°.$ $w_k = \sqrt{\sqrt[5]{2}} \operatorname{cis}\left(\frac{45° + 360°k}{5}\right)$ for $k = 0, 1, 2, 3, 4.$

$$w_k = \sqrt[10]{2} \operatorname{cis} \theta \text{ with } \theta = 9°, 81°, 153°, 225°, 297°.$$

23 $x^4 - 16 = 0 \Rightarrow x^4 = 16.$ The problem is now to find the 4 fourth roots of 16.

$16 = 16 + 0i = 16 \operatorname{cis} 0°.$ $w_k = \sqrt[4]{16} \operatorname{cis}\left(\frac{0° + 360°k}{4}\right)$ for $k = 0, 1, 2, 3.$

$w_0 = 2 \operatorname{cis} 0° = 2(1 + 0i) = 2.$ $w_1 = 2 \operatorname{cis} 90° = 2(0 + i) = 2i.$

$w_2 = 2 \operatorname{cis} 180° = 2(-1 + 0i) = -2.$ $w_3 = 2 \operatorname{cis} 270° = 2(0 - i) = -2i.$

25 $x^6 + 64 = 0 \Rightarrow x^6 = -64.$ The problem is now to find the 6 sixth roots of $-64.$

$-64 = -64 + 0i = 64 \operatorname{cis} 180°.$ $w_k = \sqrt[6]{64} \operatorname{cis}\left(\frac{180° + 360°k}{6}\right)$ for $k = 0, 1, \ldots, 5.$

$w_0 = 2 \operatorname{cis} 30° = 2\left(\frac{\sqrt{3}}{2} + \frac{1}{2}i\right) = \sqrt{3} + i.$

$w_1 = 2 \operatorname{cis} 90° = 2(0 + i) = 2i.$

$w_2 = 2 \operatorname{cis} 150° = 2\left(-\frac{\sqrt{3}}{2} + \frac{1}{2}i\right) = -\sqrt{3} + i.$

$w_3 = 2 \operatorname{cis} 210° = 2\left(-\frac{\sqrt{3}}{2} - \frac{1}{2}i\right) = -\sqrt{3} - i.$

$w_4 = 2 \operatorname{cis} 270° = 2(0 - i) = -2i.$

$w_5 = 2 \operatorname{cis} 330° = 2\left(\frac{\sqrt{3}}{2} - \frac{1}{2}i\right) = \sqrt{3} - i.$

27 $x^3 + 8i = 0 \Rightarrow x^3 = -8i$. The problem is now to find the 3 cube roots of $-8i$.

$$-8i = 0 - 8i = 8\operatorname{cis} 270°. \quad w_k = \sqrt[3]{8}\operatorname{cis}\left(\frac{270° + 360° k}{3}\right) \text{ for } k = 0, 1, 2.$$

$w_0 = 2\operatorname{cis} 90° = 2(0 + i) = 2i.$

$w_1 = 2\operatorname{cis} 210° = 2\left(-\frac{\sqrt{3}}{2} - \frac{1}{2}i\right) = -\sqrt{3} - i.$

$w_2 = 2\operatorname{cis} 330° = 2\left(\frac{\sqrt{3}}{2} - \frac{1}{2}i\right) = \sqrt{3} - i.$

29 $x^5 - 243 = 0 \Rightarrow x^5 = 243$. The problem is now to find the 5 fifth roots of 243.

$$243 = 243 + 0i = 243\operatorname{cis} 0°. \quad w_k = \sqrt[5]{243}\operatorname{cis}\left(\frac{0° + 360° k}{5}\right) \text{ for } k = 0, 1, 2, 3, 4.$$

$$w_k = 3\operatorname{cis}\theta \text{ with } \theta = 0°, 72°, 144°, 216°, 288°.$$

31 $\left[r(\cos\theta + i\sin\theta)\right]^n = \left[r(e^{i\theta})\right]^n = r^n(e^{i\theta})^n = r^n e^{i(n\theta)} = r^n(\cos n\theta + i\sin n\theta)$

Chapter 7 Review Exercises

1 We are given 2 sides of a triangle and the angle between them—so we use the law of cosines to find the third side.

$$a = \sqrt{b^2 + c^2 - 2bc\,\cos\alpha} = \sqrt{6^2 + 7^2 - 2(6)(7)\cos 60°} = \sqrt{43}.$$

$$\beta = \cos^{-1}\left(\frac{a^2 + c^2 - b^2}{2ac}\right) = \cos^{-1}\left(\frac{43 + 49 - 36}{2\sqrt{43}\,(7)}\right) = \cos^{-1}\left(\frac{4}{\sqrt{43}}\right).$$

$$\gamma = \cos^{-1}\left(\frac{a^2 + b^2 - c^2}{2ab}\right) = \cos^{-1}\left(\frac{43 + 36 - 49}{2\sqrt{43}\,(6)}\right) = \cos^{-1}\left(\frac{5}{2\sqrt{43}}\right).$$

2 $\frac{\sin\alpha}{a} = \frac{\sin\gamma}{c} \Rightarrow \alpha = \sin^{-1}\left(\frac{a\,\sin\gamma}{c}\right) = \sin^{-1}\left(\frac{2\sqrt{3}\cdot\frac{1}{2}}{2}\right) = \sin^{-1}\left(\frac{\sqrt{3}}{2}\right) = 60° \text{ or } 120°.$

There are two triangles possible since in either case $\alpha + \gamma < 180°$.

$\beta = (180° - \gamma) - \alpha = (180° - 30°) - (60° \text{ or } 120°) = 90° \text{ or } 30°.$

$$\frac{b}{\sin\beta} = \frac{c}{\sin\gamma} \Rightarrow b = \frac{c\,\sin\beta}{\sin\gamma} = \frac{2\sin(90° \text{ or } 30°)}{\sin 30°} = 4 \text{ or } 2.$$

3 $\gamma = 180° - \alpha - \beta = 180° - 60° - 45° = 75°.$

$$\frac{a}{\sin\alpha} = \frac{b}{\sin\beta} \Rightarrow a = \frac{b\,\sin\alpha}{\sin\beta} = \frac{100\sin 60°}{\sin 45°} = \frac{100\cdot(\sqrt{3}/2)}{\sqrt{2}/2}\cdot\frac{\sqrt{2}}{\sqrt{2}} = 50\sqrt{6}.$$

$$\frac{c}{\sin\gamma} = \frac{b}{\sin\beta} \Rightarrow c = \frac{b\,\sin\gamma}{\sin\beta} = \frac{100\sin(45° + 30°)}{\sqrt{2}/2} =$$

$$100\sqrt{2}\,(\sin 45° \cos 30° + \cos 45° \sin 30°) = 100\sqrt{2}\left(\frac{\sqrt{2}}{2}\cdot\frac{\sqrt{3}}{2} + \frac{\sqrt{2}}{2}\cdot\frac{1}{2}\right) =$$

$$\tfrac{100}{4}\sqrt{2}\,(\sqrt{6} + \sqrt{2}) = 25(2\sqrt{3} + 2) = 50(1 + \sqrt{3}).$$

4 $\alpha = \cos^{-1}\left(\dfrac{b^2 + c^2 - a^2}{2bc}\right) = \cos^{-1}\left(\dfrac{9 + 16 - 4}{2(3)(4)}\right) = \cos^{-1}\left(\dfrac{7}{8}\right).$

$\beta = \cos^{-1}\left(\dfrac{a^2 + c^2 - b^2}{2ac}\right) = \cos^{-1}\left(\dfrac{4 + 16 - 9}{2(2)(4)}\right) = \cos^{-1}\left(\dfrac{11}{16}\right).$

$\gamma = \cos^{-1}\left(\dfrac{a^2 + b^2 - c^2}{2ab}\right) = \cos^{-1}\left(\dfrac{4 + 9 - 16}{2(2)(3)}\right) = \cos^{-1}\left(-\dfrac{1}{4}\right).$

5 $\alpha = 180° - \beta - \gamma = 180° - 67° - 75° = 38°.$

$\dfrac{a}{\sin\alpha} = \dfrac{b}{\sin\beta} \;\Rightarrow\; a = \dfrac{b\sin\alpha}{\sin\beta} = \dfrac{12\sin 38°}{\sin 67°} \approx 8.0.$

$\dfrac{c}{\sin\gamma} = \dfrac{b}{\sin\beta} \;\Rightarrow\; c = \dfrac{b\sin\gamma}{\sin\beta} = \dfrac{12\sin 75°}{\sin 67°} \approx 12.6,\ \text{or } 13.$

6 $\dfrac{\sin\gamma}{c} = \dfrac{\sin\alpha}{a} \;\Rightarrow\; \gamma = \sin^{-1}\left(\dfrac{c\sin\alpha}{a}\right) = \sin^{-1}\left(\dfrac{125\sin 23°30'}{152}\right) \approx \sin^{-1}(0.3279) \approx$

19°10′ or 160°50′ { rounded to the nearest 10 minutes }. Reject 160°50′ because then

$\alpha + \gamma \ge 180°. \quad \beta = 180° - \alpha - \gamma \approx 180° - 23°30' - 19°10' = 137°20'.$

$\dfrac{b}{\sin\beta} = \dfrac{a}{\sin\alpha} \;\Rightarrow\; b = \dfrac{a\sin\beta}{\sin\alpha} = \dfrac{152\sin 137°20'}{\sin 23°30'} \approx 258.3,\ \text{or } 258.$

7 $b = \sqrt{a^2 + c^2 - 2ac\cos\beta} = \sqrt{(4.6)^2 + (7.3)^2 - 2(4.6)(7.3)\cos 115°} \approx \sqrt{102.8} \approx 10.1.$

$\alpha = \cos^{-1}\left(\dfrac{b^2 + c^2 - a^2}{2bc}\right) \approx \cos^{-1}(0.9116) \approx 24°.$

$\gamma = 180° - \alpha - \beta \approx 180° - 24° - 115° = 41°.$

9   Since we are given two sides and the included angle of a triangle,

we use the formula for the area of a triangle from Section 7.2.

$$\mathcal{A} = \tfrac{1}{2}bc\sin\alpha = \tfrac{1}{2}(20)(30)\sin 75° \approx 289.8,\ \text{or } 290 \text{ square units.}$$

10 Given the three sides of a triangle, we apply Heron's formula to find the area.

$s = \tfrac{1}{2}(a + b + c) = \tfrac{1}{2}(4 + 7 + 10) = 10.5.$

$$\mathcal{A} = \sqrt{s(s-a)(s-b)(s-c)} = \sqrt{(10.5)(6.5)(3.5)(0.5)} \approx 10.9 \text{ square units.}$$

11 (a) $\mathbf{a} = \langle -4, 5\rangle$ and $\mathbf{b} = \langle 2, -8\rangle \;\Rightarrow$

     $\mathbf{a} + \mathbf{b} = \langle -4 + 2,\, 5 + (-8)\rangle = \langle -2, -3\rangle.$

(b) $\mathbf{a} - \mathbf{b} = \langle -4 - 2,\, 5 - (-8)\rangle = \langle -6, 13\rangle.$

(c) $2\mathbf{a} = 2\langle -4, 5\rangle = \langle -8, 10\rangle.$

(d) $-\tfrac{1}{2}\mathbf{b} = -\tfrac{1}{2}\langle 2, -8\rangle = \langle -1, 4\rangle.$

Terminal points are $(-2, -3),\ (-6, 13),\ (-8, 10),\ (-1, 4).$

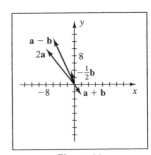

Figure 11

13 S50°E is the same as 320°, or −40°, on the xy-plane.

$$\langle 14\cos(-40°),\, 14\sin(-40°)\rangle = \langle 14\cos 40°,\, -14\sin 40°\rangle \approx \langle 10.72, -9.00\rangle.$$

15 $-2\mathbf{a} = -2(8\mathbf{i} - 6\mathbf{j}) = -16\mathbf{i} + 12\mathbf{j}$

17 $\|\mathbf{r} - \mathbf{a}\| = c \implies \|\langle x, y \rangle - \langle a_1, a_2 \rangle\| = c \implies \|\langle x - a_1, y - a_2 \rangle\| = c \implies$

$\sqrt{(x - a_1)^2 + (y - a_2)^2} = c \implies (x - a_1)^2 + (y - a_2)^2 = c^2$.

This is a circle with center (a_1, a_2) and radius c.

19 $\mathbf{p} = \langle 400 \cos 10°, 400 \sin 10° \rangle \approx \langle 393.92, 69.46 \rangle$.

$\mathbf{r} = \langle 390 \cos 0°, 390 \sin 0° \rangle = \langle 390, 0 \rangle$.

$\mathbf{w} = \mathbf{r} - \mathbf{p} \approx \langle -3.92, -69.46 \rangle$ and $\|\mathbf{w}\| \approx 69.57$, or 70 mi/hr.

$\tan \theta \approx \frac{-69.46}{-3.92} \implies \theta \approx 267°$, or in the direction of 183°.

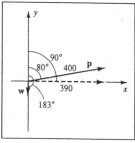

Figure 19

21 (a) $(2\mathbf{a} - 3\mathbf{b}) \cdot \mathbf{a} = [2(6\mathbf{i} - 2\mathbf{j}) - 3(\mathbf{i} + 3\mathbf{j})] \cdot (6\mathbf{i} - 2\mathbf{j})$

$= (9\mathbf{i} - 13\mathbf{j}) \cdot (6\mathbf{i} - 2\mathbf{j}) = 54 + 26 = 80$.

(b) $\mathbf{c} = \mathbf{a} + \mathbf{b} = (6\mathbf{i} - 2\mathbf{j}) + (\mathbf{i} + 3\mathbf{j}) = 7\mathbf{i} + \mathbf{j}$. The angle between $\mathbf{a}$ and $\mathbf{c}$ is

$$\theta = \cos^{-1}\left(\frac{\mathbf{a} \cdot \mathbf{c}}{\|\mathbf{a}\| \|\mathbf{c}\|}\right) = \cos^{-1}\left(\frac{(6\mathbf{i} - 2\mathbf{j}) \cdot (7\mathbf{i} + \mathbf{j})}{\|6\mathbf{i} - 2\mathbf{j}\| \|7\mathbf{i} + \mathbf{j}\|}\right) = \cos^{-1}\left(\frac{40}{\sqrt{40}\sqrt{50}}\right) \approx 26°34'.$$

(c) $\text{comp}_{\mathbf{a}}(\mathbf{a} + \mathbf{b}) = \text{comp}_{\mathbf{a}}\mathbf{c} = \frac{\mathbf{c} \cdot \mathbf{a}}{\|\mathbf{a}\|} = \frac{40}{\sqrt{40}} = \sqrt{40} = 2\sqrt{10} \approx 6.32$.

23 $z = -10 + 10i \implies r = \sqrt{(-10)^2 + 10^2} = \sqrt{200} = 10\sqrt{2}$.

$\tan \theta = \frac{10}{-10} = -1$ and θ in QII $\implies \theta = \frac{3\pi}{4}$. $z = 10\sqrt{2}\operatorname{cis}\frac{3\pi}{4}$.

25 $z = -17 \implies r = 17$. θ on the negative x-axis $\implies \theta = \pi$. $z = 17\operatorname{cis}\pi$.

27 $z = -5\sqrt{3} - 5i \implies r = \sqrt{(-5\sqrt{3})^2 + (-5)^2} = \sqrt{100} = 10$.

$\tan \theta = \frac{-5}{-5\sqrt{3}} = \frac{1}{\sqrt{3}}$ and θ in QIII $\implies \theta = \frac{7\pi}{6}$. $z = 10\operatorname{cis}\frac{7\pi}{6}$.

28 $z = 4 + 5i \implies r = \sqrt{4^2 + 5^2} = \sqrt{41}$. $\tan \theta = \frac{5}{4}$ and θ in QI $\implies \theta = \tan^{-1}\frac{5}{4}$.

$$z = \sqrt{41}\operatorname{cis}\left(\tan^{-1}\tfrac{5}{4}\right).$$

29 $20\left(\cos\frac{11\pi}{6} + i\sin\frac{11\pi}{6}\right) = 20\left(\frac{\sqrt{3}}{2} - \frac{1}{2}i\right) = 10\sqrt{3} - 10i$

31 $z_1 = -3\sqrt{3} - 3i = 6\operatorname{cis}\frac{7\pi}{6}$ and $z_2 = 2\sqrt{3} + 2i = 4\operatorname{cis}\frac{\pi}{6}$.

$z_1 z_2 = 6 \cdot 4\operatorname{cis}\left(\frac{7\pi}{6} + \frac{\pi}{6}\right) = 24\operatorname{cis}\frac{4\pi}{3} = 24\left(-\frac{1}{2} - \frac{\sqrt{3}}{2}i\right) = -12 - 12\sqrt{3}\,i$.

$$\frac{z_1}{z_2} = \frac{6}{4}\operatorname{cis}\left(\frac{7\pi}{6} - \frac{\pi}{6}\right) = \frac{3}{2}\operatorname{cis}\pi = \frac{3}{2}(-1 + 0i) = -\frac{3}{2}.$$

33 $(-\sqrt{3} + i)^9 = (2\operatorname{cis}\frac{5\pi}{6})^9 = 2^9\operatorname{cis}\left(9 \cdot \frac{5\pi}{6}\right) = 2^9\operatorname{cis}\frac{15\pi}{2} = 512\operatorname{cis}\frac{3\pi}{2} = 512(0 - i) = -512i$

35 $(3 - 3i)^5 = (3\sqrt{2} \operatorname{cis} \frac{7\pi}{4})^5 = (3\sqrt{2})^5 \operatorname{cis} \frac{35\pi}{4} = 972\sqrt{2} \operatorname{cis} \frac{3\pi}{4} = 972\sqrt{2}\left(-\frac{\sqrt{2}}{2} + \frac{\sqrt{2}}{2}i\right) =$
$$-972 + 972i$$

37 $-27 + 0i = 27 \operatorname{cis} 180°.$ $w_k = \sqrt[3]{27} \operatorname{cis}\left(\frac{180° + 360°k}{3}\right)$ for $k = 0, 1, 2.$

$$w_0 = 3 \operatorname{cis} 60° = 3\left(\frac{1}{2} + \frac{\sqrt{3}}{2}i\right) = \frac{3}{2} + \frac{3\sqrt{3}}{2}i.$$

$$w_1 = 3 \operatorname{cis} 180° = 3(-1 + 0i) = -3.$$

$$w_2 = 3 \operatorname{cis} 300° = 3\left(\frac{1}{2} - \frac{\sqrt{3}}{2}i\right) = \frac{3}{2} - \frac{3\sqrt{3}}{2}i.$$

39 $x^5 - 32 = 0 \Rightarrow x^5 = 32.$ The problem is now to find the 5 fifth roots of 32.

$$32 = 32 + 0i = 32 \operatorname{cis} 0°. \quad w_k = \sqrt[5]{32} \operatorname{cis}\left(\frac{0° + 360°k}{5}\right) \text{ for } k = 0, 1, 2, 3, 4.$$

$$w_k = 2 \operatorname{cis} \theta \text{ with } \theta = 0°, 72°, 144°, 216°, 288°.$$

41 Let a be the Earth–Venus distance, b be the Earth–sun distance, and c be the Venus–sun distance. Then, by the law of cosines (with a, b, and c in millions),
$$a^2 = b^2 + c^2 - 2bc \cos \alpha = 93^2 + 67^2 - 2(93)(67) \cos 34° \approx 2807 \Rightarrow$$
$a \approx 53$—that is, 53,000,000 miles.

43 (a) $\angle LAS = 180° - 47.2° - 66.4° = 66.4°.$ $\dfrac{\overline{AL}}{\sin 47.2°} = \dfrac{41}{\sin 66.4°} \Rightarrow$

$$\overline{AL} = \frac{41 \sin 47.2°}{\sin 66.4°} \approx 32.83, \text{ or } 33 \text{ miles.} \quad \overline{AS} = \overline{LS} = 41 \text{ since } \triangle LAS \text{ is isosceles.}$$

(b) Let $\overline{AP}$ be perpendicular to $\overline{LS}.$ $\sin 47.2° = \dfrac{\overline{AP}}{\overline{AS}} \Rightarrow \quad \overline{AP} \approx 30.08, \text{ or } 30 \text{ miles.}$

45 If d denotes the distance each girl walks before losing contact with each other, then $d = 5t$, where t is in hours. Using the law of cosines,
$$10^2 = d^2 + d^2 - 2(d)(d) \cos 105° \Rightarrow 100 = 2d^2(1 - \cos 105°) \Rightarrow d \approx 6.30 \Rightarrow$$
$$t = d/5 \approx 1.26 \text{ hours, or 1 hour and 16 minutes.}$$

47 (a) Let d denote the length of the rescue tunnel. Using the law of cosines,
$$d^2 = 45^2 + 50^2 - 2(45)(50) \cos 78° \Rightarrow d \approx 59.91 \text{ ft. Now using the law of sines,}$$
$$\frac{\sin \theta}{45} = \frac{\sin 78°}{d} \Rightarrow \theta = \sin^{-1}\left(\frac{45 \sin 78°}{d}\right) \approx 47.28°, \text{ or } 47°.$$

(b) If x denotes the number of hours needed, then
$$d \text{ ft} = (3 \text{ ft/hr})(x \text{ hr}) \Rightarrow x = \tfrac{1}{3}d = \tfrac{1}{3}(59.91) \approx 20 \text{ hr.}$$

1 (a) $\dfrac{a}{\sin \alpha} = \dfrac{c}{\sin \gamma} \Rightarrow \dfrac{a}{c} = \dfrac{\sin \alpha}{\sin \gamma}$ and $\dfrac{b}{\sin \beta} = \dfrac{c}{\sin \gamma} \Rightarrow \dfrac{b}{c} = \dfrac{\sin \beta}{\sin \gamma}$.

Adding the equations yields $\dfrac{a}{c} + \dfrac{b}{c} = \dfrac{\sin \alpha}{\sin \gamma} + \dfrac{\sin \beta}{\sin \gamma} \Rightarrow \dfrac{a+b}{c} = \dfrac{\sin \alpha + \sin \beta}{\sin \gamma}$.

(b) $\dfrac{a+b}{c} = \dfrac{\sin \alpha + \sin \beta}{\sin \gamma} \Rightarrow \dfrac{a+b}{c} = \dfrac{[\text{S1}]\ 2 \sin \frac{1}{2}(\alpha + \beta)\, \cos \frac{1}{2}(\alpha - \beta)}{2 \sin \frac{1}{2}\gamma\, \cos \frac{1}{2}\gamma}$.

Now $\gamma = 180° - (\alpha + \beta) \Rightarrow \frac{1}{2}\gamma = \left[90° - \frac{1}{2}(\alpha + \beta)\right]$ and

$\sin \frac{1}{2}(\alpha + \beta) = \cos \left[90° - \frac{1}{2}(\alpha + \beta)\right] = \cos \frac{1}{2}\gamma$. Thus, $\dfrac{a+b}{c} = \dfrac{\cos \frac{1}{2}(\alpha - \beta)}{\sin \frac{1}{2}\gamma}$.

Note: This is an interesting result and gives an answer to the question, "How can I check these triangle problems?" Some of my students have written programs for their graphing calculators to utilize this check.

3 Example 4 in Section 7.6 illustrates the case $a = 1$.

Algebraic: $\sqrt[3]{a},\ \sqrt[3]{a}\ \text{cis}\ \frac{2\pi}{3},\ \sqrt[3]{a}\ \text{cis}\ \frac{4\pi}{3}$

Geometric: All roots lie on a circle of radius $\sqrt[3]{a}$, they are all 120° apart,

one is on the real axis, one is on $\theta = \frac{2\pi}{3}$, and one is on $\theta = \frac{4\pi}{3}$

5 (a) $\mathbf{c} = \mathbf{b} + \mathbf{a} = (\|\mathbf{b}\| \cos \alpha\, \mathbf{i} + \|\mathbf{b}\| \sin \alpha\, \mathbf{j}) + (\|\mathbf{a}\| \cos (-\beta)\, \mathbf{i} + \|\mathbf{a}\| \sin (-\beta)\, \mathbf{j})$

$= \|\mathbf{b}\| \cos \alpha\, \mathbf{i} + \|\mathbf{b}\| \sin \alpha\, \mathbf{j} + \|\mathbf{a}\| \cos \beta\, \mathbf{i} - \|\mathbf{a}\| \sin \beta\, \mathbf{j}$

$= (\|\mathbf{b}\| \cos \alpha + \|\mathbf{a}\| \cos \beta)\mathbf{i} + (\|\mathbf{b}\| \sin \alpha - \|\mathbf{a}\| \sin \beta)\mathbf{j}$

(b) $\|\mathbf{c}\|^2 = (\|\mathbf{b}\| \cos \alpha + \|\mathbf{a}\| \cos \beta)^2 + (\|\mathbf{b}\| \sin \alpha - \|\mathbf{a}\| \sin \beta)^2$

$= \|\mathbf{b}\|^2 \cos^2\alpha + 2\|\mathbf{a}\|\|\mathbf{b}\| \cos \alpha\, \cos \beta + \|\mathbf{a}\|^2 \cos^2\beta +$

$\|\mathbf{b}\|^2 \sin^2\alpha - 2\|\mathbf{a}\|\|\mathbf{b}\| \sin \alpha\, \sin \beta + \|\mathbf{a}\|^2 \sin^2\beta$

$= (\|\mathbf{b}\|^2 \cos^2\alpha + \|\mathbf{b}\|^2 \sin^2\alpha) + (\|\mathbf{a}\|^2 \cos^2\beta + \|\mathbf{a}\|^2 \sin^2\beta) +$

$2\|\mathbf{a}\|\|\mathbf{b}\| \cos \alpha\, \cos \beta - 2\|\mathbf{a}\|\|\mathbf{b}\| \sin \alpha\, \sin \beta$

$= \|\mathbf{b}\|^2 + \|\mathbf{a}\|^2 + 2\|\mathbf{a}\|\|\mathbf{b}\| (\cos \alpha\, \cos \beta - \sin \alpha\, \sin \beta)$

$= \|\mathbf{a}\|^2 + \|\mathbf{b}\|^2 + 2\|\mathbf{a}\|\|\mathbf{b}\| \cos (\alpha + \beta)$

$= \|\mathbf{a}\|^2 + \|\mathbf{b}\|^2 + 2\|\mathbf{a}\|\|\mathbf{b}\| \cos (\pi - \gamma)\ \{\alpha + \beta + \gamma = \pi\}$

$= \|\mathbf{a}\|^2 + \|\mathbf{b}\|^2 - 2\|\mathbf{a}\|\|\mathbf{b}\| \cos \gamma\ \{\cos (\pi - \gamma) = -\cos \gamma\}$

(c) From part (a), we let $(\|\mathbf{b}\| \sin \alpha - \|\mathbf{a}\| \sin \beta) = 0$.

Thus, $\|\mathbf{b}\| \sin \alpha = \|\mathbf{a}\| \sin \beta$, and $\dfrac{\sin \alpha}{\|\mathbf{a}\|} = \dfrac{\sin \beta}{\|\mathbf{b}\|}$.

Chapter 8: Systems of Equations and Inequalities

Note: The notation E_1 and E_2 refers to the first equation and the second equation.

$\boxed{1}$ Substituting y in E_2 into E_1 yields $2x - 1 = x^2 - 4 \Rightarrow x^2 - 2x - 3 = 0 \Rightarrow$

$(x - 3)(x + 1) = 0 \Rightarrow x = 3, -1$. Substituting $x = 3$ into E_2 gives us $y = 5$,

so $(3, 5)$ is a solution of the system. Similarly, $(-1, -3)$ is a solution.

$\boxed{3}$ Solving E_2 for x, $x = 1 - 2y$, and substituting into E_1 yields $y^2 = 1 - (1 - 2y) \Rightarrow$

$y^2 - 2y = 0 \Rightarrow y(y - 2) = 0 \Rightarrow y = 0, 2; x = 1, -3.$ ★ $(1, 0), (-3, 2)$

$\boxed{5}$ Substituting y in E_2 into E_1 yields $2(4x^3) = x^2 \Rightarrow 8x^3 = x^2 \Rightarrow$

$8x^3 - x^2 = 0 \Rightarrow x^2(8x - 1) = 0 \Rightarrow x = 0, \frac{1}{8}; y = 0, \frac{1}{128}.$ ★ $(0, 0), (\frac{1}{8}, \frac{1}{128})$

$\boxed{7}$ Solving E_1 for x, $x = -2y - 1$, and substituting into E_2 yields

$2(-2y - 1) - 3y = 12 \Rightarrow -7y = 14 \Rightarrow y = -2; x = 3.$ ★ $(3, -2)$

$\boxed{9}$ Solving E_1 for x, $2x = 3y + 1 \Rightarrow x = \frac{3}{2}y + \frac{1}{2}$, and substituting into E_2 yields

$-6(\frac{3}{2}y + \frac{1}{2}) + 9y = 4 \Rightarrow -9y - 3 + 9y = 4 \Rightarrow -3 = 4$, a contradiction.

There are *no solutions.*

$\boxed{11}$ Solving E_1 for x, $x = 5 - 3y$, and substituting into E_2 yields $(5 - 3y)^2 + y^2 = 25 \Rightarrow$

$25 - 30y + 9y^2 + y^2 = 25 \Rightarrow 10y^2 - 30y = 0 \Rightarrow 10y(y - 3) = 0 \Rightarrow$

$y = 0, 3; x = 5, -4.$ ★ $(-4, 3), (5, 0)$

$\boxed{13}$ Solving E_2 for y, $y = x + 4$, and substituting into E_1 yields $x^2 + (x + 4)^2 = 8 \Rightarrow$

$x^2 + x^2 + 8x + 16 = 8 \Rightarrow 2x^2 + 8x + 8 = 0 \Rightarrow x^2 + 4x + 4 = 0 \Rightarrow$

$(x + 2)^2 = 0 \Rightarrow x = -2; y = 2.$ ★ $(-2, 2)$

$\boxed{15}$ Solving E_2 for y, $y = 3x + 2$, and substituting into E_1 yields

$x^2 + (3x + 2)^2 = 9 \Rightarrow x^2 + 9x^2 + 12x + 4 = 9 \Rightarrow 10x^2 + 12x - 5 = 0 \Rightarrow$

$x = \dfrac{-12 \pm \sqrt{344}}{20} = \dfrac{-12 \pm \sqrt{4 \cdot 86}}{20} = \dfrac{-12 \pm 2\sqrt{86}}{20} = \dfrac{-6 \pm \sqrt{86}}{10} = -\dfrac{3}{5} \pm \dfrac{1}{10}\sqrt{86}.$

Now we'll substitute the values of x into $y = 3x + 2$ to get the corresponding values

of y. $y = 3\left(\dfrac{-6 \pm \sqrt{86}}{10}\right) + 2 = \dfrac{-18 \pm 3\sqrt{86}}{10} + \dfrac{20}{10} = \dfrac{2 \pm 3\sqrt{86}}{10} = \dfrac{1}{5} \pm \dfrac{3}{10}\sqrt{86}.$

★ $(-\frac{3}{5} + \frac{1}{10}\sqrt{86}, \frac{1}{5} + \frac{3}{10}\sqrt{86}), (-\frac{3}{5} - \frac{1}{10}\sqrt{86}, \frac{1}{5} - \frac{3}{10}\sqrt{86})$

$\boxed{17}$ Solving E_2 for x, $x = 2y - 4$, and substituting into E_1 yields $(2y - 4)^2 + y^2 = 16 \Rightarrow$

$4y^2 - 16y + 16 + y^2 = 16 \Rightarrow 5y^2 - 16y = 0 \Rightarrow y(5y - 16) = 0 \Rightarrow$

$y = 0, \frac{16}{5}; x = -4, \frac{12}{5}.$ ★ $(-4, 0), (\frac{12}{5}, \frac{16}{5})$

19 Solving E_2 for x, $x = 1 - y$, and substituting into E_1 yields

$$(1 - y - 1)^2 + (y + 2)^2 = 10 \;\Rightarrow\; y^2 + (y^2 + 4y + 4) = 10 \;\Rightarrow$$

$$2y^2 + 4y - 6 = 0 \;\Rightarrow\; 2(y^2 + 2y - 3) = 0 \;\Rightarrow\; 2(y + 3)(y - 1) = 0 \;\Rightarrow$$

$$y = -3, 1; \; x = 4, 0. \hspace{3cm} \bigstar \; (0, 1), \, (4, -3)$$

21 Substituting y in E_1 into E_2 yields $20/x^2 = 9 - x^2$ {multiply by x^2} $\;\Rightarrow$

$$20 = 9x^2 - x^4 \;\Rightarrow\; x^4 - 9x^2 + 20 = 0 \;\Rightarrow\; (x^2 - 4)(x^2 - 5) = 0 \;\Rightarrow$$

$$x^2 = 4, 5 \;\Rightarrow\; x = \pm 2, \, \pm\sqrt{5}. \text{ If } x^2 = 4, \text{ then } y = \frac{20}{4} = 5.$$

If $x^2 = 5$, then $y = \frac{20}{5} = 4$. Thus, we get the *four* solutions $(\pm 2, 5)$ and $(\pm\sqrt{5}, 4)$.

23 Solving E_1 for y^2, $y^2 = 4x^2 + 4$, and substituting into E_2 yields

$$9(4x^2 + 4) + 16x^2 = 140 \;\Rightarrow\; 52x^2 = 104 \;\Rightarrow\; x^2 = 2 \;\Rightarrow\; x = \pm\sqrt{2}; \, y = \pm 2\sqrt{3}.$$

There are *four* solutions. $\hspace{2cm} \bigstar \; (\sqrt{2}, \, \pm 2\sqrt{3}), \, (-\sqrt{2}, \, \pm 2\sqrt{3})$

25 Solving E_1 for x^2 and substituting into E_2 yields $(y^2 + 4) + y^2 = 12 \;\Rightarrow$

$$2y^2 = 8 \;\Rightarrow\; y^2 = 4 \;\Rightarrow\; y = \pm 2; \, x = \pm\sqrt{8} = \pm 2\sqrt{2}.$$

There are *four* solutions. $\hspace{2cm} \bigstar \; (2\sqrt{2}, \, \pm 2), \, (-2\sqrt{2}, \, \pm 2)$

27 Solving E_2 for y, $y = 2x + z - 9$, and substituting into E_1 and E_3 yields

$$\begin{cases} x + 2(2x + z - 9) - z & = -1 \\ x + 3(2x + z - 9) + 3z & = 6 \end{cases} \;\Rightarrow\; \begin{cases} 5x + z & = 17 \quad (E_4) \\ 7x + 6z & = 33 \quad (E_5) \end{cases}$$

Solving E_4 for z, $z = 17 - 5x$, and substituting into E_5 yields

$$7x + 6(17 - 5x) = 33 \;\Rightarrow\; 7x + 102 - 30x = 33 \;\Rightarrow$$

$$-23x = -69 \;\Rightarrow\; x = 3. \text{ Now } z = 17 - 5x = 17 - 5(3) = 2 \text{ and}$$

$$y = 2x + z - 9 = 2(3) + 2 - 9 = -1. \hspace{2cm} \bigstar \; (3, -1, 2)$$

29 Solving E_3 for y, $y = 1 - z$, and substituting into E_2 yields $\begin{cases} x^2 + z^2 & = 5 \quad (E_1) \\ 2x - z & = 0 \quad (E_4) \end{cases}$

Now solve E_4 for z, $z = 2x$, and substitute into E_1 yielding $x^2 + (2x)^2 = 5 \;\Rightarrow$

$$x^2 + 4x^2 = 5 \;\Rightarrow\; 5x^2 = 5 \;\Rightarrow\; x^2 = 1 \;\Rightarrow\; x = \pm 1;$$

$$z = 2x = \pm 2; \, y = 1 - z = -1, 3. \hspace{2cm} \bigstar \; (1, -1, 2), \, (-1, 3, -2)$$

31 Using $P = 40 = 2l + 2w$ and $A = 96 = lw$, we have $\begin{cases} 2l + 2w & = 40 \quad (E_1) \\ lw & = 96 \quad (E_2) \end{cases}$

Solving E_1 for l, $l = 20 - w$, and substituting into E_2 yields $(20 - w)w = 96 \;\Rightarrow$

$$20w - w^2 = 96 \;\Rightarrow\; w^2 - 20w + 96 = 0 \;\Rightarrow\; (w - 8)(w - 12) = 0 \;\Rightarrow\; w = 8, 12;$$

$l = 12, 8.$ In either case, the rectangle is 12 inches $\times$ 8 inches.

[33] From the graph of $y = x$ and $y = 2^{-x}$, it is clear that there is a point of intersection and therefore a single solution between 0 and 1.

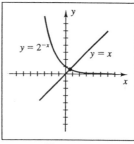

Figure 33

[35] The area of the rectangular sheet is width $\times$ height $= 200$. When rolled into a cylindrical tube, the width becomes the circumference of the circular base; that is, width $= 2\pi r$. Area of rectangular sheet $= A = 200 = (2\pi r)h \ \Rightarrow \ \pi rh = 100$.

Volume of cylindrical tube $= V = 200 = \pi r^2 h \ \Rightarrow \ h = \dfrac{200}{\pi r^2}$.

Substituting h into $\pi rh = 100$ yields $\pi r \cdot \dfrac{200}{\pi r^2} = 100 \ \Rightarrow \ \dfrac{200}{r} = 100 \ \Rightarrow \ r = 2$ inches;

$h = \dfrac{200}{\pi(2)^2} = \dfrac{200}{4\pi} = \dfrac{50}{\pi} \approx 15.9$ inches.

[37] (a) We have $R = aS/(S + b)$ and for 1998, let $S = 40{,}000$ and $R = 60{,}000$.

Then for 1999, let $S = 60{,}000$ and $R = 72{,}000$.

$$\begin{cases} 60{,}000 = (40{,}000a)/(40{,}000 + b) \\ 72{,}000 = (60{,}000a)/(60{,}000 + b) \end{cases} \Rightarrow \begin{cases} 120{,}000 + 3b = 2a \quad (E_1) \\ 360{,}000 + 6b = 5a \quad (E_2) \end{cases}$$

Solving E_1 for a and substituting into E_2 yields

$$360{,}000 + 6b = 5(60{,}000 + \tfrac{3}{2}b) \ \Rightarrow \ 60{,}000 = \tfrac{3}{2}b \ \Rightarrow \ b = 40{,}000; \ a = 120{,}000.$$

(b) Now let $S = 72{,}000$ and thus $R = \dfrac{(120{,}000)(72{,}000)}{72{,}000 + 40{,}000} = \dfrac{540{,}000}{7} \approx 77{,}143.$

[39] Let R_1 and R_2 equal 0. The system is then $\begin{cases} 0 = 0.01x(50 - x - y) \\ 0 = 0.02y(100 - y - 0.5x) \end{cases}$

The first equation is zero if $x = 0$ or if $50 - x - y = 0$. The second equation is zero if $y = 0$ or if $100 - y - 0.5x = 0$. Hence, there are 4 possible solutions.

(1) A solution is $x = 0$ and $y = 0$, or $(0, 0)$.

(2) A second solution is $x = 0$ and $100 - y - 0.5x = 0 \ \{y = 100\}$, or $(0, 100)$.

(3) A third solution is $y = 0$ and $50 - x - y = 0 \ \{x = 50\}$, or $(50, 0)$.

(4) A fourth solution occurs if $50 - x - y = 0$ and $100 - y - 0.5x = 0$.

Solve the first equation for $y \ \{y = 50 - x\}$ and substitute into the second equation: $100 - (50 - x) - 0.5x = 0 \ \Rightarrow \ 50 = -\tfrac{1}{2}x \ \Rightarrow \ x = -100; \ y = 150.$

This solution is meaningless for this problem since x and y are nonnegative.

41 Since we have an *open* top instead of a closed top, we use $3xy$ instead of $4xy$ { as in

Example 5 } for the surface area formula. $\begin{cases} x^2y = 2 & Volume \\ 2x^2 + 3xy = 8 & Surface\ Area \end{cases}$

Solving E_1 for y and substituting into E_2 yields $2x^2 + 3x(2/x^2) = 8 \Rightarrow$

$2x^2 + (6/x) = 8$ { multiply by x } $\Rightarrow 2x^3 - 8x + 6 = 0 \Rightarrow 2(x^3 - 4x + 3) = 0$.

We know that $x = 1$ is a solution of $x^3 - 4x + 3 = 0$ since the sum of the coefficients

is zero. Continuing, $2(x-1)(x^2 + x - 3) = 0 \Rightarrow \{x > 0\}\ x = 1, \dfrac{-1 + \sqrt{13}}{2}$.

There are two solutions: 1 ft × 1 ft × 2 ft or

$$\frac{\sqrt{13}-1}{2} \text{ ft} \times \frac{\sqrt{13}-1}{2} \text{ ft} \times \frac{8}{(\sqrt{13}-1)^2} \text{ ft} \approx 1.30 \text{ ft} \times 1.30 \text{ ft} \times 1.18 \text{ ft}.$$

43 We eliminate n from the equations to determine all intersection points.

(a) $x^2 + y^2 = n^2$ and $y = n - 1$ { so $n = y + 1$ } $\Rightarrow x^2 + y^2 = (y+1)^2 \Rightarrow$

$x^2 + y^2 = y^2 + 2y + 1 \Rightarrow x^2 = 2y + 1 \Rightarrow 2y = x^2 - 1 \Rightarrow y = \frac{1}{2}x^2 - \frac{1}{2}$.

The points are on the parabola $y = \frac{1}{2}x^2 - \frac{1}{2}$.

(b) As in part (a), $x^2 + y^2 = (y+2)^2 \Rightarrow x^2 = 4y + 4 \Rightarrow y = \frac{1}{4}x^2 - 1$.

45 (a) The slope of the line from $(-4, -3)$ to the origin is $\frac{3}{4}$ so the slope of the tangent

line (which is perpendicular to the line to the origin) is $-\frac{4}{3}$. An equation of the

line through $(-4, -3)$ with slope $-\frac{4}{3}$ is $y + 3 = -\frac{4}{3}(x+4)$, or, equivalently,

$4x + 3y = -25$. Letting $y = -50$, we find that $x = 31.25$.

(b) The slope of the line from an arbitrary point (x, y) on the circle to the origin is

$\frac{y}{x}$, so the slope of the tangent line is $-\frac{x}{y}$. The line through $(0, -50)$ is

$y + 50 = \left(-\frac{x}{y}\right)(x - 0)$, or, equivalently, $y^2 + 50y = -x^2$. The equation of the

circle is $x^2 + y^2 = 25$. Substituting $x^2 = 25 - y^2$ into $y^2 + 50y = -x^2$ gives us

$50y = -25$, or $y = -\frac{1}{2}$. Substituting $y = -\frac{1}{2}$ into $x^2 = 25 - y^2$ gives us $x^2 = \frac{99}{4}$

and hence $x = \pm\frac{3}{2}\sqrt{11}$. The positive solution corresponds to releasing the

hammer from a clockwise spin, whereas the negative solution corresponds to

releasing the hammer from a counterclockwise spin as depicted in the figures.

Hence, the hammer should be released at $\left(-\frac{3}{2}\sqrt{11}, -\frac{1}{2}\right) \approx (-4.975, -0.5)$.

47 Graphically: $x^2 + y^2 = 4 \Rightarrow y = \pm\sqrt{4-x^2}$ and $x + y = 1 \Rightarrow y = 1 - x$.
Graph $Y_1 = \sqrt{4-x^2}$, $Y_2 = -Y_1$, and $Y_3 = 1 - x$. There are two points of
intersection at approximately $(-0.82, 1.82)$ and $(1.82, -0.82)$.

Algebraically: $y = 1 - x$ and $x^2 + y^2 = 4 \Rightarrow x^2 + (1-x)^2 = 4 \Rightarrow$

$$x^2 + 1 - 2x + x^2 = 4 \Rightarrow 2x^2 - 2x - 3 = 0. \text{ Using the quadratic formula,}$$

$$x = \frac{2 \pm \sqrt{4 - 4(2)(-3)}}{4} = \frac{1 \pm \sqrt{7}}{2}. \text{ To find the value of } y \text{ we use}$$

$$y = 1 - x \Rightarrow y = 1 - \frac{1 \pm \sqrt{7}}{2} = \frac{1 \mp \sqrt{7}}{2}. \text{ The points of intersection are}$$

$\left(\frac{1}{2} \pm \frac{\sqrt{7}}{2}, \frac{1}{2} \mp \frac{\sqrt{7}}{2} \right)$. The graphical solution approximates the algebraic solution.

$[-6, 6]$ by $[-4, 4]$

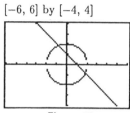

Figure 47

49 After zooming-in near the region of interest in the second quadrant, we see that the
cubic intersects the circle twice. Due to the symmetry, we know there are two more
points of intersection in the fourth quadrant. Thus, the six points of intersection are
approximately $(\mp 0.56, \pm 1.92)$, $(\mp 0.63, \pm 1.90)$, and $(\pm 1.14, \pm 1.65)$.

$[-6, 6]$ by $[-4, 4]$ $[-3, 3]$ by $[-2, 2]$

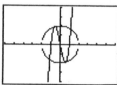   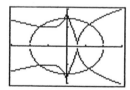

Figure 49   *Figure 51*

51 $|x + \ln|x|| - y^2 = 0 \Rightarrow y^2 = |x + \ln|x|| \Rightarrow y = \pm\sqrt{|x + \ln|x||}$ and

$$\frac{x^2}{4} + \frac{y^2}{2.25} = 1 \Rightarrow \qquad \frac{y^2}{2.25} = 1 - \frac{x^2}{4} \Rightarrow \qquad y^2 = 2.25(1 - x^2/4) \Rightarrow$$

$y = \pm 1.5\sqrt{1 - x^2/4}$. The graph is symmetric with respect to x-axis. There are 8
points of intersection. Their coordinates are approximately

$$(-1.44, \pm 1.04), (-0.12, \pm 1.50), (0.10, \pm 1.50), \text{ and } (1.22, \pm 1.19).$$

$\boxed{53}$ If $f(x) = ae^{-bx}$ and $x = 1$, then $f(1) = ae^{-b} = 0.80487 \Rightarrow a = 0.80487e^b$. When
$x = 2$, $f(2) = ae^{-2b} = 0.53930 \Rightarrow a = 0.53930e^{2b}$. Let $a = y$ and $b = x$ and then
graph $Y_1 = 0.80487e^x$ and $Y_2 = 0.53930e^{2x}$. The graphs intersect at approximately
$(0.4004, 1.2012) = (b, a)$. Thus, $a \approx 1.2012$, $b \approx 0.4004$, and $f(x) = 1.2012e^{-0.4004x}$.
The function f is also accurate at $x = 3, 4$.

[0, 3] by [0, 2]

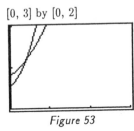

Figure 53

[0, 4] by [0, 4]

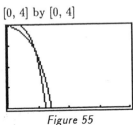

Figure 55

$\boxed{55}$ If $f(x) = ax^2 + e^{bx}$ and $x = 2$, then $f(2) = 4a + e^{2b} = 17.2597 \Rightarrow$
$a = (17.2597 - e^{2b})/4$. When $x = 3$, $f(3) = 9a + e^{3b} = 40.1058 \Rightarrow$
$a = (40.1058 - e^{3b})/9$. Graph $Y_1 = (17.2597 - e^{2x})/4$ and $Y_2 = (40.1058 - e^{3x})/9$.
The graphs intersect at approximately $(0.9002, 2.8019)$. Thus, $a \approx 2.8019$,
$b \approx 0.9002$, and $f(x) = 2.8019x^2 + e^{0.9002x}$. The function f is also accurate at $x = 4$.

8.2 Exercises

Note: The notation E_1 and E_2 refers to equation 1 and equation 2. $3\,E_2$ symbolizes "3
times equation 2". After the value of one variable is found, the value(s) of the
other variable(s) will be stated and can be found by substituting the known
value(s) back into the original equation(s).

$\boxed{1}$ The easiest choice for multipliers is to take -2 times the second equation and
eliminate x. $-2\,E_2 + E_1 \Rightarrow 4y + 3y = -16 + 2 \Rightarrow 7y = -14 \Rightarrow y = -2$; $x = 4$.
Another way to write this is as follows:
$$\begin{cases} 2x + 3y = 2 & (E_1) \\ x - 2y = 8 & (E_2) \end{cases} \Rightarrow \begin{cases} 2x + 3y = 2 & (E_1) \\ -2x + 4y = -16 & (-2\,E_2) \end{cases}$$

Alternatively, if we wanted to eliminate y instead of x, we could add 2 times
the first equation to 3 times the second equation, giving us $2\,E_1 + 3\,E_2 \Rightarrow$
$4x + 3x = 4 + 24 \Rightarrow 7x = 28 \Rightarrow x = 4$; $y = -2$ $\bigstar$ $(4, -2)$

$\boxed{3}$ $3\,E_1 - 2\,E_2 \Rightarrow 15y + 14y = 48 - 48 \Rightarrow 29y = 0 \Rightarrow y = \frac{0}{29} = 0$. From the first
equation, we see that $2x + 5(0) = 16 \Rightarrow 2x = 16 \Rightarrow x = 8$. $\bigstar$ $(8, 0)$

$\boxed{5}$ $2\,E_2 + E_1 \Rightarrow 2r + 3r = -8 + 3 \Rightarrow 5r = -5 \Rightarrow r = -1$. From the second
equation, we see that $-1 - 2s = -4 \Rightarrow -2s = -3 \Rightarrow s = \frac{3}{2}$. $\bigstar$ $(-1, \frac{3}{2})$

[7] $3\,E_1 - 5\,E_2 \;\Rightarrow\; -53y = -28 \;\Rightarrow\; y = \frac{28}{53}$. Instead of substituting into one of the equations to find the value of the other variable, it is usually easier to pick different multipliers and re-solve the system for the other variable, as in Example 4.

$7\,E_1 + 6\,E_2 \;\Rightarrow\; 53x = 76 \;\Rightarrow\; x = \frac{76}{53}$. $\qquad\qquad$ ★ $\left(\frac{76}{53}, \frac{28}{53}\right)$

[9] We will first eliminate all fractions by multiplying both sides of each equation by its

lcd. $\begin{cases} 6\,E_1 \\ 3\,E_2 \end{cases} \Rightarrow \begin{cases} 2c + 3d \;= 30 & (E_3) \\ 3c - 2d \;= -3 & (E_4) \end{cases}$

$\qquad 3\,E_3 - 2\,E_4 \;\Rightarrow\; 13d = 96 \;\Rightarrow\; d = \frac{96}{13}$.

$\qquad 2\,E_3 + 3\,E_4 \;\Rightarrow\; 13c = 51 \;\Rightarrow\; c = \frac{51}{13}$ $\qquad$ ★ $\left(\frac{51}{13}, \frac{96}{13}\right)$

[11] The least common multiple of 2 and 3 is 6, so we will multiply the first equation by $\sqrt{3}$ and the second equation by $\sqrt{2}$ so that the coefficients of y are $\sqrt{6}$ and $-\sqrt{6}$.

$\sqrt{3}\,E_1 + \sqrt{2}\,E_2 \;\Rightarrow\; 7x = 8 \;\Rightarrow\; x = \frac{8}{7}$. We will now re-solve the system to obtain a value for y. $2\sqrt{2}\,E_1 - \sqrt{3}\,E_2 \;\Rightarrow\; -7y = 3\sqrt{6} \;\Rightarrow\; y = -\frac{3}{7}\sqrt{6}$.

$\qquad\qquad\qquad\qquad\qquad\qquad\qquad\qquad\qquad\qquad$ ★ $\left(\frac{8}{7}, -\frac{3}{7}\sqrt{6}\right)$

[13] $3\,E_1 + E_2 \;\Rightarrow\; 0 = 27$; this statement is never true so there are *no solutions*.

[15] $2\,E_1 + E_2 \;\Rightarrow\; 0 = 0$; this statement is always true so the solution is all ordered pairs $(m, \, n)$ such that $3m - 4n = 2$. We could also say that the solution is all ordered pairs $(m, \, n)$ such that $-6m + 8n = -4$. Either statement is correct.

[17] $3\,E_1 - 2\,E_2 \;\Rightarrow\; -23x = 0 \;\Rightarrow\; x = 0; \; y = 0$ $\qquad\qquad$ ★ $(0, \, 0)$

[19] Using the hint, we let $u = 1/x$ and $v = 1/y$, and the system is $\begin{cases} 2u + 3v \;= -2 & (E_3) \\ 4u - 5v \;= 1 & (E_4) \end{cases}$

$\qquad -2\,E_3 + E_4 \;\Rightarrow\; -11v = 5 \;\Rightarrow\; v = -\frac{5}{11}$.

$\qquad 5\,E_3 + 3\,E_4 \;\Rightarrow\; 22u = -7 \;\Rightarrow\; u = -\frac{7}{22}$.

Resubstituting, we have $x = 1/u = -\frac{22}{7}$ and $y = 1/v = -\frac{11}{5}$. $\qquad$ ★ $\left(-\frac{22}{7}, -\frac{11}{5}\right)$

[21] Let x denote the number of \$3.00 tickets and y the number of \$4.50 tickets.

$$\begin{cases} x + y = 450 & \textit{quantity} \\ 3.00x + 4.50y = 1555.50 & \textit{value} \end{cases}$$

$E_2 - 3\,E_1 \;\Rightarrow\; 4.50y - 3y = 1555.50 - 1350 \;\Rightarrow\; 1.5y = 205.50 \;\Rightarrow$

$y = \dfrac{205.50}{1.5} = 137$. Since $x + y = 450$, $x = 450 - 137 = 313$.

23 The volume for a cylinder is $\pi r^2 h$ and the volume for a cone is $\frac{1}{3}\pi r^2 h$.

The radius of the cylinder is $\frac{1}{2}$ cm. $\begin{cases} x + y = 8 & \text{length} \\ \pi(\frac{1}{2})^2 x + \frac{1}{3}\pi(\frac{1}{2})^2 y = 5 & \text{volume} \end{cases}$

Solving E_1 for y, $y = 8 - x$, and substituting into E_2 yields

$\frac{\pi}{4}x + \frac{\pi}{12}(8 - x) = 5 \Rightarrow \frac{\pi}{4}x + \frac{8\pi}{12} - \frac{\pi}{12}x = \frac{15}{3} \Rightarrow \frac{3\pi}{12}x - \frac{\pi}{12}x = \frac{15}{3} - \frac{2\pi}{3} \Rightarrow$

$\frac{\pi}{6}x = \frac{15 - 2\pi}{3} \Rightarrow x = \frac{6}{\pi} \cdot \frac{15 - 2\pi}{3} = \frac{30 - 4\pi}{\pi} = \frac{30}{\pi} - 4 \approx 5.55$ cm.

$y = 8 - x = 8 - \left(\frac{30 - 4\pi}{\pi}\right) = \frac{8\pi}{\pi} - \frac{30}{\pi} + \frac{4\pi}{\pi} = \frac{12\pi - 30}{\pi} = 12 - \frac{30}{\pi} \approx 2.45$ cm.

25 The perimeter is composed of 2 sides of the rectangular portion of the table, $2l$, and 2 edges of the semicircular regions, $2 \cdot \frac{1}{2}(2\pi r) = 2\pi r$. Since the radius is $\frac{1}{2}w$, the system can be represented by $\begin{cases} 2l + 2\pi(\frac{1}{2}w) = 40 & \text{perimeter} \\ lw = 2\left[\pi(\frac{1}{2}w)^2\right] & \text{area} \end{cases}$

Solving E_1 for l, $l = \frac{40 - \pi w}{2}$, and substituting into E_2 yields

$\left(\frac{40 - \pi w}{2}\right)w = \frac{\pi w^2}{2} \Rightarrow (40 - \pi w)w = \pi w^2 \Rightarrow 40w - \pi w^2 = \pi w^2 \Rightarrow$

$40w - 2\pi w^2 = 0 \Rightarrow 2w(20 - \pi w) = 0 \Rightarrow w = 0, \frac{20}{\pi}$.

We discard $w = 0$ and use $w = 20/\pi \approx 6.37$ ft. Thus, $l = \frac{40 - \pi(20/\pi)}{2} = \frac{20}{2} = 10$ ft.

27 Let x denote the number of adults and y the number of kittens. Thus,

$(\frac{1}{2}x)$ is the number of adult females. $\begin{cases} x + y = 6000 & \text{total} \\ y = 3(\frac{1}{2}x) & \text{3 kittens per adult female} \end{cases}$

Substituting y from E_2 into E_1 yields

$x + \frac{3}{2}x = 6000 \Rightarrow \frac{5}{2}x = 6000 \Rightarrow x = \frac{2}{5} \cdot 6000 = 2400; y = 3600.$

29 Let x denote the number of grams of the 35% alloy and y the number of grams of the

60% alloy. $\begin{cases} x + y = 100 & \text{quantity} \\ 0.35x + 0.60y = (0.50)(100) & \text{quality} \end{cases}$

Multiply the second equation by 100 to obtain $35x + 60y = 5000$ (E_3).

$\qquad\qquad\qquad E_3 - 35\,E_1 \Rightarrow 25y = 1500 \Rightarrow y = 60; x = 40$

31 Let x denote the speed of the plane and y the speed of the wind. Use $d = rt$.

$\begin{cases} 1200 = (x + y)(2) & \text{with the wind} \\ 1200 = (x - y)(2\frac{1}{2}) & \text{against the wind} \end{cases} \Rightarrow \begin{cases} 600 = x + y \\ 480 = x - y \end{cases}$

$\qquad\qquad\qquad E_1 + E_2 \Rightarrow 2x = 1080 \Rightarrow x = 540$ mi/hr; $y = 60$ mi/hr

33 $v(t) = v_0 + at$ • $\begin{cases} v(2) = 16 \\ v(5) = 25 \end{cases}$ $\Rightarrow$ $\begin{cases} 16 = v_0 + 2a \quad (E_1) \\ 25 = v_0 + 5a \quad (E_2) \end{cases}$

$$E_2 - E_1 \;\Rightarrow\; 9 = 3a \;\Rightarrow\; a = 3;\; v_0 = 10$$

35 Let x denote the number of sofas produced and y the number of recliners produced.

$$\begin{cases} 8x + 6y = 340 & \text{\textit{labor hours}} \\ 60x + 35y = 2250 & \text{\textit{cost of materials}} \end{cases}$$

$$6\,E_2 - 35\,E_1 \;\Rightarrow\; 360x - 280x = 13{,}500 - 11{,}900 \;\Rightarrow\; 80x = 1600 \;\Rightarrow$$
$$x = 20;\; y = 30$$

37 (a) The expression $6x + 5y$ represents the total bill for the plumber's business. This should equal the plumber's income, which is the plumber's number of hours times his/her hourly wage—that is, $(6 + 4)(x) = 10x$. The expression $4x + 6y$ represents the total bill for the electrician's business. This should equal the electrician's income, that is, $(5 + 6)(y) = 11y$.

$$\begin{cases} 6x + 5y = 10x \\ 4x + 6y = 11y \end{cases} \;\Rightarrow\; \begin{cases} 5y = 4x \\ 4x = 5y \end{cases} \;\Rightarrow\; y = \tfrac{4}{5}x, \text{ or equivalently, } y = 0.80x$$

(b) The electrician should charge 80% of what the plumber charges—80% of $20 per hour is $16 per hour.

39 Let $t = 0$ correspond to the year 1891. The average daily maximum can then be approximated by the linear equation $y_1 = 0.011t + 15.1$ and the average daily minimum by the linear equation $y_2 = 0.019t + 5.8$. We must determine t when y_1 and y_2 differ by 9. $y_1 - y_2 = 9 \;\Rightarrow\; (0.011t + 15.1) - (0.019t + 5.8) = 9 \;\Rightarrow$ $-0.008t + 9.3 = 9 \;\Rightarrow\; -0.008t = -0.3 \;\Rightarrow\; t = 37.5.$ $1891 + 37.5 = 1928.5$, or during the year 1928. Now we find the corresponding average maximum temperature: $t = 37.5 \;\Rightarrow\; y_1 = 0.011(37.5) + 15.1 = 15.5125 \approx 15.5°C$.

41 Let x and y denote the amount of time in hours at the LP and SLP speeds, respectively. 5 hours and 20 minutes $= 5\tfrac{1}{3}$ hours, so $\dfrac{x}{5\frac{1}{3}}$ is the portion of the tape used at the LP speed. $\dfrac{y}{8}$ is the portion of the tape used at the SLP speed, and together, these fractions add up to 1 whole tape.

$$\begin{cases} x + y = 6 & \text{\textit{total time}} \\ \dfrac{x}{5\frac{1}{3}} + \dfrac{y}{8} = 1 & \text{\textit{portions of tape}} \end{cases} \;\Rightarrow\; \begin{cases} x + y = 6 & (E_1) \\ \tfrac{3}{16}x + \tfrac{1}{8}y = 1 & (E_2) \end{cases}$$

$$16\,E_2 - 2\,E_1 \;\Rightarrow\; 3x - 2x = 16 - 12 \;\Rightarrow\; x = 4;\; y = 2$$

Note: If you convert hours to minutes, the equations can be written as $x + y = 360$ and $x/320 + y/480 = 1$ with the solution $x = 240$ and $y = 120$.

43 $\begin{cases} a e^{3x} + b e^{-3x} & = 0 \quad (E_1) \\ a(3e^{3x}) + b(-3e^{-3x}) & = e^{3x} \quad (E_2) \end{cases}$

Following the hint in the text, think of this system as a system of two equations in the variables a and b. In E_1, we have "e^{3x}" times a, and in E_2, we have "$3e^{3x}$" times a. Hence, we will multiply E_1 by -3 and add this equation to E_2 to eliminate a and then solve the system for b. $-3E_1 + E_2 \Rightarrow -3be^{-3x} - 3be^{-3x} = e^{3x} \Rightarrow -6be^{-3x} = e^{3x} \Rightarrow b = -\frac{1}{6}e^{6x}$. Substituting back into E_1 yields $a e^{3x} + (-\frac{1}{6}e^{6x})e^{-3x} = 0 \Rightarrow a e^{3x} = \frac{1}{6}e^{3x} \Rightarrow a = \frac{1}{6}$.

Note: The solution for Exercise 45 is on page 336 at the end of this chapter.

8.3 Exercises

---

1 $3x - 2y < 6 \Leftrightarrow y > \frac{3}{2}x - 3$. Sketch the graph of $y = \frac{3}{2}x - 3$ with dashes. The point $(0, 0)$ is clearly on one side of the line, so we will substitute $x = 0$ and $y = 0$ into the inequality $y > \frac{3}{2}x - 3$. Checking, we have $0 > -3$, a *true* statement. Hence, we shade *all* points that are on the *same* side of the line as $(0, 0)$.

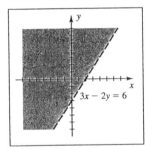

Figure 1

3 $2x + 3y \geq 2y + 1 \Leftrightarrow y \geq -2x + 1$. Sketch the graph of $y = -2x + 1$ as a solid line. Checking the test point $(0, 0)$ in $y \geq -2x + 1$ gives us $0 \geq 1$, a *false* statement. Hence, we shade *all* points that are on the *opposite* side of the line as $(0, 0)$.

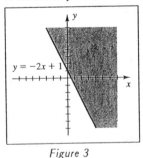

Figure 3

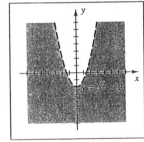

Figure 5

5 $y + 2 < x^2 \Leftrightarrow y < x^2 - 2$. Sketch the graph of the parabola $y = x^2 - 2$ with dashes. Checking the test point $(0, 0)$ in $y < x^2 - 2$ gives us $0 < -2$, a *false* statement. Hence, we shade *all* points that are on the *opposite* side of the parabola as $(0, 0)$.

[7] $x^2 + 1 \le y$ ⇔ $y \ge x^2 + 1$. Sketch the graph of the parabola $y = x^2 + 1$ with a solid curve. Checking the test point $(0, 0)$ in $y \ge x^2 + 1$ gives us $0 \ge 1$, a *false* statement. Hence, we shade *all* points that are on the *opposite* side of the parabola as $(0, 0)$.

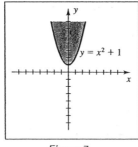

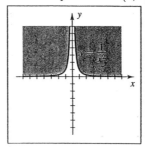

Figure 7   Figure 9

[9] $yx^2 \ge 1$ ⇔ $y \ge 1/x^2$ $\{x \ne 0\}$. Sketch the graph of $y = 1/x^2$ with a solid curve. The point $(1, 0)$ is clearly not on the graph. Substituting $x = 1$ and $y = 0$ into $y \ge 1/x^2$ yields $0 \ge 1$, a false statement. Hence, we do not shade the region containing $(1, 0)$, but we do shade the regions in the first and second quadrants that are above the graph.

Note: The will use the notation V @ (a, b), (c, d), ... to denote the intersection point(s) of the solution region of the graph. These can be found by solving the system of *equalities* that correspond to the given system of inequalities.

[11] $\begin{cases} 3x + y < 3 \\ 4 - y < 2x \end{cases}$ ⇔ $\begin{cases} y < -3x + 3 \\ y > -2x + 4 \end{cases}$ V @ $(-1, 6)$

Solving the system $y = -3x + 3$ and $y = -2x + 4$ gives us the solution $(-1, 6)$. Testing the point $(0, 0)$ in both inequalities, we see that we need to shade under $y = -3x + 3$ and above $y = -2x + 4$, as shown in *Figure 11*.

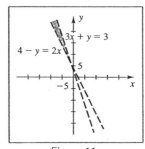

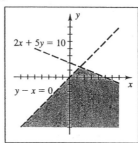

Figure 11   Figure 13

[13] $\begin{cases} y - x < 0 \\ 2x + 5y < 10 \end{cases}$ ⇔ $\begin{cases} y < x \\ y < -\frac{2}{5}x + 2 \end{cases}$ V @ $(\frac{10}{7}, \frac{10}{7})$

Solving the system $y = x$ and $y = -\frac{2}{5}x + 2$ gives us the solution $(\frac{10}{7}, \frac{10}{7})$.
We need to shade under $y = x$ and under $y = -\frac{2}{5}x + 2$.

$\boxed{15}$ $\begin{cases} 3x + y \le 6 \\ y - 2x \ge 1 \\ x \ge -2 \\ y \le 4 \end{cases}$ $\Leftrightarrow$ $\begin{cases} y \le -3x + 6 \\ y \ge 2x + 1 \\ x \ge -2 \\ y \le 4 \end{cases}$ V @ $(-2, -3)$, $(-2, 4)$, $(\frac{2}{3}, 4)$, $(1, 3)$

We need to shade under $y = -3x + 6$, above $y = 2x + 1$, to the right of $x = -2$, and

under $y = 4$. Include all the boundaries in the solution region.

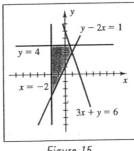

Figure 15

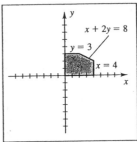

Figure 17

$\boxed{17}$ $\begin{cases} x + 2y \le 8 \\ 0 \le x \le 4 \\ 0 \le y \le 3 \end{cases}$ • V @ $(0, 0)$, $(4, 0)$, $(4, 2)$, $(2, 3)$, $(0, 3)$

We can write $x + 2y \le 8$ as $y \le -\frac{1}{2}x + 4$. Shade under $y = -\frac{1}{2}x + 4$, between $x = 0$

and $x = 4$, and between $y = 0$ and $y = 3$. Include all the boundaries in the solution

region.

$\boxed{19}$ $|x| \ge 2$ $\Leftrightarrow$ $x \ge 2$ or $x \le -2$. This region is bounded on the left by the vertical

line $x = 2$ and on the right by $x = -2$, including the lines.

$|y| < 3$ $\Leftrightarrow$ $-3 < y < 3$. This region is bounded below by the horizontal line

$y = -3$ and above by $y = 3$, excluding the lines.

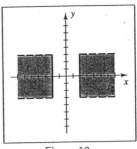

Figure 19

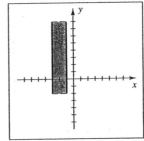

Figure 21

$\boxed{21}$ $|x + 2| \le 1$ $\Leftrightarrow$ $-1 \le x + 2 \le 1$ $\Leftrightarrow$ $-3 \le x \le -1$.

This region is bounded by the vertical lines $x = -3$ and $x = -1$, including the lines.

$|y - 3| < 5$ $\Leftrightarrow$ $-5 < y - 3 < 5$ $\Leftrightarrow$ $-2 < y < 8$.

This region is bounded by the horizontal lines $y = -2$ and $y = 8$, excluding the lines.

23 $\begin{cases} x^2 + y^2 \le 4 \\ x + y \ge 1 \end{cases}$ $\Leftrightarrow$ $\begin{cases} x^2 + y^2 \le 2^2 \\ y \ge -x + 1 \end{cases}$ $V @ (\frac{1}{2} \mp \frac{1}{2}\sqrt{7}, \frac{1}{2} \pm \frac{1}{2}\sqrt{7})$

Shade inside the circle $x^2 + y^2 = 4$ and above the line $y = -x + 1$.

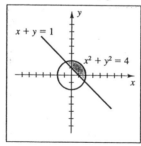

Figure 23

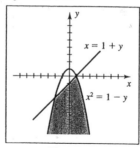

Figure 25

25 $\begin{cases} x^2 \le 1 - y \\ x \ge 1 + y \end{cases}$ $\Leftrightarrow$ $\begin{cases} y \le -x^2 + 1 \\ y \le x - 1 \end{cases}$ $V @ (-2, -3), (1, 0)$

Shade under { or inside } the parabola $y = -x^2 + 1$ and under the line $y = x - 1$.

27   The shaded region lies between $x = 0$ and $x = 3$, including $x = 0$ and excluding $x = 3$. This region is described by $0 \le x < 3$.

The dashed line goes through $(0, 4)$ and $(4, 0)$. Hence, its slope is -1, its y-intercept is 4, and an equation for it is $y = -x + 4$. Substituting $x = 0$ and $y = 0$ into $y = -x + 4$ gives us $0 = 4$. We replace $=$ by $<$ { the line is dashed } to make this statement a true statement. Hence, we have the inequality $y < -x + 4$.

The solid line goes through $(0, -4)$ and $(4, 0)$. Hence, its slope is 1, its y-intercept is -4, and an equation for it is $y = x - 4$. Substituting $x = 0$ and $y = 0$ into $y = x - 4$ gives us $0 = -4$. We replace $=$ by $\ge$ { the line is solid } to make this statement a true statement. Hence, we have the inequality $y \ge x - 4$. Thus, the graph may be described by the system $\begin{cases} 0 \le x < 3 \\ y < -x + 4 \\ y \ge x - 4 \end{cases}$

29   The shaded region lies *inside* the circle $x^2 + y^2 = 3^2$, so we can describe it with the inequality $x^2 + y^2 \le 9$.

The dashed line goes through $(0, 4)$ and $(2, 0)$. Hence, its slope is -2, its y-intercept is 4, and an equation for it is $y = -2x + 4$. We want the region shaded *above* the line, so replace $=$ by $>$ { the line is dashed } to make this statement a true statement. Hence, we have the inequality $y > -2x + 4$.

Thus, the graph may be described by the system

$$\begin{cases} x^2 + y^2 \le 9 \\ y > -2x + 4 \end{cases}$$

31 The center of the circle is (2, 2) and it passes through (0, 0). The radius is the distance from (0, 0) to (2, 2), which is $\sqrt{8}$. An equation of the circle is $(x-2)^2 + (y-2)^2 = 8$. The center is clearly inside the circle, and if we substitute $x = 2$ and $y = 2$ in the equation of the circle, we get $0 = 8$. Since we want to make this a true statement and include the boundary, we use $(x-2)^2 + (y-2)^2 \le 8$ to describe the shaded circular region and its boundary.

For the dashed line $y = x$, test the point (1, 0). This test gives us $0 = 1$, which we can make true by replacing $=$ by $<$. Hence $y < x$ describes the shaded region under the line $y = x$. The other inequality may be found as in Exercise 27. Thus, the graph may be described by the system $\begin{cases} y < x \\ y \le -x + 4 \\ (x-2)^2 + (y-2)^2 \le 8 \end{cases}$

33   For the dashed line through $(-4, 0)$ and (4, 1), we can use the point-slope form to find an equation of the line. $y - 0 = \frac{1-0}{4-(-4)}(x-(-4)) \Leftrightarrow y = \frac{1}{8}x + \frac{1}{2}$. Checking the point (0, 1) { which is in the shaded region } gives us $1 = \frac{1}{2}$, so we change $=$ to $>$ to obtain $y > \frac{1}{8}x + \frac{1}{2}$.

Similarly, for the solid line with x-intercept -4 and y-intercept 4, the shaded region can be described by $y \le x + 4$.

Lastly, the solid line passing through (4, 1) has slope $-\frac{3}{4}$, y-intercept 4, and its shaded region can be described by $y \le -\frac{3}{4}x + 4$. Thus, the system is $\begin{cases} y > \frac{1}{8}x + \frac{1}{2} \\ y \le x + 4 \\ y \le -\frac{3}{4}x + 4 \end{cases}$

35 Let x and y denote the number of sets of brand A and brand B, respectively. "Necessary to stock at least twice as many sets of brand A as of brand B" may be symbolized as $x \ge 2y$. "Necessary to have on hand at least 10 sets of brand B" may be symbolized as $y \ge 10$. Consequently, $x \ge 20$ from the first inequality. "Room for not more than 100 sets in the store" may be symbolized as $x + y \le 100$. We now sketch the system of inequalities and shade the region that they have in common. The graph is the region bounded by the triangle with vertices (20, 10), (90, 10), and $(\frac{200}{3}, \frac{100}{3})$. See *Figure 35* on the next page.

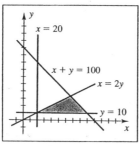

Figure 35

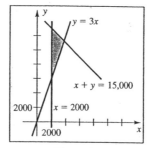

Figure 37

37 If x and y denote the amount placed in the high-risk and low-risk investment, respectively, then a system is $x \geq 2000$, $y \geq 3x$, $x + y \leq 15{,}000$. The graph is the region bounded by the triangle with vertices $(2000,\ 6000)$, $(2000,\ 13{,}000)$, and $(3750,\ 11{,}250)$.

39 A system is $x + y \leq 9$, $y \geq x$, $x \geq 1$. To justify the condition $y \geq x$, start with

$$\frac{\text{cylinder volume}}{\text{total volume}} \geq 0.75 \quad \Rightarrow \quad \frac{\pi r^2 y}{\pi r^2 y + \frac{1}{3}\pi r^2 x} \geq \frac{3}{4} \quad \Rightarrow \quad 4\pi r^2 y \geq 3\pi r^2 y + \pi r^2 x \quad \Rightarrow$$

$\pi r^2 y \geq \pi r^2 x \ \Rightarrow \ y \geq x$. The graph is the region bounded by the triangle with vertices $(1,\ 1)$, $(1,\ 8)$, and $\left(\frac{9}{2},\ \frac{9}{2}\right)$.

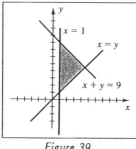

Figure 39

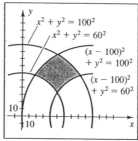

Figure 41

41 If the plant is located at $(x,\ y)$, then a system is $(60)^2 \leq x^2 + y^2 \leq (100)^2$, $(60)^2 \leq (x - 100)^2 + y^2 \leq (100)^2$, $y \geq 0$. The graph is the region in the first quadrant that lies between the two concentric circles with center $(0,\ 0)$ and radii 60 and 100, and also between the two concentric circles with center $(100,\ 0)$ and radii 60 and 100.

Equating the different circle equations, we obtain the vertices of the solution region $(50,\ 50\sqrt{3})$, $(50,\ 10\sqrt{11})$, $(18,\ 6\sqrt{91})$, and $(82,\ 6\sqrt{91})$. For example, to determine where the circle $x^2 + y^2 = 100^2$ intersects the circle $(x - 100)^2 + y^2 = 60^2$, we subtract the first equation from the second to obtain $[(x - 100)^2 + y^2] - (x^2 + y^2) = 60^2 - 100^2 \ \Rightarrow \ -200x + 10{,}000 = -6400 \ \Rightarrow$ $200x = 16{,}400 \ \Rightarrow \ x = 82$. Substituting $x = 82$ into $x^2 + y^2 = 100^2$ gives us $y^2 = 100^2 - 82^2 = 3276 \ \Rightarrow \ y = \pm\sqrt{3276} = \pm 6\sqrt{91} \approx \pm 57.2$.

$\boxed{43}$ $64y^3 - x^3 \le e^{1-2x}$ $\Rightarrow$ $64y^3 \le e^{1-2x} + x^3$ $\Rightarrow$ $y^3 \le \frac{1}{64}(e^{1-2x} + x^3)$ $\Rightarrow$
$y \le \frac{1}{4}(e^{1-2x} + x^3)^{1/3}$. Graph the curve $y = \frac{1}{4}(e^{1-2x} + x^3)^{1/3}$.

The solution includes the curve and the region below it.

[−3.5, 4] by [−1, 4]

[−1.5, 1.5, 0.5] by [−1, 1, 0.5]

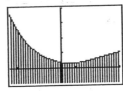

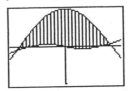

Figure 43

Figure 45

$\boxed{45}$ $5^{1-y} \ge x^4 + x^2 + 1$ $\Rightarrow$ $1 - y \ge \log_5(x^4 + x^2 + 1)$ $\Rightarrow$ $y \le 1 - \log_5(x^4 + x^2 + 1)$.
$x + 3y \ge x^{5/3}$ $\Rightarrow$ $y \ge \frac{1}{3}(x^{5/3} - x)$. Graph $y = 1 - \log_5(x^4 + x^2 + 1)$ $\{Y_1\}$ and
$y = \frac{1}{3}(x^{5/3} - x)$ $\{Y_2\}$. The curves intersect at approximately $(-1.32, -0.09)$ and
$(1.21, 0.05)$. The solution is located between the points of intersection. It includes
the curves and the region below the first curve and above the second curve.

$\boxed{47}$ $x^4 - 2x < 3y$ $\Rightarrow$ $y > \frac{1}{3}(x^4 - 2x)$. $x + 2y < x^3 - 5$ $\Rightarrow$ $y < \frac{1}{2}(x^3 - x - 5)$.   Graph
$y = \frac{1}{3}(x^4 - 2x)$ and $y = \frac{1}{2}(x^3 - x - 5)$. The curves do not intersect. The solution must
be above the first curve and below the second curve. Since the regions do not
intersect, there is no solution.

[−4.5, 4.5] by [−3, 3]

[33, 80, 5] by [0, 50, 5]

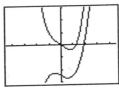

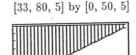

Figure 47

Figure 49

$\boxed{49}$ (a) $29T - 39P < 450$ $\Rightarrow$ $29(37) - 39(21.2) < 450$ $\Rightarrow$ $246.2 < 450$ (true).

Yes, forests can grow.

(b) $29T - 39P < 450$ $\Rightarrow$ $-39P < -29T + 450$ $\Rightarrow$ $P > \frac{29}{39}T - \frac{150}{13}$.

Graph $Y_1 = \frac{29}{39}x - \frac{150}{13}$.

(c) Forests can grow whenever the point (T, P) lies above the line $P = \frac{29}{39}T - \frac{150}{13}$.

8.4 Exercises

1 We substitute the x and y values of each vertex into $C = 3x + 2y + 5$ and summarize the results in the following table. We see that there is a maximum of 27 at $(6, 2)$, and a minimum of 9 at $(0, 2)$.

(x, y)	$(0, 2)$	$(0, 4)$	$(3, 5)$	$(6, 2)$	$(5, 0)$	$(2, 0)$
C	9 ■	13	24	27 ■	20	11

3 The region R is sketched in *Figure 3*. The vertices are found by obtaining the intersection points of the system of equations corresponding to the given inequalities.

(x, y)	$(0, 0)$	$(0, 3)$	$(4, 6)$	$(6, 3)$	$(5, 0)$
C	0	3	18	21 ■	15

$C = 3x + y$;

maximum of 21 at $(6, 3)$

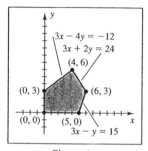

Figure 3

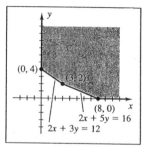

Figure 5

5

(x, y)	$(8, 0)$	$(3, 2)$	$(0, 4)$
C	24	21 ■	24

$C = 3x + 6y$;

minimum of 21 at $(3, 2)$

[7]

(x, y)	(0, 0)	(0, 4)	(2, 5)	(6, 3)	(8, 0)
C	0	16	24 ∎	24 ∎	16

As in Example 2, $C = 2x + 4y$ has the maximum value 24 for *any* point on the line segment from (2, 5) to (6, 3). To show that the last statement is true, we will substitute $6 - \frac{1}{2}x$ for y in C. $C = 2x + 4y = 2x + 4(6 - \frac{1}{2}x) = 2x + 24 - 2x = 24$.

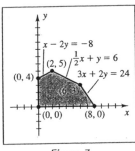

Figure 7

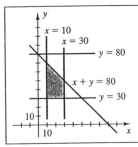

Figure 9

Note: **For the linear programming application exercises, the solution outline is as follows:**

(1) Define the variables used in the problem.

(2) Define the function to be maximized or minimized.

(3) List the system of inequalities that determine the solution region R.

(4) A table of intersection points and values of the function in (2) at those points is given. The maximum or minimum value is denoted by a ∎.

(5) A summarizing statement is given.

(6) A figure is shown for the system of inequalities.

[9] Let x and y denote the number of oversized and standard rackets, respectively.

Profit function: $P = 15x + 8y$

(x, y)	(10, 30)	(30, 30)	(30, 50)	(10, 70)
P	390	690	850 ∎	710

$$\begin{cases} 30 \leq y \leq 80 \\ 10 \leq x \leq 30 \\ x + y \leq 80 \end{cases}$$

The maximum profit of \$850 per day occurs when 30 oversized rackets and 50 standard rackets are manufactured.

[11] Let x and y denote the number of pounds of S and T, respectively.

Cost function: $C = 3x + 4y$

$$\begin{cases} 2x + 2y \geq 9 & \text{amount of } I \\ 4x + 6y \geq 20 & \text{amount of } G \\ x \geq 0 \\ y \geq 0 \end{cases}$$

(x, y)	$(0, 4.5)$	$(3.5, 1)$	$(5, 0)$
C	18	14.5 ■	15

The minimum cost of \$14.50 occurs when 3.5 pounds of S and 1 pound of T are used.

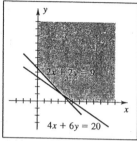

Figure 11

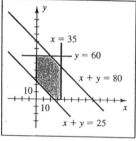

Figure 13

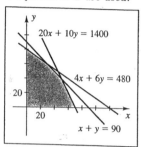

Figure 15

[13] Let x and y denote the number of units sent to A and B, respectively, from W_1.

Cost function: $C = 12x + 10(35 - x) + 16y + 12(60 - y) = 2x + 4y + 1070$

The points are the same as those in Example 4.

$$\begin{cases} 0 \leq x \leq 35 \\ 0 \leq y \leq 60 \\ x + y \leq 80 \\ x + y \geq 25 \end{cases}$$

(x, y)	$(0, 25)$	$(0, 60)$	$(20, 60)$	$(35, 45)$	$(35, 0)$	$(25, 0)$
C	1170	1310	1350	1320	1140	1120 ■

To minimize the shipping costs, send 25 units from W_1 to A and 0 from W_1 to B.

Send 10 $\{35 - 25\}$ units from W_2 to A and 60 $\{60 - 0\}$ units from W_2 to B.

[15] Let x and y denote the number of acres planted with alfalfa and corn, respectively.

Profit function: $P = 110x - 4x - 20x + 150y - 6y - 10y = 86x + 134y$

$$\begin{cases} 4x + 6y \leq 480 & \text{seed cost} \\ 20x + 10y \leq 1400 & \text{labor cost} \\ x + y \leq 90 & \text{area} \\ x, y \geq 0 \end{cases}$$

(x, y)	$(70, 0)$	$(50, 40)$	$(30, 60)$	$(0, 80)$	$(0, 0)$
P	6020	9660	10,620	10,720 ■	0

The maximum profit of \$10,720 occurs when

0 acres of alfalfa are planted and 80 acres of corn are planted.

[17] Let x, y, and z denote the number of ounces of X, Y, and Z, respectively.

Cost function: $C = 0.25x + 0.35y + 0.50z$

$$= 0.25x + 0.35y + 0.50(20 - x - y) = \underline{10 - 0.25x - 0.15y}$$

$$\left\{ \begin{array}{l} 0.20x + 0.20y + 0.10z \geq 0.14(20) \quad amount\ of\ \text{A} \\ 0.10x + 0.40y + 0.20z \geq 0.16(20) \quad amount\ of\ \text{B} \\ 0.25x + 0.15y + 0.25z \geq 0.20(20) \quad amount\ of\ \text{C} \end{array} \right. \Rightarrow \left\{ \begin{array}{r} x + y \geq 8 \\ x - 2y \leq 8 \\ y \leq 10 \\ x + y \leq 20 \\ 0 \leq x,\ y \leq 20 \end{array} \right.$$

The new restrictions are found by substituting $z = 20 - x - y$ into the 3 inequalities, simplifying, and adding the last 2 inequalities.

(x, y)	$(8, 0)$	$(16, 4)$	$(10, 10)$	$(0, 10)$	$(0, 8)$
C	8.00	5.40 ■	6.00	8.50	8.80 ■■

The minimum cost of $5.40 requires 16 oz of X, 4 oz of Y, and 0 oz of Z.

The maximum cost of $8.80 requires 0 oz of X, 8 oz of Y, and 12 oz of Z.

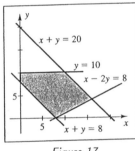

Figure 17

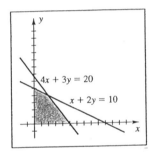

Figure 19

[19] Let x and y denote the number of vans and buses purchased, respectively.

$$\left\{ \begin{array}{l} 10{,}000x + 20{,}000y \leq 100{,}000 \quad purchase \\ 100x + 75y \leq 500 \quad maintenance \\ x \geq 0 \\ y \geq 0 \end{array} \right. \Rightarrow \left\{ \begin{array}{r} x + 2y \leq 10 \\ 4x + 3y \leq 20 \\ x \geq 0 \\ y \geq 0 \end{array} \right.$$

(x, y)	$(5, 0)$	$(2, 4)$	$(0, 5)$	$(0, 0)$
P	75	130 ■	125	0

Passenger capacity function: $P = 15x + 25y$

The maximum passenger capacity of 130 would occur if the community purchases 2 vans and 4 buses.

21 Let x and y denote the number of trout and bass, respectively.

Pound function: $P = 3x + 4y$

$$\begin{cases} x + y \le 5000 & \textit{number of fish} \\ 0.50x + 0.75y \le 3000 & \textit{cost} \\ x \ge 0 \\ y \ge 0 \end{cases} \Rightarrow \begin{cases} x + y \le 5000 \\ 2x + 3y \le 12,000 \\ x \ge 0 \\ y \ge 0 \end{cases}$$

(x, y)	(5000, 0)	(3000, 2000)	(0, 4000)	(0, 0)
P	15,000	17,000 ■	16,000	0

The total number of pounds of fish will be a maximum of 17,000 if 3000 trout and 2000 bass are purchased.

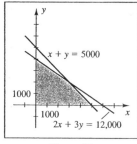

Figure 21

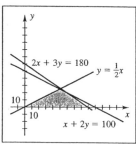

Figure 23

23 Let x and y denote the number of basic and deluxe units constructed, respectively.

$$\begin{cases} 300x + 600y \le 30,000 & \textit{cost} \\ x \ge 2y & \textit{ratio} \\ 80x + 120y \le 7200 & \textit{area} \\ x, y \ge 0 \end{cases} \Rightarrow \begin{cases} x + 2y \le 100 \\ y \le \frac{1}{2}x \\ 2x + 3y \le 180 \\ x, y \ge 0 \end{cases}$$

(x, y)	(90, 0)	(60, 20)	(50, 25)	(0, 0)
R	3600	3900 ■	3875	0

Revenue function: $R = 40x + 75y$

The maximum monthly revenue of $3900 occurs if 60 basic units and 20 deluxe units are constructed.

8.5 Exercises

Note: Most systems are solved using the back substitution method. The solution for Exercise 7 uses the reduced echelon method. To avoid fractions, some solutions include linear combinations of rows—that is, $m\mathrm{R}_i + n\mathrm{R}_j$.

$$\boxed{1} \begin{bmatrix} 1 & -2 & -3 & -1 \\ 2 & 1 & 1 & 6 \\ 1 & 3 & -2 & 13 \end{bmatrix} \begin{array}{l} R_2 - 2\,R_1 \to R_2 \\ R_3 - R_1 \to R_3 \end{array}$$

$$\begin{bmatrix} 1 & -2 & -3 & -1 \\ 0 & 5 & 7 & 8 \\ 0 & 5 & 1 & 14 \end{bmatrix} R_3 - R_2 \to R_3$$

$$\begin{bmatrix} 1 & -2 & -3 & -1 \\ 0 & 5 & 7 & 8 \\ 0 & 0 & -6 & 6 \end{bmatrix} -\tfrac{1}{6}R_3 \to R_3$$

R_3: $-6z = 6 \Rightarrow z = -1$

R_2: $5y + 7z = 8 \Rightarrow 5y + 7(-1) = 8 \Rightarrow 5y = 15 \Rightarrow y = 3$ ★ $(2, 3, -1)$

R_1: $x - 2y - 3z = -1 \Rightarrow x - 2(3) - 3(-1) = -1 \Rightarrow x = 2$

$$\boxed{3} \begin{bmatrix} 1 & -2 & 2 & 0 \\ 5 & 2 & -1 & -7 \\ 0 & 3 & 1 & 17 \end{bmatrix} R_2 - 5\,R_1 \to R_2$$

$$\begin{bmatrix} 1 & -2 & 2 & 0 \\ 0 & 12 & -11 & -7 \\ 0 & 3 & 1 & 17 \end{bmatrix} 4\,R_3 - R_2 \to R_3$$

$$\begin{bmatrix} 1 & -2 & 2 & 0 \\ 0 & 12 & -11 & -7 \\ 0 & 0 & 15 & 75 \end{bmatrix} \tfrac{1}{15}R_3 \to R_3$$

R_3: $15z = 75 \Rightarrow z = 5$

R_2: $12y - 11z = -7 \Rightarrow 12y - 11(5) = -7 \Rightarrow 12y = 48 \Rightarrow y = 4$ ★ $(-2, 4, 5)$

R_1: $x - 2y + 2z = 0 \Rightarrow x - 2(4) + 2(5) = 0 \Rightarrow x = -2$

$$\boxed{5} \begin{bmatrix} 2 & 6 & -4 & 1 \\ 1 & 3 & -2 & 4 \\ 2 & 1 & -3 & -7 \end{bmatrix} R_1 \leftrightarrow R_2$$

$$\begin{bmatrix} 1 & 3 & -2 & 4 \\ 2 & 6 & -4 & 1 \\ 2 & 1 & -3 & -7 \end{bmatrix} \begin{array}{l} R_2 - 2\,R_1 \to R_2 \\ R_3 - 2\,R_1 \to R_3 \end{array}$$

$$\begin{bmatrix} 1 & 3 & -2 & 4 \\ 0 & 0 & 0 & -7 \\ 0 & -5 & 1 & -15 \end{bmatrix}$$

The second row, $0x + 0y + 0z = -7$, has *no solution*.

7 $\begin{bmatrix} 2 & -3 & 2 & -3 \\ -3 & 2 & 1 & 1 \\ 4 & 1 & -3 & 4 \end{bmatrix}$ $R_1 \leftrightarrow R_2$ and then $R_2 \leftrightarrow R_3$

$\begin{bmatrix} -3 & 2 & 1 & 1 \\ 4 & 1 & -3 & 4 \\ 2 & -3 & 2 & -3 \end{bmatrix}$ $R_1 + R_2 \rightarrow R_1$

$\begin{bmatrix} 1 & 3 & -2 & 5 \\ 4 & 1 & -3 & 4 \\ 2 & -3 & 2 & -3 \end{bmatrix}$ $\begin{aligned} R_2 - 4\,R_1 \rightarrow R_2 \\ R_3 - 2\,R_1 \rightarrow R_3 \end{aligned}$

$\begin{bmatrix} 1 & 3 & -2 & 5 \\ 0 & -11 & 5 & -16 \\ 0 & -9 & 6 & -13 \end{bmatrix}$ $4\,R_2 - 5\,R_3 \rightarrow R_2$ ($*$ see note)

$\begin{bmatrix} 1 & 3 & -2 & 5 \\ 0 & 1 & -10 & 1 \\ 0 & -9 & 6 & -13 \end{bmatrix}$ $R_3 + 9\,R_2 \rightarrow R_3$

$\begin{bmatrix} 1 & 3 & -2 & 5 \\ 0 & 1 & -10 & 1 \\ 0 & 0 & -84 & -4 \end{bmatrix}$ $-\frac{1}{84}R_3 \rightarrow R_3$

R_3: $-84z = -4 \;\Rightarrow\; z = \frac{1}{21}$

R_2: $y - 10z = 1 \;\Rightarrow\; y = \frac{31}{21}$ $\qquad\qquad\qquad\qquad$ $\star\ \left(\frac{2}{3}, \frac{31}{21}, \frac{1}{21}\right)$

R_1: $x + 3y - 2z = 5 \;\Rightarrow\; x = \frac{14}{21} = \frac{2}{3}$

($*$) We could have just used $-\frac{1}{11}R_2 \rightarrow R_2$, but we would then have to work with many fractions and increase our chance of making a mistake. We will solve this system again, but this time we will obtain the reduced echelon form. The first 4 matrices are exactly the same. Starting with the fourth matrix we have:

$\begin{bmatrix} 1 & 3 & -2 & 5 \\ 0 & 1 & -10 & 1 \\ 0 & -9 & 6 & -13 \end{bmatrix}$ $\begin{aligned} R_1 - 3\,R_2 \rightarrow R_1 \\ R_3 + 9\,R_2 \rightarrow R_3 \end{aligned}$

$\begin{bmatrix} 1 & 0 & 28 & 2 \\ 0 & 1 & -10 & 1 \\ 0 & 0 & -84 & -4 \end{bmatrix}$ $-\frac{1}{84}R_3 \rightarrow R_3$

$\begin{bmatrix} 1 & 0 & 28 & 2 \\ 0 & 1 & -10 & 1 \\ 0 & 0 & 1 & \frac{1}{21} \end{bmatrix}$ $\begin{aligned} R_1 - 28\,R_3 \rightarrow R_1 \\ R_2 + 10\,R_3 \rightarrow R_2 \end{aligned}$

$\begin{bmatrix} 1 & 0 & 0 & \frac{14}{21} \\ 0 & 1 & 0 & \frac{31}{21} \\ 0 & 0 & 1 & \frac{1}{21} \end{bmatrix}$

R_1: $x = \frac{14}{21} = \frac{2}{3}$; R_2: $y = \frac{31}{21}$; R_3: $z = \frac{1}{21}$

Note: Exer. 9–16: There are other forms for the answers; c is any real number.

9 $\begin{bmatrix} 1 & 3 & 1 & 0 \\ 1 & 1 & -1 & 0 \\ 1 & -2 & -4 & 0 \end{bmatrix}$ $\begin{matrix} \\ R_2 - R_1 \to R_2 \\ R_3 - R_1 \to R_3 \end{matrix}$

$\begin{bmatrix} 1 & 3 & 1 & 0 \\ 0 & -2 & -2 & 0 \\ 0 & -5 & -5 & 0 \end{bmatrix}$ $\begin{matrix} -\frac{1}{2} R_2 \to R_2 \\ \\ -\frac{1}{5} R_3 \to R_3 \end{matrix}$

$\begin{bmatrix} 1 & 3 & 1 & 0 \\ 0 & 1 & 1 & 0 \\ 0 & 1 & 1 & 0 \end{bmatrix}$ $\begin{matrix} R_1 - 3 R_2 \to R_1 \\ \\ R_3 - R_2 \to R_3 \end{matrix}$

$\begin{bmatrix} 1 & 0 & -2 & 0 \\ 0 & 1 & 1 & 0 \\ 0 & 0 & 0 & 0 \end{bmatrix}$

R_1: $x - 2z = 0 \implies x = 2z$

R_2: $y + z = 0 \implies y = -z$ ★ $(2c, -c, c)$

11 $\begin{bmatrix} 2 & 1 & 1 & 0 \\ 1 & -2 & -2 & 0 \\ 1 & 1 & 1 & 0 \end{bmatrix}$ $R_1 \leftrightarrow R_2$

$\begin{bmatrix} 1 & -2 & -2 & 0 \\ 2 & 1 & 1 & 0 \\ 1 & 1 & 1 & 0 \end{bmatrix}$ $\begin{matrix} \\ R_2 - 2 R_1 \to R_2 \\ R_3 - R_1 \to R_3 \end{matrix}$

$\begin{bmatrix} 1 & -2 & -2 & 0 \\ 0 & 5 & 5 & 0 \\ 0 & 3 & 3 & 0 \end{bmatrix}$ $\begin{matrix} \frac{1}{5} R_2 \to R_2 \\ \\ \frac{1}{3} R_3 \to R_3 \end{matrix}$

$\begin{bmatrix} 1 & -2 & -2 & 0 \\ 0 & 1 & 1 & 0 \\ 0 & 1 & 1 & 0 \end{bmatrix}$ $\begin{matrix} R_1 + 2 R_2 \to R_1 \\ \\ R_3 - R_2 \to R_3 \end{matrix}$

$\begin{bmatrix} 1 & 0 & 0 & 0 \\ 0 & 1 & 1 & 0 \\ 0 & 0 & 0 & 0 \end{bmatrix}$

R_1: $x = 0$

R_2: $y + z = 0 \implies y = -z$ ★ $(0, -c, c)$

13 $\begin{bmatrix} 1 & 4 & -1 & -2 \\ 3 & -2 & 5 & 7 \end{bmatrix}$ $R_2 - 3 R_1 \to R_2$

$\begin{bmatrix} 1 & 4 & -1 & -2 \\ 0 & -14 & 8 & 13 \end{bmatrix}$

R_2: $-14y + 8z = 13 \implies y = \frac{4}{7}z - \frac{13}{14}$

R_1: $x + 4y - z = -2 \implies x = -4\left(\frac{4}{7}z - \frac{13}{14}\right) + z - 2 = -\frac{9}{7}z + \frac{12}{7}$

★ $\left(\frac{12}{7} - \frac{9}{7}c, \frac{4}{7}c - \frac{13}{14}, c\right)$

15 $\begin{bmatrix} 4 & -2 & 1 & 5 \\ 3 & 1 & -4 & 0 \end{bmatrix}$ $R_1 - R_2 \to R_1$

$\begin{bmatrix} 1 & -3 & 5 & 5 \\ 3 & 1 & -4 & 0 \end{bmatrix}$ $R_2 - 3\,R_1 \to R_2$

$\begin{bmatrix} 1 & -3 & 5 & 5 \\ 0 & 10 & -19 & -15 \end{bmatrix}$

$R_2: 10y - 19z = -15 \;\Rightarrow\; y = \frac{19}{10}z - \frac{3}{2}$

$R_1: x - 3y + 5z = 5 \;\Rightarrow\; x = 3(\frac{19}{10}z - \frac{3}{2}) - 5z + 5 = \frac{7}{10}z + \frac{1}{2}$ $\qquad \bigstar \; (\frac{7}{10}c + \frac{1}{2}, \; \frac{19}{10}c - \frac{3}{2}, \; c)$

17 $\begin{bmatrix} 5 & 0 & 2 & 1 \\ 0 & 1 & -3 & 2 \\ 2 & 1 & 0 & 3 \end{bmatrix}$ $R_1 - 2\,R_3 \to R_1$

$\begin{bmatrix} 1 & -2 & 2 & -5 \\ 0 & 1 & -3 & 2 \\ 2 & 1 & 0 & 3 \end{bmatrix}$ $R_3 - 2\,R_1 \to R_3$

$\begin{bmatrix} 1 & -2 & 2 & -5 \\ 0 & 1 & -3 & 2 \\ 0 & 5 & -4 & 13 \end{bmatrix}$ $R_3 - 5\,R_2 \to R_3$

$\begin{bmatrix} 1 & -2 & 2 & -5 \\ 0 & 1 & -3 & 2 \\ 0 & 0 & 11 & 3 \end{bmatrix}$ $\frac{1}{11}R_3 \to R_3$

$R_3: z = \frac{3}{11}$

$R_2: y - 3z = 2 \;\Rightarrow\; y = \frac{31}{11}$

$R_1: x - 2y + 2z = -5 \;\Rightarrow\; x = \frac{1}{11}$ $\qquad\qquad \bigstar \; (\frac{1}{11}, \frac{31}{11}, \frac{3}{11})$

19 $\begin{bmatrix} 4 & -3 & 1 \\ 2 & 1 & -7 \\ -1 & 1 & -1 \end{bmatrix}$ $R_1 + 3\,R_3 \to R_1$

$\begin{bmatrix} 1 & 0 & -2 \\ 2 & 1 & -7 \\ -1 & 1 & -1 \end{bmatrix}$ $\begin{array}{l} R_2 - 2\,R_1 \to R_2 \\ R_3 + R_1 \to R_3 \end{array}$

$\begin{bmatrix} 1 & 0 & -2 \\ 0 & 1 & -3 \\ 0 & 1 & -3 \end{bmatrix}$ $R_3 - R_2 \to R_3$

$\begin{bmatrix} 1 & 0 & -2 \\ 0 & 1 & -3 \\ 0 & 0 & 0 \end{bmatrix}$ $\qquad R_2: y = -3; \; R_1: x = -2$ $\qquad \bigstar \; (-2, -3)$

21
$$\begin{bmatrix} 2 & 3 & 5 \\ 1 & -3 & 4 \\ 1 & 1 & -2 \end{bmatrix} \qquad R_1 \leftrightarrow R_2$$

$$\begin{bmatrix} 1 & -3 & 4 \\ 2 & 3 & 5 \\ 1 & 1 & -2 \end{bmatrix} \qquad \begin{matrix} R_2 - 2\,R_1 \rightarrow R_2 \\ R_3 - R_1 \rightarrow R_3 \end{matrix}$$

$$\begin{bmatrix} 1 & -3 & 4 \\ 0 & 9 & -3 \\ 0 & 4 & -6 \end{bmatrix} \qquad R_2 - 2\,R_3 \rightarrow R_2$$

$$\begin{bmatrix} 1 & -3 & 4 \\ 0 & 1 & 9 \\ 0 & 4 & -6 \end{bmatrix} \qquad \begin{matrix} R_1 + 3\,R_2 \rightarrow R_1 \\ R_3 - 4\,R_2 \rightarrow R_3 \end{matrix}$$

$$\begin{bmatrix} 1 & 0 & 31 \\ 0 & 1 & 9 \\ 0 & 0 & -42 \end{bmatrix}$$

There is a contradiction in row 3, therefore there is *no solution.*

23 Let x, y, and z denote the number of liters of the 10% acid, 30% acid, and 50% acid.

$$\begin{cases} x + y + z = 50 \\ 0.10x + 0.30y + 0.50z = (0.32)(50) \\ z = 2y \end{cases} \qquad \begin{matrix} quantity \\ quality \\ constraint \end{matrix} \qquad \begin{matrix} (E_1) \\ (E_2) \\ (E_3) \end{matrix}$$

Substitute $z = 2y$ into E_1 and $100\,E_2$ to obtain

$$\begin{cases} x + 3y & = 50 & (E_4) \\ 10x + 130y & = 1600 & (E_5) \end{cases}$$

$$E_5 - 10\,E_4 \;\Rightarrow\; 100y = 1100 \;\Rightarrow\; y = 11;\; x = 17;\; z = 22$$

25 Let x, y, and z denote the number of hours needed for A, B, and C, respectively, to produce 1000 items. In one hour, A, B, and C produce $\frac{1000}{x}$, $\frac{1000}{y}$, and $\frac{1000}{z}$ items, respectively. In 6 hours, A and B produce $\frac{6000}{x}$ and $\frac{6000}{y}$ items. From the table, this sum must equal 4500. The system of equations is then:

$$\begin{cases} \dfrac{6000}{x} + \dfrac{6000}{y} & = 4500 \\[2mm] \dfrac{8000}{x} \phantom{{}+{}} + \dfrac{8000}{z} & = 3600 \\[2mm] \dfrac{7000}{y} + \dfrac{7000}{z} & = 4900 \end{cases}$$

To simplify, let $a = 1/x$, $b = 1/y$, $c = 1/z$ and divide each equation by its greatest common factor $\{1500, 400, \text{ and } 700\}$.

$$\begin{cases} 4a + 4b & = 3 & (E_1) \\ 20a + {}+ 20c & = 9 & (E_2) \\ 10b + 10c & = 7 & (E_3) \end{cases} \qquad E_2 - 5\,E_1 \;\Rightarrow\; 20c - 20b = -6 \;\; (E_4)$$

$$E_4 + 2\,E_3 \;\Rightarrow\; 40c = 8 \;\Rightarrow\; c = \tfrac{1}{5};\; b = \tfrac{1}{2};\; a = \tfrac{1}{4}.$$

Resubstituting, $x = 4$, $y = 2$, and $z = 5$.

27 Let x, y, and z denote the amounts of G_1, G_2, and G_3, respectively.

$$\begin{cases} x + y + z = 600 & \textit{quantity} & (E_1) \\ 0.30x + 0.20y + 0.15z = (0.25)(600) & \textit{quality} & (E_2) \\ z = 100 + y & \textit{constraint} & (E_3) \end{cases}$$

Substitute $z = 100 + y$ into E_1 and $100\,E_2$ to obtain

$$\begin{cases} x + 2y = 500 & (E_4) \\ 30x + 35y = 13{,}500 & (E_5) \end{cases}$$

$$E_5 - 30\,E_4 \Rightarrow -25y = -1500 \Rightarrow y = 60;\ z = 160;\ x = 380.$$

29 (a) Let $R_1 = R_2 = R_3 = 3$ in the given system of equations.

$$\begin{cases} I_1 - I_2 + I_3 = 0 \\ 3I_1 + 3I_2 = 6 \\ 3I_2 + 3I_3 = 12 \end{cases} \Rightarrow \begin{cases} I_1 - I_2 + I_3 = 0 & (E_1) \\ I_1 + I_2 = 2 & (E_2) \\ I_2 + I_3 = 4 & (E_3) \end{cases}$$

Substitute $I_1 = I_2 - I_3$ from E_1 into E_2 to obtain $2I_2 - I_3 = 2$ (E_4).

$$\text{Now } E_4 + E_3 \Rightarrow 3I_2 = 6 \Rightarrow I_2 = 2;\ I_3 = 2;\ I_1 = 0.$$

(b) Let $R_1 = 4$, $R_2 = 1$, and $R_3 = 4$ in the given system of equations.

$$\begin{cases} I_1 - I_2 + I_3 = 0 & (E_1) \\ 4I_1 + 1I_2 = 6 & (E_2) \\ 1I_2 + 4I_3 = 12 & (E_3) \end{cases}$$

Substitute $I_1 = I_2 - I_3$ from E_1 into E_2 to obtain $4(I_2 - I_3) + I_2 = 6 \Rightarrow$

$$5I_2 - 4I_3 = 6 \ (E_4). \text{ Now } E_4 + E_3 \Rightarrow 6I_2 = 18 \Rightarrow I_2 = 3;\ I_3 = \tfrac{9}{4};\ I_1 = \tfrac{3}{4}.$$

31 Let x, y, and z denote the amount of Colombian, Brazilian, and Kenyan coffee used, respectively.

$$\begin{cases} x + y + z = 1 & \textit{quantity} & (E_1) \\ 10x + 6y + 8z = (8.50)(1) & \textit{quality} & (E_2) \\ x = 3y & \textit{constraint} & (E_3) \end{cases}$$

Substitute $x = 3y$ into E_1 and E_2 to obtain

$$\begin{cases} 4y + z = 1 & (E_4) \\ 36y + 8z = 8.5 & (E_5) \end{cases} \quad E_5 - 8\,E_4 \Rightarrow 4y = \tfrac{1}{2} \Rightarrow y = \tfrac{1}{8};\ z = \tfrac{1}{2};\ x = \tfrac{3}{8}.$$

33 (a) A: $x_1 + x_4 = 50 + 25 = 75$, B: $x_1 + x_2 = 100 + 50 = 150$,

C: $x_2 + x_3 = 150 + 75 = 225$, D: $x_3 + x_4 = 100 + 50 = 150$

(b) From C, $x_3 = 100 \Rightarrow x_2 = 225 - 100 = 125$.

From D, $x_3 = 100 \Rightarrow x_4 = 150 - 100 = 50$.

From A, $x_4 = 50 \Rightarrow x_1 = 75 - 50 = 25$.

(c) From D in part (a), $x_3 = 150 - x_4 \Rightarrow x_3 \le 150$ since $x_4 \ge 0$. From C,

$x_3 = 225 - x_2 = 225 - (150 - x_1)\ \{\text{from B}\} = 75 + x_1 \Rightarrow x_3 \ge 75$ since $x_1 \ge 0$.

$\boxed{35}$ $t = 2070 - 1990 = 80.$ $rt = (0.025)(80) = 2$ and $A = 800$ for E_1,

$(0.015)(80) = 1.2$ and $A = 560$ for E_2, and $(0.01)(0) = 0$ and $A = 340$ for E_3.

Substituting into $A = a + ct + ke^{rt}$ and summarizing as a system, we have:

$$\begin{cases} a + 80c + \quad e^2 k = 800 & (E_1) \\ a + 80c + \quad e^{1.2} k = 560 & (E_2) \\ a \quad + \quad k = 340 & (E_3) \end{cases}$$

We want to find t when A has doubled—that is, $A = 2(340)$. First we find c and k.

$E_1 - E_2 \Rightarrow e^2 k - e^{1.2} k = 240 \Rightarrow k = \dfrac{240}{e^2 - e^{1.2}} \approx 58.98.$

Substituting into E_3 gives $a = 340 - k \approx 281.02.$

Substituting into E_1 gives $c = \dfrac{800 - a - e^2 k}{80} \approx \dfrac{800 - 281.02 - (58.98)e^2}{80} \approx 1.04.$

Thus, $A = 281.02 + 1.04t + 58.98e^{rt}$. If $A = 680$ and $r = 0.01$, then

$680 = 281.02 + 1.04t + 58.98e^{0.01t} \Rightarrow 1.04t + 58.98e^{0.01t} - 398.98 = 0.$ Graphing

$y = 1.04t + 58.98e^{0.01t} - 398.98$, we see there is an x-intercept at $x \approx 144.08$.

$1990 + 144.08 = 2134.08$, or during the year 2134.

$[0, 1050, 100]$ by $[-350, 350, 100]$

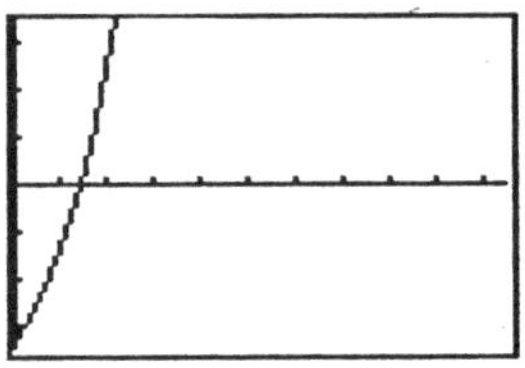

Figure 35

$\boxed{37}$ The circle has an equation of the form $x^2 + y^2 + ax + by + c = 0.$

Substituting the x and y values of P, Q, and R into this equation yields:

$$\begin{cases} 2a + \quad b + \quad c = -5 & P \quad (E_1) \\ -a - \quad 4b + \quad c = -17 & Q \quad (E_2) \\ 3a \quad + \quad c = -9 & R \quad (E_3) \end{cases}$$

Solving E_3 for c $\{c = -9 - 3a\}$ and substituting into E_1 and E_2 yields:

$$\begin{cases} -a + \quad b = 4 & (E_4) \\ -4a - \quad 4b = -8 & (E_5) \end{cases} \Rightarrow \begin{cases} -a + \quad b = 4 & (E_6) \\ a + \quad b = 2 & (E_7) \end{cases}$$

$E_6 + E_7 \Rightarrow 2b = 6 \Rightarrow b = 3;\ a = -1;\ c = -6.$

The equation is $x^2 + y^2 - x + 3y - 6 = 0.$

39 For $y = f(x) = ax^3 + bx^2 + cx + d$ and the points $(-1,\ 2)$, $(0.5,\ 2)$, $(1,\ 3)$, and $(2,\ 4.5)$, we obtain the system:

$$\begin{cases} -a + & b - & c + & d = & 2 & (-1,\ 2) \\ 0.125a + & 0.25b + & 0.5c + & d = & 2 & (0.5,\ 2) \\ a + & b + & c + & d = & 3 & (1,\ 3) \\ 8a + & 4b + & 2c + & d = & 4.5 & (2,\ 4.5) \end{cases}$$

Solving the system yields $a = -\frac{4}{9}$, $b = \frac{11}{9}$, $c = \frac{17}{18}$, and $d = \frac{23}{18}$.

8.6 Exercises

Note: For Exercises 1–8, $(A + B)$ and $(A - B)$ are possible only if A and B have the same number of rows and columns. {#7 & #8 are not possible} $(2A)$ and $(-3B)$ are always possible by simply multiplying the elements of A and B by 2 and -3, respectively.

1 $A + B = \begin{bmatrix} 5 & -2 \\ 1 & 3 \end{bmatrix} + \begin{bmatrix} 4 & 1 \\ -3 & 2 \end{bmatrix} = \begin{bmatrix} 5+4 & -2+1 \\ 1+(-3) & 3+2 \end{bmatrix} = \begin{bmatrix} 9 & -1 \\ -2 & 5 \end{bmatrix}$

$A - B = \begin{bmatrix} 5 & -2 \\ 1 & 3 \end{bmatrix} - \begin{bmatrix} 4 & 1 \\ -3 & 2 \end{bmatrix} = \begin{bmatrix} 5-4 & -2-1 \\ 1-(-3) & 3-2 \end{bmatrix} = \begin{bmatrix} 1 & -3 \\ 4 & 1 \end{bmatrix}$

$2A = 2 \begin{bmatrix} 5 & -2 \\ 1 & 3 \end{bmatrix} = \begin{bmatrix} 2(5) & 2(-2) \\ 2(1) & 2(3) \end{bmatrix} = \begin{bmatrix} 10 & -4 \\ 2 & 6 \end{bmatrix}$

$-3B = -3 \begin{bmatrix} 4 & 1 \\ -3 & 2 \end{bmatrix} = \begin{bmatrix} -3(4) & -3(1) \\ -3(-3) & -3(2) \end{bmatrix} = \begin{bmatrix} -12 & -3 \\ 9 & -6 \end{bmatrix}$

3 $A + B = \begin{bmatrix} 6 & -1 \\ 2 & 0 \\ -3 & 4 \end{bmatrix} + \begin{bmatrix} 3 & 1 \\ -1 & 5 \\ 6 & 0 \end{bmatrix} = \begin{bmatrix} 6+3 & -1+1 \\ 2+(-1) & 0+5 \\ -3+6 & 4+0 \end{bmatrix} = \begin{bmatrix} 9 & 0 \\ 1 & 5 \\ 3 & 4 \end{bmatrix}$

$A - B = \begin{bmatrix} 6 & -1 \\ 2 & 0 \\ -3 & 4 \end{bmatrix} - \begin{bmatrix} 3 & 1 \\ -1 & 5 \\ 6 & 0 \end{bmatrix} = \begin{bmatrix} 6-3 & -1-1 \\ 2-(-1) & 0-5 \\ -3-6 & 4-0 \end{bmatrix} = \begin{bmatrix} 3 & -2 \\ 3 & -5 \\ -9 & 4 \end{bmatrix}$

$2A = 2 \begin{bmatrix} 6 & -1 \\ 2 & 0 \\ -3 & 4 \end{bmatrix} = \begin{bmatrix} 2(6) & 2(-1) \\ 2(2) & 2(0) \\ 2(-3) & 2(4) \end{bmatrix} = \begin{bmatrix} 12 & -2 \\ 4 & 0 \\ -6 & 8 \end{bmatrix}$

$-3B = -3 \begin{bmatrix} 3 & 1 \\ -1 & 5 \\ 6 & 0 \end{bmatrix} = \begin{bmatrix} -3(3) & -3(1) \\ -3(-1) & -3(5) \\ -3(6) & -3(0) \end{bmatrix} = \begin{bmatrix} -9 & -3 \\ 3 & -15 \\ -18 & 0 \end{bmatrix}$

5 $A + B = \begin{bmatrix} 4 & -3 & 2 \end{bmatrix} + \begin{bmatrix} 7 & 0 & -5 \end{bmatrix} = \begin{bmatrix} 11 & -3 & -3 \end{bmatrix}$,

$A - B = \begin{bmatrix} -3 & -3 & 7 \end{bmatrix}$, $2A = \begin{bmatrix} 8 & -6 & 4 \end{bmatrix}$, $-3B = \begin{bmatrix} -21 & 0 & 15 \end{bmatrix}$

7 $A + B$ and $A - B$ are not possible since A and B are different sizes.

$2A = 2 \begin{bmatrix} 3 & -2 & 2 \\ 0 & 1 & -4 \\ -3 & 2 & -1 \end{bmatrix} = \begin{bmatrix} 6 & -4 & 4 \\ 0 & 2 & -8 \\ -6 & 4 & -2 \end{bmatrix}$, $-3B = -3 \begin{bmatrix} 4 & 0 \\ 2 & -1 \\ -1 & 3 \end{bmatrix} = \begin{bmatrix} -12 & 0 \\ -6 & 3 \\ 3 & -9 \end{bmatrix}$

⑨ To find c_{21} in Exercise 15, use the second row of A and the first column of B.

$$c_{21} = (-5)(2) + (2)(0) + (2)(-4) = -10 + 0 - 8 = -18.$$

Note: For Exercises 11–22, AB is possible only if the number of columns of A equal the number of rows of B. { BA is not possible in #21, AB is not possible in #22 } Multiplying an $\boxed{m} \times n$ matrix with an $n \times \boxed{k}$ matrix will result in an $m \times k$ matrix.

⑪ $AB = \begin{bmatrix} 2 & 6 \\ 3 & -4 \end{bmatrix} \begin{bmatrix} 5 & -2 \\ 1 & 7 \end{bmatrix}$

$= \begin{bmatrix} 2(5) + 6(1) & 2(-2) + 6(7) \\ 3(5) + (-4)(1) & 3(-2) + (-4)(7) \end{bmatrix} = \begin{bmatrix} 10 + 6 & -4 + 42 \\ 15 - 4 & -6 - 28 \end{bmatrix} = \begin{bmatrix} 16 & 38 \\ 11 & -34 \end{bmatrix}$

$BA = \begin{bmatrix} 5 & -2 \\ 1 & 7 \end{bmatrix} \begin{bmatrix} 2 & 6 \\ 3 & -4 \end{bmatrix}$

$= \begin{bmatrix} 5(2) + (-2)(3) & 5(6) + (-2)(-4) \\ 1(2) + 7(3) & 1(6) + 7(-4) \end{bmatrix} = \begin{bmatrix} 4 & 38 \\ 23 & -22 \end{bmatrix}$

⑬ $AB = \begin{bmatrix} 3 & 0 & -1 \\ 0 & 4 & 2 \\ 5 & -3 & 1 \end{bmatrix} \begin{bmatrix} 1 & -5 & 0 \\ 4 & 1 & -2 \\ 0 & -1 & 3 \end{bmatrix}$

$= \begin{bmatrix} 3(1) + 0(4) + (-1)(0) & 3(-5) + 0(1) + (-1)(-1) & 3(0) + 0(-2) + (-1)(3) \\ 0(1) + 4(4) + 2(0) & 0(-5) + 4(1) + 2(-1) & 0(0) + 4(-2) + 2(3) \\ 5(1) + (-3)(4) + 1(0) & 5(-5) + (-3)(1) + 1(-1) & 5(0) + (-3)(-2) + 1(3) \end{bmatrix}$

$= \begin{bmatrix} 3 & -14 & -3 \\ 16 & 2 & -2 \\ -7 & -29 & 9 \end{bmatrix}$

$BA = \begin{bmatrix} 1 & -5 & 0 \\ 4 & 1 & -2 \\ 0 & -1 & 3 \end{bmatrix} \begin{bmatrix} 3 & 0 & -1 \\ 0 & 4 & 2 \\ 5 & -3 & 1 \end{bmatrix}$

$= \begin{bmatrix} 1(3) + (-5)(0) + 0(5) & 1(0) + (-5)(4) + 0(-3) & 1(-1) + (-5)(2) + 0(1) \\ 4(3) + 1(0) + (-2)(5) & 4(0) + 1(4) + (-2)(-3) & 4(-1) + 1(2) + (-2)(1) \\ 0(3) + (-1)(0) + 3(5) & 0(0) + (-1)(4) + 3(-3) & 0(-1) + (-1)(2) + 3(1) \end{bmatrix}$

$= \begin{bmatrix} 3 & -20 & -11 \\ 2 & 10 & -4 \\ 15 & -13 & 1 \end{bmatrix}$

15 $AB = \begin{bmatrix} 4 & -3 & 1 \\ -5 & 2 & 2 \end{bmatrix} \begin{bmatrix} 2 & 1 \\ 0 & 1 \\ -4 & 7 \end{bmatrix}$

$= \begin{bmatrix} 4(2) + (-3)(0) + 1(-4) & 4(1) + (-3)(1) + 1(7) \\ -5(2) + 2(0) + 2(-4) & -5(1) + 2(1) + 2(7) \end{bmatrix} = \begin{bmatrix} 4 & 8 \\ -18 & 11 \end{bmatrix}$

$BA = \begin{bmatrix} 2 & 1 \\ 0 & 1 \\ -4 & 7 \end{bmatrix} \begin{bmatrix} 4 & -3 & 1 \\ -5 & 2 & 2 \end{bmatrix}$

$= \begin{bmatrix} 2(4) + 1(-5) & 2(-3) + 1(2) & 2(1) + 1(2) \\ 0(4) + 1(-5) & 0(-3) + 1(2) & 0(1) + 1(2) \\ -4(4) + 7(-5) & -4(-3) + 7(2) & -4(1) + 7(2) \end{bmatrix} = \begin{bmatrix} 3 & -4 & 4 \\ -5 & 2 & 2 \\ -51 & 26 & 10 \end{bmatrix}$

17 $AB = \begin{bmatrix} 1 & 2 & 3 \\ 4 & 5 & 6 \\ 7 & 8 & 9 \end{bmatrix} \begin{bmatrix} 1 & 0 & 0 \\ 0 & 1 & 0 \\ 0 & 0 & 1 \end{bmatrix} = \begin{bmatrix} 1 & 2 & 3 \\ 4 & 5 & 6 \\ 7 & 8 & 9 \end{bmatrix}, BA = \begin{bmatrix} 1 & 2 & 3 \\ 4 & 5 & 6 \\ 7 & 8 & 9 \end{bmatrix}$

19 $AB = \begin{bmatrix} -3 & 7 & 2 \end{bmatrix} \begin{bmatrix} 1 \\ 4 \\ -5 \end{bmatrix} = \begin{bmatrix} -3(1) + 7(4) + 2(-5) \end{bmatrix} = \begin{bmatrix} 15 \end{bmatrix}$

$BA = \begin{bmatrix} 1 \\ 4 \\ -5 \end{bmatrix} \begin{bmatrix} -3 & 7 & 2 \end{bmatrix} = \begin{bmatrix} 1(-3) & 1(7) & 1(2) \\ 4(-3) & 4(7) & 4(2) \\ -5(-3) & -5(7) & -5(2) \end{bmatrix} = \begin{bmatrix} -3 & 7 & 2 \\ -12 & 28 & 8 \\ 15 & -35 & -10 \end{bmatrix}$

21 $AB = \begin{bmatrix} 2 & 0 & 1 \\ -1 & 2 & 0 \end{bmatrix} \begin{bmatrix} 1 & -1 & 2 \\ 3 & 1 & 0 \\ 0 & 2 & 1 \end{bmatrix}$

$= \begin{bmatrix} 2(1) + 0(3) + 1(0) & 2(-1) + 0(1) + 1(2) & 2(2) + 0(0) + 1(1) \\ -1(1) + 2(3) + 0(0) & -1(-1) + 2(1) + 0(2) & -1(2) + 2(0) + 0(1) \end{bmatrix}$

$= \begin{bmatrix} 2 & 0 & 5 \\ 5 & 3 & -2 \end{bmatrix}$. BA is a 3×3 matrix times a 2×3 matrix. Since the number

of columns of B, 3, is not equal to the number of rows of A, 2, BA is not possible.

23 $AB = \begin{bmatrix} 4 & -2 \\ 0 & 3 \\ -7 & 5 \end{bmatrix} \begin{bmatrix} 3 \\ 4 \end{bmatrix} = \begin{bmatrix} 4(3) + (-2)(4) \\ 0(3) + 3(4) \\ (-7)(3) + 5(4) \end{bmatrix} = \begin{bmatrix} 4 \\ 12 \\ -1 \end{bmatrix}$

25 $AB = \begin{bmatrix} 2 & 1 & 0 & -3 \\ -7 & 0 & -2 & 4 \end{bmatrix} \begin{bmatrix} 4 & -2 & 0 \\ 1 & 1 & -2 \\ 0 & 0 & 5 \\ -3 & -1 & 0 \end{bmatrix} = \begin{bmatrix} c_{11} & c_{12} & c_{13} \\ c_{21} & c_{22} & c_{23} \end{bmatrix}$, where

$c_{11} = 2(4) + 1(1) + 0(0) + (-3)(-3)$ $c_{12} = 2(-2) + 1(1) + 0(0) + (-3)(-1)$

$c_{13} = 2(0) + 1(-2) + 0(5) + (-3)(0)$ $c_{21} = -7(4) + 0(1) + (-2)(0) + 4(-3)$

$c_{22} = -7(-2) + 0(1) + (-2)(0) + 4(-1)$ $c_{23} = -7(0) + 0(-2) + (-2)(5) + 4(0)$

Hence, $AB = \begin{bmatrix} 18 & 0 & -2 \\ -40 & 10 & -10 \end{bmatrix}$.

27
$$A + B = \begin{bmatrix} 1 & 2 \\ 0 & -3 \end{bmatrix} + \begin{bmatrix} 2 & -1 \\ 3 & 1 \end{bmatrix} = \begin{bmatrix} 3 & 1 \\ 3 & -2 \end{bmatrix}$$

$$A - B = \begin{bmatrix} 1 & 2 \\ 0 & -3 \end{bmatrix} - \begin{bmatrix} 2 & -1 \\ 3 & 1 \end{bmatrix} = \begin{bmatrix} -1 & 3 \\ -3 & -4 \end{bmatrix}$$

So $(A + B)(A - B) = \begin{bmatrix} 3 & 1 \\ 3 & -2 \end{bmatrix} \begin{bmatrix} -1 & 3 \\ -3 & -4 \end{bmatrix} = \begin{bmatrix} -6 & 5 \\ 3 & 17 \end{bmatrix}$.

$$A^2 = AA = \begin{bmatrix} 1 & 2 \\ 0 & -3 \end{bmatrix} \begin{bmatrix} 1 & 2 \\ 0 & -3 \end{bmatrix} = \begin{bmatrix} 1 & -4 \\ 0 & 9 \end{bmatrix}$$

$$B^2 = BB = \begin{bmatrix} 2 & -1 \\ 3 & 1 \end{bmatrix} \begin{bmatrix} 2 & -1 \\ 3 & 1 \end{bmatrix} = \begin{bmatrix} 1 & -3 \\ 9 & -2 \end{bmatrix}$$

So $A^2 - B^2 = \begin{bmatrix} 1 & -4 \\ 0 & 9 \end{bmatrix} - \begin{bmatrix} 1 & -3 \\ 9 & -2 \end{bmatrix} = \begin{bmatrix} 0 & -1 \\ -9 & 11 \end{bmatrix}$.

Thus, $(A + B)(A - B) \neq A^2 - B^2$ since $\begin{bmatrix} -6 & 5 \\ 3 & 17 \end{bmatrix} \neq \begin{bmatrix} 0 & -1 \\ -9 & 11 \end{bmatrix}$.

29 $A(B + C) = \begin{bmatrix} 1 & 2 \\ 0 & -3 \end{bmatrix} \begin{bmatrix} 5 & 0 \\ 1 & 1 \end{bmatrix} = \begin{bmatrix} 7 & 2 \\ -3 & -3 \end{bmatrix}$;

$AB + AC = \begin{bmatrix} 8 & 1 \\ -9 & -3 \end{bmatrix} + \begin{bmatrix} -1 & 1 \\ 6 & 0 \end{bmatrix} = \begin{bmatrix} 7 & 2 \\ -3 & -3 \end{bmatrix}$

31 $m(A + B) = m \begin{bmatrix} a + p & b + q \\ c + r & d + s \end{bmatrix}$

$= \begin{bmatrix} m(a + p) & m(b + q) \\ m(c + r) & m(d + s) \end{bmatrix}$

$= \begin{bmatrix} ma + mp & mb + mq \\ mc + mr & md + ms \end{bmatrix}$

$= \begin{bmatrix} ma & mb \\ mc & md \end{bmatrix} + \begin{bmatrix} mp & mq \\ mr & ms \end{bmatrix}$

$= m \begin{bmatrix} a & b \\ c & d \end{bmatrix} + m \begin{bmatrix} p & q \\ r & s \end{bmatrix}$

$= mA + mB$

33 $A(B + C) = \begin{bmatrix} a & b \\ c & d \end{bmatrix} \begin{bmatrix} p + w & q + x \\ r + y & s + z \end{bmatrix}$

$$= \begin{bmatrix} a(p + w) + b(r + y) & a(q + x) + b(s + z) \\ c(p + w) + d(r + y) & c(q + x) + d(s + z) \end{bmatrix}$$

$$= \begin{bmatrix} ap + aw + br + by & aq + ax + bs + bz \\ cp + cw + dr + dy & cq + cx + ds + dz \end{bmatrix}$$

$$= \begin{bmatrix} ap + br & aq + bs \\ cp + dr & cq + ds \end{bmatrix} + \begin{bmatrix} aw + by & ax + bz \\ cw + dy & cx + dz \end{bmatrix}$$

$$= \begin{bmatrix} a & b \\ c & d \end{bmatrix} \begin{bmatrix} p & q \\ r & s \end{bmatrix} + \begin{bmatrix} a & b \\ c & d \end{bmatrix} \begin{bmatrix} w & x \\ y & z \end{bmatrix} = AB + AC$$

Note: For Exercises 35–38, $A = \begin{bmatrix} 3 & -3 & 7 \\ 2 & 6 & -2 \\ 4 & 2 & 5 \end{bmatrix}$ and $B = \begin{bmatrix} -9 & 5 & -8 \\ 3 & -7 & 1 \\ -1 & 2 & 6 \end{bmatrix}$.

35 $A^2 = AA = \begin{bmatrix} 3 & -3 & 7 \\ 2 & 6 & -2 \\ 4 & 2 & 5 \end{bmatrix} \begin{bmatrix} 3 & -3 & 7 \\ 2 & 6 & -2 \\ 4 & 2 & 5 \end{bmatrix} = \begin{bmatrix} 31 & -13 & 62 \\ 10 & 26 & -8 \\ 36 & 10 & 49 \end{bmatrix}$

$B^2 = BB = \begin{bmatrix} -9 & 5 & -8 \\ 3 & -7 & 1 \\ -1 & 2 & 6 \end{bmatrix} \begin{bmatrix} -9 & 5 & -8 \\ 3 & -7 & 1 \\ -1 & 2 & 6 \end{bmatrix} = \begin{bmatrix} 104 & -96 & 29 \\ -49 & 66 & -25 \\ 9 & -7 & 46 \end{bmatrix}$

$A^2 + B^2 = \begin{bmatrix} 31 & -13 & 62 \\ 10 & 26 & -8 \\ 36 & 10 & 49 \end{bmatrix} + \begin{bmatrix} 104 & -96 & 29 \\ -49 & 66 & -25 \\ 9 & -7 & 46 \end{bmatrix} = \begin{bmatrix} 135 & -109 & 91 \\ -39 & 92 & -33 \\ 45 & 3 & 95 \end{bmatrix}$

37 $A^2 - 5B = \begin{bmatrix} 31 & -13 & 62 \\ 10 & 26 & -8 \\ 36 & 10 & 49 \end{bmatrix} - \begin{bmatrix} -45 & 25 & -40 \\ 15 & -35 & 5 \\ -5 & 10 & 30 \end{bmatrix} = \begin{bmatrix} 76 & -38 & 102 \\ -5 & 61 & -13 \\ 41 & 0 & 19 \end{bmatrix}$

39 (a) For the inventory matrix, we have 5 colors and 3 sizes of towels, and for each size of towel, we have 1 price. We choose a 5×3 matrix A and a 3×1 matrix B.

$$\text{inventory matrix } A = \begin{bmatrix} 400 & 550 & 500 \\ 400 & 450 & 500 \\ 300 & 500 & 600 \\ 250 & 200 & 300 \\ 100 & 100 & 200 \end{bmatrix}, \text{ price matrix } B = \begin{bmatrix} \$1.39 \\ \$2.99 \\ \$4.99 \end{bmatrix}$$

(b) $C = AB = \begin{bmatrix} 400 & 550 & 500 \\ 400 & 450 & 500 \\ 300 & 500 & 600 \\ 250 & 200 & 300 \\ 100 & 100 & 200 \end{bmatrix} \begin{bmatrix} \$1.39 \\ \$2.99 \\ \$4.99 \end{bmatrix} = \begin{bmatrix} \$4695.50 \\ \$4396.50 \\ \$4906.00 \\ \$2442.50 \\ \$1436.00 \end{bmatrix}$

(c) The $1436.00 represents the amount the store would receive if all the yellow towels were sold.

Note: Exer. 1–12: Let A denote the given matrix.

[1] $AB = \begin{bmatrix} 5 & 7 \\ 2 & 3 \end{bmatrix} \begin{bmatrix} 3 & -7 \\ -2 & 5 \end{bmatrix} = \begin{bmatrix} 1 & 0 \\ 0 & 1 \end{bmatrix} = I_2$ and

$BA = \begin{bmatrix} 3 & -7 \\ -2 & 5 \end{bmatrix} \begin{bmatrix} 5 & 7 \\ 2 & 3 \end{bmatrix} = \begin{bmatrix} 1 & 0 \\ 0 & 1 \end{bmatrix} = I_2.$

Since $AB = I_2$ and $BA = I_2$, B is the inverse of A.

[3] $\left[\begin{array}{cc|cc} 2 & -4 & 1 & 0 \\ 1 & 3 & 0 & 1 \end{array}\right] R_1 - R_2 \rightarrow R_1 \;\Rightarrow\; \left[\begin{array}{cc|cc} 1 & -7 & 1 & -1 \\ 1 & 3 & 0 & 1 \end{array}\right] R_2 - R_1 \rightarrow R_2$

$\left[\begin{array}{cc|cc} 1 & -7 & 1 & -1 \\ 0 & 10 & -1 & 2 \end{array}\right] \tfrac{1}{10}R_2 \rightarrow R_2 \;\Rightarrow\; \left[\begin{array}{cc|cc} 1 & -7 & 1 & -1 \\ 0 & 1 & -\tfrac{1}{10} & \tfrac{2}{10} \end{array}\right] R_1 + 7R_2 \rightarrow R_1$

$\left[\begin{array}{cc|cc} 1 & 0 & \tfrac{3}{10} & \tfrac{4}{10} \\ 0 & 1 & -\tfrac{1}{10} & \tfrac{2}{10} \end{array}\right] \;\Rightarrow\; A^{-1} = \tfrac{1}{10}\begin{bmatrix} 3 & 4 \\ -1 & 2 \end{bmatrix}$

[5] $\left[\begin{array}{cc|cc} 2 & 4 & 1 & 0 \\ 4 & 8 & 0 & 1 \end{array}\right] \tfrac{1}{2}R_1 \rightarrow R_1 \;\Rightarrow\; \left[\begin{array}{cc|cc} 1 & 2 & \tfrac{1}{2} & 0 \\ 4 & 8 & 0 & 1 \end{array}\right] R_2 - 4R_1 \rightarrow R_2$

$\left[\begin{array}{cc|cc} 1 & 2 & \tfrac{1}{2} & 0 \\ 0 & 0 & -2 & 1 \end{array}\right]$

Since the identity matrix cannot be obtained on the left, *no inverse exists.*

[7] $\left[\begin{array}{ccc|ccc} 3 & -1 & 0 & 1 & 0 & 0 \\ 2 & 2 & 0 & 0 & 1 & 0 \\ 0 & 0 & 4 & 0 & 0 & 1 \end{array}\right] R_1 - R_2 \rightarrow R_1$

$\left[\begin{array}{ccc|ccc} 1 & -3 & 0 & 1 & -1 & 0 \\ 2 & 2 & 0 & 0 & 1 & 0 \\ 0 & 0 & 4 & 0 & 0 & 1 \end{array}\right] R_2 - 2R_1 \rightarrow R_2$

$\left[\begin{array}{ccc|ccc} 1 & -3 & 0 & 1 & -1 & 0 \\ 0 & 8 & 0 & -2 & 3 & 0 \\ 0 & 0 & 4 & 0 & 0 & 1 \end{array}\right] \begin{array}{l} (1/8)R_2 \rightarrow R_2 \\ (1/4)R_3 \rightarrow R_3 \end{array}$

$\left[\begin{array}{ccc|ccc} 1 & -3 & 0 & 1 & -1 & 0 \\ 0 & 1 & 0 & -\tfrac{2}{8} & \tfrac{3}{8} & 0 \\ 0 & 0 & 1 & 0 & 0 & \tfrac{1}{4} \end{array}\right] R_1 + 3R_2 \rightarrow R_1$

$\left[\begin{array}{ccc|ccc} 1 & 0 & 0 & \tfrac{2}{8} & \tfrac{1}{8} & 0 \\ 0 & 1 & 0 & -\tfrac{2}{8} & \tfrac{3}{8} & 0 \\ 0 & 0 & 1 & 0 & 0 & \tfrac{2}{8} \end{array}\right] \qquad A^{-1} = \tfrac{1}{8}\begin{bmatrix} 2 & 1 & 0 \\ -2 & 3 & 0 \\ 0 & 0 & 2 \end{bmatrix}$

[9] $\begin{bmatrix} -2 & 2 & 3 & | & 1 & 0 & 0 \\ 1 & -1 & 0 & | & 0 & 1 & 0 \\ 0 & 1 & 4 & | & 0 & 0 & 1 \end{bmatrix}$ $R_1 + 2R_2 \leftrightarrow R_2$

$\begin{bmatrix} 1 & -1 & 0 & | & 0 & 1 & 0 \\ 0 & 0 & 3 & | & 1 & 2 & 0 \\ 0 & 1 & 4 & | & 0 & 0 & 1 \end{bmatrix}$ $R_1 + R_3 \rightarrow R_1$

$\begin{bmatrix} 1 & 0 & 4 & | & 0 & 1 & 1 \\ 0 & 0 & 3 & | & 1 & 2 & 0 \\ 0 & 1 & 4 & | & 0 & 0 & 1 \end{bmatrix}$ $\frac{1}{3}R_2 \leftrightarrow R_3$

$\begin{bmatrix} 1 & 0 & 4 & | & 0 & 1 & 1 \\ 0 & 1 & 4 & | & 0 & 0 & 1 \\ 0 & 0 & 1 & | & \frac{1}{3} & \frac{2}{3} & 0 \end{bmatrix}$ $\begin{array}{l} R_1 - 4R_3 \rightarrow R_1 \\ R_2 - 4R_3 \rightarrow R_2 \end{array}$

$\begin{bmatrix} 1 & 0 & 0 & | & -\frac{4}{3} & -\frac{5}{3} & 1 \\ 0 & 1 & 0 & | & -\frac{4}{3} & -\frac{8}{3} & 1 \\ 0 & 0 & 1 & | & \frac{1}{3} & \frac{2}{3} & 0 \end{bmatrix}$ $\qquad A^{-1} = \frac{1}{3}\begin{bmatrix} -4 & -5 & 3 \\ -4 & -8 & 3 \\ 1 & 2 & 0 \end{bmatrix}$

[11] $\begin{bmatrix} 2 & 0 & 0 & | & 1 & 0 & 0 \\ 0 & 4 & 0 & | & 0 & 1 & 0 \\ 0 & 0 & 6 & | & 0 & 0 & 1 \end{bmatrix}$ $\begin{array}{l} (1/2)R_1 \rightarrow R_1 \\ (1/4)R_2 \rightarrow R_2 \\ (1/6)R_3 \rightarrow R_3 \end{array}$

$\begin{bmatrix} 1 & 0 & 0 & | & \frac{1}{2} & 0 & 0 \\ 0 & 1 & 0 & | & 0 & \frac{1}{4} & 0 \\ 0 & 0 & 1 & | & 0 & 0 & \frac{1}{6} \end{bmatrix}$ $\qquad A^{-1} = \frac{1}{12}\begin{bmatrix} 6 & 0 & 0 \\ 0 & 3 & 0 \\ 0 & 0 & 2 \end{bmatrix}$

[13] $\begin{bmatrix} a & 0 & | & 1 & 0 \\ 0 & b & | & 0 & 1 \end{bmatrix}$ $\begin{array}{l} (1/a)R_1 \rightarrow R_1 \\ (1/b)R_2 \rightarrow R_2 \end{array}$ $\Rightarrow$ $\begin{bmatrix} 1 & 0 & | & 1/a & 0 \\ 0 & 1 & | & 0 & 1/b \end{bmatrix}$

The inverse is the matrix with main diagonal elements $(1/a)$ and $(1/b)$.

The required conditions are that a and b are nonzero to avoid division by zero.

[15] $AI_3 = \begin{bmatrix} a_{11} & a_{12} & a_{13} \\ a_{21} & a_{22} & a_{23} \\ a_{31} & a_{32} & a_{33} \end{bmatrix}\begin{bmatrix} 1 & 0 & 0 \\ 0 & 1 & 0 \\ 0 & 0 & 1 \end{bmatrix} = \begin{bmatrix} a_{11} & a_{12} & a_{13} \\ a_{21} & a_{22} & a_{23} \\ a_{31} & a_{32} & a_{33} \end{bmatrix} = A$

$I_3 A = \begin{bmatrix} 1 & 0 & 0 \\ 0 & 1 & 0 \\ 0 & 0 & 1 \end{bmatrix}\begin{bmatrix} a_{11} & a_{12} & a_{13} \\ a_{21} & a_{22} & a_{23} \\ a_{31} & a_{32} & a_{33} \end{bmatrix} = \begin{bmatrix} a_{11} & a_{12} & a_{13} \\ a_{21} & a_{22} & a_{23} \\ a_{31} & a_{32} & a_{33} \end{bmatrix} = A$

[17] (a) $X = A^{-1}B = \frac{1}{10}\begin{bmatrix} 3 & 4 \\ -1 & 2 \end{bmatrix}\begin{bmatrix} 3 \\ 1 \end{bmatrix} = \frac{1}{10}\begin{bmatrix} 13 \\ -1 \end{bmatrix}; \quad (\frac{13}{10}, -\frac{1}{10})$

(b) $X = A^{-1}B = \frac{1}{10}\begin{bmatrix} 3 & 4 \\ -1 & 2 \end{bmatrix}\begin{bmatrix} -2 \\ 5 \end{bmatrix} = \frac{1}{10}\begin{bmatrix} 14 \\ 12 \end{bmatrix}; \quad (\frac{7}{5}, \frac{6}{5})$

19 (a) $X = A^{-1}B = \frac{1}{3}\begin{bmatrix} -4 & -5 & 3 \\ -4 & -8 & 3 \\ 1 & 2 & 0 \end{bmatrix}\begin{bmatrix} 1 \\ 3 \\ -2 \end{bmatrix} = \frac{1}{3}\begin{bmatrix} -25 \\ -34 \\ 7 \end{bmatrix};$ $(-\frac{25}{3}, -\frac{34}{3}, \frac{7}{3})$

(b) $X = A^{-1}B = \frac{1}{3}\begin{bmatrix} -4 & -5 & 3 \\ -4 & -8 & 3 \\ 1 & 2 & 0 \end{bmatrix}\begin{bmatrix} -1 \\ 0 \\ 4 \end{bmatrix} = \frac{1}{3}\begin{bmatrix} 16 \\ 16 \\ -1 \end{bmatrix};$ $(\frac{16}{3}, \frac{16}{3}, -\frac{1}{3})$

21 The inverse should be found using some type of computational device. If you are using a TI-83/86, enter the 9 values into the matrix [A]. Change Float to 5 via MODE. Now find $[A]^{-1}$ { be sure to use the $\boxed{x^{-1}}$ key }. Use the right arrow key to see the rightmost elements in [A].

$A = \begin{bmatrix} 2 & -5 & 8 \\ 3 & 7 & -1 \\ 0 & 2 & 1 \end{bmatrix} \Rightarrow A^{-1} \approx \begin{bmatrix} 0.11111 & 0.25926 & -0.62963 \\ -0.03704 & 0.02469 & 0.32099 \\ 0.07407 & -0.04938 & 0.35802 \end{bmatrix}$

23 $A = \begin{bmatrix} 2 & -1 & 1 & 4 \\ 7 & 1.2 & -8 & 0 \\ 2.5 & 0 & 1.9 & 7.9 \\ 1 & -1 & 3 & 1 \end{bmatrix} \Rightarrow A^{-1} \approx \begin{bmatrix} -0.22278 & 0.12932 & 0.06496 & 0.37796 \\ -1.17767 & 0.09503 & 0.55936 & 0.29171 \\ -0.37159 & 0.00241 & 0.14074 & 0.37447 \\ 0.15987 & -0.04150 & 0.07218 & -0.20967 \end{bmatrix}$

25 (a) $AX = B \Leftrightarrow \begin{bmatrix} 4.0 & 7.1 \\ 2.2 & -4.9 \end{bmatrix}\begin{bmatrix} x \\ y \end{bmatrix} = \begin{bmatrix} 6.2 \\ 2.9 \end{bmatrix}$

(b) On your calculator, Find $[A]^{-1}$ and STOre this matrix into matrix [B] to use in part (c).

$$A^{-1} \approx \begin{bmatrix} 0.1391 & 0.2016 \\ 0.0625 & -0.1136 \end{bmatrix}$$

(c) Following the instructions in part (b), enter the 3 values into [C] { a 3×1 matrix }, and then evaluate [B]*[C].

$$X = A^{-1}B \approx \begin{bmatrix} 0.1391 & 0.2016 \\ 0.0625 & -0.1136 \end{bmatrix}\begin{bmatrix} 6.2 \\ 2.9 \end{bmatrix} \approx \begin{bmatrix} 1.4472 \\ 0.0579 \end{bmatrix}.$$

27 (a) $AX = B \Leftrightarrow \begin{bmatrix} 3.1 & 6.7 & -8.7 \\ 4.1 & -5.1 & 0.2 \\ 0.6 & 1.1 & -7.4 \end{bmatrix}\begin{bmatrix} x \\ y \\ z \end{bmatrix} = \begin{bmatrix} 1.5 \\ 2.1 \\ 3.9 \end{bmatrix}$

(b) $A^{-1} \approx \begin{bmatrix} 0.1474 & 0.1572 & -0.1691 \\ 0.1197 & -0.0696 & -0.1426 \\ 0.0297 & 0.0024 & -0.1700 \end{bmatrix}$

(c) $X = A^{-1}B \approx \begin{bmatrix} 0.1474 & 0.1572 & -0.1691 \\ 0.1197 & -0.0696 & -0.1426 \\ 0.0297 & 0.0024 & -0.1700 \end{bmatrix} \begin{bmatrix} 1.5 \\ 2.1 \\ 3.9 \end{bmatrix} \approx \begin{bmatrix} -0.1081 \\ -0.5227 \\ -0.6135 \end{bmatrix}$

29 (a) $f(2) = 4a + 2b + c = 19$; $f(8) = 64a + 8b + c = 59$; $f(11) = 121a + 11b + c = 26$

Solve the 3×3 linear system using the inverse method.

$$\begin{bmatrix} 4 & 2 & 1 \\ 64 & 8 & 1 \\ 121 & 11 & 1 \end{bmatrix} \begin{bmatrix} a \\ b \\ c \end{bmatrix} = \begin{bmatrix} 19 \\ 59 \\ 26 \end{bmatrix} \Rightarrow \begin{bmatrix} a \\ b \\ c \end{bmatrix} \approx \begin{bmatrix} -1.9630 \\ 26.2963 \\ -25.7407 \end{bmatrix}$$

Thus, let $f(x) = -1.9630x^2 + 26.2963x - 25.7407$.

(b)

[1, 12] by [−15, 70, 5]

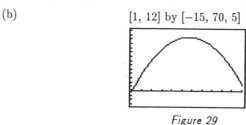

Figure 29

(c) For June, $f(6) \approx 61°F$ and for October, $f(10) \approx 41°F$.

8.8 Exercises

Note: The minor M_{ij} and the cofactor A_{ij} are equal if $(i + j)$ is even and of opposite sign if $(i + j)$ is odd.

1 The minor M_{11} is obtained by deleting the first row and first column from

$A = \begin{bmatrix} 7 & -1 \\ 5 & 0 \end{bmatrix}$. Thus, $M_{11} = 0 = A_{11}$. Similarly, $M_{12} = 5$ and $A_{12} = -5$;

$M_{21} = -1$ and $A_{21} = 1$; and $M_{22} = 7 = A_{22}$.

3 Let $A = \begin{bmatrix} 2 & 4 & -1 \\ 0 & 3 & 2 \\ -5 & 7 & 0 \end{bmatrix}$.

Be sure you understand the note preceding the solution for Exercise 1.

$M_{11} = \begin{vmatrix} 3 & 2 \\ 7 & 0 \end{vmatrix} = (3)(0) - (7)(2) = 0 - 14 = -14 = A_{11}$;

$M_{12} = \begin{vmatrix} 0 & 2 \\ -5 & 0 \end{vmatrix} = (0)(0) - (-5)(2) = 0 - (-10) = 10$; $A_{12} = -10$;

$M_{13} = \begin{vmatrix} 0 & 3 \\ -5 & 7 \end{vmatrix} = 15 = A_{13}$; $\qquad M_{21} = \begin{vmatrix} 4 & -1 \\ 7 & 0 \end{vmatrix} = 7$; $A_{21} = -7$;

$M_{22} = \begin{vmatrix} 2 & -1 \\ -5 & 0 \end{vmatrix} = -5 = A_{22}$; $\qquad M_{23} = \begin{vmatrix} 2 & 4 \\ -5 & 7 \end{vmatrix} = 34$; $A_{23} = -34$; (cont.)

$$M_{31} = \begin{vmatrix} 4 & -1 \\ 3 & 2 \end{vmatrix} = 11 = A_{31}; \qquad M_{32} = \begin{vmatrix} 2 & -1 \\ 0 & 2 \end{vmatrix} = 4; \; A_{32} = -4;$$

$$M_{33} = \begin{vmatrix} 2 & 4 \\ 0 & 3 \end{vmatrix} = 6 = A_{33}.$$

Note: Exercises 5–20: Let A denote the given matrix.

$\boxed{5}$ $\begin{vmatrix} 7 & -1 \\ 5 & 0 \end{vmatrix} = (7)(0) - (-1)(5) = 0 + 5 = 5$

$\boxed{7}$ Expanding $|A|$ by the first column and using the cofactor values from Exercise 3,

we obtain $|A| = a_{11}A_{11} + a_{21}A_{21} + a_{31}A_{31} = 2(-14) + 0(A_{21}) - 5(11) = -83.$

$\boxed{9}$ By the definition of the determinant of a 2×2 matrix A (on page 652),

$$|A| = \begin{vmatrix} -5 & 4 \\ -3 & 2 \end{vmatrix} = (-5)(2) - (4)(-3) = -10 + 12 = 2.$$

$\boxed{11}$ $|A| = (a)(-b) - (-a)(b) = -ab + ab = 0$

$\boxed{13}$ Expand by the first row.

$|A| = a_{11}A_{11} + a_{12}A_{12} + a_{13}A_{13}$

$$= (3)(-1)^{1+1} \begin{vmatrix} 2 & 5 \\ 3 & -1 \end{vmatrix} + (1)(-1)^{1+2} \begin{vmatrix} 4 & 5 \\ -6 & -1 \end{vmatrix} + (-2)(-1)^{1+3} \begin{vmatrix} 4 & 2 \\ -6 & 3 \end{vmatrix}$$

$$= (3)(1)(-17) + (1)(-1)(26) + (-2)(1)(24)$$

$$= -51 - 26 - 48 = -125$$

$\boxed{15}$ Expand by the third row.

$|A| = a_{31}A_{31} + a_{32}A_{32} + a_{33}A_{33} = 2(30) + 0(A_{32}) + 6(-2) = 48$

$\boxed{17}$ Expand $|A|$ by the third row. $|A| = 6\,A_{32} = -6\,M_{32} = -6 \begin{vmatrix} 3 & 2 & 0 \\ 4 & -3 & 5 \\ 1 & -4 & 2 \end{vmatrix}.$

Expand M_{32} by the first row.

$M_{32} = 3(14) + 2(-3) + 0(-13) = 36 \;\Rightarrow\; |A| = -6(36) = -216.$

$\boxed{19}$ Expand by the first row. $|A| = -b \begin{vmatrix} 0 & c & 0 \\ a & 0 & 0 \\ 0 & 0 & d \end{vmatrix}.$

Expand again by the first row. $|A| = (-b)(-c) \begin{vmatrix} a & 0 \\ 0 & d \end{vmatrix} = bc(ad - 0) = abcd.$

$\boxed{21}$ LS $= ad - bc;$ RS $= -(bc - ad) = ad - bc$

$\boxed{23}$ LS $= adk - bck;$ RS $= k(ad - bc) = adk - bck$

$\boxed{25}$ LS $= ad - bc;$ RS $= abk + ad - abk - bc = ad - bc$

$\boxed{27}$ LS $= ad - bc + af - ce;$ RS $= ad + af - bc - ce$

29 Consider the matrix in Exercise 20. If we expanded along the first column, we would obtain a times its cofactor. Expanding along the first column again, we obtain ab times another cofactor. This exercise is similar since all elements in A *above* {rather than below} the main diagonal are zero. We can evaluate the determinant using n expansions by the first row, and obtain $|A| = a_{11}a_{22}\cdots a_{nn}$.

31 (a) $A - xI = \begin{bmatrix} 1 & 2 \\ 3 & 2 \end{bmatrix} - x \begin{bmatrix} 1 & 0 \\ 0 & 1 \end{bmatrix} = \begin{bmatrix} 1-x & 2 \\ 3 & 2-x \end{bmatrix}$.

$f(x) = |A - xI| = \begin{vmatrix} 1-x & 2 \\ 3 & 2-x \end{vmatrix}$

$= (1-x)(2-x) - (3)(2) = (2 - 3x + x^2) - 6 = x^2 - 3x - 4.$

(b) $x^2 - 3x - 4 = 0 \Rightarrow (x-4)(x+1) = 0 \Rightarrow x = -1, 4$

33 (a) $f(x) = |A - xI| = \begin{vmatrix} -3-x & -2 \\ 2 & 2-x \end{vmatrix} = x^2 + x - 2$

(b) $x^2 + x - 2 = 0 \Rightarrow (x+2)(x-1) = 0 \Rightarrow x = -2, 1$

35 (a) $f(x) = \begin{vmatrix} 1-x & 0 & 0 \\ 1 & 0-x & -2 \\ -1 & 1 & -3-x \end{vmatrix}$ {Expand by the first row.}

$= (1-x)[(-x)(-3-x) - (-2)]$

$= (1-x)(x^2 + 3x + 2)$

$= (1-x)(x+1)(x+2)$ or

$\qquad\qquad (-x^3 - 2x^2 + x + 2)$

(b) $(1-x)(x+1)(x+2) = 0 \Rightarrow x = -2, -1, 1$

37 (a) $f(x) = \begin{vmatrix} 0-x & 2 & -2 \\ -1 & 3-x & 1 \\ -3 & 3 & 1-x \end{vmatrix}$ {Expand by the first row.}

$= (-x)[(3-x)(1-x) - 3] - 2[(x-1) + 3] - 2[-3 + 3(3-x)]$

$= (-x)[(3 - 4x + x^2) - 3] - 2(x+2) - 2(-3x + 6)$

$= (-x^3 + 4x^2) - 2x - 4 + 6x - 12$

$= -x^3 + 4x^2 + 4x - 16$

(b) By trying possible rational roots, we determine that 2 is a zero of f.

Thus, $-x^3 + 4x^2 + 4x - 16 = 0 \Rightarrow (x-2)(-x^2 + 2x + 8) = 0 \Rightarrow$

$\qquad\qquad (x+2)(x-2)(-x+4) = 0 \Rightarrow x = -2, 2, 4.$

39 Expand the determinant by the first row.

$\begin{vmatrix} i & j & k \\ 2 & -1 & 6 \\ -3 & 5 & 1 \end{vmatrix} = i \begin{vmatrix} -1 & 6 \\ 5 & 1 \end{vmatrix} - j \begin{vmatrix} 2 & 6 \\ -3 & 1 \end{vmatrix} + k \begin{vmatrix} 2 & -1 \\ -3 & 5 \end{vmatrix} = -31i - 20j + 7k$

$$\boxed{41} \quad \begin{vmatrix} i & j & k \\ 5 & -6 & -1 \\ 3 & 0 & 1 \end{vmatrix} = i \begin{vmatrix} -6 & -1 \\ 0 & 1 \end{vmatrix} - j \begin{vmatrix} 5 & -1 \\ 3 & 1 \end{vmatrix} + k \begin{vmatrix} 5 & -6 \\ 3 & 0 \end{vmatrix} = -6i - 8j + 18k$$

$$\boxed{43} \quad A = \begin{bmatrix} 29 & -17 & 90 \\ -34 & 91 & -34 \\ 48 & 7 & 10 \end{bmatrix} \Rightarrow |A| = -359{,}284.$$

$$\boxed{45} \quad A = \begin{bmatrix} 4 & -7 & -3 & 13 \\ -17 & -0.8 & 5 & 0.9 \\ 1.1 & 0.2 & 10 & -4 \\ 3 & -6 & 2 & 1 \end{bmatrix} \Rightarrow |A| = 10{,}739.92.$$

$\boxed{47}$ (a) $f(x) = |A - xI| = \begin{vmatrix} 1-x & 0 & 1 \\ 0 & 2-x & 1 \\ 1 & 1 & -2-x \end{vmatrix} = -x^3 + x^2 + 6x - 7$

(b) The characteristic values of A are equal to the zeros of f. From the graph,

we see that the zeros are approximately -2.51, 1.22, and 2.29.

$[-10, 11]$ by $[-12, 2]$

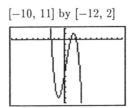

Figure 47

8.9 Exercises

$\boxed{1}$ R_2 and R_3 are interchanged. The determinant value is multiplied by -1.

$\boxed{3}$ R_3 is replaced by $(R_3 - R_1)$. There is no change in the determinant value.

$\boxed{5}$ The number 2 can be factored out of R_1, yielding $2 \begin{vmatrix} 1 & 2 & 1 \\ 1 & 2 & 4 \\ 2 & 6 & 4 \end{vmatrix}$.

Next, the number 2 can be factored out of R_3, yielding $4 \begin{vmatrix} 1 & 2 & 1 \\ 1 & 2 & 4 \\ 1 & 3 & 2 \end{vmatrix}$.

$\boxed{7}$ R_1 and R_3 are identical. The determinant is 0.

$\boxed{9}$ The number -1 can be factored out of R_2, yielding the determinant on the right side.

$\boxed{11}$ Every number in C_2 is 0. The determinant is zero.

$\boxed{13}$ C_3 is replaced by $(2C_1 + C_3)$. There is no change in the determinant value.

Note: The notation $\{R_i\,(C_i)\}$ means expand the determinant by the ith row (column).

[15] There are many possibilities for introducing zeros. In this case, obtaining a zero in the third row, second column would lead to an easy evaluation using the second column.

$$\begin{vmatrix} 3 & 1 & 0 \\ -2 & 0 & 1 \\ 1 & 3 & -1 \end{vmatrix} \begin{matrix} \\ R_3 - 3R_1 \to R_3 \\ \end{matrix} = \begin{vmatrix} 3 & 1 & 0 \\ -2 & 0 & 1 \\ -8 & 0 & -1 \end{vmatrix} \{C_2\}$$

$$= (-1)\begin{vmatrix} -2 & 1 \\ -8 & -1 \end{vmatrix} = (-1)(2 + 8) = -10$$

[17]
$$\begin{vmatrix} 5 & 4 & 3 \\ -3 & 2 & 1 \\ 0 & 7 & -2 \end{vmatrix} \begin{matrix} R_1 - 3R_2 \to R_1 \\ = \\ R_3 + 2R_2 \to R_3 \end{matrix} \begin{vmatrix} 14 & -2 & 0 \\ -3 & 2 & 1 \\ -6 & 11 & 0 \end{vmatrix} \{C_3\}$$

$$= (-1)\begin{vmatrix} 14 & -2 \\ -6 & 11 \end{vmatrix} = (-1)(154 - 12) = -142$$

[19]
$$\begin{vmatrix} 2 & 2 & -3 \\ 3 & 6 & 9 \\ -2 & 5 & 4 \end{vmatrix} \{3 \text{ is a common factor of } R_2\} = (3)\begin{vmatrix} 2 & 2 & -3 \\ 1 & 2 & 3 \\ -2 & 5 & 4 \end{vmatrix}$$

Since there is a "1" in the second row, first column, we will obtain zeros in the other two locations of the first column.

$$(3)\begin{vmatrix} 2 & 2 & -3 \\ 1 & 2 & 3 \\ -2 & 5 & 4 \end{vmatrix} \begin{matrix} R_1 - 2R_2 \to R_1 \\ \\ R_3 + 2R_2 \to R_3 \end{matrix} = (3)\begin{vmatrix} 0 & -2 & -9 \\ 1 & 2 & 3 \\ 0 & 9 & 10 \end{vmatrix} \{C_1\}$$

$$= (3)(-1)\begin{vmatrix} -2 & -9 \\ 9 & 10 \end{vmatrix} = (-3)(-20 + 81) = -183$$

[21]
$$\begin{vmatrix} 3 & 1 & -2 & 2 \\ 2 & 0 & 1 & 4 \\ 0 & 1 & 3 & 5 \\ -1 & 2 & 0 & -3 \end{vmatrix} \begin{matrix} \\ \\ R_3 - R_1 \to R_3 \\ R_4 - 2R_1 \to R_4 \end{matrix} = \begin{vmatrix} 3 & 1 & -2 & 2 \\ 2 & 0 & 1 & 4 \\ -3 & 0 & 5 & 3 \\ -7 & 0 & 4 & -7 \end{vmatrix} \{C_2\}$$

$$= (-1)\begin{vmatrix} 2 & 1 & 4 \\ -3 & 5 & 3 \\ -7 & 4 & -7 \end{vmatrix} \begin{matrix} R_2 - 5R_1 \to R_2 \\ R_3 - 4R_1 \to R_3 \end{matrix} = (-1)\begin{vmatrix} 2 & 1 & 4 \\ -13 & 0 & -17 \\ -15 & 0 & -23 \end{vmatrix} \{C_2\}$$

$$= (-1)(-1)\begin{vmatrix} -13 & -17 \\ -15 & -23 \end{vmatrix} = (1)(299 - 255) = 44$$

$$
\boxed{23} \quad
\begin{vmatrix}
2 & -2 & 0 & 0 & -3 \\
3 & 0 & 3 & 2 & -1 \\
0 & 1 & -2 & 0 & 2 \\
-1 & 2 & 0 & 3 & 0 \\
0 & 4 & 1 & 0 & 0
\end{vmatrix}
\begin{array}{l} C_2 - 4C_3 \to C_2 \end{array}
=
\begin{vmatrix}
2 & -2 & 0 & 0 & -3 \\
3 & -12 & 3 & 2 & -1 \\
0 & 9 & -2 & 0 & 2 \\
-1 & 2 & 0 & 3 & 0 \\
0 & 0 & 1 & 0 & 0
\end{vmatrix}
\{R_5\}
$$

$$
= (1)
\begin{vmatrix}
2 & -2 & 0 & -3 \\
3 & -12 & 2 & -1 \\
0 & 9 & 0 & 2 \\
-1 & 2 & 3 & 0
\end{vmatrix}
\begin{array}{l} R_1 + 2R_4 \to R_1 \\ R_2 + 3R_4 \to R_2 \end{array}
=
\begin{vmatrix}
0 & 2 & 6 & -3 \\
0 & -6 & 11 & -1 \\
0 & 9 & 0 & 2 \\
-1 & 2 & 3 & 0
\end{vmatrix}
\{C_1\}
$$

$$
= (1)
\begin{vmatrix}
2 & 6 & -3 \\
-6 & 11 & -1 \\
9 & 0 & 2
\end{vmatrix}
\begin{array}{l} R_1 - 3R_2 \to R_1 \\ R_3 + 2R_2 \to R_3 \end{array}
=
\begin{vmatrix}
20 & -27 & 0 \\
-6 & 11 & -1 \\
-3 & 22 & 0
\end{vmatrix}
\{C_3\}
$$

$$
= (1)
\begin{vmatrix}
20 & -27 \\
-3 & 22
\end{vmatrix}
= (1)(440 - 81) = 359
$$

$$
\boxed{25} \quad
\begin{vmatrix}
1 & 1 & 1 \\
a & b & c \\
a^2 & b^2 & c^2
\end{vmatrix}
\begin{array}{l} C_1 - C_2 \to C_1 \\ C_3 - C_2 \to C_3 \end{array}
$$

$$
=
\begin{vmatrix}
0 & 1 & 0 \\
a-b & b & c-b \\
a^2 - b^2 & b^2 & c^2 - b^2
\end{vmatrix}
\begin{array}{l} a-b \text{ is a common factor of } C_1 \\ c-b \text{ is a common factor of } C_3 \end{array}
$$

$$
= (a-b)(c-b)
\begin{vmatrix}
0 & 1 & 0 \\
1 & b & 1 \\
a+b & b^2 & c+b
\end{vmatrix}
\{R_1\}
$$

$$
= (a-b)(c-b)(-1)
\begin{vmatrix}
1 & 1 \\
a+b & c+b
\end{vmatrix}
$$

$$
= (a-b)(b-c)(c+b-a-b) = \underline{(a-b)(b-c)(c-a)}
$$

$$
\boxed{27} \quad
\begin{vmatrix}
a_{11} & a_{12} & a_{13} & a_{14} \\
0 & a_{22} & a_{23} & a_{24} \\
0 & 0 & a_{33} & a_{34} \\
0 & 0 & 0 & a_{44}
\end{vmatrix}
\{C_1\} = (a_{11})
\begin{vmatrix}
a_{22} & a_{23} & a_{24} \\
0 & a_{33} & a_{34} \\
0 & 0 & a_{44}
\end{vmatrix}
\{C_1\}
$$

$$
= (a_{11})(a_{22})
\begin{vmatrix}
a_{33} & a_{34} \\
0 & a_{44}
\end{vmatrix}
= (a_{11} a_{22})(a_{33} a_{44} - 0) = a_{11} a_{22} a_{33} a_{44}
$$

$$
\boxed{29} \quad |AB| =
\begin{vmatrix}
a_{11}b_{11} + a_{12}b_{21} & a_{11}b_{12} + a_{12}b_{22} \\
a_{21}b_{11} + a_{22}b_{21} & a_{21}b_{12} + a_{22}b_{22}
\end{vmatrix}
$$

$$
= (a_{11}b_{11} + a_{12}b_{21})(a_{21}b_{12} + a_{22}b_{22}) - (a_{11}b_{12} + a_{12}b_{22})(a_{21}b_{11} + a_{22}b_{21})
$$

$$
= \quad a_{11}b_{11}a_{21}b_{12} + a_{11}b_{11}a_{22}b_{22} + a_{12}b_{21}a_{21}b_{12} + a_{12}b_{21}a_{22}b_{22}
$$

$$
\quad - a_{11}b_{12}a_{21}b_{11} - a_{11}b_{12}a_{22}b_{21} - a_{12}b_{22}a_{21}b_{11} - a_{12}b_{22}a_{22}b_{21}
$$

$$
= a_{11}a_{22}b_{11}b_{22} - a_{11}a_{22}b_{21}b_{12} - a_{21}a_{12}b_{11}b_{22} + a_{21}a_{12}b_{21}b_{12}
$$

$$
= (a_{11}a_{22} - a_{21}a_{12})(b_{11}b_{22} - b_{21}b_{12}) = |A| \, |B|
$$

$\boxed{31}$ Expanding by the first row yields $Ax + By + C = 0$ {an equation of a line} where A, B, and C are constants. To show that the line contains (x_1, y_1) and (x_2, y_2), we must show that these points are solutions of the equation. Substituting x_1 for x and y_1 for y, we obtain two identical rows and the determinant is zero. Hence, (x_1, y_1) is a solution of the equation and a similar argument can be made for (x_2, y_2).

$\boxed{33}$ For the system $\begin{cases} 2x + 3y = 2 \\ x - 2y = 8 \end{cases}$, $|D| = \begin{vmatrix} 2 & 3 \\ 1 & -2 \end{vmatrix} = -4 - 3 = -7$.

Since $|D| = -7 \ne 0$, we may solve the system using Cramer's rule.

$|D_x| = \begin{vmatrix} 2 & 3 \\ 8 & -2 \end{vmatrix} = -4 - 24 = -28$. $|D_y| = \begin{vmatrix} 2 & 2 \\ 1 & 8 \end{vmatrix} = 16 - 2 = 14$.

$x = \dfrac{|D_x|}{|D|} = \dfrac{-28}{-7} = 4$ and $y = \dfrac{|D_y|}{|D|} = \dfrac{14}{-7} = -2$. ★ $(4, -2)$

$\boxed{35}$ For the system $\begin{cases} 2x + 5y = 16 \\ 3x - 7y = 24 \end{cases}$, $|D| = \begin{vmatrix} 2 & 5 \\ 3 & -7 \end{vmatrix} = -14 - 15 = -29$.

Since $|D| = -29 \ne 0$, we may solve the system using Cramer's rule.

$|D_x| = \begin{vmatrix} 16 & 5 \\ 24 & -7 \end{vmatrix} = -112 - 120 = -232$. $|D_y| = \begin{vmatrix} 2 & 16 \\ 3 & 24 \end{vmatrix} = 48 - 48 = 0$.

$x = \dfrac{|D_x|}{|D|} = \dfrac{-232}{-29} = 8$ and $y = \dfrac{|D_y|}{|D|} = \dfrac{0}{-29} = 0$. ★ $(8, 0)$

$\boxed{37}$ $|D| = \begin{vmatrix} 2 & -3 \\ -6 & 9 \end{vmatrix} = 18 - 18 = 0$, so Cramer's rule cannot be used.

$\boxed{39}$ $|D| = \begin{vmatrix} 1 & -2 & -3 \\ 2 & 1 & 1 \\ 1 & 3 & -2 \end{vmatrix}$ $\{R_1\} = 1(-5) - (-2)(-5) + (-3)(5) = -30$

$|D_x| = \begin{vmatrix} -1 & -2 & -3 \\ 6 & 1 & 1 \\ 13 & 3 & -2 \end{vmatrix}$ $\{R_1\} = (-1)(-5) - (-2)(-25) + (-3)(5) = -60$

$|D_y| = \begin{vmatrix} 1 & -1 & -3 \\ 2 & 6 & 1 \\ 1 & 13 & -2 \end{vmatrix}$ $\{R_1\} = 1(-25) - (-1)(-5) + (-3)(20) = -90$

$|D_z| = \begin{vmatrix} 1 & -2 & -1 \\ 2 & 1 & 6 \\ 1 & 3 & 13 \end{vmatrix}$ $\{R_1\} = 1(-5) - (-2)(20) + (-1)(5) = 30$

$x = \dfrac{|D_x|}{|D|} = \dfrac{-60}{-30} = 2$; $y = \dfrac{|D_y|}{|D|} = \dfrac{-90}{-30} = 3$; $z = \dfrac{|D_z|}{|D|} = \dfrac{30}{-30} = -1$.

★ $(2, 3, -1)$

41 $|D| = \begin{vmatrix} 5 & 2 & -1 \\ 1 & -2 & 2 \\ 0 & 3 & 1 \end{vmatrix} \{R_3\} = -3(11) + 1(-12) = -45$

$|D_x| = \begin{vmatrix} -7 & 2 & -1 \\ 0 & -2 & 2 \\ 17 & 3 & 1 \end{vmatrix} \{C_1\} = -7(-8) + 17(2) = 90$

$|D_y| = \begin{vmatrix} 5 & -7 & -1 \\ 1 & 0 & 2 \\ 0 & 17 & 1 \end{vmatrix} \{R_2\} = -1(10) - 2(85) = -180$

$|D_z| = \begin{vmatrix} 5 & 2 & -7 \\ 1 & -2 & 0 \\ 0 & 3 & 17 \end{vmatrix} \{C_1\} = 5(-34) - 1(55) = -225$

$x = \dfrac{|D_x|}{|D|} = \dfrac{90}{-45} = -2; \quad y = \dfrac{|D_y|}{|D|} = \dfrac{-180}{-45} = 4; \quad z = \dfrac{|D_z|}{|D|} = \dfrac{-225}{-45} = 5.$

$\star \ (-2, 4, 5)$

8.10 Exercises

Note: The general outline for the solutions in this section is as follows:

1st line)   The expression is shown on the left side of the equation and its decomposition is on the right side.

2nd line)   The equation in the first line is multiplied by its least common denominator and left in factored form.

3rd line and beyond)   Values are substituted into the equation in the second line and the coefficients are found by solving the resulting equations. It will be stated when the method of equating coefficients is used.

1 $\dfrac{8x - 1}{(x - 2)(x + 3)} = \dfrac{A}{x - 2} + \dfrac{B}{x + 3}$

$8x - 1 = A(x + 3) + B(x - 2)$

$x = -3: \ -25 = -5B \ \Rightarrow \ B = 5$

$x = 2: \ 15 = 5A \ \Rightarrow \ A = 3$

$\star \ \dfrac{3}{x - 2} + \dfrac{5}{x + 3}$

3 $\dfrac{x + 34}{(x - 6)(x + 2)} = \dfrac{A}{x - 6} + \dfrac{B}{x + 2}$

$x + 34 = A(x + 2) + B(x - 6)$

$x = -2: \ 32 = -8B \ \Rightarrow \ B = -4$

$x = 6: \ 40 = 8A \ \Rightarrow \ A = 5$

$\star \ \dfrac{5}{x - 6} - \dfrac{4}{x + 2}$

[5] $\dfrac{4x^2 - 15x - 1}{(x-1)(x+2)(x-3)} = \dfrac{A}{x-1} + \dfrac{B}{x+2} + \dfrac{C}{x-3}$

$4x^2 - 15x - 1 = A(x+2)(x-3) + B(x-1)(x-3) + C(x-1)(x+2)$

$x = -2:\ 45 = 15B \ \Rightarrow \ B = 3$

$x = 1:\ -12 = -6A \ \Rightarrow \ A = 2$ $\qquad\qquad$ ★ $\dfrac{2}{x-1} + \dfrac{3}{x+2} - \dfrac{1}{x-3}$

$x = 3:\ -10 = 10C \ \Rightarrow \ C = -1$

[7] $\dfrac{4x^2 - 5x - 15}{x(x-5)(x+1)} = \dfrac{A}{x} + \dfrac{B}{x-5} + \dfrac{C}{x+1}$

$4x^2 - 5x - 15 = A(x-5)(x+1) + Bx(x+1) + Cx(x-5)$

$x = -1:\ -6 = 6C \ \Rightarrow \ C = -1$

$x = 0:\ -15 = -5A \ \Rightarrow \ A = 3$ $\qquad\qquad$ ★ $\dfrac{3}{x} + \dfrac{2}{x-5} - \dfrac{1}{x+1}$

$x = 5:\ 60 = 30B \ \Rightarrow \ B = 2$

[9] $\dfrac{2x+3}{(x-1)^2} = \dfrac{A}{x-1} + \dfrac{B}{(x-1)^2}$

$2x + 3 = A(x-1) + B$

$x = 1:\ 5 = B$ $\qquad\qquad\qquad\qquad\qquad\qquad\qquad$ ★ $\dfrac{2}{x-1} + \dfrac{5}{(x-1)^2}$

$x = 0:\ 3 = -A + B \ \Rightarrow \ A = 2$

[11] $\dfrac{19x^2 + 50x - 25}{x^2(3x-5)} = \dfrac{A}{x} + \dfrac{B}{x^2} + \dfrac{C}{3x-5}$

$19x^2 + 50x - 25 = Ax(3x-5) + B(3x-5) + Cx^2$

$x = \frac{5}{3}:\ \frac{1000}{9} = \frac{25}{9}C \ \Rightarrow \ C = 40$

$x = 0:\ -25 = -5B \ \Rightarrow \ B = 5$ $\qquad\qquad$ ★ $-\dfrac{7}{x} + \dfrac{5}{x^2} + \dfrac{40}{3x-5}$

$x = 1:\ 44 = -2A - 2B + C \ \Rightarrow \ A = -7$

[13] $\dfrac{x^2 - 6}{(x+2)^2(2x-1)} = \dfrac{A}{x+2} + \dfrac{B}{(x+2)^2} + \dfrac{C}{2x-1}$

$x^2 - 6 = A(x+2)(2x-1) + B(2x-1) + C(x+2)^2$

$x = -2:\ -2 = -5B \ \Rightarrow \ B = \frac{2}{5}$

$x = \frac{1}{2}:\ -\frac{23}{4} = \frac{25}{4}C \ \Rightarrow \ C = -\frac{23}{25}$ $\qquad\qquad$ ★ $\dfrac{\frac{24}{25}}{x+2} + \dfrac{\frac{2}{5}}{(x+2)^2} - \dfrac{\frac{23}{25}}{2x-1}$

$x = 1:\ -5 = 3A + B + 9C \ \Rightarrow \ A = \frac{24}{25}$

$\boxed{15}$ $\dfrac{3x^3 + 11x^2 + 16x + 5}{x(x+1)^3} = \dfrac{A}{x} + \dfrac{B}{x+1} + \dfrac{C}{(x+1)^2} + \dfrac{D}{(x+1)^3}$

$3x^3 + 11x^2 + 16x + 5 = A(x+1)^3 + Bx(x+1)^2 + Cx(x+1) + Dx$

$x = -1: -3 = -D \;\Rightarrow\; D = 3$

$x = 0: 5 = A$

$x = 1: 35 = 8A + 4B + 2C + D \quad (\text{E}_1)$

$x = -2: -7 = -A - 2B + 2C - 2D \quad (\text{E}_2)$

Substituting the values for A and D into E_1 and E_2 yields

$\begin{cases} 4B + 2C & = -8 \\ -2B + 2C & = 4 \end{cases} \;\Rightarrow\; \begin{cases} 2B + C & = -4 \quad (\text{E}_3) \\ -B + C & = 2 \quad (\text{E}_4) \end{cases}$ $\bigstar \; \dfrac{5}{x} - \dfrac{2}{x+1} + \dfrac{3}{(x+1)^3}$

$\text{E}_3 + 2\,\text{E}_4 \;\Rightarrow\; 3C = 0 \;\Rightarrow\; C = 0; \; B = -2$

$\boxed{17}$ $\dfrac{x^2 + x - 6}{(x^2+1)(x-1)} = \dfrac{Ax + B}{x^2+1} + \dfrac{C}{x-1}$

$x^2 + x - 6 = (Ax + B)(x - 1) + C(x^2 + 1)$

$x = 1: -4 = 2C \;\Rightarrow\; C = -2$

$x = 0: -6 = -B + C \;\Rightarrow\; B = 4$ $\bigstar \; -\dfrac{2}{x-1} + \dfrac{3x+4}{x^2+1}$

$x = 2: 0 = 2A + B + 5C \;\Rightarrow\; A = 3$

$\boxed{19}$ $\dfrac{9x^2 - 3x + 8}{x(x^2+2)} = \dfrac{A}{x} + \dfrac{Bx + C}{x^2+2}$

$9x^2 - 3x + 8 = A(x^2 + 2) + (Bx + C)x$

$x = 0: 8 = 2A \;\Rightarrow\; A = 4$

$x = 1: 14 = 3A + B + C \quad (\text{E}_1)$

$x = -1: 20 = 3A + B - C \quad (\text{E}_2)$ $\bigstar \; \dfrac{4}{x} + \dfrac{5x-3}{x^2+2}$

$\text{E}_1 - \text{E}_2 \;\Rightarrow\; -6 = 2C \;\Rightarrow\; C = -3; \; B = 5$

$\boxed{21}$ $\dfrac{4x^3 - x^2 + 4x + 2}{(x^2+1)^2} = \dfrac{Ax + B}{x^2+1} + \dfrac{Cx + D}{(x^2+1)^2}$

$\begin{aligned} 4x^3 - x^2 + 4x + 2 &= (Ax + B)(x^2 + 1) + Cx + D \\ &= Ax^3 + Bx^2 + (A + C)x + (B + D) \end{aligned}$

Equating coefficients, that is, the coefficient of x^3 on the left side must equal the coefficient of x^3 on the right side, we have the following:

$\begin{array}{lll} x^3 & : A = 4 \\ x^2 & : B = -1 \\ x & : A + C = 4 \;\Rightarrow\; C = 0 \\ \text{constant} & : B + D = 2 \;\Rightarrow\; D = 3 \end{array}$ $\bigstar \; \dfrac{4x-1}{x^2+1} + \dfrac{3}{(x^2+1)^2}$

$\boxed{23}$ The degree of the numerator is not lower than the degree of the denominator.

Thus, we must first use long division and then decompose the remaining expression.

Hence, by first dividing and then factoring, we have the following:

$$2x + \frac{4x^2 - 3x + 1}{(x^2 + 1)(x - 1)} = 2x + \frac{Ax + B}{x^2 + 1} + \frac{C}{x - 1}$$

$4x^2 - 3x + 1 = (Ax + B)(x - 1) + C(x^2 + 1)$

$x = 1$: $2 = 2C \Rightarrow C = 1$

$x = 0$: $1 = -B + C \Rightarrow B = 0$

$x = -1$: $8 = 2A - 2B + 2C \Rightarrow A = 3$

$\star\ 2x + \dfrac{1}{x - 1} + \dfrac{3x}{x^2 + 1}$

$\boxed{25}$ By first dividing and then factoring, we have the following:

$$3 + \frac{12x - 16}{x(x - 4)} = 3 + \frac{A}{x} + \frac{B}{x - 4}$$

$12x - 16 = A(x - 4) + Bx$

$x = 0$: $-16 = -4A \Rightarrow A = 4$

$x = 4$: $32 = 4B \Rightarrow B = 8$

$\star\ 3 + \dfrac{4}{x} + \dfrac{8}{x - 4}$

$\boxed{27}$ By first dividing and then factoring, we have the following:

$$2x + 3 + \frac{x + 5}{(2x + 1)(x - 1)} = 2x + 3 + \frac{A}{2x + 1} + \frac{B}{x - 1}$$

$x + 5 = A(x - 1) + B(2x + 1)$

$x = 1$: $6 = 3B \Rightarrow B = 2$

$x = -\frac{1}{2}$: $\frac{9}{2} = -\frac{3}{2}A \Rightarrow A = -3$

$\star\ 2x + 3 + \dfrac{2}{x - 1} - \dfrac{3}{2x + 1}$

Chapter 8 Review Exercises

$\boxed{1}$ $4\,E_1 + 3\,E_2 \Rightarrow 8x + 15x = 16 + 3 \Rightarrow 23x = 19 \Rightarrow x = \frac{19}{23}$;

$-5\,E_1 + 2\,E_2 \Rightarrow 15y + 8y = -20 + 2 \Rightarrow 23y = -18 \Rightarrow y = -\frac{18}{23}$

$\star\ \left(\frac{19}{23}, -\frac{18}{23}\right)$

$\boxed{3}$ Solve E_2 for y, $y = -2x - 1$, and substitute into E_1 to yield $-2x - 1 + 4 = x^2 \Rightarrow$

$x^2 + 2x - 3 = 0 \Rightarrow (x + 3)(x - 1) = 0 \Rightarrow x = -3, 1$ and $y = 5, -3$.

$\star\ (-3, 5),\ (1, -3)$

$\boxed{5}$ $4\,E_2 + E_1 \Rightarrow 4x^2 + 9x^2 = 16 + 140 \Rightarrow 13x^2 = 156 \Rightarrow x^2 = 12 \Rightarrow$

$x = \pm\sqrt{12} = \pm 2\sqrt{3}$. Substituting 12 for x^2 in E_2 gives us $12 - 4y^2 = 4 \Rightarrow$

$8 = 4y^2 \Rightarrow 2 = y^2 \Rightarrow y = \pm\sqrt{2}$.

There are four solutions: $(2\sqrt{3},\ \pm\sqrt{2}),\ (-2\sqrt{3},\ \pm\sqrt{2})$.

$\boxed{7}$ $-4E_1 + E_2 \Rightarrow -12/y - 2/y = -28 + 1 \Rightarrow -14/y = -27 \Rightarrow y = \frac{14}{27};$

$2E_1 + 3E_2 \Rightarrow 2/x + 12/x = 14 + 3 \Rightarrow 14/x = 17 \Rightarrow x = \frac{14}{17}$ $\qquad$ ★ $(\frac{14}{17}, \frac{14}{27})$

$\boxed{8}$ Treat 3^{y+1} as $3 \cdot 3^y$ and 2^{x+1} as $2 \cdot 2^x$.

Thus, E_1 is $1(2^x) + 3(3^y) = 10$ and E_2 is $2(2^x) - 1(3^y) = 5$.

Now $E_1 + 3E_2 \Rightarrow 7 \cdot 2^x = 25 \Rightarrow 2^x = \frac{25}{7} \Rightarrow x = \log_2 \frac{25}{7} = \dfrac{\log \frac{25}{7}}{\log 2} \approx 1.84.$

Resolving the original system for y, we have

$$-2E_1 + E_2 \Rightarrow -7 \cdot 3^y = -15 \Rightarrow 3^y = \frac{15}{7} \Rightarrow y = \log_3 \frac{15}{7} = \dfrac{\log \frac{15}{7}}{\log 3} \approx 0.69.$$

$\boxed{9}$ Solve E_3 for z, $z = 4x + 5y + 2$, and substitute into E_1 and E_2 to yield

$$\begin{cases} 3x + y - 2(4x + 5y + 2) = -1 \\ 2x - 3y + (4x + 5y + 2) = 4 \end{cases} \Rightarrow \begin{cases} -5x - 9y = 3 & (E_4) \\ 6x + 2y = 2 & (E_5) \end{cases}$$

$6E_4 + 5E_5 \Rightarrow -44y = 28 \Rightarrow y = -\frac{7}{11};$

$2E_4 + 9E_5 \Rightarrow 44x = 24 \Rightarrow x = \frac{6}{11}; z = 1$ $\qquad$ ★ $(\frac{6}{11}, -\frac{7}{11}, 1)$

$\boxed{11}$ Solve E_2 for x, $x = y + z$, and substitute into E_1 and E_3 to yield

$$\begin{cases} 4(y + z) - 3y - z = 0 \\ 3(y + z) - y + 3z = 0 \end{cases} \Rightarrow \begin{cases} y + 3z = 0 & (E_4) \\ 2y + 6z = 0 & (E_5) \end{cases}$$

Now E_5 is $2E_4$, hence $y = -3z$ and $x = y + z = -3z + z = -2z$.

$\qquad\qquad$ The general solution is $(-2c, -3c, c)$ for any real number c.

$\boxed{13}$ $E_1 - E_2 \Rightarrow x - 5z = -1 \Rightarrow x = 5z - 1;$

Substitute this value into E_1 to obtain $y = \dfrac{-19z + 5}{2}; \left(5c - 1, \dfrac{-19c + 5}{2}, c \right)$ is

$\qquad\qquad$ the general solution, where c is any real number.

$\boxed{15}$ Let $a = 1/x$, $b = 1/y$, and $c = 1/z$ to obtain the system

$$\begin{cases} 4a + b + 2c = 4 & (E_1) \\ 2a + 3b - c = 1 & (E_2) \\ a + b + c = 4 & (E_3) \end{cases}$$

Solving E_2 for c and substituting into E_1 and E_3 yields

$$\begin{cases} 4a + b + 2(2a + 3b - 1) = 4 \\ a + b + (2a + 3b - 1) = 4 \end{cases} \Rightarrow \begin{cases} 8a + 7b = 6 & (E_4) \\ 3a + 4b = 5 & (E_5) \end{cases}$$

$3E_4 - 8E_5 \Rightarrow -11b = -22 \Rightarrow b = 2$

$\qquad\qquad 4E_4 - 7E_5 \Rightarrow 11a = -11 \Rightarrow a = -1, c = 3; (x, y, z) = (-1, \frac{1}{2}, \frac{1}{3})$

17
$$\begin{cases} x^2 + y^2 < 16 \\ y - x^2 > 0 \end{cases} \quad \Leftrightarrow \quad \begin{cases} x^2 + y^2 < 4^2 \\ y > x^2 \end{cases}$$

$$V \text{ @ } \left(\pm \sqrt{-\tfrac{1}{2} + \tfrac{1}{2}\sqrt{65}}, \; -\tfrac{1}{2} + \tfrac{1}{2}\sqrt{65} \right) \approx (\pm 1.88, \, 3.53)$$

Shade inside the dashed circle $x^2 + y^2 = 4^2$ and inside the dashed parabola $y = x^2$.

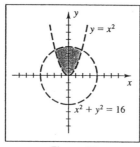

Figure 17

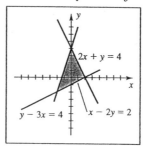

Figure 19

19
$$\begin{cases} x - 2y \le 2 \\ y - 3x \le 4 \\ 2x + y \le 4 \end{cases} \quad \Leftrightarrow \quad \begin{cases} y \ge \tfrac{1}{2}x - 1 \\ y \le 3x + 4 \\ y \le -2x + 4 \end{cases} \qquad V \text{ @ } (-2, -2), \, (0, 4), \, (2, 0)$$

Shade above the solid line $y = \tfrac{1}{2}x - 1$, under the solid line $y = 3x + 4$, and under the solid line $y = -2x + 4$.

21
$$\begin{bmatrix} 2 & -1 & 0 \\ 3 & 0 & -2 \end{bmatrix} \begin{bmatrix} 2 & -1 & 3 \\ 0 & 3 & 0 \\ 1 & 4 & 2 \end{bmatrix} =$$

$$= \begin{bmatrix} 2(2) + (-1)(0) + 0(1) & 2(-1) + (-1)(3) + 0(4) & 2(3) + (-1)(0) + 0(2) \\ 3(2) + 0(0) + (-2)(1) & 3(-1) + 0(3) + (-2)(4) & 3(3) + 0(0) + (-2)(2) \end{bmatrix}$$

$$= \begin{bmatrix} 4 & -5 & 6 \\ 4 & -11 & 5 \end{bmatrix}$$

23
$$\begin{bmatrix} 2 & 0 \\ 1 & 4 \\ -2 & 3 \end{bmatrix} \begin{bmatrix} 0 & 2 & -3 \\ 4 & 5 & 1 \end{bmatrix} =$$

$$\begin{bmatrix} 2(0) + 0(4) & 2(2) + 0(5) & 2(-3) + 0(1) \\ 1(0) + 4(4) & 1(2) + 4(5) & 1(-3) + 4(1) \\ -2(0) + 3(4) & -2(2) + 3(5) & -2(-3) + 3(1) \end{bmatrix} = \begin{bmatrix} 0 & 4 & -6 \\ 16 & 22 & 1 \\ 12 & 11 & 9 \end{bmatrix}$$

25
$$2 \begin{bmatrix} 0 & -1 & -4 \\ 3 & 2 & 1 \end{bmatrix} - 3 \begin{bmatrix} 4 & -2 & 1 \\ 0 & 5 & -1 \end{bmatrix} =$$

$$\begin{bmatrix} 0 & -2 & -8 \\ 6 & 4 & 2 \end{bmatrix} + \begin{bmatrix} -12 & 6 & -3 \\ 0 & -15 & 3 \end{bmatrix} = \begin{bmatrix} -12 & 4 & -11 \\ 6 & -11 & 5 \end{bmatrix}$$

$$\boxed{27} \begin{bmatrix} a & 0 \\ 0 & b \end{bmatrix} \begin{bmatrix} 1 & 3 \\ 2 & 4 \end{bmatrix} = \begin{bmatrix} a & 3a \\ 2b & 4b \end{bmatrix}$$

$$\boxed{29} \begin{bmatrix} 1 & 2 \\ 3 & 4 \end{bmatrix} \left\{ \begin{bmatrix} 2 & -4 \\ 3 & 7 \end{bmatrix} + \begin{bmatrix} 1 & 5 \\ -2 & -3 \end{bmatrix} \right\} = \begin{bmatrix} 1 & 2 \\ 3 & 4 \end{bmatrix} \begin{bmatrix} 3 & 1 \\ 1 & 4 \end{bmatrix} = \begin{bmatrix} 5 & 9 \\ 13 & 19 \end{bmatrix}$$

$\boxed{30}$ Don't multiply, just remember that $A\,A^{-1} = I_3$.

Note: Let A denote each of the matrices in Exercises 31–46.

$\boxed{31} \begin{bmatrix} 5 & -4 & | & 1 & 0 \\ -3 & 2 & | & 0 & 1 \end{bmatrix} 2\,R_1 + 3\,R_2 \rightarrow R_1 \Rightarrow \begin{bmatrix} 1 & -2 & | & 2 & 3 \\ -3 & 2 & | & 0 & 1 \end{bmatrix} R_2 + 3\,R_1 \rightarrow R_2$

$\begin{bmatrix} 1 & -2 & | & 2 & 3 \\ 0 & -4 & | & 6 & 10 \end{bmatrix} -\tfrac{1}{4}R_2 \rightarrow R_2 \quad \Rightarrow \begin{bmatrix} 1 & -2 & | & 2 & 3 \\ 0 & 1 & | & -\tfrac{3}{2} & -\tfrac{5}{2} \end{bmatrix} R_1 + 2\,R_2 \rightarrow R_1$

$\begin{bmatrix} 1 & 0 & | & -1 & -2 \\ 0 & 1 & | & -\tfrac{3}{2} & -\tfrac{5}{2} \end{bmatrix} \qquad\qquad A^{-1} = -\tfrac{1}{2}\begin{bmatrix} 2 & 4 \\ 3 & 5 \end{bmatrix}$

$\boxed{33} \begin{bmatrix} 1 & 0 & 0 & | & 1 & 0 & 0 \\ 0 & 4 & 7 & | & 0 & 1 & 0 \\ 0 & 1 & 2 & | & 0 & 0 & 1 \end{bmatrix} R_3 \leftrightarrow R_2$

$\begin{bmatrix} 1 & 0 & 0 & | & 1 & 0 & 0 \\ 0 & 1 & 2 & | & 0 & 0 & 1 \\ 0 & 4 & 7 & | & 0 & 1 & 0 \end{bmatrix} R_3 - 4\,R_2 \rightarrow R_3$

$\begin{bmatrix} 1 & 0 & 0 & | & 1 & 0 & 0 \\ 0 & 1 & 2 & | & 0 & 0 & 1 \\ 0 & 0 & -1 & | & 0 & 1 & -4 \end{bmatrix} -R_3 \rightarrow R_3$

$\begin{bmatrix} 1 & 0 & 0 & | & 1 & 0 & 0 \\ 0 & 1 & 2 & | & 0 & 0 & 1 \\ 0 & 0 & 1 & | & 0 & -1 & 4 \end{bmatrix} R_2 - 2\,R_3 \rightarrow R_2$

$\begin{bmatrix} 1 & 0 & 0 & | & 1 & 0 & 0 \\ 0 & 1 & 0 & | & 0 & 2 & -7 \\ 0 & 0 & 1 & | & 0 & -1 & 4 \end{bmatrix} \qquad A^{-1} = \begin{bmatrix} 1 & 0 & 0 \\ 0 & 2 & -7 \\ 0 & -1 & 4 \end{bmatrix}$

$\boxed{35}\ X = A^{-1}B = -\tfrac{1}{2}\begin{bmatrix} 2 & 4 \\ 3 & 5 \end{bmatrix}\begin{bmatrix} 30 \\ -16 \end{bmatrix} = -\tfrac{1}{2}\begin{bmatrix} -4 \\ 10 \end{bmatrix} = \begin{bmatrix} 2 \\ -5 \end{bmatrix};\ (x,\,y) = (2,\,-5)$

$\boxed{37}\ A = \begin{bmatrix} -6 \end{bmatrix} \ \Rightarrow\ |\,A\,| = -6.$

$\boxed{39}\ A = \begin{bmatrix} 3 & -4 \\ 6 & 8 \end{bmatrix} \ \Rightarrow\ |\,A\,| = \begin{vmatrix} 3 & -4 \\ 6 & 8 \end{vmatrix} = 3(8) - 6(-4) = 24 + 24 = 48$

$\boxed{41}\ |\,A\,| = \begin{vmatrix} 2 & -3 & 5 \\ -4 & 1 & 3 \\ 3 & 2 & -1 \end{vmatrix} \{R_1\} = 2(-7) + 3(-5) + 5(-11) = -84$

$\boxed{43}$ From Exercise 29 of §8.8, the determinant of A is the product of the main diagonal elements of A—that is, $|\,A\,| = (5)(-3)(-4)(2) = 120.$

44
$$\begin{vmatrix} 1 & 2 & 0 & 3 & 1 \\ -2 & -1 & 4 & 1 & 2 \\ 3 & 0 & -1 & 0 & -1 \\ 2 & -3 & 2 & -4 & 2 \\ -1 & 1 & 0 & 1 & 3 \end{vmatrix} \begin{matrix} R_2 + 4R_3 \to R_2 \\ \\ R_4 + 2R_3 \to R_4 \end{matrix} = \begin{vmatrix} 1 & 2 & 0 & 3 & 1 \\ 10 & -1 & 0 & 1 & -2 \\ 3 & 0 & -1 & 0 & -1 \\ 8 & -3 & 0 & -4 & 0 \\ -1 & 1 & 0 & 1 & 3 \end{vmatrix} \{C_3\}$$

$$= (-1) \begin{vmatrix} 1 & 2 & 3 & 1 \\ 10 & -1 & 1 & -2 \\ 8 & -3 & -4 & 0 \\ -1 & 1 & 1 & 3 \end{vmatrix} \begin{matrix} R_2 + 2R_1 \to R_2 \\ \\ R_4 - 3R_1 \to R_4 \end{matrix} = (-1) \begin{vmatrix} 1 & 2 & 3 & 1 \\ 12 & 3 & 7 & 0 \\ 8 & -3 & -4 & 0 \\ -4 & -5 & -8 & 0 \end{vmatrix} \{C_4\}$$

$$= (-1)(-1) \begin{vmatrix} 12 & 3 & 7 \\ 8 & -3 & -4 \\ -4 & -5 & -8 \end{vmatrix} \{4 \text{ is a common factor of } C_1 \text{ and } -1 \text{ of } R_3\}$$

$$= (-4) \begin{vmatrix} 3 & 3 & 7 \\ 2 & -3 & -4 \\ 1 & 5 & 8 \end{vmatrix} \begin{matrix} R_1 - 3R_3 \to R_1 \\ R_2 - 2R_3 \to R_2 \end{matrix} = (-4) \begin{vmatrix} 0 & -12 & -17 \\ 0 & -13 & -20 \\ 1 & 5 & 8 \end{vmatrix} \{C_1\}$$

$$= (-4) \begin{vmatrix} -12 & -17 \\ -13 & -20 \end{vmatrix} = (-4)(240 - 221) = -76$$

45 C_2 and C_4 are identical, so $|A| = 0$.

47 $\begin{vmatrix} 2-x & 3 \\ 1 & -4-x \end{vmatrix} = 0 \Rightarrow (2-x)(-4-x) - 3 = 0 \Rightarrow$

$$x^2 + 2x - 11 = 0 \Rightarrow x = \frac{-2 \pm \sqrt{4+44}}{2} = \frac{-2 \pm 4\sqrt{3}}{2} = -1 \pm 2\sqrt{3}$$

49 2 is a common factor of R_1, 2 is a common factor of C_2,

and 3 is a common factor of C_3.

51 This is an extension of Exercise 29 of §8.8. Expanding by C_1, only a_{11} is not 0.

Expanding by the new C_1 again, only a_{22} is not 0. Repeating this process yields

$|A| = a_{11}a_{22}a_{33}\cdots a_{nn}$, the product of the main diagonal elements.

52 $\begin{vmatrix} 1 & a & b+c \\ 1 & b & a+c \\ 1 & c & a+b \end{vmatrix} C_3 + C_2 \to C_2$

$$= \begin{vmatrix} 1 & a & a+b+c \\ 1 & b & a+b+c \\ 1 & c & a+b+c \end{vmatrix} C_3 - (a+b+c)C_1 \to C_3$$

$$= \begin{vmatrix} 1 & a & 0 \\ 1 & b & 0 \\ 1 & c & 0 \end{vmatrix} = 0, \text{ since } C_3 \text{ consists of all zeros.}$$

53 For the system $\begin{cases} 5x - 6y = 4 \\ 3x + 7y = 8 \end{cases}$, $|D| = \begin{vmatrix} 5 & -6 \\ 3 & 7 \end{vmatrix} = 35 + 18 = 53.$

Since $|D| = 53 \neq 0$, we may solve the system using Cramer's rule.

$|D_x| = \begin{vmatrix} 4 & -6 \\ 8 & 7 \end{vmatrix} = 28 + 48 = 76.$　$|D_y| = \begin{vmatrix} 5 & 4 \\ 3 & 8 \end{vmatrix} = 40 - 12 = 28.$

$x = \dfrac{|D_x|}{|D|} = \dfrac{76}{53}$ and $y = \dfrac{|D_y|}{|D|} = \dfrac{28}{53}.$　　　　★ $\left(\dfrac{76}{53}, \dfrac{28}{53}\right)$

55 $\dfrac{4x^2 + 54x + 134}{(x+3)(x^2+4x-5)} = \dfrac{A}{x+3} + \dfrac{B}{x+5} + \dfrac{C}{x-1}$

$4x^2 + 54x + 134 = A(x+5)(x-1) + B(x+3)(x-1) + C(x+3)(x+5)$

$x = 1: 192 = 24C \;\Rightarrow\; C = 8$

$x = -3: 8 = -8A \;\Rightarrow\; A = -1$　　　★ $\dfrac{8}{x-1} - \dfrac{3}{x+5} - \dfrac{1}{x+3}$

$x = -5: -36 = 12B \;\Rightarrow\; B = -3$

57 $\dfrac{x^2 + 14x - 13}{x^3 + 5x^2 + 4x + 20} = \dfrac{A}{x+5} + \dfrac{Bx+C}{x^2+4}$

$x^2 + 14x - 13 = A(x^2+4) + (Bx+C)(x+5)$

$x = -5: -58 = 29A \;\Rightarrow\; A = -2$

$x = 0: -13 = 4A + 5C \;\Rightarrow\; C = -1$　　　★ $-\dfrac{2}{x+5} + \dfrac{3x-1}{x^2+4}$

$x = 1: 2 = 5A + 6B + 6C \;\Rightarrow\; B = 3$

59 Let x and y denote the length and width, respectively, of the rectangle. A diagonal of the field is 100 ft. We can use the Pythagorean theorem to formulate an equation that relates the sides and a diagonal.

$\begin{cases} xy = 4000 & \text{area} & (E_1) \\ x^2 + y^2 = 100^2 & \text{diagonal} & (E_2) \end{cases}$

Solve E_1 for y, $y = 4000/x$, and substitute into E_2.

$x^2 + \dfrac{4000^2}{x^2} = 100^2 \;\Rightarrow\; x^4 - 10{,}000x^2 + 16{,}000{,}000 = 0 \;\Rightarrow$

$(x^2 - 2000)(x^2 - 8000) = 0 \;\Rightarrow\; x = 20\sqrt{5},\, 40\sqrt{5}$ and $y = 40\sqrt{5},\, 20\sqrt{5}.$

The dimensions are $20\sqrt{5}$ ft $\times\, 40\sqrt{5}$ ft.

61 Let x and y denote the total amount of taxes paid and bonus money, respectively.

$\begin{cases} x = 0.40(50{,}000 - y) & \text{taxes} \\ y = 0.10(50{,}000 - x) & \text{bonuses} \end{cases} \;\Rightarrow\; \begin{cases} 10x + 4y = 200{,}000 & (E_1) \\ x + 10y = 50{,}000 & (E_2) \end{cases}$

$E_1 - 10\,E_2 \;\Rightarrow\; -96y = -300{,}000 \;\Rightarrow\; y = \$3{,}125; \; x = \$18{,}750$

63 Let x, y, and z denote the number of ft^3/hr flowing through pipes A, B, and C, respectively.

$$\begin{cases} 10x + 10y + 10z = 1000 & \text{all 3 working} \\ 20x + 20y = 1000 & \text{A and B only} \\ 12.5x + 12.5z = 1000 & \text{A and C only} \end{cases} \Rightarrow \begin{cases} x + y + z = 100 & (E_1) \\ x + y = 50 & (E_2) \\ x + z = 80 & (E_3) \end{cases}$$

$$E_1 - E_2 \Rightarrow z = 50; \; E_1 - E_3 \Rightarrow y = 20; \text{ from } E_2, \; x = 30.$$

65 If x and y denote the length and the width, respectively, then a system is $x \le 12$, $y \le 8$, $y \ge \frac{1}{2}x$. The graph is the region bounded by the quadrilateral with vertices $(0, 0)$, $(0, 8)$, $(12, 8)$, and $(12, 6)$.

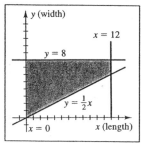

Figure 65

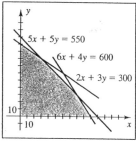

Figure 67

67 Let x and y denote the number of lawn mowers and edgers produced, respectively.

Profit function: $P = 100x + 80y$

$$\begin{cases} 6x + 4y \le 600 & \text{machining} \\ 2x + 3y \le 300 & \text{welding} \\ 5x + 5y \le 550 & \text{assembly} \\ x, y \ge 0 \end{cases}$$

(x, y)	$(100, 0)$	$(80, 30)$	$(30, 80)$	$(0, 100)$	$(0, 0)$
P	10,000	10,400 ■	9400	8000	0

The maximum weekly profit of $10,400 occurs when 80 lawn mowers and 30 edgers are produced.

Chapter 8 Discussion Exercises

1 (a) For $b = 1.99$, we get $x = 204$ and $y = -100$. For $b = 1.999$, we get $x = 2004$ and $y = -1000$.

(b) Solving $\begin{cases} x + 2y = 4 \\ x + by = 5 \end{cases}$ for y gives us $y = \dfrac{1}{b-2}$ and then solving for x we

obtain $x = \dfrac{4b - 10}{b - 2}$. Both x and y are rational functions of b and as $b \to 2^-$,

$x \to \infty$ and $y \to -\infty$.

(c) If b gets very large, (x, y) gets close to $(4, 0)$.

$\boxed{3}$ If we let A be an $m \times n$ matrix $(m \neq n)$, then B would have to be an $n \times m$ matrix so that AB and BA are both defined. But then $AB = I_m$ and $BA = I_n$, different orders of the identity matrix, which we can't have (they must be the same).

$\boxed{5}$ Synthetically dividing $x^4 + ax^2 + bx + c$ by $x + 1$, $x - 2$, and $x - 3$ yields the remainders $a - b + c + 1$, $4a + 2b + c + 16$, and $9a + 3b + c + 81$, respectively. Setting each of the remainders equal to 0 gives us the following system of equations in matrix form.

$$AX = B \iff \begin{bmatrix} 1 & -1 & 1 \\ 4 & 2 & 1 \\ 9 & 3 & 1 \end{bmatrix} \begin{bmatrix} a \\ b \\ c \end{bmatrix} = \begin{bmatrix} -1 \\ -16 \\ -81 \end{bmatrix} \implies X = A^{-1}B = \begin{bmatrix} -15 \\ 10 \\ 24 \end{bmatrix}.$$

Hence, $a = -15$, $b = 10$, $c = 24$, and we graph $Y_1 = x^4 - 15x^2 + 10x + 24$.

The roots of Y_1 are -1, 2, 3, and -4.

Note: The next solution is for Exercise 45 in Section 8.2.

$\boxed{45}$
$$\begin{cases} a \cos x + b \sin x & = 0 \quad (E_1) \\ -a \sin x + b \cos x & = \tan x \quad (E_2) \end{cases}$$

$\sin x$ (E_1) and $\cos x$ (E_2) yield $\begin{cases} a \sin x \cos x + b \sin^2 x & = 0 \quad (E_3) \\ -a \sin x \cos x + b \cos^2 x & = \sin x \quad (E_4) \end{cases}$

$E_3 + E_4 \implies$

$b \sin^2 x + b \cos^2 x = \sin x \implies b(\sin^2 x + \cos^2 x) = \sin x \implies b(1) = \sin x \implies$

$b = \sin x$. Substituting back into E_1 yields $a \cos x + \sin^2 x = 0 \implies$

$$a = -\frac{\sin^2 x}{\cos x} = -\frac{1 - \cos^2 x}{\cos x} = -\frac{1}{\cos x} + \frac{\cos^2 x}{\cos x} = -\sec x + \cos x = \cos x - \sec x.$$

Chapter 9: Sequences, Series, and Probability

Note: For Exercises 1–16, the answers are listed in the order a_1, a_2, a_3, a_4; and a_8.

Simply substitute 1, 2, 3, 4, and 8 for n in the formula for a_n to obtain the results.

1. $a_n = 12 - 3n$ •

$a_2 = 12 - 3(2) = 6,$

$a_4 = 12 - 3(4) = 0;$

$a_1 = 12 - 3(1) = 9,$

$a_3 = 12 - 3(3) = 3,$

$a_8 = 12 - 3(8) = -12$ ★ 9, 6, 3, 0; −12

3. $a_n = \dfrac{3n - 2}{n^2 + 1}$ •

$a_1 = \dfrac{3(1) - 2}{(1)^2 + 1} = \dfrac{1}{2},$ $a_2 = \dfrac{3(2) - 2}{(2)^2 + 1} = \dfrac{4}{5},$

$a_3 = \dfrac{3(3) - 2}{(3)^2 + 1} = \dfrac{7}{10},$ $a_4 = \dfrac{3(4) - 2}{(4)^2 + 1} = \dfrac{10}{17};$ $a_8 = \dfrac{3(8) - 2}{(8)^2 + 1} = \dfrac{22}{65}$ ★ $\dfrac{1}{2}, \dfrac{4}{5}, \dfrac{7}{10}, \dfrac{10}{17}; \dfrac{22}{65}$

5. $a_n = 9$ • $a_{(\text{any allowable value})} = 9$ ★ 9, 9, 9, 9; 9

7. $a_n = 2 + (-0.1)^n$ • $a_1 = 2 + (-0.1)^1 = 2 - 0.1 = 1.9,$

$a_2 = 2 + (-0.1)^2 = 2 + 0.01 = 2.01,$

$a_3 = 2 + (-0.1)^3 = 2 - 0.001 = 1.999,$

$a_4 = 2 + (-0.1)^4 = 2 + 0.0001 = 2.0001;$

$a_8 = 2 + (-0.1)^8 = 2 + 0.00000001 = 2.00000001$

★ 1.9, 2.01, 1.999, 2.0001; 2.00000001

9. $a_n = (-1)^{n-1} \left(\dfrac{n + 7}{2n} \right)$ •

$a_1 = (-1)^{1-1} \left(\dfrac{1 + 7}{2(1)} \right) = (-1)^0 \left(\tfrac{8}{2} \right) = (1)(4) = 4,$

$a_2 = (-1)^{2-1} \left(\dfrac{2 + 7}{2(2)} \right) = (-1)^1 \left(\tfrac{9}{4} \right) = -\dfrac{9}{4},$

$a_3 = (-1)^{3-1} \left(\dfrac{3 + 7}{2(3)} \right) = (-1)^2 \left(\tfrac{10}{6} \right) = \dfrac{5}{3},$

$a_4 = (-1)^{4-1} \left(\dfrac{4 + 7}{2(4)} \right) = (-1)^3 \left(\tfrac{11}{8} \right) = -\dfrac{11}{8};$

$a_8 = (-1)^{8-1} \left(\dfrac{8 + 7}{2(8)} \right) = (-1)^7 \left(\tfrac{15}{16} \right) = -\dfrac{15}{16}$ ★ $4, -\dfrac{9}{4}, \dfrac{5}{3}, -\dfrac{11}{8}; -\dfrac{15}{16}$

11. $a_n = 1 + (-1)^{n+1}$ • $a_1 = 1 + (-1)^{1+1} = 1 + (-1)^2 = 1 + 1 = 2,$

$a_2 = 1 + (-1)^{2+1} = 1 + (-1)^3 = 1 - 1 = 0,$

$a_3 = 1 + (-1)^{3+1} = 1 + (-1)^4 = 1 + 1 = 2,$

$a_4 = 1 + (-1)^{4+1} = 1 + (-1)^5 = 1 - 1 = 0;$

$a_8 = 1 + (-1)^{8+1} = 1 + (-1)^9 = 1 - 1 = 0$

★ 2, 0, 2, 0; 0

$\boxed{13}$ $a_n = \dfrac{2^n}{n^2 + 2}$ • $a_1 = \dfrac{2^1}{1^2 + 2} = \dfrac{2}{3}$, $a_2 = \dfrac{2^2}{2^2 + 2} = \dfrac{4}{6} = \dfrac{2}{3}$, $a_3 = \dfrac{2^3}{3^2 + 2} = \dfrac{8}{11}$,

$a_4 = \dfrac{2^4}{4^2 + 2} = \dfrac{16}{18} = \dfrac{8}{9}$; $a_8 = \dfrac{2^8}{8^2 + 2} = \dfrac{256}{66} = \dfrac{128}{33}$ ★ $\dfrac{2}{3}, \dfrac{2}{3}, \dfrac{8}{11}, \dfrac{8}{9}; \dfrac{128}{33}$

$\boxed{15}$ a_n is the number of decimal places in $(0.1)^n$. •

$(0.1)^1 = 0.1$ { 1 decimal place } $\Rightarrow$ $a_1 = 1$,

$(0.1)^2 = 0.01$ { 2 decimal places } $\Rightarrow$ $a_2 = 2$,

$(0.1)^3 = 0.001$ { 3 decimal places } $\Rightarrow$ $a_3 = 3$,

$(0.1)^4 = 0.0001$ { 4 decimal places } $\Rightarrow$ $a_4 = 4$;

$(0.1)^8 = 0.00000001$ { 8 decimal places } $\Rightarrow$ $a_8 = 8$ ★ 1, 2, 3, 4; 8

$\boxed{17}$ $\left\{ \dfrac{1}{\sqrt{n}} \right\} = \dfrac{1}{\sqrt{1}}, \dfrac{1}{\sqrt{2}}, \dfrac{1}{\sqrt{3}}, \dfrac{1}{\sqrt{4}}, \dfrac{1}{\sqrt{5}}, \ldots \approx 1, 0.71, 0.58, 0.5, 0.45, \ldots$

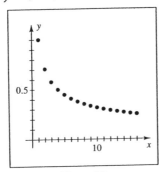

Figure 17

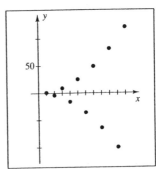

Figure 19

$\boxed{19}$ $\left\{ (-1)^{n+1} n^2 \right\} = 1 \cdot 1^2, -1 \cdot 2^2, 1 \cdot 3^2, -1 \cdot 4^2, \ldots = 1, -4, 9, -16, \ldots$

$\boxed{21}$ $a_1 = 2$, $a_{k+1} = 3a_k - 5$ •

$a_2 = a_{1+1}$ { $k = 1$ } $= 3a_1 - 5 = 3(2) - 5 = 1$, $a_3 = 3a_2 - 5 = 3(1) - 5 = -2$,

$a_4 = 3a_3 - 5 = 3(-2) - 5 = -11$, $a_5 = 3a_4 - 5 = 3(-11) - 5 = -38$

$\boxed{23}$ $a_1 = -3$, $a_{k+1} = a_k^2$ •

$a_2 = a_{1+1}$ { $k = 1$ } $= a_1^2 = (-3)^2 = 3^2 = 9$, $a_3 = a_2^2 = (3^2)^2 = 3^4$,

$a_4 = a_3^2 = (3^4)^2 = 3^8$, $a_5 = a_4^2 = (3^8)^2 = 3^{16}$

$\boxed{25}$ $a_1 = 5$, $a_{k+1} = ka_k$ •

$a_2 = a_{1+1}$ { $k = 1$ } $= 1\,a_1 = 1(5) = 5$, $a_3 = 2\,a_2 = 2(5) = 10$,

$a_4 = 3\,a_3 = 3(10) = 30$, $a_5 = 4\,a_4 = 4(30) = 120$

$\boxed{27}$ $a_1 = 2$, $a_{k+1} = (a_k)^k$ •

$a_2 = a_{1+1}$ { $k = 1$ } $= (a_1)^1 = (2)^1 = 2$, $a_3 = (a_2)^2 = (2)^2 = 4$,

$a_4 = (a_3)^3 = (4)^3 = 4^3$ or 64, $a_5 = (a_4)^4 = (4^3)^4 = 4^{12}$ or 16,777,216

29 $a_n = 3 + \frac{1}{2}n$ • $S_1 = a_1 = 3 + \frac{1}{2} = \frac{7}{2}$. $S_2 = S_1 + a_2 = \frac{7}{2} + 4 = \frac{15}{2}$. Alternatively, to find S_2 we could use $a_1 + a_2$. However, in most cases, it is easier to compute S_{n+1} by using $S_n + a_{n+1}$. $S_3 = S_2 + a_3 = \frac{15}{2} + \frac{9}{2} = 12$. $S_4 = S_3 + a_4 = 12 + 5 = 17$.

31 $\{(-1)^n n^{-1/2}\} = \left\{(-1)^n \frac{1}{\sqrt{n}}\right\}$ • $S_1 = a_1 = -1$.

$$S_2 = S_1 + a_2 = -1 + \frac{1}{\sqrt{2}}. \qquad\qquad S_3 = S_2 + a_3 = -1 + \frac{1}{\sqrt{2}} - \frac{1}{\sqrt{3}}.$$

$$S_4 = S_3 + a_4 = -1 + \frac{1}{\sqrt{2}} - \frac{1}{\sqrt{3}} + \frac{1}{2} = -\frac{1}{2} + \frac{1}{\sqrt{2}} - \frac{1}{\sqrt{3}}.$$

33 Substitute $k = 1, 2, 3, 4$, and 5 into $2k - 7$ and then add the results.

$$\sum_{k=1}^{5} (2k - 7) = (-5) + (-3) + (-1) + 1 + 3 = -5$$

35 $\displaystyle\sum_{k=1}^{4} (k^2 - 5) = (1^2 - 5) + (2^2 - 5) + (3^2 - 5) + (4^2 - 5) = (-4) + (-1) + 4 + 11 = 10$

37 $\displaystyle\sum_{k=0}^{5} k(k - 2) = 0(-2) + 1(-1) + 2(0) + 3(1) + 4(2) + 5(3)$
$$= 0 + (-1) + 0 + 3 + 8 + 15 = 25$$

39 $\displaystyle\sum_{k=3}^{6} \frac{k-5}{k-1} = \frac{-2}{2} + \frac{-1}{3} + \frac{0}{4} + \frac{1}{5} = -\frac{15}{15} - \frac{5}{15} + \frac{3}{15} = -\frac{17}{15}$

41 $\displaystyle\sum_{k=1}^{5} (-3)^{k-1} = (-3)^0 + (-3)^1 + (-3)^2 + (-3)^3 + (-3)^4$
$$= 1 + (-3) + 9 + (-27) + 81 = 61$$

43 Using part (1) of the theorem on the sum of a constant, we have

$$\sum_{k=1}^{100} 100 = 100(100) = 10{,}000.$$

45 Using part (2) of the theorem on the sum of a constant, we have

$$\sum_{k=253}^{571} \frac{1}{3} = (571 - 253 + 1)\left(\frac{1}{3}\right) = 319\left(\frac{1}{3}\right) = \frac{319}{3}.$$

47 Note that j, not k, is the summation variable. Hence, we again use part (1) of the theorem on the sum of a constant, obtaining $\displaystyle\sum_{j=1}^{7} \frac{1}{2}k^2 = 7\left(\frac{1}{2}k^2\right) = \frac{7}{2}k^2$.

49 $\displaystyle\sum_{k=1}^{n} (a_k - b_k) = (a_1 - b_1) + (a_2 - b_2) + \cdots + (a_n - b_n)$
$$= (a_1 + a_2 + \cdots + a_n) + (-b_1 - b_2 - \cdots - b_n)$$
$$= (a_1 + a_2 + \cdots + a_n) - (b_1 + b_2 + \cdots + b_n)$$
$$= \sum_{k=1}^{n} a_k - \sum_{k=1}^{n} b_k$$

51 $a_1 = 5$, $a_2 = 5^{1/2}$, $a_3 = 5^{1/4}$, $a_4 = 5^{1/8}$, ...

As k increases, the exponent gets closer to zero and the terms approach 1.

$\boxed{53}$ $a_1 = 0.4$, $a_k = 0.1(3 \cdot 2^{k-2} + 4)$ $\Rightarrow$

$a_2 = 0.1(3 \cdot 2^{2-2} + 4) = 0.1(3 \cdot 2^0 + 4) = 0.1(3 \cdot 1 + 4) = 0.1(7) = 0.7$

$a_3 = 0.1(3 \cdot 2^{3-2} + 4) = 0.1(3 \cdot 2^1 + 4) = 0.1(3 \cdot 2 + 4) = 0.1(10) = 1$

$a_4 = 0.1(3 \cdot 2^{4-2} + 4) = 0.1(3 \cdot 2^2 + 4) = 0.1(3 \cdot 4 + 4) = 0.1(16) = 1.6$

$a_5 = 0.1(3 \cdot 2^{5-2} + 4) = 0.1(3 \cdot 2^3 + 4) = 0.1(3 \cdot 8 + 4) = 0.1(28) = 2.8$

$\boxed{55}$ (a) $a_1 = 1$, $a_2 = 1$, $a_{k+1} = a_k + a_{k-1}$ for $k \geq 2$. So

$a_3 = a_2 + a_1 = 1 + 1 = 2$,

$a_4 = a_3 + a_2 = 2 + 1 = 3$,

$a_5 = a_4 + a_3 = 3 + 2 = 5$,

$a_6 = a_5 + a_4 = 5 + 3 = 8$,

$a_7 = a_6 + a_5 = 8 + 5 = 13$,

$a_8 = a_7 + a_6 = 13 + 8 = 21$,

$a_9 = a_8 + a_7 = 21 + 13 = 34$, and

$a_{10} = a_9 + a_8 = 34 + 21 = 55$.

The following screens show the assignment needed to generate the Fibonacci

sequence on the TI-83 Plus and a listing of the first 14 Fibonacci numbers.

```
Plot1 Plot2 Plot3
nMin=1
·u(n)Bu(n-1)+u(n
-2)
u(nMin)B{1,1}
\v(n)=
v(nMin)=
\w(n)=
```

```
u(1,7)
{1  1  2  3  5  8  13}
u(8,11)
      {21  34  55  89}
u(12,14)
      {144  233  377}
```

(b) $r_1 = \frac{1}{1} = 1$, $r_2 = \frac{2}{1} = 2$, $r_3 = \frac{3}{2} = 1.5$,

$r_4 = \frac{5}{3} = 1.\overline{6}$, $r_5 = \frac{8}{5} = 1.6$, $r_6 = \frac{13}{8} = 1.625$,

$r_7 = \frac{21}{13} \approx 1.6153846$, $r_8 = \frac{34}{21} \approx 1.6190476$, $r_9 = \frac{55}{34} \approx 1.6176471$,

and $r_{10} = \frac{89}{55} \approx 1.6181818$.

$\boxed{57}$ (a) Since the amount of chlorine decreases by a factor of 0.20 each day {retain

80%}, $a_n = 0.8a_{n-1}$, where a_0 is the initial amount of chlorine in the pool.

(b) Let $a_0 = 7$ and $a_n = 0.8a_{n-1}$.

Day (n)	0	1	2	3	4	5
Chlorine (a_n)	7.00	5.60	4.48	3.58	2.87	2.29

The chlorine level will drop below 3 ppm during the fourth day.

59 $N = 5$ and $x_1 = \frac{5}{2}$ $\Rightarrow$ $x_2 = \frac{1}{2}\left(x_1 + \frac{N}{x_1}\right) = \frac{1}{2}\left(2.5 + \frac{5}{2.5}\right) = 2.25$ $\Rightarrow$

$x_3 = \frac{1}{2}\left(2.25 + \frac{5}{2.25}\right) \approx 2.236111$ $\Rightarrow$ $x_4 \approx 2.236068$ $\Rightarrow$ $x_5 \approx 2.236068$.

Thus, $\sqrt{5} \approx 2.236068$.

Note: TI-users—see the key sequence at the end of the solution of #63.

61 $x_1 = 2$ and $x_2 = \frac{1}{3}\sqrt[3]{x_1} + 2$ $\Rightarrow$ $x_2 \approx 2.419974$ $\Rightarrow$ $x_3 \approx 2.447523$ $\Rightarrow$

$x_4 \approx 2.449215$ $\Rightarrow$ $x_5 \approx 2.449319$ $\Rightarrow$ $x_6 \approx 2.449325$. The root is approximately

2.4493. *Note:* TI-users—see the key sequence at the end of the solution of #63.

63 (a) $f(1) = (1 - 2 + \log 1) = (1 - 2 + 0) = -1 < 0$ and

$f(2) = (2 - 2 + \log 2) = \log 2 \approx 0.30 > 0$.

Thus, f assumes both positive and negative values on $[1, 2]$.

(b) Solving $\underline{x - 2 + \log x = 0}$ for x in terms of x gives us $\underline{x = 2 - \log x}$.

$x_1 = 1.5$ $\Rightarrow$ $x_2 = 2 - \log x_1 = 2 - \log 1.5 \approx 1.823909$ $\Rightarrow$ $x_3 \approx 1.738997$ $\Rightarrow$

$x_4 \approx 1.759701$ $\Rightarrow$ $x_5 \approx 1.754561$ $\Rightarrow$ $x_6 \approx 1.755832$ $\Rightarrow$ $x_7 \approx 1.755517$.

The zero is approximately 1.76. *Note:* If you are using a TI-8x, type 1.5 and

press the ENTER key to store 1.5 in the memory location ANS. Now type

$2 - \text{LOG ANS}$ and then successively press the ENTER key to obtain the

approximations for x_1, x_2, ... (see the display below).

1.5 | ENTER | 2 | $-$ | LOG | ANS | ENTER | ENTER | ...

Similar key sequences for **Exercises 59 and 61**, respectively, are:

2.5 | ENTER | $.5$ | (| ANS | $+$ | 5 | $\div$ | ANS |) | ENTER | ENTER | ...

2 | ENTER | MATH | 4 | ANS | $\div$ | 3 | $+$ | 2 | ENTER | ENTER | ...

Alternatively, we can use the memory location X instead of ANS by

using the | STO ▷ | key as shown in *Figure 63*.

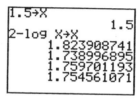

Figure 63

65 Graph $y = \left(1 + \frac{1}{x} + \frac{1}{2x^2}\right)^x$ on the interval $[1, 100]$.

The graph approaches the horizontal asymptote $y \approx 2.718 \approx e$.

For increasing values of n, the terms of the sequence appear to approximate e.

$[1, 100, 10]$ by $[0, 3]$ $[1, 100, 10]$ by $[0, 3]$

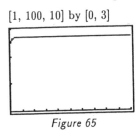 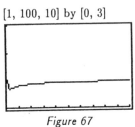

Figure 65   *Figure 67*

67 Graph $y = \left(\frac{1}{x}\right)^{1/x}$ on the interval $[1, 100]$.

The graph approaches the horizontal asymptote $y \approx 1$ from below.

For increasing values of n, the terms of the sequence appear to approximate 1.

69 *Note:* **TI-82 users:** Under MODE, make sure you are in Seq and Dot mode.

Under $Y =$, assign $1.7\text{U}n - 1 + .5$ to $\text{U}n$ $\left\{\text{U}n - 1 \text{ is } \boxed{\text{2nd}}\ \boxed{7}\right\}$.

Under WINDOW, let $\text{U}n\text{Start} = .25$, $n\text{Start} = 1$, $n\text{Min} = 1$, $n\text{Max} = 20$,

and the rest of the window parameters as shown in *Figure 69*.

TI-83 users: Under MODE, make sure you are in Seq mode.

Under $Y =$, let $n\text{Min} = 1$, $\text{u}(n\text{Min}) = .25$, and $\text{u}(n) = 1.7\text{u}(n-1) + .5$

$\left\{\text{u is } \boxed{\text{2nd}}\ \boxed{7}\ \text{and } n \text{ is } \boxed{\text{X,T,}\theta\text{,}n}\right\}$.

Under WINDOW, let $n\text{Min} = 1$, $n\text{Max} = 20$, PlotStart $= 1$, PlotStep $= 1$,

and the rest of the window parameters as shown in the figure.

By tracing the graph we see that $a_9 \approx 66.55$ and $a_{10} \approx 113.64$. Thus, $k = 10$.

$[0, 20, 5]$ by $[0, 125, 25]$ $[0, 20, 5]$ by $[0, 300, 50]$

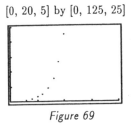 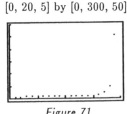

Figure 69   *Figure 71*

71 By tracing the graph we see that $a_{18} \approx 50.39$ and $a_{19} \approx 255.96$. Thus, $k = 19$.

73 (a) Since $c = 0.5$, let the sequence be defined by $a_{k+1} = 0.5a_k(1 - a_k)$. Then, $a_1 = 0.25$, $a_2 = 0.5a_1(1 - a_1) = 0.5(0.25)(1 - 0.25) = 0.09375$. In a similar manner, $a_3 \approx 0.04248$, $a_4 \approx 0.02034$, ..., $a_{10} \approx 0.00031$, $a_{11} \approx 0.00015$, $a_{12} \approx 0.00008$. The insect population is initially $1000a_1 = 1000(0.25) = 250$. It then becomes approximately 94, 42, 20, and so on, until the population decreases to zero.

(b) The sequence determined is $a_1 = 0.25$, $a_2 \approx 0.28125$, $a_3 \approx 0.30322$, $a_4 \approx 0.31692$, ..., $a_{18} \approx 0.33333$, $a_{19} \approx 0.33333$, $a_{20} \approx 0.33333$. The insect population is initially 250. It then becomes approximately 281, 303, 317, and so on, until the population stabilizes at 333.

(c) The sequence determined is $a_1 = 0.25$, $a_2 \approx 0.51563$, $a_3 \approx 0.68683$, $a_4 \approx 0.59151$, ..., $a_{40} \approx 0.63636$, $a_{41} \approx 0.63636$, $a_{42} \approx 0.63636$. The insect population is initially 250. It then becomes approximately 516, 687, 592, and so on, until the population stabilizes at 636.

9.2 Exercises

1 To show that the given sequence, -6, -2, 2, ..., $4n - 10$, ..., is arithmetic, we must show that $a_{k+1} - a_k$ is equal to some constant, which is the common difference. $a_n = 4n - 10 \Rightarrow$

$$a_{k+1} - a_k = \left[4(k+1) - 10\right] - \left[4(k) - 10\right] = 4k + 4 - 10 - 4k + 10 = 4$$

Note: For Exercises 3–10, we will find the nth term first, and then use that term to find a_5 and a_{10}.

3 The common difference can be found by *subtracting* any term from its successor.

$d = 6 - 2 = 4$; $a_n = a_1 + (n-1)d = 2 + (n-1)(4) = 2 + 4n - 4 = 4n - 2$;

$$a_5 = 4(5) - 2 = 18; \quad a_{10} = 4(10) - 2 = 38$$

5 $d = 2.7 - 3 = -0.3$; $a_n = a_1 + (n-1)d = 3 + (n-1)(-0.3) = -0.3n + 3.3$;

$$a_5 = -0.3(5) + 3.3 = 1.8; \quad a_{10} = -0.3(10) + 3.3 = 0.3$$

7 $d = -3.9 - (-7) = 3.1$; $a_n = a_1 + (n-1)d = -7 + (n-1)(3.1) = 3.1n - 10.1$;

$$a_5 = 3.1(5) - 10.1 = 5.4; \quad a_{10} = 3.1(10) - 10.1 = 20.9$$

9 An equivalent sequence is $\ln 3$, $\ln 3^2$, $\ln 3^3$, $\ln 3^4$, ..., which is also equivalent to the sequence $\ln 3$, $2\ln 3$, $3\ln 3$, $4\ln 3$, ...; $d = 2\ln 3 - \ln 3 = \ln 3$;

$a_n = \ln 3 + (n-1)(\ln 3) = n\ln 3$ or $\ln 3^n$; $a_5 = 5\ln 3$ or $\ln 3^5$; $a_{10} = 10\ln 3$ or $\ln 3^{10}$

11 $a_6 = a_1 + 5d$ and $a_2 = a_1 + d \Rightarrow a_6 - a_2 = (a_1 + 5d) - (a_1 + d) = 4d$.

But $a_6 - a_2 = -11 - 21 = -32$. Hence, $4d = -32$ and $d = -8$.

13 Given a_1 and a_2, we can find the difference d. $d = a_2 - a_1 = 7.5 - 9.1 = -1.6$.

The twelfth term is equal to the first term plus eleven differences.

$$a_{12} = 9.1 + (11)(-1.6) = -8.5.$$

15 $d = a_7 - a_6 = 5.2 - 2.7 = 2.5;\ a_6 = a_1 + 5d\ \Rightarrow\ 2.7 = a_1 + 5(2.5)\ \Rightarrow\ a_1 = -9.8$

17 $a_3 = a_1 + 2d$ and $a_{20} = a_1 + 19d\ \Rightarrow\ a_{20} - a_3 = (a_1 + 19d) - (a_1 + 2d) = 17d$.

$a_3 = 7$ and $a_{20} = 43\ \Rightarrow\ 17d = 43 - 7\ \Rightarrow\ d = \frac{36}{17}$.

$$a_{15} = a_3 + 12d = 7 + 12(\tfrac{36}{17}) = \tfrac{551}{17}.$$

Note: To find the sums in Exercises 19–26, we use the sum formulas

$$\textbf{(1)}\ S_n = \tfrac{n}{2}\big[2a_1 + (n-1)d\big] \quad \text{and} \quad \textbf{(2)}\ S_n = \tfrac{n}{2}(a_1 + a_n).$$

19 $a_1 = 40,\ d = -3,\ n = 30\ \Rightarrow\ S_{30} = \tfrac{30}{2}\big[2(40) + (29)(-3)\big] = -105$.

21 $a_1 = -9,\ a_{10} = 15,\ n = 10\ \Rightarrow\ S_{10} = \tfrac{10}{2}(-9 + 15) = 30$.

23 $\displaystyle\sum_{k=1}^{20} (3k - 5)$ • Using the second formula for S_n with $n = 20$, $a_1 = 3(1) - 5 = -2$,

and $a_{20} = 3(20) - 5 = 55$, we obtain $S_{20} = \tfrac{20}{2}(-2 + 55) = 530$.

25 $\displaystyle\sum_{k=1}^{18} (\tfrac{1}{2}k + 7)$ • $a_1 = \tfrac{15}{2},\ a_{18} = 16,\ n = 18\ \Rightarrow\ S_{18} = \tfrac{18}{2}(\tfrac{15}{2} + 16) = \tfrac{423}{2}$.

27 $1 + 3 + 5 + 7$. Since the difference in terms is 2, the general term is

$$a_1 + (n-1)d = 1 + (n-1)2 = 2n - 1. \qquad \sum_{n=1}^{4} (2n - 1)$$

29 $1 + 3 + 5 + \cdots + 73$. From Exercise 27, the general term is $2n - 1$ with

n starting at 1. $2n - 1 = 73\ \Rightarrow\ n = 37$, the largest value. $\displaystyle\sum_{n=1}^{37} (2n - 1)$

31 $\frac{3}{7} + \frac{6}{11} + \frac{9}{15} + \frac{12}{19} + \frac{15}{23} + \frac{18}{27}$. The numerators increase by 3, the denominators increase

by 4. The general terms are $3 + (n-1)3 = 3n$ and $7 + (n-1)4 = 4n + 3$. $\displaystyle\sum_{n=1}^{6} \frac{3n}{4n+3}$

33 $a_1 = -2,\ d = \tfrac{1}{4},\ S = 21,$ and $S_n = \tfrac{n}{2}\big[2a_1 + (n-1)(d)\big]\ \Rightarrow$

$21 = \tfrac{n}{2}\big[2(-2) + (n-1)(\tfrac{1}{4})\big]\ \Rightarrow\ 42 = n(\tfrac{1}{4}n - \tfrac{17}{4})\ \Rightarrow\ 168 = n^2 - 17n\ \Rightarrow$

$n^2 - 17n - 168 = 0\ \Rightarrow\ (n - 24)(n + 7) = 0\ \Rightarrow\ n = 24,\ -7$.

Since n can not be negative, $n = 24$.

35 If we insert five arithmetic means between 2 and 10, there will be 6 differences that

span the distance from 2 to 10. Hence, $6d = 10 - 2\ \Rightarrow\ d = \tfrac{4}{3}$;

The terms are $2,\ \tfrac{10}{3},\ \tfrac{14}{3},\ 6,\ \tfrac{22}{3},\ \tfrac{26}{3},\ 10$.

37 (a) The first integer greater than 32 that is divisible by 6 is 36 $\{6 \cdot 6\}$

and the last integer less than 395 that is divisible by 6 is 390 $\{65 \cdot 6\}$.

The number of terms is then $65 - 6 + 1 = 60$.

(b) The sum is $S_{60} = \frac{60}{2}(36 + 390) = 12{,}780$.

39 There are $(24 - 10 + 1) = 15$ layers. Model this problem as an arithmetic sequence

with $a_1 = 10$ and $a_{15} = 24$. $S_{15} = \frac{15}{2}(10 + 24) = 255$.

41 This is similar to inserting 9 arithmetic means between 4 and 24.

$10d = 20 \;\Rightarrow\; d = 2$. The circumference of each ring is πD with $D = 4, 6, 8, \ldots, 24$.

$$a_1 = 4\pi \text{ and } a_{11} = 24\pi \;\Rightarrow\; S_{11} = \frac{11}{2}(4\pi + 24\pi) = 154\pi \text{ ft.}$$

43 $n = 5$, $S_5 = 5000$, $d = -100 \;\Rightarrow$

$$5000 = \frac{5}{2}\big[2a_1 + 4(-100)\big] \;\Rightarrow\; 2000 = 2a_1 - 400 \;\Rightarrow\; a_1 = \$1200$$

45 The sequence $16, 48, 80, 112, \ldots$, is an arithmetic sequence with $a_1 = 16$ and

$d = 48 - 16 = 32$. The total distance traveled in n seconds is

$$a_1 + a_2 + \cdots + a_n = \frac{n}{2}\big[2a_1 + (n-1)d\big] = \frac{n}{2}\big[2(16) + (n-1)(32)\big] = \frac{n}{2}(32n) = 16n^2.$$

47 If the nth term is $\frac{1}{x_n}$ and $x_{n+1} = \frac{x_n}{1 + x_n}$, then the $(n+1)$st term is

$$\frac{1}{x_{n+1}} = \frac{1}{\dfrac{x_n}{1 + x_n}} = \frac{1 + x_n}{x_n} = \frac{1}{x_n} + \frac{x_n}{x_n} = 1 + \frac{1}{x_n},$$

which is 1 greater than the nth term and therefore the sequence is arithmetic.

49 (a) $T_8 = 1 + 2 + \cdots + 8 = 36$. $A_k = \frac{n - k + 1}{T_n} \;\Rightarrow\; A_1 = \frac{8 - 1 + 1}{36} = \frac{8}{36}$.

$$A_2 = \frac{7}{36}, \; A_3 = \frac{6}{36}, \; A_4 = \frac{5}{36}, \; A_5 = \frac{4}{36}, \; A_6 = \frac{3}{36}, \; A_7 = \frac{2}{36}, \; A_8 = \frac{1}{36}.$$

(b) $d = A_{k+1} - A_k = -\frac{1}{36}$ for $k = 1, 2, \ldots, 7$. $S_8 = \sum_{k=1}^{8} A_k = \frac{8}{36} + \frac{7}{36} + \cdots + \frac{1}{36} = 1$.

(c) $\$1000\left(\frac{8}{36} + \frac{7}{36} + \frac{6}{36} + \frac{5}{36}\right) \approx \$722.22$

9.3 Exercises

1 To show that the given sequence, $5, -\frac{5}{4}, \frac{5}{16}, \ldots, 5\left(-\frac{1}{4}\right)^{n-1}, \ldots$, is geometric,

we must show that $\frac{a_{k+1}}{a_k}$ is equal to some constant, which is the common ratio.

$$a_n = a_1 r^{n-1} = 5\left(-\frac{1}{4}\right)^{n-1} \;\Rightarrow\; \frac{a_{k+1}}{a_k} = \frac{5\left(-\frac{1}{4}\right)^{(k+1)-1}}{5\left(-\frac{1}{4}\right)^{k-1}} = -\frac{1}{4}$$

Note: For Exercises 3–14, we will find the nth term first and then use that term to find

a_5 and a_8.

3. The common ratio can be found by *dividing* any term by its successor. $r = \frac{4}{8} = \frac{1}{2}$;

$$a_n = a_1 r^{n-1} = 8(\tfrac{1}{2})^{n-1} = 2^3 (2^{-1})^{n-1} = 2^3 2^{1-n} = 2^{4-n};$$

$$a_5 = 2^{4-5} = 2^{-1} = \tfrac{1}{2}; \qquad\qquad a_8 = 2^{4-8} = 2^{-4} = \tfrac{1}{16}$$

5. $r = \frac{-30}{300} = -0.1$; $a_n = a_1 r^{n-1} = 300(-0.1)^{n-1}$;

$$a_5 = 300(-0.1)^4 = 0.03; \qquad\qquad a_8 = 300(-0.1)^7 = -0.00003$$

7. $r = \frac{25}{5} = 5$; $a_n = 5(5)^{n-1} = 5^n$; $a_5 = 5^5 = 3125$; $a_8 = 5^8 = 390{,}625$

9. $r = \frac{-6}{4} = -1.5$; $a_n = 4(-1.5)^{n-1}$; $a_5 = 4(-1.5)^4 = 20.25$; $a_8 = 4(-1.5)^7 = -68.34375$

11. $r = \frac{-x^2}{1} = -x^2$; $a_n = a_1 r^{n-1} = 1(-x^2)^{n-1} = (-1)^{n-1} x^{2n-2}$;

$$a_5 = (-1)^4 x^{10-2} = x^8; \qquad\qquad a_8 = (-1)^7 x^{16-2} = -x^{14}$$

13. $r = \frac{2^{x+1}}{2} = 2^x$;

$$a_n = 2(2^x)^{n-1} = 2^{\left[x(n-1)\right]} \cdot 2^1 = 2^{(n-1)x + 1}; \; a_5 = 2^{4x+1}; \; a_8 = 2^{7x+1}$$

15. $\frac{a_6}{a_4} = \frac{9}{3} = 3$ and $\frac{a_6}{a_4} = \frac{a_1 r^5}{a_1 r^3} = r^2$. Hence, $r^2 = 3$ and $r = \pm\sqrt{3}$.

17. $r = \frac{6}{4} = \frac{3}{2}$; $a_6 = a_1 r^{n-1} = 4(\tfrac{3}{2})^5 = \frac{243}{8}$

19. $\frac{a_7}{a_4} = \frac{12}{4} = 3$ and $\frac{a_7}{a_4} = \frac{a_1 r^6}{a_1 r^3} = r^3$. Hence, $r^3 = 3$ and $r = \sqrt[3]{3}$. Since the tenth term

    can be obtained by multiplying the seventh term by the common ratio three times,

$$a_{10} = a_7 r^3 = 12(3) = 36.$$

21. Using the formula for S_n with $n = 10$, $r = 3$, and $a_1 = 3^1 = 3$, we get

$$\sum_{k=1}^{10} 3^k = 3 \cdot \frac{1 - 3^{10}}{1 - 3} = 3 \cdot \frac{-59{,}048}{-2} = 88{,}572.$$

23. $\displaystyle\sum_{k=0}^{9} (-\tfrac{1}{2})^{k+1} = \sum_{k=1}^{10} (-\tfrac{1}{2})^k = -\frac{1}{2} \cdot \frac{1 - (-\tfrac{1}{2})^{10}}{1 - (-\tfrac{1}{2})} = -\frac{1}{2} \cdot \frac{\frac{1023}{1024}}{\frac{3}{2}} = -\frac{1023}{3072} = -\frac{341}{1024}$

25. $2 + 4 + 8 + 16 + 32 + 64 + 128 = 2^1 + 2^2 + 2^3 + 2^4 + 2^5 + 2^6 + 2^7 = \displaystyle\sum_{n=1}^{7} 2^n$

27. $\frac{1}{4} - \frac{1}{12} + \frac{1}{36} - \frac{1}{108} = \frac{1}{4} - \frac{1}{4} \cdot \frac{1}{3^1} + \frac{1}{4} \cdot \frac{1}{3^2} - \frac{1}{4} \cdot \frac{1}{3^3} = \displaystyle\sum_{n=1}^{4} (-1)^{n+1} \frac{1}{4}\left(\frac{1}{3}\right)^{n-1}$

29. $1 - \frac{1}{2} + \frac{1}{4} - \frac{1}{8} + \cdots$ • $a_1 = 1$, $r = -\frac{1}{2}$, $S = \frac{a}{1-r} = \frac{1}{1 - (-\tfrac{1}{2})} = \frac{1}{\frac{3}{2}} = \frac{2}{3}$

31. $1.5 + 0.015 + 0.00015 + \cdots$ •

$$a_1 = 1.5, \; r = \frac{0.015}{1.5} = 0.01, \; S = \frac{1.5}{1 - 0.01} = \frac{1.5}{0.99} = \frac{150}{99} = \frac{50}{33}$$

33 $\sqrt{2} - 2 + \sqrt{8} - 4 + \cdots$ •

The ratio $r = \dfrac{-2}{\sqrt{2}} = -\sqrt{2}$, but since $|r| = \sqrt{2} > 1$, the sum does not exist.

35 $256 + 192 + 144 + 108 + \cdots$ • $a_1 = 256$, $r = \dfrac{192}{256} = \dfrac{3}{4}$, $S = \dfrac{a}{1-r} = \dfrac{256}{1 - \frac{3}{4}} = 1024$

37 $0.\overline{23}$ • If we write $0.\overline{23}$ as $\dfrac{23}{100} + \dfrac{23}{10{,}000} + \cdots$, we see that the first term is $\dfrac{23}{100}$ and

the common ratio is $\dfrac{1}{100}$. $a_1 = 0.23$, $r = 0.01$, $S = \dfrac{0.23}{1 - 0.01} = \dfrac{23}{99}$

39 $2.4\overline{17} = 2.4171717\ldots = 2.4 + \dfrac{17}{1000} + \dfrac{17}{1000} \cdot \dfrac{1}{100} + \dfrac{17}{1000} \cdot \left(\dfrac{1}{100}\right)^2 + \cdots.$

$a_1 = \dfrac{17}{1000}$ and $r = \dfrac{1}{100}$ $\Rightarrow$ $S = \dfrac{\frac{17}{1000}}{1 - \frac{1}{100}} = \dfrac{\frac{17}{1000}}{\frac{99}{100}} = \dfrac{17}{990}$, so

$$2.4\overline{17} = 2.4 + \dfrac{17}{990} = \dfrac{2376}{990} + \dfrac{17}{990} = \dfrac{2393}{990}.$$

41 $5.\overline{146}$ • In this problem, we work only with $0.\overline{146}$ and add its rational number representation to 5. $a_1 = 0.146$, $r = 0.001$,

$$S = \dfrac{0.146}{1 - 0.001} = \dfrac{146}{999}; \quad 5.\overline{146} = 5 + \dfrac{146}{999} = \dfrac{5141}{999}$$

43 $1.\overline{6124}$ • $a_1 = 0.6124$, $r = 0.0001$, $S = \dfrac{0.6124}{1 - 0.0001} = \dfrac{6124}{9999};$

$$1.\overline{6124} = 1 + \dfrac{6124}{9999} = \dfrac{16{,}123}{9999}$$

45 The geometric mean of 12 and 48 is $\sqrt{12 \cdot 48} = \sqrt{576} = 24$.

47 Inserting 2 geometric means results in a sequence that looks like 4, x, y, 500.

Now $4r = x$, $4r^2 = y$, and $4r^3 = 500$. Thus, $4r^3 = 500$ $\Rightarrow$ $r^3 = \dfrac{500}{4} = 125$ $\Rightarrow$

$r = 5$. The terms are 4, 20, 100, and 500.

49 Let $a_1 = x$ { the original amount of air in the container } and $r = \frac{1}{2}$.

$a_{11} = x\left(\frac{1}{2}\right)^{10} = \dfrac{1}{1024}x$. This is $\left(\dfrac{1}{1024} \cdot 100\right)\%$ or $\dfrac{25}{256}\%$ or approximately 0.1% of x.

51 Let $a_1 = 10{,}000$ and $r = 1.2$, i.e., 120% every hour.

(a) $N(1) = a_2 = 10{,}000(1.2)^1$, $N(2) = a_3 = 10{,}000(1.2)^2$, ...,

$$N(t) = a_{t+1} = 10{,}000(1.2)^t$$

(b) $N(10) = a_{11} = 10{,}000(1.2)^{10} \approx 61{,}917$.

53 Distance$_{\text{total}}$ = Distance$_{\text{down}}$ + Distance$_{\text{up}}$ = $60 + 2 \cdot$ Distance$_{\text{up}}$

$$= 60 + 2\left[60\left(\tfrac{2}{3}\right) + 60\left(\tfrac{2}{3}\right)^2 + \cdots\right] = 60 + 2\left(\dfrac{60\left(\frac{2}{3}\right)}{1 - \frac{2}{3}}\right) = 60 + 2(120) = 300 \text{ ft.}$$

55 Total amount of local spending $= 2{,}000{,}000(0.60) + 2{,}000{,}000(0.60)^2 + \cdots$. We have

$a_1 = 2{,}000{,}000(0.60) = 1{,}200{,}000$ and $r = 0.60$, so $S = \dfrac{1{,}200{,}000}{1 - 0.60} = \$3{,}000{,}000$.

[57] (a) A half-life of 2 hours means there will be $(\frac{1}{2})(\frac{1}{2}D) = \frac{1}{4}D$ after 4 hours. The amount remaining after n doses {not hours} for a given dose is $a_n = D(\frac{1}{4})^{n-1}$. Since D mg are administered every 4 hours, the amount of the drug in the bloodstream after n doses is $\sum_{k=1}^{n} a_k = \sum_{k=1}^{n} D(\frac{1}{4})^{k-1} = D + \frac{1}{4}D + \cdots + (\frac{1}{4})^{n-1}D$. Since $r = \frac{1}{4} < 1$, S_n may be approximated by $S = \frac{a_1}{1-r}$ for large n.

$$S = \frac{D}{1 - \frac{1}{4}} = \frac{4}{3}D.$$

(b) Since the amount of the drug in the bloodstream is given by $\frac{4}{3}D$, and this amount must be less than or equal to 500 mg, we have $\frac{4}{3}D \leq 500$ mg, or, equivalently, $D \leq 375$ mg.

[59] (a) From the figure on the right, we see that

$$(\tfrac{1}{4}a_k)^2 + (\tfrac{3}{4}a_k)^2 = (a_{k+1})^2 \;\Rightarrow\; \tfrac{10}{16}a_k^2 = a_{k+1}^2 \;\Rightarrow\; a_{k+1} = \tfrac{1}{4}\sqrt{10}\,a_k.$$

(b) From part (a), the common ratio $r = \frac{a_{k+1}}{a_k} = \frac{1}{4}\sqrt{10}$, so $a_n = a_1 r^{n-1} = a_1(\frac{1}{4}\sqrt{10})^{n-1}$. For the square S_{k+1}, the area $A_{k+1} = a_{k+1}^2$. But from part (a), $a_{k+1}^2 = \frac{10}{16}a_k^2$. Since a_k^2 is the area A_k of square S_k, we have $A_{k+1} = \frac{5}{8}A_k$, and hence $A_n = (\frac{5}{8})^{n-1}A_1$. Similarly, for the perimeter of the square S_k, $P_{k+1} = 4a_{k+1} = 4 \cdot \frac{1}{4}\sqrt{10}\,a_k = \sqrt{10}\,a_k = \sqrt{10}(\frac{1}{4}P_k)$, and hence $P_n = (\frac{1}{4}\sqrt{10})^{n-1}P_1$.

(c) $\sum_{n=1}^{\infty} P_n$ is an infinite geometric series with first term P_1 and $r = \frac{1}{4}\sqrt{10}$.

$$S = \frac{P_1}{1 - \frac{1}{4}\sqrt{10}} \cdot \frac{4}{4} = \frac{4P_1}{4 - \sqrt{10}} = \frac{16a_1}{4 - \sqrt{10}}.$$

[61] (a) The sequence is 1, 3, 9, 27, 81, ... Thus, $a_k = 3^{k-1}$ for $k = 1, 2, 3, \ldots$.

(b) $a_{15} = 3^{14} = 4{,}782{,}969$

(c) The area of the triangle removed first is $\frac{1}{4}$. During the next step, 3 triangles with an area of $\frac{1}{16}$ are removed. Then, 9 triangles with an area of $\frac{1}{64}$ are removed. At each step the number of triangles increase by a factor of 3, while the area of each triangle decreases by a factor of 4. Thus, $b_k = \frac{3^{k-1}}{4^k} = \frac{1}{4}\left(\frac{3}{4}\right)^{k-1}$.

(d) $b_7 = \frac{1}{4}\left(\frac{3}{4}\right)^6 = \frac{729}{16{,}384} \approx 0.0445 = 4.45\%$.

63 Let $a_k = 100\left(1 + \frac{0.06}{12}\right)^k = 100(1.005)^k$, where k represents the number of compounding periods for each deposit. For the first deposit, $k = 18 \cdot 12 = 216$. For the last deposit, $k = 1$.　　$S_{216} = a_1 + a_2 + \cdots + a_{216}$

$$= 100(1.005)^1 + 100(1.005)^2 + \cdots + 100(1.005)^{216}$$

$$= 100(1.005)\left(\frac{1 - (1.005)^{216}}{1 - (1.005)}\right) \approx \$38,929.00$$

65 Use $A = P\left(\frac{12}{r} + 1\right)\left[\left(1 + \frac{r}{12}\right)^n - 1\right]$ with $P = 100$, $r = 0.08$, and $n = 60$ to get

$$A = 100\left(\frac{12}{0.08} + 1\right)\left[\left(1 + \frac{0.08}{12}\right)^{60} - 1\right] \approx \$7396.67.$$

67 (a) $A_1 = \frac{2}{5}\left(1 - \frac{2}{5}\right)^{1-1} = \frac{2}{5}\left(\frac{3}{5}\right)^0 = \frac{2}{5}$.

$A_2 = \frac{2}{5}\left(\frac{3}{5}\right)^1 = \frac{6}{25}$, $A_3 = \frac{2}{5}\left(\frac{3}{5}\right)^2 = \frac{18}{125}$, $A_4 = \frac{2}{5}\left(\frac{3}{5}\right)^3 = \frac{54}{625}$, $A_5 = \frac{2}{5}\left(\frac{3}{5}\right)^4 = \frac{162}{3125}$.

(b) $r = A_{k+1}/A_k = \frac{3}{5}$ for $k = 1, 2, 3, 4$.

$$S_5 = \sum_{k=1}^{5} A_k = A_1 + A_2 + A_3 + A_4 + A_5 = \frac{2}{5} \cdot \frac{1 - \left(\frac{3}{5}\right)^5}{1 - \frac{3}{5}} = 1 - \left(\frac{3}{5}\right)^5 = \frac{2882}{3125} = 0.92224.$$

(c) $\$25,000\left(\frac{2}{5} + \frac{6}{25}\right) = \$25,000\left(\frac{16}{25}\right) = \$16,000$

9.4 Exercises

1 (1) P_1 is true, since $2(1) = 1(1 + 1) = 2$.

(2) Assume P_k is true:

$2 + 4 + 6 + \cdots + 2k = k(k + 1)$. Hence,

$$2 + 4 + 6 + \cdots + 2k + 2(k + 1) = k(k + 1) + 2(k + 1)$$
$$= (k + 1)(k + 2)$$
$$= (k + 1)(k + 1 + 1).$$

Thus, P_{k+1} is true, and the proof is complete.

3 (1) P_1 is true, since $2(1) - 1 = (1)^2 = 1$.

(2) Assume P_k is true:

$1 + 3 + 5 + \cdots + (2k - 1) = k^2$. Hence,

$$1 + 3 + 5 + \cdots + (2k - 1) + 2(k + 1) - 1 = k^2 + 2(k + 1) - 1$$
$$= k^2 + 2k + 1$$
$$= (k + 1)^2.$$

Thus, P_{k+1} is true, and the proof is complete.

$\boxed{5}$ (1) P_1 is true, since $5(1) - 3 = \frac{1}{2}(1)[5(1) - 1] = 2$.

(2) Assume P_k is true:

$$2 + 7 + 12 + \cdots + (5k - 3) = \tfrac{1}{2}k(5k - 1). \text{ Hence,}$$

$$2 + 7 + 12 + \cdots + (5k - 3) + 5(k + 1) - 3 = \tfrac{1}{2}k(5k - 1) + 5(k + 1) - 3$$
$$= \tfrac{5}{2}k^2 + \tfrac{9}{2}k + 2$$
$$= \tfrac{1}{2}(5k^2 + 9k + 4)$$
$$= \tfrac{1}{2}(k + 1)(5k + 4)$$
$$= \tfrac{1}{2}(k + 1)[5(k + 1) - 1].$$

Thus, P_{k+1} is true, and the proof is complete.

$\boxed{7}$ (1) P_1 is true, since $1 \cdot 2^{1-1} = 1 + (1 - 1) \cdot 2^1 = 1$.

(2) Assume P_k is true:

$$1 + 2 \cdot 2 + 3 \cdot 2^2 + \cdots + k \cdot 2^{k-1} = 1 + (k - 1) \cdot 2^k. \text{ Hence,}$$

$$1 + 2 \cdot 2 + 3 \cdot 2^2 + \cdots + k \cdot 2^{k-1} + (k + 1) \cdot 2^k = 1 + (k - 1) \cdot 2^k + (k + 1) \cdot 2^k$$
$$= 1 + k \cdot 2^k - 2^k + k \cdot 2^k + 2^k$$
$$= 1 + k \cdot 2^1 \cdot 2^k$$
$$= 1 + [(k + 1) - 1] \cdot 2^{k+1}.$$

Thus, P_{k+1} is true, and the proof is complete.

$\boxed{9}$ (1) P_1 is true, since $(1)^1 = \dfrac{1(1 + 1)[2(1) + 1]}{6} = 1$.

(2) Assume P_k is true:

$$1^2 + 2^2 + 3^2 + \cdots + k^2 = \frac{k(k + 1)(2k + 1)}{6}. \text{ Hence,}$$

$$1^2 + 2^2 + 3^2 + \cdots + k^2 + (k + 1)^2 = \frac{k(k + 1)(2k + 1)}{6} + (k + 1)^2$$
$$= (k + 1)\left[\frac{k(2k + 1)}{6} + \frac{6(k + 1)}{6}\right]$$
$$= \frac{(k + 1)(2k^2 + 7k + 6)}{6}$$
$$= \frac{(k + 1)(k + 2)(2k + 3)}{6}.$$

Thus, P_{k+1} is true, and the proof is complete.

11 (1) P_1 is true, since $\dfrac{1}{1(1+1)} = \dfrac{1}{1+1} = \dfrac{1}{2}$.

(2) Assume P_k is true:

$$\dfrac{1}{1\cdot 2}+\dfrac{1}{2\cdot 3}+\dfrac{1}{3\cdot 4}+\cdots+\dfrac{1}{k(k+1)} = \dfrac{k}{k+1}. \quad \text{Hence,}$$

$$\begin{aligned}
\dfrac{1}{1\cdot 2}+\dfrac{1}{2\cdot 3}+\dfrac{1}{3\cdot 4}+\cdots+\dfrac{1}{k(k+1)}+\dfrac{1}{(k+1)(k+2)} &= \dfrac{k}{k+1}+\dfrac{1}{(k+1)(k+2)} \\
&= \dfrac{k}{k+1}+\dfrac{1}{(k+1)(k+2)} \\
&= \dfrac{k(k+2)+1}{(k+1)(k+2)} \\
&= \dfrac{k^2+2k+1}{(k+1)(k+2)} \\
&= \dfrac{k+1}{(k+1)+1}.
\end{aligned}$$

Thus, P_{k+1} is true, and the proof is complete.

13 (1) P_1 is true, since $3^1 = \tfrac{3}{2}(3^1-1) = 3$.

(2) Assume P_k is true:

$$3+3^2+3^3+\cdots+3^k = \tfrac{3}{2}(3^k-1). \quad \text{Hence,}$$

$$\begin{aligned}
3+3^2+3^3+\cdots+3^k+3^{k+1} &= \tfrac{3}{2}(3^k-1)+3^{k+1} \\
&= \tfrac{3}{2}\cdot 3^k - \tfrac{3}{2}+3\cdot 3^k \\
&= \tfrac{9}{2}\cdot 3^k - \tfrac{3}{2} \\
&= \tfrac{3}{2}(3\cdot 3^k-1) \\
&= \tfrac{3}{2}(3^{k+1}-1).
\end{aligned}$$

Thus, P_{k+1} is true, and the proof is complete.

15 (1) P_1 is true, since $1 < 2^1$.

(2) Assume P_k is true: $k < 2^k$. Now $k+1 < k+k = 2(k)$ for $k > 1$.

From P_k, we see that $2(k) < 2(2^k) = 2^{k+1}$ and conclude that $k+1 < 2^{k+1}$.

Thus, P_{k+1} is true, and the proof is complete.

17 (1) P_1 is true, since $1 < \tfrac{1}{8}[2(1)+1]^2 = \tfrac{9}{8}$.

(2) Assume P_k is true: $1+2+3+\cdots+k < \tfrac{1}{8}(2k+1)^2$. Hence,

$$\begin{aligned}
1+2+3+\cdots+k+(k+1) &< \tfrac{1}{8}(2k+1)^2+(k+1) \\
&= \tfrac{1}{2}k^2 + \tfrac{3}{2}k + \tfrac{9}{8} \\
&= \tfrac{1}{8}(4k^2+12k+9) \\
&= \tfrac{1}{8}(2k+3)^2 \\
&= \tfrac{1}{8}[2(k+1)+1]^2.
\end{aligned}$$

Thus, P_{k+1} is true, and the proof is complete.

19 (1) For $n = 1$, $n^3 - n + 3 = 3$ and 3 is a factor of 3.

(2) Assume 3 is a factor of $k^3 - k + 3$. The $(k+1)$st term is

$$(k+1)^3 - (k+1) + 3 = k^3 + 3k^2 + 2k + 3$$
$$= (k^3 - k + 3) + 3k^2 + 3k$$
$$= (k^3 - k + 3) + 3(k^2 + k).$$

By the induction hypothesis, 3 is a factor of $k^3 - k + 3$ and 3 is a factor of $3(k^2 + k)$, so 3 is a factor of the $(k+1)$st term. Thus, P_{k+1} is true, and the proof is complete.

21 (1) For $n = 1$, $5^n - 1 = 4$ and 4 is a factor of 4.

(2) Assume 4 is a factor of $5^k - 1$. The $(k+1)$st term is

$$5^{k+1} - 1 = 5 \cdot 5^k - 1$$
$$= 5 \cdot 5^k - 5 + 4$$
$$= 5(5^k - 1) + 4.$$

By the induction hypothesis, 4 is a factor of $5^k - 1$ and 4 is a factor of 4, so 4 is a factor of the $(k+1)$st term. Thus, P_{k+1} is true, and the proof is complete.

23 (1) If $a > 1$, then $a^1 = a > 1$, so P_1 is true.

(2) Assume P_k is true: $a^k > 1$.

Multiply both sides by a to obtain $a^{k+1} > a$, but since $a > 1$, we have $a^{k+1} > 1$.

Thus, P_{k+1} is true, and the proof is complete.

25 (1) For $n = 1$, $a - b$ is a factor of $a^1 - b^1$.

(2) Assume $a - b$ is a factor of $a^k - b^k$. Following the hint for the $(k+1)$st term, $a^{k+1} - b^{k+1} = a^k \cdot a - b \cdot a^k + b \cdot a^k - b^k \cdot b = a^k(a-b) + (a^k - b^k)b$. Since $(a-b)$ is a factor of $a^k(a-b)$ and since by the induction hypothesis $a - b$ is a factor of $(a^k - b^k)$, it follows that $a - b$ is a factor of the $(k+1)$st term. Thus, P_{k+1} is true, and the proof is complete.

Note: For Exercises 27–32 in this section and Exercises 47–48 in the Chapter Review, there are several ways to find j. Possibilities include: solve the inequality, sketch the graphs of functions representing each side, and trial and error. Trial and error may be the easiest to use.

27 For j: $n^2 \geq n + 12 \Rightarrow n^2 - n - 12 \geq 0 \Rightarrow (n-4)(n+3) \geq 0 \Rightarrow n \geq 4 \ \{n > 0\}$

(1) P_4 is true, since $4 + 12 \leq 4^2$.

(2) Assume P_k is true: $k + 12 \leq k^2$. Hence,

$$(k+1) + 12 = (k+12) + 1 \leq (k^2) + 1 < k^2 + 2k + 1 = (k+1)^2.$$

Thus, P_{k+1} is true, and the proof is complete.

29 For j: By sketching $y = 5 + \log_2 x$ and $y = x$, we see that the solution for $x > 1$

must be larger than 5. By trial and error, $j = 8$.

(1) P_8 is true, since $5 + \log_2 8 \le 8$.

(2) Assume P_k is true: $5 + \log_2 k \le k$. Hence,

$$5 + \log_2 (k + 1) < 5 + \log_2 (k + k)$$
$$= 5 + \log_2 2k$$
$$= 5 + \log_2 2 + \log_2 k$$
$$= (5 + \log_2 k) + 1$$
$$\le k + 1.$$

Thus, P_{k+1} is true, and the proof is complete.

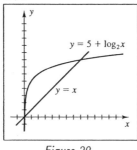

Figure 29

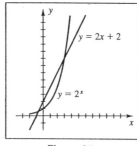

Figure 31

31 For j: By sketching $y = 2x + 2$ and $y = 2^x$, we see there is one positive solution.

By trial and error, $j = 3$.

(1) P_3 is true, since $2(3) + 2 \le 2^3$.

(2) Assume P_k is true: $2k + 2 \le 2^k$. Hence,

$$2(k + 1) + 2 = (2k + 2) + 2 \le 2^k + 2^k = 2 \cdot 2^k = 2^{k+1}.$$

Thus, P_{k+1} is true, and the proof is complete.

33 Following the hint in the text:

$$\sum_{k=1}^{n} (k^2 + 3k + 5) = \sum_{k=1}^{n} k^2 + 3 \sum_{k=1}^{n} k + \sum_{k=1}^{n} 5$$
$$= \frac{n(n+1)(2n+1)}{6} + 3\left[\frac{n(n+1)}{2}\right] + 5n$$
$$= \frac{n(n+1)(2n+1) + 9n(n+1) + 30n}{6}$$
$$= \frac{2n^3 + 12n^2 + 40n}{6}$$
$$= \frac{n^3 + 6n^2 + 20n}{3}$$

[35] $\displaystyle\sum_{k=1}^{n} (2k - 3)^2 = \sum_{k=1}^{n} (4k^2 - 12k + 9) = 4\sum_{k=1}^{n} k^2 - 12\sum_{k=1}^{n} k + \sum_{k=1}^{n} 9$

$$= 4\left[\frac{n(n+1)(2n+1)}{6}\right] - 12\left[\frac{n(n+1)}{2}\right] + 9n = \frac{4n^3 - 12n^2 + 11n}{3}$$

[37] (a) $n = 1 \;\Rightarrow\; a(1)^3 + b(1)^2 + c(1) = 1^2 \;\Rightarrow\; a + b + c = 1$

$n = 2 \;\Rightarrow\; a(2)^3 + b(2)^2 + c(2) = 1^2 + 2^2 \;\Rightarrow\; 8a + 4b + 2c = 5$

$n = 3 \;\Rightarrow\; a(3)^3 + b(3)^2 + c(3) = 1^2 + 2^2 + 3^2 \;\Rightarrow\; 27a + 9b + 3c = 14$

$$AX = B \;\Rightarrow\; \begin{bmatrix} 1 & 1 & 1 \\ 8 & 4 & 2 \\ 27 & 9 & 3 \end{bmatrix} \begin{bmatrix} a \\ b \\ c \end{bmatrix} = \begin{bmatrix} 1 \\ 5 \\ 14 \end{bmatrix} \;\Rightarrow\; X = A^{-1}B = \begin{bmatrix} 1/3 \\ 1/2 \\ 1/6 \end{bmatrix}.$$

(b) $a = \frac{1}{3}, \, b = \frac{1}{2}, \, c = \frac{1}{6} \;\Rightarrow$

$$1^2 + 2^2 + 3^2 + \cdots + n^2 = \tfrac{1}{3}n^3 + \tfrac{1}{2}n^2 + \tfrac{1}{6}n = \frac{n(n+1)(2n+1)}{6},$$

which is the formula found in Exercise 9. This method does not verify the formula for all n but only for $n = 1$, 2, 3. Mathematical induction should be used to verify the formula for all n as in Exercise 9.

Note: The solutions for Exercises 39 and 41 are on page 374 at the end of the chapter.

9.5 Exercises

[1] $2!\, 6! = (2 \cdot 1) \cdot (6 \cdot 5 \cdot 4 \cdot 3 \cdot 2 \cdot 1) = 2 \cdot 720 = 1440$

[3] $7!\, 0! = (7 \cdot 6 \cdot 5 \cdot 4 \cdot 3 \cdot 2 \cdot 1) \cdot (1) \, \{\text{Remember that } 0! = 1.\} = 5040$

[5] $\dfrac{8!}{5!} = \dfrac{8 \cdot 7 \cdot 6 \cdot 5!}{5!} \, \{\text{cancel } 5!\} = 8 \cdot 7 \cdot 6 = 336$

[7] $\dbinom{5}{5} = \dfrac{5!}{5!\, 0!} \, \{\text{cancel } 5!\} = \dfrac{1}{0!} = \dfrac{1}{1} = 1$

[9] $\dbinom{7}{5} = \dfrac{7!}{5!\, 2!} \, \{\text{cancel } 5!\} = \dfrac{7 \cdot 6}{2} = 21$

[11] $\dbinom{13}{4} = \dfrac{13!}{4!\, 9!} \, \{\text{cancel } 9!\} = \dfrac{13 \cdot 12 \cdot 11 \cdot 10}{4 \cdot 3 \cdot 2} = 715$

[13] $\dfrac{n!}{(n-2)!} = \dfrac{n(n-1)(n-2)!}{(n-2)!} \, \{\text{cancel } (n-2)!\} = n(n-1)$

[15] $\dfrac{(2n+2)!}{(2n)!} = \dfrac{(2n+2)(2n+1)(2n)!}{(2n)!} \, \{\text{cancel } (2n)!\} = (2n+2)(2n+1)$

[17] We use the binomial theorem formula with $a = 4x$, $b = -y$, and $n = 3$.

$$(4x - y)^3 = \dbinom{3}{0}(4x)^3(-y)^0 + \dbinom{3}{1}(4x)^2(-y)^1 + \dbinom{3}{2}(4x)^1(-y)^2 + \dbinom{3}{3}(4x)^0(-y)^3$$

$$= (1)(64x^3)(1) + (3)(16x^2)(-y) + (3)(4x)(y^2) + (1)(1)(-y^3)$$

$$= 64x^3 - 48x^2y + 12xy^2 - y^3$$

19 We use the binomial theorem formula with $a = a$, $b = b$, and $n = 6$.

$$(a+b)^6 = a^6 + \binom{6}{1}a^5b^1 + \binom{6}{2}a^4b^2 + \binom{6}{3}a^3b^3 + \binom{6}{4}a^2b^4 + \binom{6}{5}a^1b^5 + b^6$$

$$= a^6 + 6a^5b + 15a^4b^2 + 20a^3b^3 + 15a^2b^4 + 6ab^5 + b^6$$

21 We use the binomial theorem formula with $a = a$, $b = -b$, and $n = 7$.

$$(a-b)^7 = a^7 + \binom{7}{1}a^6(-b)^1 + \binom{7}{2}a^5(-b)^2 + \binom{7}{3}a^4(-b)^3 + \binom{7}{4}a^3(-b)^4 +$$

$$\binom{7}{5}a^2(-b)^5 + \binom{7}{6}a^1(-b)^6 + (-b)^7$$

$$= a^7 - 7a^6b + 21a^5b^2 - 35a^4b^3 + 35a^3b^4 - 21a^2b^5 + 7ab^6 - b^7$$

23 We use the binomial theorem formula with $a = 3x$, $b = -5y$, and $n = 4$.

$$(3x - 5y)^4 = \binom{4}{0}(3x)^4(-5y)^0 + \binom{4}{1}(3x)^3(-5y)^1 + \binom{4}{2}(3x)^2(-5y)^2 +$$

$$\binom{4}{3}(3x)^1(-5y)^3 + \binom{4}{4}(3x)^0(-5y)^4$$

$$= (1)(81x^4)(1) + (4)(27x^3)(-5y) + (6)(9x^2)(25y^2) +$$

$$(4)(3x)(-125y^3) + (1)(1)(625y^4)$$

$$= 81x^4 - 540x^3y + 1350x^2y^2 - 1500xy^3 + 625y^4$$

25 We use the binomial theorem formula with $a = \frac{1}{3}x$, $b = y^2$, and $n = 5$.

$$(\tfrac{1}{3}x + y^2)^5 = \binom{5}{0}(\tfrac{1}{3}x)^5(y^2)^0 + \binom{5}{1}(\tfrac{1}{3}x)^4(y^2)^1 + \binom{5}{2}(\tfrac{1}{3}x)^3(y^2)^2 +$$

$$\binom{5}{3}(\tfrac{1}{3}x)^2(y^2)^3 + \binom{5}{4}(\tfrac{1}{3}x)^1(y^2)^4 + \binom{5}{5}(\tfrac{1}{3}x)^0(y^2)^5$$

$$= (1)(\tfrac{1}{243}x^5)(1) + (5)(\tfrac{1}{81}x^4)(y^2) + (10)(\tfrac{1}{27}x^3)(y^4) +$$

$$(10)(\tfrac{1}{9}x^2)(y^6) + (5)(\tfrac{1}{3}x)(y^8) + (1)(1)(y^{10})$$

$$= \tfrac{1}{243}x^5 + \tfrac{5}{81}x^4y^2 + \tfrac{10}{27}x^3y^4 + \tfrac{10}{9}x^2y^6 + \tfrac{5}{3}xy^8 + y^{10}$$

27 We use the binomial theorem formula with $a = x^{-2}$, $b = 3x$, and $n = 6$.

$$\left(\frac{1}{x^2} + 3x\right)^6 = (x^{-2} + 3x)^6$$

$$= \binom{6}{0}(x^{-2})^6(3x)^0 + \binom{6}{1}(x^{-2})^5(3x)^1 + \binom{6}{2}(x^{-2})^4(3x)^2 + \binom{6}{3}(x^{-2})^3(3x)^3 +$$

$$\binom{6}{4}(x^{-2})^2(3x)^4 + \binom{6}{5}(x^{-2})^1(3x)^5 + \binom{6}{6}(x^{-2})^0(3x)^6$$

$$= (1)(x^{-12})(1) + (6)(x^{-10})(3x^1) + (15)(x^{-8})(9x^2) + (20)(x^{-6})(27x^3) +$$

$$(15)(x^{-4})(81x^4) + (6)(x^{-2})(243x^5) + (1)(1)(729x^6)$$

$$= x^{-12} + 18x^{-9} + 135x^{-6} + 540x^{-3} + 1215 + 1458x^3 + 729x^6$$

[29] We use the binomial theorem formula with $a = x^{1/2}$, $b = -x^{-1/2}$, and $n = 5$.

$$\left(\sqrt{x} - \frac{1}{\sqrt{x}}\right)^5 = (x^{1/2} - x^{-1/2})^5$$

$$= \binom{5}{0}(x^{1/2})^5(-x^{-1/2})^0 + \binom{5}{1}(x^{1/2})^4(-x^{-1/2})^1 + \binom{5}{2}(x^{1/2})^3(-x^{-1/2})^2 +$$

$$\binom{5}{3}(x^{1/2})^2(-x^{-1/2})^3 + \binom{5}{4}(x^{1/2})^1(-x^{-1/2})^4 + \binom{5}{5}(x^{1/2})^0(-x^{-1/2})^5$$

$$= (1)(x^{5/2})(1) + (5)(x^2)(-x^{-1/2}) + (10)(x^{3/2})(x^{-1}) +$$

$$(10)(x^1)(-x^{-3/2}) + (5)(x^{1/2})(x^{-2}) + (1)(1)(-x^{-5/2})$$

$$= x^{5/2} - 5x^{3/2} + 10x^{1/2} - 10x^{-1/2} + 5x^{-3/2} - x^{-5/2}$$

[31] For the binomial expression $(3c^{2/5} + c^{4/5})^{25}$, the first three terms are

$$\sum_{k=0}^{2} \binom{25}{k}(3c^{2/5})^{25-k}(c^{4/5})^k$$

$$= \binom{25}{0}(3c^{2/5})^{25}(c^{4/5})^0 + \binom{25}{1}(3c^{2/5})^{24}(c^{4/5})^1 + \binom{25}{2}(3c^{2/5})^{23}(c^{4/5})^2$$

$$= (1)(3^{25}c^{10})(1) + (25)(3^{24}c^{48/5})(c^{4/5}) + (300)(3^{23}c^{46/5})(c^{8/5})$$

$$= 3^{25}c^{10} + 25 \cdot 3^{24}c^{52/5} + 300 \cdot 3^{23}c^{54/5}$$

[33] For the binomial expression $(4b^{-1} - 3b)^{15}$, the last three terms are

$$\sum_{k=13}^{15} \binom{15}{k}(4b^{-1})^{15-k}(-3b)^k$$

$$= \binom{15}{13}(4b^{-1})^2(-3b)^{13} + \binom{15}{14}(4b^{-1})^1(-3b)^{14} + \binom{15}{15}(4b^{-1})^0(-3b)^{15}$$

$$= (105)(16b^{-2})(-3^{13}b^{13}) + (15)(4b^{-1})(3^{14}b^{14}) + (1)(1)(-3^{15}b^{15})$$

$$= -1680 \cdot 3^{13}b^{11} + 60 \cdot 3^{14}b^{13} - 3^{15}b^{15}$$

Note: For the following exercises, the general formula for the

$(k+1)$st term of the expansion of $(a+b)^n$ is $\boxed{\binom{n}{k}(a)^{n-k}(b)^k}$.

[35] $\left(\frac{3}{c} + \frac{c^2}{4}\right)^7$; sixth term $\{k = 5\} = \binom{7}{5}\left(\frac{3}{c}\right)^2\left(\frac{c^2}{4}\right)^5 = 21\left(\frac{9}{c^2}\right)\left(\frac{c^{10}}{1024}\right) = \frac{189}{1024}c^8$

[37] $(\frac{1}{3}u + 4v)^8$; seventh term $\{k = 6\} = \binom{8}{6}(\frac{1}{3}u)^2(4v)^6 = 28\left(\frac{u^2}{9}\right)(4096v^6) = \frac{114,688}{9}u^2v^6$

[39] Since there are 9 terms, the middle term is the fifth term $\left\{\frac{9+1}{2} = 5\right\}$.

Using the formula in the *Note* above, we obtain the fifth term of $(x^{1/2} + y^{1/2})^8$,

$$\binom{8}{4}(x^{1/2})^4(y^{1/2})^4 = 70x^2y^2.$$

[41] $(2y + x^2)^8$; term that contains x^{10} •

Consider only the variable x in the expansion: $(x^2)^k = x^{10} \Rightarrow 2k = 10 \Rightarrow k = 5$;

$$\text{sixth term} = \binom{8}{5}(2y)^3(x^2)^5 = 448y^3x^{10}$$

43 $(3b^3 - 2a^2)^4$; term that contains b^9 •

Consider only the variable b in the expansion:

$$(b^3)^{4-k} = b^9 \implies 12 - 3k = 9 \implies k = 1; \text{2nd term} = \binom{4}{1}(3b^3)^3(-2a^2)^1 = -216b^9a^2$$

45 $\left(3x - \dfrac{1}{4x}\right)^6$; term that does not contain x •

Consider only the variable x in the expansion:

$$x^{6-k}(x^{-1})^k = x^0 \implies x^{6-2k} = x^0 \implies k = 3; \text{4th term} = \binom{6}{3}(3x)^3\left(-\frac{1}{4x}\right)^3 = -\frac{135}{16}$$

47 The first three terms in the expansion of $(1 + 0.2)^{10}$ are

$$\sum_{k=0}^{2}\binom{10}{k}(1)^{10-k}(0.2)^k = \binom{10}{0}(1)^{10}(0.2)^0 + \binom{10}{1}(1)^9(0.2)^1 + \binom{10}{2}(1)^8(0.2)^2$$

$$= (1)(1)(1) + (10)(1)(0.2) + (45)(1)(0.04) = 1 + 2 + 1.8 = 4.8.$$

The calculator result for $(1.2)^{10}$ is approximately 6.19.

49 $\dfrac{(x+h)^4 - x^4}{h} = \dfrac{(x^4 + 4x^3h + 6x^2h^2 + 4xh^3 + h^4) - x^4}{h} = \dfrac{h(4x^3 + 6x^2h + 4xh^2 + h^3)}{h} =$

$$4x^3 + 6x^2h + 4xh^2 + h^3$$

51 $\dbinom{n}{1} = \dfrac{n!}{(n-1)!\,1!} = n$ and $\dbinom{n}{n-1} = \dfrac{n!}{[n-(n-1)]!\,(n-1)!} = \dfrac{n!}{1!\,(n-1)!} = n$

9.6 Exercises

1 $P(7, 3) = \dfrac{7!}{(7-3)!} = \dfrac{7!}{4!} = \dfrac{7\cdot 6\cdot 5\cdot 4!}{4!}\,\{\text{cancel } 4!\} = 7\cdot 6\cdot 5 = 210$

3 $P(9, 6) = \dfrac{9!}{(9-6)!} = \dfrac{9!}{3!} = \dfrac{9\cdot 8\cdot 7\cdot 6\cdot 5\cdot 4\cdot 3!}{3!}\,\{\text{cancel } 3!\} = 9\cdot 8\cdot 7\cdot 6\cdot 5\cdot 4 = 60{,}480$

5 $P(5, 5) = \dfrac{5!}{(5-5)!} = \dfrac{5!}{0!} = \dfrac{5\cdot 4\cdot 3\cdot 2\cdot 1}{1} = 120$

7 $P(6, 1) = \dfrac{6!}{(6-1)!} = \dfrac{6!}{5!} = \dfrac{6\cdot 5!}{5!} = 6$

9 $P(n, 0) = \dfrac{n!}{(n-0)!} = \dfrac{n!}{n!} = 1$

11 $P(n, n-1) = \dfrac{n!}{[n-(n-1)]!} = \dfrac{n!}{1!} = \dfrac{n!}{1} = n!$

13 (a) We can think of the three-digit numbers as filling 3 slots. There are 5 digits to pick from for filling the first slot. There are 4 remaining digits to pick from to fill the second slot. There are 3 remaining digits to pick from to fill the third slot. By the fundamental counting principle, there are a total of $5\cdot 4\cdot 3 = 60$ three-digit numbers.

(b) The difference between parts (a) and (b) is that we can use any of the 5 digits for all 3 slots. Hence, there are a total of $5\cdot 5\cdot 5 = 125$ three-digit numbers if repetitions are allowed.

15 There are 4 one digit numbers; $4 \cdot 3 = 12$ two digit numbers;

$4 \cdot 3 \cdot 2 = 24$ three digit numbers; $4 \cdot 3 \cdot 2 \cdot 1 = 24$ four digit numbers.

The total number of numbers is $4 + 12 + 24 + 24 = 64$.

17 We want eight teams taken three at a time. $P(8, 3) = \dfrac{8!}{(8-3)!} = \dfrac{8!}{5!} = 8 \cdot 7 \cdot 6 = 336$

19 By the fundamental counting principle, $4 \cdot 6 = 24$.

21 (a) By the fundamental counting principle, $26 \cdot 9 \cdot 10^4 = 2,340,000$.

(b) By the fundamental counting principle, $24 \cdot 9 \cdot 10^4 = 2,160,000$.

23 (a) $P(10, 6) = \dfrac{10!}{(10-6)!} = \dfrac{10!}{4!} = 10 \cdot 9 \cdot 8 \cdot 7 \cdot 6 \cdot 5 = 151,200$

(b) $\boxed{\text{Boy}}$- girl seatings: $\boxed{6} \cdot \underline{4} \cdot \boxed{5} \cdot \underline{3} \cdot \boxed{4} \cdot \underline{2} = 2880$.

$\underline{\text{Girl}} \text{-} \boxed{\text{boy}}$ seatings: $\underline{4} \cdot \boxed{6} \cdot \underline{3} \cdot \boxed{5} \cdot \underline{2} \cdot \boxed{4} = 2880$.

The total number of seatings is $2880 + 2880 = 5760$.

25 There are 2 choices for each of the 10 questions. 2 times itself 10 times $= 2^{10} = 1024$

27 We want eight people taken eight at a time. $P(8, 8) = \dfrac{8!}{(8-8)!} = \dfrac{8!}{0!} = 8! = 40,320$

29 We want six flags taken three at a time. $P(6, 3) = \dfrac{6!}{(6-3)!} = \dfrac{6!}{3!} = 6 \cdot 5 \cdot 4 = 120$

31 (a) The number of choices for each letter are: $\underline{2} \cdot \underline{25} \cdot \underline{24} \cdot \underline{23} = 27,600$

(b) The number of choices for each letter are: $\underline{2} \cdot \underline{26} \cdot \underline{26} \cdot \underline{26} = 35,152$

33 We have a choice of 9 digits for the first number and a choice of 10 digits for the other numbers. By the fundamental counting principle, $9 \cdot 10^6 = 9,000,000$.

35 $P(4, 4) = \dfrac{4!}{(4-4)!} = \dfrac{4!}{0!} = 4! = 24$

37 There are 3! ways to choose the ordering of the couples and 2 ways for each couple to sit. By the fundamental counting principle, $3! \cdot 2^3 = 48$.

39 Two ways to sum these integers are

$1 + 2 + 3 + 4 + 5 + 6 + 7 + 8 + 9 + 10$ and $2 + 1 + 4 + 3 + 6 + 5 + 8 + 7 + 10 + 9$.

$$P(10, 10) = \dfrac{10!}{(10-10)!} = \dfrac{10!}{0!} = 10! = 3,628,800$$

41 (a) There are 9 choices for the first digit { can't have 0 }, 10 for the second, 10 for the third, and 1 for the fourth and fifth. $9 \cdot 10 \cdot 10 \cdot 1 \cdot 1 = 900$

(b) If n is even, we need to select the first $\frac{n}{2}$ digits. There are 9 choices for the first digit and 10 choices for each of the next $\left(\frac{n}{2} - 1\right)$ digits, so the total number of even n-digit palindromes is $9 \cdot 10^{(n/2)-1}$. (continued)

If n is odd, we need to select the first $\left(\frac{n}{2}+\frac{1}{2}\right)$ digits. There are 9 choices for the first digit and 10 choices for each of the next $\left(\frac{n}{2}+\frac{1}{2}-1\right)=\left(\frac{n}{2}-\frac{1}{2}\right)=\frac{n-1}{2}$ digits, so the total number of odd n-digit palindromes is $9\cdot 10^{(n-1)/2}$.

[43] (a) There is a horizontal asymptote of $y=1$.

(b) $\dfrac{n!\,e^n}{n^n\sqrt{2\pi n}}\approx 1 \;\Rightarrow\; n!\approx \dfrac{n^n\sqrt{2\pi n}}{e^n}$.

Example: $50!\approx \dfrac{50^{50}\sqrt{2\pi(50)}}{e^{50}}\approx 3.0363\times 10^{64}$.

The actual value is closer to 3.0414×10^{64}.

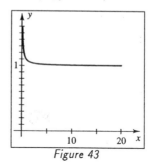

Figure 43

9.7 Exercises

[1] $C(7,\,3)=\dfrac{7!}{(7-3)!\,3!}=\dfrac{7!}{4!\,3!}=\dfrac{7\cdot 6\cdot 5\cdot 4!}{4!\,3!}=\dfrac{7\cdot 6\cdot 5}{3!}=\dfrac{7\cdot 6\cdot 5}{6}=7\cdot 5=35$

[3] $C(9,\,8)=\dfrac{9!}{1!\,8!}=\dfrac{9\cdot 8!}{1\cdot 8!}=9$

[5] $C(n,\,n-1)=\dfrac{n!}{[n-(n-1)]!\,(n-1)!}=\dfrac{n!}{1!\,(n-1)!}=n$

[7] $C(7,\,0)=\dfrac{7!}{7!\,0!}=\dfrac{1}{0!}=\dfrac{1}{1}=1$

[9] There are 12! total permutations. We want the number of *distinguishable* permutations of the 12 disks. Using the second theorem on distinguishable permutations, we have $\dfrac{(5+3+2+2)!}{5!\,3!\,2!\,2!}=\dfrac{12!}{5!\,3!\,2!\,2!}=166{,}320$.

[11] In the 10-letter word *bookkeeper*, there are 3 e's, 2 o's, and 2 k's. Using the second theorem on distinguishable permutations, we have $\dfrac{10!}{3!\,2!\,2!\,1!\,1!\,1!}=151{,}200$.

[13] There are $C(10,\,5)$ ways to pick the first team.

The second team is determined once the first team is selected. $C(10,\,5)=252$

[15] Two points determine a unique line. $C(8,\,2)=28$

[17] There are 3! ways to order the categories. $(5!\cdot 4!\cdot 8!)\cdot 3!=696{,}729{,}600$

[19] Pick the center, $C(3,\,1)$; two guards, $C(10,\,2)$;

two tackles from the 8 remaining linemen, $C(8,\,2)$; two ends, $C(4,\,2)$;

two halfbacks, $C(6,\,2)$; the quarterback, $C(3,\,1)$; and the fullback, $C(4,\,1)$.

$$3\cdot C(10,\,2)\cdot C(8,\,2)\cdot C(4,\,2)\cdot C(6,\,2)\cdot 3\cdot 4=4{,}082{,}400$$

21 There are $C(12, 3) = 220$ ways to pick the men and $C(8, 2) = 28$ ways to pick the women. By the fundamental counting principle, the total number of ways to pick the committee is $220 \cdot 28 = 6160$.

23 We need 3 U's out of 8 moves. $C(8, 3) = 56$

25 (a) We want 6 of the 49 numbers and order is not important. $C(49, 6) = 13{,}983{,}816$

(b) There are 24 even numbers from 1 to 49. $C(24, 6) = 134{,}596$

27 Let n denote the number of players. $C(n, 2) = 45 \Rightarrow \dfrac{n!}{(n-2)!\,2!} = 45 \Rightarrow$

$$n(n-1) = 90 \Rightarrow (n-10)(n+9) = 0 \Rightarrow \{\, n > 0 \,\}\ n = 10.$$

29 Each team must win 3 of the first 6 games for the series to be extended to a

7th game. $C(6, 3) = 20$

31 They may have computed $C(31, 3)$, which is 4495.

33 (a) The amounts received are the same, so the order of selection *is not* important,

and we use a combination. $C(1000, 30) = \dfrac{1000!}{970!\,30!} \approx 2.43 \times 10^{57}$

(b) The amounts received are different, so the order of selection *is* important,

and we use a permutation. $P(1000, 30) = \dfrac{1000!}{970!} \approx 6.44 \times 10^{89}$

35 We want to select 3 of the 4 kings and 2 of the remaining 48 cards. The order of selection is not important, so we use combinations. Any of the groups of 3 kings can be selected with any pair of the remaining 48 cards, so by the fundamental counting principle the total number of hands is $C(4, 3) \cdot C(48, 2) = 4 \cdot 1128 = 4512$.

37 (a) $S_1 = \binom{1}{1} + \binom{1}{3} + \binom{1}{5} + \cdots = 1 + 0 + 0 + \cdots = 1.$

$S_2 = \binom{2}{1} + \binom{2}{3} + \binom{2}{5} + \cdots = 2 + 0 + 0 + \cdots = 2.$

$S_3 = 3 + 1 + 0 + \cdots = 4.\quad S_4 = 4 + 4 + 0 + \cdots = 8.$

$S_5 = 16,\ S_6 = 32,\ S_7 = 64,\ S_8 = 128,\ S_9 = 256,\ S_{10} = 512.$

(b) It appears that $S_n = 2^{n-1}$.

39 (a) Graph the values of $C(10, 1), C(10, 2), C(10, 3), \ldots, C(10, 10)$.

(b) The maximum value of $C(10, r)$ is 252 and occurs at $r = 5$.

[0, 10] by [0, 300, 50]   [0, 19] by [0, 10^5, 10^4]

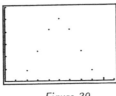

Figure 39

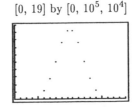

Figure 41

41 (a) Graph the values of $C(19, 1), C(19, 2), C(19, 3), \ldots, C(19, 19)$.

(b) The maximum value of $C(19, r)$ is 92,378 and occurs at $r = 9, 10$.

9.8 Exercises

1. (a) The sample space is the 52-card deck, thus, $n(S) = 52$.

    There are 4 kings, so the probability is $P(E) = \dfrac{n(E)}{n(S)} = \dfrac{4}{52} = \dfrac{1}{13}$.

    The odds are $n(E)$ to $n(E')$, which are 4 to 48, or 1 to 12.

    (b) $\frac{4}{52} + \frac{4}{52} = \frac{8}{52} = \frac{2}{13}$; $O(E)$ are 8 to 44, or 2 to 11

    (c) $\frac{4}{52} + \frac{4}{52} + \frac{4}{52} = \frac{12}{52} = \frac{3}{13}$; $O(E)$ are 12 to 40, or 3 to 10

3. $S = \{1, 2, 3, 4, 5, 6\}$ and $n(S) = 6$.

    (a) $P(\text{rolling a 4}) = \frac{1}{6}$. There's 1 way to get a 4 and 5 ways to not get a 4,

    so $O(E)$ are 1 to 5.

    (b) $P(\text{rolling a 6}) = \frac{1}{6}$; $O(E)$ are 1 to 5.

    (c) $\frac{1}{6} + \frac{1}{6} = \frac{2}{6} = \frac{1}{3}$; $O(E)$ are 2 to 4, or 1 to 2

5. $n(S) = 5 + 6 + 4 = 15$

    (a) $P(\text{drawing a red ball}) = \frac{5}{15} = \frac{1}{3}$; $O(E)$ are 5 to 10, or 1 to 2.

    (b) $P(\text{drawing a green ball}) = \frac{6}{15} = \frac{2}{5}$; $O(E)$ are 6 to 9, or 2 to 3.

    (c) $\frac{5}{15} + \frac{4}{15} = \frac{9}{15} = \frac{3}{5}$; $O(E)$ are 9 to 6, or 3 to 2

7. *Note:* The following table summarizes the results for the sum of two dice being

    tossed. Notice the symmetry about the sum of 7 in the # of ways to obtain.

Sum of two dice	2	3	4	5	6	7	8	9	10	11	12
# of ways to obtain	1	2	3	4	5	6	5	4	3	2	1

    (a) $\frac{2}{36} = \frac{1}{18}$; $O(E)$ are 2 to 34, or 1 to 17

    (b) $\frac{5}{36}$; $O(E)$ are 5 to 31

    (c) $\frac{2}{36} + \frac{5}{36} = \frac{7}{36}$; $O(E)$ are 7 to 29

9. $n(S) = 6 \times 6 \times 6 = 216$.

    There are 6 ways to make a sum of 5 (3 with 1, 1, 3 and 3 with 1, 2, 2). $\frac{6}{216} = \frac{1}{36}$

11. There are 2 choices for each of the 3 coins, so $n(S) = 2 \cdot 2 \cdot 2 = 8$. Getting two heads

    to turn up is the same as getting one tail to turn up. There are 3 ways to get one

    tail: THH, HTH, HHT. So the probability is $\dfrac{3}{2^3} = \dfrac{3}{8}$.

13. If $P(E) = \frac{5}{7}$, then $O(E)$ are 5 to 2 and $O(E')$ are 2 to 5.

15. If $O(E)$ are 9 to 5, then $O(E')$ are 5 to 9 and $P(E) = \frac{9}{5+9} = \frac{9}{14}$.

17. $P(E) \approx 0.659 = \frac{659}{1000}$, so $n(E) = 659$ and $n(E') = 1000 - 659 = 341$ and

    $O(E)$ are 659 to 341. Divide both numbers by 341 to get odds of 1.93 to 1.

Note: For Exercises 19–24, there are $C(52, 5) = 2,598,960$ ways to draw 5 cards.

19 There are 13 denominations to pick from and any one of them could be combined

with any one of the remaining 48 cards. $\dfrac{48 \cdot 13}{C(52, 5)} = \dfrac{1}{4165} \approx 0.00024$

21 Pick 4 of the 13 diamonds and 1 of the 13 spades.
$$\frac{C(13, 4) \cdot C(13, 1)}{C(52, 5)} = \frac{143}{39,984} \approx 0.00358$$

23 Pick 5 of the 13 cards in one suit. There are 4 suits. $\dfrac{C(13, 5) \cdot 4}{C(52, 5)} = \dfrac{33}{16,660} \approx 0.00198$

25 Let E_1 be the event that the number is odd, E_2 that the number is prime.

$E_1 = \{1, 3, 5\}$ and $E_2 = \{2, 3, 5\}$.
$$P(E_1 \cup E_2) = P(E_1) + P(E_2) - P(E_1 \cap E_2) = \tfrac{3}{6} + \tfrac{3}{6} - \tfrac{2}{6} = \tfrac{4}{6} = \tfrac{2}{3}.$$

27 The probability that the player does *not* get a hit is $1 - 0.326 = 0.674$. The probability that the player does not get a hit four times in a row is $(0.674) \cdot (0.674) \cdot (0.674) \cdot (0.674) = (0.674)^4 \approx 0.2064$.

29 (a) $P(E_2) = P(2) + P(3) + P(4) = 0.10 + 0.15 + 0.20 = 0.45$

(b) $P(E_1 \cap E_2) = P(2) = 0.10$

(c) $P(E_1 \cup E_2) = P(E_1) + P(E_2) - P(E_1 \cap E_2) = 0.35 + 0.45 - 0.10 = 0.70$

(d) $P(E_2 \cup E_3') = P(E_2) + P(E_3') - P(E_2 \cap E_3')$. $E_3' = \{1, 2, 3, 5\}$ and
$$E_2 \cap E_3' = \{2, 3\} \quad \Rightarrow \quad P(E_2 \cup E_3') = 0.45 + 0.75 - 0.25 = 0.95.$$

Note: For Exercises 31–32, there are $C(60, 5) = 5,461,512$ ways to draw 5 chips.

31 (a) We want 5 blue and 0 non-blue. There are 20 blue and 40 non-blue chips in the
$$\text{box.} \quad \frac{C(20, 5) \cdot C(40, 0)}{C(60, 5)} = \frac{34}{11,977} \approx 0.0028.$$

(b) $P(\text{at least 1 green}) = 1 - P(\text{no green}) =$
$$1 - \frac{C(30, 0) \cdot C(30, 5)}{C(60, 5)} = 1 - \frac{117}{4484} = \frac{4367}{4484} \approx 0.9739.$$

(c) $P(\text{at most 1 red}) = P(0 \text{ red}) + P(1 \text{ red}) =$
$$\frac{C(10, 0) \cdot C(50, 5)}{C(60, 5)} + \frac{C(10, 1) \cdot C(50, 4)}{C(60, 5)} = \frac{26,320}{32,509} \approx 0.8096.$$

33 (a) $\dfrac{C(8, 8)}{2^8} = \dfrac{1}{256} \approx 0.00391$ (b) $\dfrac{C(8, 7)}{2^8} = \dfrac{1}{32} = 0.03125$

(c) $\dfrac{C(8, 6)}{2^8} = \dfrac{7}{64} = 0.109375$ (d) $\dfrac{C(8, 6) + C(8, 7) + C(8, 8)}{2^8} = \dfrac{37}{256} \approx 0.14453$

35 $P(\text{obtaining at least one ace}) = 1 - P(\text{no aces}) =$
$$1 - \frac{C(48, 5)}{C(52, 5)} = 1 - \frac{35,673}{54,145} = \frac{18,472}{54,145} \approx 0.34116$$

$\boxed{37}$ (a) We may use ordered pairs to represent the outcomes of the sample space S of the experiment. A representative outcome is (nine of clubs, 3). The number of outcomes in the sample space S is $n(S) = 52 \cdot 6 = 312$.

(b) For each integer k, where $2 \leq k \leq 6$, there are 4 ways to obtain an outcome of the form $(k,\ k)$ since there are 4 suits. Because there are 5 values of k, $n(E_1) = 5 \cdot 4 = 20$. $n(E_1') = n(S) - n(E_1) = 312 - 20 = 292$.

$$P(E_1) = \frac{n(E_1)}{n(S)} = \frac{20}{312} = \frac{5}{78}.$$

(c) No, if E_2 or E_3 occurs, then the other event may occur.

Yes, the occurrence of either E_2 or E_3 has no effect on the other event.

$$P(E_2) = \frac{n(E_2)}{n(S)} = \frac{12 \cdot 6}{312} = \frac{72}{312} = \frac{3}{13}. \qquad P(E_3) = \frac{n(E_3)}{n(S)} = \frac{52 \cdot 3}{312} = \frac{156}{312} = \frac{1}{2}.$$

Since E_2 and E_3 are indep., $P(E_2 \cap E_3) = P(E_2) \cdot P(E_3) = \frac{3}{13} \cdot \frac{1}{2} = \frac{3}{26} = \frac{36}{312}$.

$P(E_2 \cup E_3) = P(E_2) + P(E_3) - P(E_2 \cap E_3) = \frac{72}{312} + \frac{156}{312} - \frac{36}{312} = \frac{192}{312} = \frac{8}{13}$.

(d) Yes, if E_1 or E_2 occurs, then the other event cannot occur. No, the occurrence of either E_1 or E_2 influences the occurrence of the other event.

Remember, (non-empty) *mutually exclusive events* **cannot** *be independent events.*

Since E_1 and E_2 are mutually exclusive, $P(E_1 \cap E_2) = 0$ and

$$P(E_1 \cup E_2) = P(E_1) + P(E_2) = \frac{20}{312} + \frac{72}{312} = \frac{92}{312} = \frac{23}{78}.$$

$\boxed{39}$ Let k denote the sum.

$$P(k > 5) = 1 - P(k \leq 5) = 1 - \left(\frac{1}{36} + \frac{2}{36} + \frac{3}{36} + \frac{4}{36} \right) = 1 - \frac{10}{36} = \frac{26}{36} = \frac{13}{18}.$$

$\boxed{41}$ (a) Each birth is independent of the others and the probability of having a boy is $\frac{1}{2}$.

$P(\text{all boys}) = \frac{1}{2} \cdot \frac{1}{2} \cdot \frac{1}{2} \cdot \frac{1}{2} \cdot \frac{1}{2} = \frac{1}{32} = 0.03125$.

(b) $P(\text{at least one girl}) = 1 - P(\text{all boys}) = 1 - \frac{1}{32} = \frac{31}{32} = 0.96875$.

$\boxed{43}$ There are 4! ways to place the cards, so $n(S) = 4!$.

(a) $\dfrac{C(4,\ 4)}{4!} = \dfrac{1}{24} \approx 0.04167$ \qquad (b) $\dfrac{C(4,\ 2)}{4!} = \dfrac{1}{4} = 0.25$

$\boxed{45}$ (a) The 3, 4, or 5 on the left die would need to combine with a 4, 3, or 2 on the right die to sum to 7, but the right die only has 1, 5, or 6. The probability is 0.

(b) To obtain 8, we would need a 3 on the left die and a 5 on the right.

$$\tfrac{1}{3} \cdot \tfrac{1}{3} = \tfrac{1}{9} = 0.\overline{1}$$

$\boxed{47}$ There are 3273 smoking-related deaths that are not in any of the 3 categories.

(a) $P(\text{cardiovascular disease or cancer}) = \dfrac{179{,}820 + 151{,}322}{418{,}890} = \dfrac{331{,}142}{418{,}890} \approx 0.791$

(b) $P(not \text{ respiratory}) = 1 - P(\text{respiratory}) = \dfrac{418{,}890 - 84{,}475}{418{,}890} = \dfrac{334{,}415}{418{,}890} \approx 0.798$

49 (a) The ball must take 4 "lefts". $\frac{1}{2} \cdot \frac{1}{2} \cdot \frac{1}{2} \cdot \frac{1}{2} = \frac{1}{16} = 0.0625$

   (b) We need two "lefts". $\frac{C(4,\,2)}{2^4} = \frac{6}{16} = \frac{3}{8} = 0.375$

51 For one ticket, $P(E) = \frac{n(E)}{n(S)} = \frac{C(6,\,6)}{C(54,\,6)} = \frac{1}{25,827,165}.$

          For two tickets, $P(E) = \frac{2 \times 1}{25,827,165}$, or about 1 chance in 13 million.

53 The probability that the first bulb is not defective is $\frac{195}{200}$ since 195 of the 200 bulbs
are not defective. If the first bulb is not replaced and not defective, then there are
199 bulbs left, and 194 of them are not defective. The probability that the second
bulb is not defective is then $\frac{194}{199}$. Thus, the probability that both bulbs are not
defective is $\frac{195}{200} \times \frac{194}{199} = \frac{37,830}{39,800} \approx 0.9505$. The event that either light bulb is defective is
the complement of the event that neither bulb is defective. The probability that the
sample will be rejected is $1 - \frac{37,830}{39,800} = \frac{1970}{39,800} \approx 0.0495$.

55 (a) $P(7 \text{ or } 11) = P(7) + P(11) = \frac{6}{36} + \frac{2}{36} = \frac{8}{36}$

   (b) To win with a 4 on the first roll, we must first get a 4, and then get another 4
       before a 7. The probability of getting a 4 is $\frac{3}{36}$. The probability of getting
       another 4 before a 7 is $\frac{3}{3+6}$ since there are 3 ways to get a 4, 6 ways to get a 7,
       and numbers other than 4 and 7 are immaterial.

                   Thus, $P(\text{winning with } 4) = \frac{3}{36} \cdot \frac{3}{3+6} = \frac{1}{36}.$

   (c) Let $P(k)$ denote the probability of winning a pass line bet with the number k.
       We first note that $P(4) = P(10)$, $P(5) = P(9)$, and $P(6) = P(8)$.
       $P(\text{winning}) = 2 \cdot P(4) + 2 \cdot P(5) + 2 \cdot P(6) + P(7) + P(11)$

$$= 2 \cdot \frac{3}{36} \cdot \frac{3}{3+6} + 2 \cdot \frac{4}{36} \cdot \frac{4}{4+6} + 2 \cdot \frac{5}{36} \cdot \frac{5}{5+6} + \frac{6}{36} + \frac{2}{36}$$

$$= 2 \cdot \frac{1}{36} + 2 \cdot \frac{2}{45} + 2 \cdot \frac{25}{396} + \frac{1}{6} + \frac{1}{18} = \frac{488}{990} = \frac{244}{495} \approx 0.4929$$

57 (a) $p = P((S_1 \cap S_2) \cup (S_3 \cap S_4))$

       $= P(S_1 \cap S_2) + P(S_3 \cap S_4) - P((S_1 \cap S_2) \cap (S_3 \cap S_4))$

       $= P(S_1) \cdot P(S_2) + P(S_3) \cdot P(S_4) - P(S_1 \cap S_2) \cdot P(S_3 \cap S_4)$

       $= P(S_1) \cdot P(S_2) + P(S_3) \cdot P(S_4) - P(S_1) \cdot P(S_2) \cdot P(S_3) \cdot P(S_4)$

       Let $P(S_k) = x$. Then $p = x \cdot x + x \cdot x - x \cdot x \cdot x \cdot x = -x^4 + 2x^2$.

                                 $x = 0.9 \;\; \Rightarrow \;\; p = 0.9639.$

$[-2.25, 2.25, 0.5]$ by $[-2, 1, 0.5]$

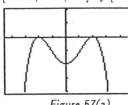

Figure 57(a)

$[0.96, 1.05, 0.5]$ by $[-0.03, 0.04, 0.5]$

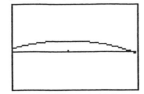

Figure 57(b)

(b) $p = 0.99 \Rightarrow -x^4 + 2x^2 = 0.99$. The graph of $y = -x^4 + 2x^2 - 0.99$ is shown in Figure 57(a). The region near $x = 1$ is enlarged in Figure 57(b) to show that the graph is above the x-axis for some values of x. The approximate x-intercepts are ± 0.95, ± 1.05. Since $0 \le P(S_k) \le 1$, $P(S_k) = 0.95$.

59 (a) The number of ways that n people can all have a different birthday is $P(365, n)$. The number of outcomes in the sample space is 365^n.

$$\text{Thus, } p = \frac{P(365, n)}{365^n} = \frac{365!}{365^n (365 - n)!}.$$

(b) $n = 32 \Rightarrow p = \dfrac{365!}{365^{32}\, 333!} \Rightarrow \ln p = \ln\dfrac{365!}{365^{32}\, 333!} =$

$\ln 365! - \ln 365^{32} - \ln 333! \approx (365 \ln 365 - 365) - (32 \ln 365) - (333 \ln 333 - 333) \approx$ -1.45. Thus, $p \approx e^{-1.45} \approx 0.24$.

The probability that two or more people have the same birthday is $1 - p \approx 0.76$.

61 From Exercise 55(c), the payoff amount of $2 has a probability of $\frac{244}{495}$. Hence, $EV = 2 \cdot \frac{244}{495} = \frac{488}{495} \approx \$0.986$, or about $0.99. Of course, after paying $1 to play, you can expect to lose about $7 for every 495 one-dollar pass line bets.

63 $EV = 1,000,000 \cdot \frac{1}{20,000,000} + 100,000 \cdot \frac{10}{20,000,000} + 10,000 \cdot \frac{100}{20,000,000} + 1000 \cdot \frac{1000}{20,000,000}$

$= \$0.20$ { less than the cost of a first class stamp }

Chapter 9 Review Exercises

1 $a_n = \dfrac{5n}{3 - 2n^2}$ •

$a_1 = \dfrac{5(1)}{3 - 2(1)^2} = \dfrac{5}{1} = 5,$

$a_2 = \dfrac{5(2)}{3 - 2(2)^2} = \dfrac{10}{-5} = -2,$

$a_3 = \dfrac{5(3)}{3 - 2(3)^2} = \dfrac{15}{-15} = -1,$

$a_4 = \dfrac{5(4)}{3 - 2(4)^2} = \dfrac{20}{-29} = -\dfrac{20}{29},$

$a_7 = \dfrac{5(7)}{3 - 2(7)^2} = \dfrac{35}{-95} = -\dfrac{7}{19}.$

★ $5, -2, -1, -\dfrac{20}{29}; -\dfrac{7}{19}$

$\boxed{3}$ $a_n = 1 + (-\frac{1}{2})^{n-1}$ •

$a_1 = 1 + (-\frac{1}{2})^0 = 1 + 1 = 2,$ $a_2 = 1 + (-\frac{1}{2})^1 = 1 - \frac{1}{2} = \frac{1}{2},$

$a_3 = 1 + (-\frac{1}{2})^2 = 1 + \frac{1}{4} = \frac{5}{4},$ $a_4 = 1 + (-\frac{1}{2})^3 = 1 - \frac{1}{8} = \frac{7}{8},$

$a_7 = 1 + (-\frac{1}{2})^6 = 1 + \frac{1}{64} = \frac{65}{64}$ $\star$ $2, \frac{1}{2}, \frac{5}{4}, \frac{7}{8}, \frac{65}{64}$

$\boxed{5}$ $a_1 = 10, a_{k+1} = 1 + (1/a_k)$ •

$a_2 = a_{1+1} \{k=1\} = 1 + 1/a_1 = 1 + 1/10 = \frac{10}{10} + \frac{1}{10} = \frac{11}{10}$

$a_3 = a_{2+1} = 1 + 1/a_2 = 1 + 1/\frac{11}{10} = 1 + \frac{10}{11} = \frac{11}{11} + \frac{10}{11} = \frac{21}{11}$

$a_4 = a_{3+1} = 1 + 1/a_3 = 1 + 1/\frac{21}{11} = 1 + \frac{11}{21} = \frac{21}{21} + \frac{11}{21} = \frac{32}{21}$

$a_5 = a_{4+1} = 1 + 1/a_4 = 1 + 1/\frac{32}{21} = 1 + \frac{21}{32} = \frac{32}{32} + \frac{21}{32} = \frac{53}{32}$ $\star$ $10, \frac{11}{10}, \frac{21}{11}, \frac{32}{21}, \frac{53}{32}$

$\boxed{7}$ $a_1 = 9, a_{k+1} = \sqrt{a_k}$ • $a_2 = a_{1+1} \{k=1\} = \sqrt{a_1} = \sqrt{9} = 3$

$a_3 = a_{2+1} = \sqrt{a_2} = \sqrt{3}$ $a_4 = a_{3+1} = \sqrt{a_3} = \sqrt{\sqrt{3}} = \sqrt[4]{3}$

$a_5 = a_{4+1} = \sqrt{a_4} = \sqrt{\sqrt[4]{3}} = \sqrt[8]{3}$ $\star$ $9, 3, \sqrt{3}, \sqrt[4]{3}, \sqrt[8]{3}$

$\boxed{9}$ $\sum\limits_{k=1}^{5} (k^2 + 4) = (1^2 + 4) + (2^2 + 4) + (3^2 + 4) + (4^2 + 4) + (5^2 + 4)$

$= 5 + 8 + 13 + 20 + 29 = 75$

$\boxed{11}$ $\sum\limits_{k=7}^{100} 10 = (100 - 7 + 1)(10) = 94(10) = 940$

$\boxed{13}$ $3 + 6 + 9 + 12 + 15 = 3(1) + 3(2) + 3(3) + 3(4) + 3(5) = \sum\limits_{n=1}^{5} (3n)$

$\boxed{14}$ We first note that $4 + 2 + 1 + \frac{1}{2} + \frac{1}{4} + \frac{1}{8}$ is the sum of six terms of a geometric sequence with first term 4 and ratio $\frac{1}{2}$—that is, $a_n = a_1 r^{n-1} = 4(\frac{1}{2})^{n-1}$.

$4 + 2 + 1 + \frac{1}{2} + \frac{1}{4} + \frac{1}{8} = \sum\limits_{n=1}^{6} 4(\frac{1}{2})^{n-1} = \sum\limits_{n=1}^{6} 2^2(2^{-1})^{n-1} = \sum\limits_{n=1}^{6} 2^2 2^{1-n} = \sum\limits_{n=1}^{6} 2^{3-n}$

$\boxed{15}$ $\frac{1}{1 \cdot 2} + \frac{1}{2 \cdot 3} + \frac{1}{3 \cdot 4} + \cdots + \frac{1}{99 \cdot 100} = \sum\limits_{n=1}^{99} \frac{1}{n(n+1)}$

$\boxed{17}$ $\frac{1}{2} + \frac{2}{5} + \frac{3}{8} + \frac{4}{11}$ • The numerators are just the positive integers, so we'll use the summation notation with n having values from 1 to 4. The denominators increase by 3, so it must be of the form $3n + k$. When $n = 1$, $3n + k = 3(1) + k = 3 + k$ and $3 + k$ must equal 2 (the first denominator). Thus, $3 + k = 2 \Rightarrow k = -1$, so the denominator is $3n - 1$ and the summation notation is $\sum\limits_{n=1}^{4} \frac{n}{3n-1}$.

$\boxed{19}$ $100 - 95 + 90 - 85 + 80$ • The terms have alternating signs and decrease by 5 in magnitude. For $n = 1$, we subtract 5 from 105 to get 100. $\sum\limits_{n=1}^{5} (-1)^{n+1}(105 - 5n)$

[20] $1 - \frac{1}{2} + \frac{1}{3} - \frac{1}{4} + \frac{1}{5} - \frac{1}{6} + \frac{1}{7}$ • The terms have alternating signs,

the numerator is 1, and the denominators increase by 1. $\sum_{n=1}^{7} (-1)^{n-1} \frac{1}{n}$

[21] $a_0 + a_1 x^4 + a_2 x^8 + \cdots + a_{25} x^{100}$ { the exponents are multiples of 4 } $= \sum_{n=0}^{25} a_n x^{4n}$

[23] $1 - \frac{x^2}{2} + \frac{x^4}{4} - \frac{x^6}{6} + \cdots + (-1)^n \frac{x^{2n}}{2n}$ • The pattern begins with the second term

and the general term is listed. $1 + \sum_{k=1}^{n} (-1)^k \frac{x^{2k}}{2k}$

[25] $d = 3 - (4 + \sqrt{3}) = -1 - \sqrt{3};\ a_{10} = (4 + \sqrt{3}) + (9)(-1 - \sqrt{3}) = -5 - 8\sqrt{3};$

$$S_{10} = \frac{10}{2}\left[(4 + \sqrt{3}) + (-5 - 8\sqrt{3})\right] = -5 - 35\sqrt{3}.$$

[27] $a_5 = a_1 + 4d = 5$ and $a_{13} = a_1 + 12d = 77.$ $a_{13} - a_5 = (a_1 + 12d) - (a_1 + 4d) = 8d$ and

$a_{13} - a_5 = 77 - 5 = 72 \Rightarrow 8d = 72 \Rightarrow d = 9.$

$a_5 = a_1 + 4d \Rightarrow 5 = a_1 + 36 \Rightarrow a_1 = -31.$ $\qquad a_{10} = -31 + 9(9) = 50.$

[28] Four arithmetic means $\Rightarrow 5d = -10 - 20 \Rightarrow d = -6.$

The terms are 20, 14, 8, 2, −4, and −10.

[29] $r = \dfrac{\frac{1}{4}}{\frac{1}{8}} = \dfrac{1 \cdot 8}{4 \cdot 1} = 2;\ a_n = \frac{1}{8}(2)^{n-1} = 2^{-3} \cdot 2^{n-1} = 2^{n-4} \Rightarrow a_{10} = 2^6 = 64$

[30] We can divide the fourth term by the third term to obtain the common ratio,

$r = \dfrac{-0.3}{3} = -0.1.$ If we multiply the third term by r five times,

we will get the eighth term. $a_8 = a_3 r^5 = 3(-0.1)^5 = -0.00003.$

[31] The geometric mean of 4 and 8 is $\sqrt{4 \cdot 8} = \sqrt{32} = 4\sqrt{2}.$

[33] $402 = \frac{12}{2}(a_1 + 50) \Rightarrow a_1 = 17.$ $a_{12} = a_1 + 11d \Rightarrow d = (50 - 17)/11 = 3.$

[34] $a_5 = a_1 r^4 \Rightarrow \frac{1}{16} = a_1 \left(\frac{3}{2}\right)^4 \Rightarrow a_1 = \frac{1}{16} \cdot \frac{16}{81} = \frac{1}{81}.$

$$S_5 = \frac{1}{81} \cdot \frac{1 - \left(\frac{3}{2}\right)^5}{1 - \frac{3}{2}} = \frac{1}{81} \cdot \frac{1 - \frac{243}{32}}{-\frac{1}{2}} = \frac{1}{81} \cdot \frac{-\frac{211}{32}}{-\frac{1}{2}} = \frac{211}{1296}.$$

[35] The sequence of terms is arithmetic.

$a_1 = 5(1) - 2 = 3$ and $a_{15} = 5(15) - 2 = 73,$ so $S_{15} = \frac{15}{2}(3 + 73) = 570.$

[37] This sum can be written as a sum of a geometric sequence and an arithmetic

sequence. $\sum_{k=1}^{10} \left(2^k - \frac{1}{2}\right) = \sum_{k=1}^{10} 2^k - \sum_{k=1}^{10} \frac{1}{2} = 2 \cdot \frac{1 - 2^{10}}{1 - 2} - 10(\frac{1}{2}) = 2046 - 5 = 2041$

[39] $a_1 = 1,\ r = -\frac{2}{5} \Rightarrow S = \dfrac{1}{1 - (-\frac{2}{5})} = \dfrac{1}{\frac{7}{5}} = \dfrac{5}{7}.$

41 (1) P_1 is true, since $3(1) - 1 = \dfrac{1[3(1)+1]}{2} = 2$.

(2) Assume P_k is true: $2 + 5 + 8 + \cdots + (3k - 1) = \dfrac{k(3k+1)}{2}$. Hence,

$$2 + 5 + 8 + \cdots + (3k - 1) + 3(k+1) - 1 = \dfrac{k(3k+1)}{2} + 3(k+1) - 1$$

$$= \dfrac{3k^2 + k + 6k + 4}{2}$$

$$= \dfrac{3k^2 + 7k + 4}{2}$$

$$= \dfrac{(k+1)(3k+4)}{2}$$

$$= \dfrac{(k+1)[3(k+1)+1]}{2}.$$

Thus, P_{k+1} is true, and the proof is complete.

43 (1) P_1 is true, since $\dfrac{1}{[2(1)-1][2(1)+1]} = \dfrac{1}{2(1)+1} = \dfrac{1}{3}$.

(2) Assume P_k is true:

$$\dfrac{1}{1 \cdot 3} + \dfrac{1}{3 \cdot 5} + \dfrac{1}{5 \cdot 7} + \cdots + \dfrac{1}{(2k-1)(2k+1)} = \dfrac{k}{2k+1}. \text{ Hence,}$$

$$\dfrac{1}{1 \cdot 3} + \dfrac{1}{3 \cdot 5} + \dfrac{1}{5 \cdot 7} + \cdots + \dfrac{1}{(2k-1)(2k+1)} + \dfrac{1}{(2k+1)(2k+3)}$$

$$= \dfrac{k}{2k+1} + \dfrac{1}{(2k+1)(2k+3)}$$

$$= \dfrac{k(2k+3)+1}{(2k+1)(2k+3)}$$

$$= \dfrac{2k^2 + 3k + 1}{(2k+1)(2k+3)}$$

$$= \dfrac{(2k+1)(k+1)}{(2k+1)(2k+3)}$$

$$= \dfrac{k+1}{2(k+1)+1}.$$

Thus, P_{k+1} is true, and the proof is complete.

45 (1) For $n = 1$, $n^3 + 2n = 3$ and 3 is a factor of 3.

(2) Assume 3 is a factor of $k^3 + 2k$. The $(k+1)$st term is

$$(k+1)^3 + 2(k+1) = k^3 + 3k^2 + 5k + 3$$

$$= (k^3 + 2k) + (3k^2 + 3k + 3)$$

$$= (k^3 + 2k) + 3(k^2 + k + 1).$$

By the induction hypothesis, 3 is a factor of $k^3 + 2k$ and 3 is a factor of $3(k^2 + k + 1)$, so 3 is a factor of the $(k+1)$st term. Thus, P_{k+1} is true, and the proof is complete.

$\boxed{47}$ For j: Examining the pattern formed by letting $n = 1, 2, 3, 4$ leads us to the
conclusion that $j = 4$.

(1) P_4 is true, since $2^4 \le 4!$.

(2) Assume P_k is true: $2^k \le k!$. Hence,

$$2^{k+1} = 2 \cdot 2^k \le 2 \cdot k! < (k+1) \cdot k! = (k+1)!.$$

Thus, P_{k+1} is true, and the proof is complete.

$\boxed{49}$ We use the binomial theorem formula with $a = x^2$, $b = -3y$, and $n = 6$.

$$(x^2 - 3y)^6 = (x^2)^6 + \binom{6}{1}(x^2)^5(-3y)^1 + \binom{6}{2}(x^2)^4(-3y)^2 + \binom{6}{3}(x^2)^3(-3y)^3$$

$$+ \binom{6}{4}(x^2)^2(-3y)^4 + \binom{6}{5}(x^2)^1(-3y)^5 + (-3y)^6$$

$$= x^{12} - 6 \cdot 3x^{10}y + 15 \cdot 9x^8y^2 - 20 \cdot 27x^6y^3$$

$$+ 15 \cdot 81x^4y^4 - 6 \cdot 243x^2y^5 + 729y^6$$

$$= x^{12} - 18x^{10}y + 135x^8y^2 - 540x^6y^3 + 1215x^4y^4 - 1458x^2y^5 + 729y^6$$

$\boxed{51}$ $(a^{2/5} + 2a^{-3/5})^{20}$; first three terms $\bullet$

$$= \binom{20}{0}(a^{2/5})^{20}(2a^{-3/5})^0 + \binom{20}{1}(a^{2/5})^{19}(2a^{-3/5})^1 + \binom{20}{2}(a^{2/5})^{18}(2a^{-3/5})^2$$

$$= 1 \cdot a^8 \cdot 1 + 20 \cdot a^{38/5} \cdot 2a^{-3/5} + 190 \cdot a^{36/5} \cdot 4a^{-6/5}$$

$$= a^8 + 40 \cdot a^{35/5} + 760 \cdot a^{30/5} = a^8 + 40a^7 + 760a^6$$

$\boxed{53}$ $(4a^2 - b)^7$; term that contains a^{10} $\bullet$

Consider only the variable a in the expansion: $(a^2)^{7-k} = a^{10} \Rightarrow 14 - 2k = 10 \Rightarrow$

$2k = 4 \Rightarrow k = 2$; $(k+1)$st term = 3rd term $= \binom{7}{2}(4a^2)^5(-b)^2 = 21{,}504a^{10}b^2$

$\boxed{55}$ (a) $S_5 = 10 \Rightarrow 10 = \frac{5}{2}(2a_1 + 4d) \Rightarrow 4 = 2a_1 + 4d \Rightarrow 4d = 4 - 2a_1 \Rightarrow$

$d = 1 - \frac{1}{2}a_1$. Since a_1 is positive, $1 - \frac{1}{2}a_1$ is less than 1 ft.

(b) $a_1 = \frac{1}{2} \Rightarrow d = 1 - \frac{1}{2}(\frac{1}{2}) = \frac{3}{4}$.

The lengths of the other four pieces are $1\frac{1}{4}$, 2, $2\frac{3}{4}$, and $3\frac{1}{2}$ ft.

$\boxed{57}$ If $s_1 = 1$, then $s_2 = f$, $s_3 = f^2$, ... There are two of each of the s_k's.

Since $0 < f < 1$, and f is the common ratio, we can sum the infinite sequence.

The sum of the s_k's is $2(1 + f + f^2 + \cdots) = 2\left(\frac{1}{1-f}\right) = \frac{2}{1-f}$.

$\boxed{59}$ (a) We want 6 digits taken 4 at a time. Order is important, so we use a

permutation. $P(6, 4) = \dfrac{6!}{(6-4)!} = \dfrac{6 \cdot 5 \cdot 4 \cdot 3 \cdot 2!}{2!} = 6 \cdot 5 \cdot 4 \cdot 3 = 360$

(b) Since repetitions are allowed, we can use any of the 6 digits for each choice:

$$6 \cdot 6 \cdot 6 \cdot 6 = 6^4 = 1296$$

$\boxed{61}$ $\dfrac{(6+5+4+2)!}{6!\,5!\,4!\,2!} = \dfrac{17!}{6!\,5!\,4!\,2!} = 85{,}765{,}680$

63 There is one way to get all heads and one way to get all tails,

so there are only two ways to get the coins to match.

(a) With two boys, $n(S) = 2 \cdot 2 = 4$, so the probability of a match is $\frac{2}{4} = \frac{1}{2}$.

(b) With three boys, $n(S) = 2 \cdot 2 \cdot 2 = 8$, so the probability of a match is $\frac{2}{8} = \frac{1}{4}$.

65 (a) $\frac{1}{1000}$ (b) $\frac{10}{1000} = \frac{1}{100}$ (c) $\frac{50}{1000} = \frac{1}{20}$

66 $P(1 \text{ head}) = \dfrac{C(4,\, 1)}{2^4} = \frac{4}{16} = \frac{1}{4} = 0.25$. Odds are 4 to 12, or 1 to 3.

67 (a) $P(\text{passing}) = P(4,\, 5, \text{ or } 6 \text{ correct}) =$

$$\frac{C(6,\, 4) + C(6,\, 5) + C(6,\, 6)}{2^6} = \frac{15 + 6 + 1}{64} = \frac{22}{64} = \frac{11}{32}$$

(b) $P(\text{failing}) = 1 - P(\text{passing}) = 1 - \frac{22}{64} = \frac{42}{64}$

69 Let O denote the event that the individual is over 60 and F denote the event that the

individual is female.

$$P(O \cup F) = P(O) + P(F) - P(O \cap F) = \tfrac{1000}{5000} + \tfrac{2000}{5000} - \tfrac{0.40(2000)}{5000} = \tfrac{2200}{5000} = 0.44$$

71 The two teams, A and B, can play as few as 4 games or as many as 7 games. Since the two teams are equally matched, the probability that either team wins any particular game is $\frac{1}{2}$.

$P(\text{team A wins in 4 games}) = \frac{1}{2} \cdot \frac{1}{2} \cdot \frac{1}{2} \cdot \frac{1}{2} = \left(\frac{1}{2}\right)^4 = \frac{1}{16} = 0.0625$.

$P(\text{team A wins in 5 games})$

$\quad = P(\text{team A wins 3 of the first 4 games } \{ \text{losing 1 game} \} \text{ and then wins game 5})$

$= \binom{4}{3}\left(\frac{1}{2}\right)^3 \cdot \left(\frac{1}{2}\right)^1 \cdot \frac{1}{2} = \binom{4}{3}\left(\frac{1}{2}\right)^5 = \frac{4}{32} = 0.125$. In a similar fashion,

$$P(\text{team A wins in 6 games}) = \binom{5}{3}\left(\tfrac{1}{2}\right)^6 = \tfrac{10}{64} = \tfrac{5}{32} = 0.15625$$

and $\qquad P(\text{team A wins in 7 games}) = \binom{6}{3}\left(\tfrac{1}{2}\right)^7 = \tfrac{20}{128} = \tfrac{5}{32} = 0.15625$.

Thus, the probability that team A wins the series is

$$0.0625 + 0.125 + 0.15625 + 0.15625 = 0.5,$$

which agrees with our common sense. Since the probabilities for team B winning the series are the same, the expected number of games is

$$4\left(2 \cdot \tfrac{1}{16}\right) + 5\left(2 \cdot \tfrac{4}{32}\right) + 6\left(2 \cdot \tfrac{5}{32}\right) + 7\left(2 \cdot \tfrac{5}{32}\right) = 5\tfrac{13}{16} = 5.8125.$$

$\boxed{1}$ Probably the easiest way to describe this sequence is by using a piecewise-defined function. If $1 \le n \le 4$, then $2n$ describes the sequence 2, 4, 6, 8. For $n = 5$, we need to obtain a as the general term—one way to symbolize this is $(n-4)a$. Hence, one possibility is $a_n = \begin{cases} 2n & \text{if } 1 \le n \le 4 \\ (n-4)a & \text{if } n \ge 5 \end{cases}$

A more sophisticated (and complicated) approach is to add a term to $2n$—call this term k—so that k is 0 for $n = 1$, 2, 3, 4, but equal to $(a-10)$ for $n = 5$ { the *minus* 10 is needed to compensate for the *plus* 10 from the $2n$ term }. We can do this by starting with the expression $(n-1)(n-2)(n-3)(n-4)$, which is zero for $n = 1$, 2, 3, 4. For $n = 5$, this expression is $4 \cdot 3 \cdot 2 \cdot 1 = 24$, so we will divide it by 24 and multiply it by $(a-10)$ to obtain $(a-10)$. Thus, another possibility is

$$a_n = 2n + \frac{(n-1)(n-2)(n-3)(n-4)(a-10)}{24}.$$

$\boxed{3}$ (a) Following the pattern from Exercises 37–38 in Section 9.4, write

$1^4 + 2^4 + 3^4 + \cdots + n^4 = an^5 + bn^4 + cn^3 + dn^2 + en$. Then, it follows that:

$n = 1 \;\Rightarrow\; a(1)^5 + b(1)^4 + c(1)^3 + d(1)^2 + e(1) = 1^4 \;\Rightarrow\; a + b + c + d + e = 1,$

$n = 2 \;\Rightarrow\; a(2)^5 + b(2)^4 + c(2)^3 + d(2)^2 + e(2) = 1^4 + 2^4 \;\Rightarrow$

$$32a + 16b + 8c + 4d + 2e = 17,$$

$n = 3 \;\Rightarrow\; a(3)^5 + b(3)^4 + c(3)^3 + d(3)^2 + e(3) = 1^4 + 2^4 + 3^4 \;\Rightarrow$

$$243a + 81b + 27c + 9d + 3e = 98,$$

$n = 4 \;\Rightarrow\; a(4)^5 + b(4)^4 + c(4)^3 + d(4)^2 + e(4) = 1^4 + 2^4 + 3^4 + 4^4 \;\Rightarrow$

$$1024a + 256b + 64c + 16d + 4e = 354,$$

$n = 5 \;\Rightarrow\; a(5)^5 + b(5)^4 + c(5)^3 + d(5)^2 + e(5) = 1^4 + 2^4 + 3^4 + 4^4 + 5^4 \;\Rightarrow$

$$3125a + 625b + 125c + 25d + 5e = 979.$$

$$AX = B \Rightarrow \begin{bmatrix} 1 & 1 & 1 & 1 & 1 \\ 32 & 16 & 8 & 4 & 2 \\ 243 & 81 & 27 & 9 & 3 \\ 1024 & 256 & 64 & 16 & 4 \\ 3125 & 625 & 125 & 25 & 5 \end{bmatrix} \begin{bmatrix} a \\ b \\ c \\ d \\ e \end{bmatrix} = \begin{bmatrix} 1 \\ 17 \\ 98 \\ 354 \\ 979 \end{bmatrix} \Rightarrow X = A^{-1}B = \begin{bmatrix} 1/5 \\ 1/2 \\ 1/3 \\ 0 \\ -1/30 \end{bmatrix}$$

Thus, $1^4 + 2^4 + 3^4 + \cdots + n^4 = \frac{1}{5}n^5 + \frac{1}{2}n^4 + \frac{1}{3}n^3 - \frac{1}{30}n$.

This formula must be verified.

(b) Let P_n be the statement that $1^4 + 2^4 + 3^4 + \cdots + n^4 = \frac{1}{5}n^5 + \frac{1}{2}n^4 + \frac{1}{3}n^3 - \frac{1}{30}n$.

(1) $1^4 = \frac{1}{5} + \frac{1}{2} + \frac{1}{3} - \frac{1}{30} = 1$ and P_1 is true.

(2) Assume that P_k is true. We must show that P_{k+1} is true.

$P_k \Rightarrow 1^4 + 2^4 + 3^4 + \cdots + k^4 = \frac{1}{5}k^5 + \frac{1}{2}k^4 + \frac{1}{3}k^3 - \frac{1}{30}k \Rightarrow$

$1^4 + 2^4 + 3^4 + \cdots + k^4 + (k+1)^4 = \frac{1}{5}k^5 + \frac{1}{2}k^4 + \frac{1}{3}k^3 - \frac{1}{30}k + (k+1)^4 \Rightarrow$

$1^4 + 2^4 + 3^4 + \cdots + k^4 + (k+1)^4 =$

$$\frac{1}{5}k^5 + \frac{1}{2}k^4 + \frac{1}{3}k^3 - \frac{1}{30}k + (k^4 + 4k^3 + 6k^2 + 4k + 1) \Rightarrow$$

$1^4 + 2^4 + 3^4 + \cdots + k^4 + (k+1)^4 = \frac{1}{5}k^5 + \frac{3}{2}k^4 + \frac{13}{3}k^3 + 6k^2 + \frac{119}{30}k + 1.$

$P_{k+1} \Rightarrow 1^4 + 2^4 + 3^4 + \cdots + k^4 + (k+1)^4$

$= \frac{1}{5}(k+1)^5 + \frac{1}{2}(k+1)^4 + \frac{1}{3}(k+1)^3 - \frac{1}{30}(k+1)$

$= \frac{1}{5}(k^5 + 5k^4 + 10k^3 + 10k^2 + 5k + 1) + \frac{1}{2}(k^4 + 4k^3 + 6k^2 + 4k + 1) +$

$$\frac{1}{3}(k^3 + 3k^2 + 3k + 1) - \frac{1}{30}(k+1)$$

$= \frac{1}{5}k^5 + \frac{3}{2}k^4 + \frac{13}{3}k^3 + 6k^2 + \frac{119}{30}k + 1.$ Thus, the formula is true for all n.

⑤ Examine the number of digits in the exponent of the value in scientific notation. The TI-82/83 can compute 69!, but not 70!, since 70! is larger than a 2-digit exponent. The TI-85/86 can compute 449!, but not 450!, since 450! is larger than a 3-digit exponent.

⑦ $\text{Time}_{\text{total}} = \text{Time}_{\text{down}} + \text{Time}_{\text{up}}$ { note $h = 10$ and $d = 10 \cdot \frac{1}{2}$ for the first rebound }

$$= \left[\frac{\sqrt{10}}{4} + \frac{\sqrt{10 \cdot \frac{1}{2}}}{4} + \frac{\sqrt{10 \cdot (\frac{1}{2})^2}}{4} + \cdots \right] + \left[\frac{\sqrt{10 \cdot \frac{1}{2}}}{4} + \frac{\sqrt{10 \cdot (\frac{1}{2})^2}}{4} + \cdots \right]$$

$$= \frac{\sqrt{10}}{4} + 2 \left[\frac{\sqrt{10 \cdot \frac{1}{2}}}{4} + \frac{\sqrt{10 \cdot (\frac{1}{2})^2}}{4} + \cdots \right].$$ The ratio of this geometric sequence

is $\sqrt{\frac{1}{2}}$ and its absolute value is less than 1, so we can find its infinite sum using

$S = a_1/(1 - r)$. Thus, the total time is $\dfrac{\sqrt{10}}{4} + 2 \cdot \dfrac{\sqrt{5}/4}{1 - \sqrt{\frac{1}{2}}} \approx 4.61$ seconds.

⑨ To the nearest penny with $r = 1.1008163$ (found by trial and error), we have the places 1st–10th:

$0.01:	237.37	215.63	195.89	177.95	161.65	146.85	133.40	121.18	110.08	100
$1.00:	237.00	216.00	196.00	178.00	162.00	147.00	133.00	121.00	110.00	100
$5.00:	240.00	215.00	195.00	180.00	160.00	145.00	135.00	120.00	110.00	100
$10.00:	240.00	220.00	200.00	180.00	160.00	140.00	130.00	120.00	110.00	100

If the amounts are to be realistic, the amounts may not be rounded to the *nearest* amount—i.e, $134 may be rounded to $140 rather than $130.

$\boxed{11}$ First, calculate the probabilities for all prizes: $n(S) = C(49, 5) \cdot 42 = 80,089,128$

$n(5\,\text{W}, 1\,\text{R}) = C(5, 5) \cdot C(44, 0) \cdot 1 \ \Rightarrow \ P(5\,\text{W}, 1\,\text{R}) = \qquad\qquad 1/80,089,128 \ (p_1)$

$n(5\,\text{W}, 0\,\text{R}) = C(5, 5) \cdot C(44, 0) \cdot 41 \ \Rightarrow \ P(5\,\text{W}, 0\,\text{R}) = \qquad\quad 41/80,089,128 \ (p_2)$

$n(4\,\text{W}, 1\,\text{R}) = C(5, 4) \cdot C(44, 1) \cdot 1 \ \Rightarrow \ P(4\,\text{W}, 1\,\text{R}) = \qquad\quad 220/80,089,128 \ (p_3)$

$n(4\,\text{W}, 0\,\text{R}) = C(5, 4) \cdot C(44, 1) \cdot 41 \ \Rightarrow \ P(4\,\text{W}, 0\,\text{R}) = \qquad 9020/80,089,128 \ (p_4)$

$n(3\,\text{W}, 1\,\text{R}) = C(5, 3) \cdot C(44, 2) \cdot 1 \ \Rightarrow \ P(3\,\text{W}, 1\,\text{R}) = \qquad 9460/80,089,128 \ (p_5)$

$n(3\,\text{W}, 0\,\text{R}) = C(5, 3) \cdot C(44, 2) \cdot 41 \ \Rightarrow \ P(3\,\text{W}, 0\,\text{R}) = \quad 387,860/80,089,128 \ (p_6)$

$n(2\,\text{W}, 1\,\text{R}) = C(5, 2) \cdot C(44, 3) \cdot 1 \ \Rightarrow \ P(2\,\text{W}, 1\,\text{R}) = \quad 132,440/80,089,128 \ (p_7)$

$n(1\,\text{W}, 1\,\text{R}) = C(5, 1) \cdot C(44, 4) \cdot 1 \ \Rightarrow \ P(1\,\text{W}, 1\,\text{R}) = \quad 678,755/80,089,128 \ (p_8)$

$n(0\,\text{W}, 1\,\text{R}) = C(5, 0) \cdot C(44, 5) \cdot 1 \ \Rightarrow \ P(0\,\text{W}, 1\,\text{R}) = \ 1,086,008/80,089,128 \ (p_9)$

(a) The probability of winning the jackpot is $\frac{1}{80,089,128}$.

(b) The probability of winning any prize is the sum of all the probabilities, that is, $\frac{2,303,805}{80,089,128}$. This gives us odds of 77,785,323 to 2,303,805 (or about 34 to 1) for *not* winning any prize.

(c) The expected value of the game without the jackpot is

$$\sum_{k=2}^{9} a_k p_k = (\$100{,}000)p_2 + (\$5000)p_3 + (\$100)p_4 + (\$100)p_5 + (\$7)p_6 +$$

$$(\$7)p_7 + (\$4)p_8 + (\$3)p_9 = \frac{16{,}663{,}144}{80{,}089{,}128} \approx 0.21$$

(d) Assuming that one ticket costs $1, we must have an expected value of 1 to be a fair game. Since the probability of winning the jackpot is $\frac{1}{80,089,128}$, we must multiply that probability by \$63,425,984 $\{\,80{,}089{,}128 - 16{,}663{,}144\,\}$ so that the numerator of the expected value in part (c) is equal to the denominator.

[13] (a) The pattern has a 1 in the denominator, an n in the numerator, and the next terms of any row of Pascal's triangle oscillate between denominator and numerator. The signs are in pairs (two positive, two negative, etc.). The power of $\tan x$ increases one with each term. Thus,

$$\tan 5x = \frac{5\tan x - 10\tan^3 x + \tan^5 x}{1 - 10\tan^2 x + 5\tan^4 x}.$$

(b) The coefficients listed in the text are in the form 1–2–1. The next identities are:

$$\cos 3x = 1\cos^3 x \qquad\qquad -3\cos x \sin^2 x$$
$$\sin 3x = \qquad 3\cos^2 x \sin x \qquad\qquad\quad \cdot -1\sin^3 x$$

$$\cos 4x = 1\cos^4 x \qquad\qquad -6\cos^2 x \sin^2 x \qquad\qquad +1\sin^4 x$$
$$\sin 4x = \qquad 4\cos^3 x \sin x \qquad\qquad -4\cos x \sin^3 x$$

Notice the pattern of coefficients: for $\cos 3x$ and $\sin 3x$ we have 1–3–3–1; for $\cos 4x$ and $\sin 4x$ we have 1–4–6–4–1. Since these are rows in Pascal's triangle, we predict the following pattern for $\cos 5x$ and $\sin 5x$ (1–5–10–10–5–1):

$$\cos 5x = 1\cos^5 x \qquad\qquad -10\cos^3 x \sin^2 x \qquad\qquad +5\cos x \sin^4 x$$
$$\sin 5x = \qquad 5\cos^4 x \sin x \qquad\qquad -10\cos^2 x \sin^3 x \qquad\qquad +1\sin^5 x$$

Note: The next two solutions are for Exercises 39 and 41 in Section 9.4.

[39] (1) For $n = 1$, $\sin(\theta + 1\pi) = \sin\theta\cos\pi + \cos\theta\sin\pi = -\sin\theta = (-1)^1\sin\theta$.

(2) Assume P_k is true: $\sin(\theta + k\pi) = (-1)^k \sin\theta$. Hence,

$$\begin{aligned}
\sin[\theta + (k+1)\pi] &= \sin[(\theta + k\pi) + \pi]\\
&= \sin(\theta + k\pi)\cos\pi + \cos(\theta + k\pi)\sin\pi\\
&= \Big[(-1)^k \sin\theta\Big]\cdot(-1) + \cos(\theta + k\pi)\cdot(0)\\
&= (-1)^{k+1} \sin\theta.
\end{aligned}$$

Thus, P_{k+1} is true, and the proof is complete.

[41] (1) For $n = 1$, $[r(\cos\theta + i\sin\theta)]^1 = r^1[\cos(1\theta) + i\sin(1\theta)]$.

(2) Assume P_k is true: $[r(\cos\theta + i\sin\theta)]^k = r^k(\cos k\theta + i\sin k\theta)$. Hence,

$$\begin{aligned}
[r(\cos\theta + i\sin\theta)]^{k+1} &= [r(\cos\theta + i\sin\theta)]^k[r(\cos\theta + i\sin\theta)]\\
&= r^k[\cos k\theta + i\sin k\theta][r(\cos\theta + i\sin\theta)]\\
&= r^{k+1}[(\cos k\theta\cos\theta - \sin k\theta\sin\theta)\\
&\qquad\qquad + i(\sin k\theta\cos\theta + \cos k\theta\sin\theta)]
\end{aligned}$$

$$\{\text{Use the addition formulas for the sine and cosine.}\}$$

$$= r^{k+1}[\cos(k+1)\theta + i\sin(k+1)\theta].$$

Thus, P_{k+1} is true, and the proof is complete.

Chapter 10: Topics from Analytic Geometry

Note: For Exercises 1–12, we will put each parabola equation in one of the forms listed on page 768—either

$$(x-h)^2 = 4p(y-k) \quad \text{or} \quad (y-k)^2 = 4p(x-h).$$

Once in one of those forms, the information concerning the vertex, focus, and directrix is easily obtainable and illustrated in the chart in the text. Let V, F, and l denote the vertex, focus, and directrix, respectively.

$\boxed{1}$ $8y = x^2 \Rightarrow y = \frac{1}{8}x^2 \Rightarrow p = \frac{1}{4a} = \frac{1}{4\left(\frac{1}{8}\right)} = \frac{1}{\frac{1}{2}} = 2.$ We know that the parabola

opens either upward or downward since the variable "x" is squared. Since p is positive, we know that the parabola opens upward and that the focus is 2 units above the vertex. The directrix is 2 units below the vertex. $\qquad V(0, 0);\ F(0, 2);\ l\!: y = -2$

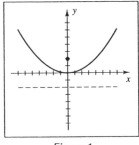

Figure 1

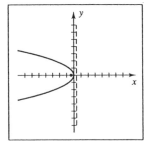

Figure 3

$\boxed{3}$ $2y^2 = -3x \Rightarrow (y-0)^2 = -\frac{3}{2}(x-0) \Rightarrow 4p = -\frac{3}{2} \Rightarrow p = -\frac{3}{8}.$ We know that the parabola opens either right or left since the variable "y" is squared. Since p is negative, we know that the parabola opens left and that the focus is $\frac{3}{8}$ unit to the left of the vertex. The directrix is $\frac{3}{8}$ unit to the right of the vertex.

$$V(0, 0);\ F\!\left(-\tfrac{3}{8}, 0\right);\ l\!: x = \tfrac{3}{8}$$

$\boxed{5}$ $(x+2)^2 = -8(y-1) \Rightarrow 4p = -8 \Rightarrow p = -2.$ The $(x+2)$ and $(y-1)$ factors indicate that we need to shift the vertex, of the parabola having equation $x^2 = -8y$, 2 units left and 1 unit up—that is, move it from $(0, 0)$ to $(-2, 1)$.

$$V(-2, 1);\ F(-2, -1);\ l\!: y = 3$$

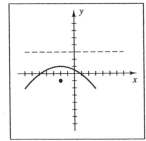

Figure 5

$\boxed{7}$ $(y-2)^2 = \frac{1}{4}(x-3)$ $\Rightarrow$ $4p = \frac{1}{4}$ $\Rightarrow$ $p = \frac{1}{16}$. "y" squared and p positive imply that the parabola opens to the right and the focus is to the right of the vertex. The $(x-3)$ and $(y-2)$ factors indicate that we need to shift the vertex 3 units right and 2 units up from $(0, 0)$ to $(3, 2)$.
$$V(3, 2);\ F(\tfrac{49}{16}, 2);\ l\text{: } x = \tfrac{47}{16}$$

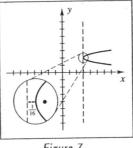

Figure 7

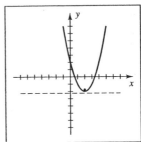

Figure 9

$\boxed{9}$ For this exercise, we need to "complete the square" in order to get the equation in proper form. The term we need to add is $\left[\frac{1}{2}(\text{coefficient of } x)\right]^2$. In this case, that value is $\left[\frac{1}{2}(-4)\right]^2 = 4$. Notice that we add and subtract the value 4 from the same side of the equation as opposed to adding 4 to both sides of the equation.

$y = x^2 - 4x + 2$ $\Rightarrow$ $y = (x^2 - 4x + \underline{4}) + 2 - \underline{4}$ $\Rightarrow$ $y = (x-2)^2 - 2$ $\Rightarrow$

$(y+2) = 1(x-2)^2$ $\Rightarrow$ $4p = 1$ $\Rightarrow$ $p = \frac{1}{4}$. $V(2, -2);\ F(2, -\frac{7}{4});\ l\text{: } y = -\frac{9}{4}$

$\boxed{11}$ $x^2 + 20y = 10$ $\Rightarrow$ $x^2 = -20y + 10$ $\Rightarrow$

$(x-0)^2 = -20(y - \frac{1}{2})$ $\Rightarrow$ $4p = -20$ $\Rightarrow$ $p = -5$.

$V(0, \frac{1}{2});\ F(0, -\frac{9}{2});\ l\text{: } y = \frac{11}{2}$

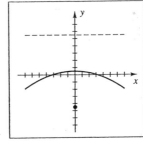

Figure 11

$\boxed{13}$ Since the vertex is at $(1, 0)$ and the parabola has a horizontal axis, the standard equation has the form $(y-0)^2 = 4p(x-1)$. The distance from the focus $F(6, 0)$ to the vertex $V(1, 0)$ is $6 - 1 = 5$, which is the value of p. Hence, an equation of the parabola is $y^2 = 20(x-1)$.

$\boxed{15}$ An equation of a parabola with vertex $V(-2, 3)$ is $(x+2)^2 = 4p(y-3)$. Since the point $P(2, 2)$ is on the graph, we'll substitute 2 for x and 2 for y and then find p.

$x = 2,\ y = 2$ $\Rightarrow$ $(2+2)^2 = 4p(2-3)$ $\Rightarrow$ $16 = 4p(-1)$ $\Rightarrow$ $p = -4$,

so an equation is $(x+2)^2 = -16(y-3)$.

17 The distance from the directrix to the focus is $2 - (-2) = 4$ units. The vertex $V(0, 0)$ is halfway between the directrix and the focus—that is, 2 units from either one. Since the focus is 2 units to the right of the vertex, p is 2. Using one of the forms of an equation of a parabola with vertex at (h, k), we have $(y - 0)^2 = 4p(x - 0)$, or, equivalently, $y^2 = 8x$.

19 $F(6, 4)$ and $l: y = -2 \Rightarrow 2p = 4 - (-2) \Rightarrow 2p = 6 \Rightarrow p = 3$ and $V(6, 1)$.
$$(x - 6)^2 = 4p(y - 1) \Rightarrow (x - 6)^2 = 12(y - 1).$$

21 $V(3, -5)$ and $l: x = 2 \Rightarrow p = 3 - 2 = 1.$
$$(y + 5)^2 = 4p(x - 3) \Rightarrow (y + 5)^2 = 4(x - 3).$$

23 $V(-1, 0)$ and $F(-4, 0) \Rightarrow p = -4 - (-1) = -3.$
$$(y - 0)^2 = 4p(x + 1) \Rightarrow y^2 = -12(x + 1).$$

25 The vertex at the origin and symmetric to the y-axis imply that the equation is of the form $y = ax^2$. Substituting $x = 2$ and $y = -3$ into that equation yields
$$-3 = a \cdot 4 \Rightarrow a = -\tfrac{3}{4}. \text{ Thus, an equation is } y = -\tfrac{3}{4}x^2, \text{ or } 3x^2 = -4y.$$

27 The vertex at $(-3, 5)$ and axis parallel to the x-axis imply that the equation is of the form $(y - 5)^2 = 4p(x + 3)$. Substituting $x = 5$ and $y = 9$ into that equation yields $16 = 4p \cdot 8 \Rightarrow p = \tfrac{1}{2}$. Thus, an equation is $(y - 5)^2 = 2(x + 3)$.

29 Refer to the definition of a parabola. The point $P(0, 5)$ is the fixed point (focus) and the line $l: y = -3$ is the fixed line (directrix). The vertex is halfway between the focus and the directrix—that is, at $V(0, 1)$. An equation is of the form $(x - h)^2 = 4p(y - k) \Rightarrow (x - 0)^2 = 4p(y - 1)$. The distance from the vertex to the focus is $p = 5 - 1 = 4$. Thus, an equation is
$$(x - 0)^2 = 4(4)(y - 1) \Rightarrow x^2 = 16(y - 1).$$

31 The point $P(-6, 3)$ is the fixed point (focus) and the line $l: x = -2$ is the fixed line (directrix). The vertex is halfway between the focus and the directrix—that is, at $V(-4, 3)$. An equation is of the form $(y - 3)^2 = 4p(x + 4)$. The distance from the vertex to the focus is $p = -6 - (-4) = -2$. Thus, an equation is $(y - 3)^2 = -8(x + 4)$.

Note: To find an equation for a lower or upper half, we need to solve for y (use $-$ or $+$ respectively). For the left or right half, solve for x (use $-$ or $+$ respectively).

33 $(y + 1)^2 = x + 3 \Rightarrow y + 1 = \pm\sqrt{x + 3} \Rightarrow y = -\sqrt{x + 3} - 1$

35 $(x + 1)^2 = y - 4 \Rightarrow x + 1 = \pm\sqrt{y - 4} \Rightarrow x = \sqrt{y - 4} - 1$

$\boxed{37}$ The parabola has an equation of the form $y = ax^2 + bx + c$. Substituting the x and y values of $P(2, 5)$, $Q(-2, -3)$, and $R(1, 6)$ into this equation yields:

$$\begin{cases} 4a + 2b + c = 5 & P \quad (E_1) \\ 4a - 2b + c = -3 & Q \quad (E_2) \\ a + b + c = 6 & R \quad (E_3) \end{cases}$$

Solving E_3 for c $\{c = 6 - a - b\}$ and substituting into E_1 and E_2 yields:

$$\begin{cases} 3a + b = -1 & (E_4) \\ 3a - 3b = -9 & (E_5) \end{cases}$$

$E_4 - E_5 \Rightarrow 4b = 8 \Rightarrow b = 2$; $a = -1$; $c = 5$. The equation is $y = -x^2 + 2x + 5$.

$\boxed{39}$ The parabola has an equation of the form $x = ay^2 + by + c$. Substituting the x and y values of $P(-1, 1)$, $Q(11, -2)$, and $R(5, -1)$ into this equation yields:

$$\begin{cases} a + b + c = -1 & P \quad (E_1) \\ 4a - 2b + c = 11 & Q \quad (E_2) \\ a - b + c = 5 & R \quad (E_3) \end{cases}$$

Solving E_3 for c $\{c = 5 - a + b\}$ and substituting into E_1 and E_2 yields:

$$\begin{cases} 2b = -6 & (E_4) \\ 3a - b = 6 & (E_5) \end{cases}$$

$E_4 \Rightarrow b = -3$; $a = 1$; $c = 1$. The equation is $x = y^2 - 3y + 1$.

$\boxed{41}$ A cross section of the mirror is a parabola with $V(0, 0)$ and passing through $P(4, 1)$. The incoming light will collect at the focus F. A general equation of this form of a parabola is $y = ax^2$. Substituting $x = 4$ and $y = 1$ gives us $1 = a(4)^2 \Rightarrow a = \frac{1}{16}$. $p = 1/(4a) = 1/(\frac{1}{4}) = 4$. The light will collect 4 inches from the center of the mirror.

$\boxed{43}$ If we set up a coordinate system with a parabola opening upward and the vertex at the origin, then the phrase "3 feet across at the opening and 1 foot deep" implies that the points $(\pm \frac{3}{2}, 1)$ are on the parabola.

$$y = ax^2 \Rightarrow 1 = a(\tfrac{3}{2})^2 \Rightarrow a = \tfrac{4}{9}. \quad p = 1/(4a) = 1/(\tfrac{16}{9}) = \tfrac{9}{16} \text{ ft.}$$

$\boxed{45}$ $p = 5 \Rightarrow a = 1/(4p) = 1/(4 \cdot 5) = \frac{1}{20}$.

$$y = ax^2 \; \{y = 2 \text{ ft} = 24 \text{ inches}\} \Rightarrow 24 = \tfrac{1}{20}x^2 \Rightarrow x^2 = 480 \Rightarrow x = \sqrt{480}.$$

The width is twice the value of x. Width $= 2\sqrt{480} \approx 43.82$ in.

$\boxed{47}$ (a) Let the parabola have the equation $x^2 = 4py$. Since the point (r, h) is on the parabola, we can substitute r for x and h for y, giving us $r^2 = 4ph$. Solving for p we have $p = \dfrac{r^2}{4h}$.

(b) $p = 10$ and $h = 5 \Rightarrow r^2 = 4(10)(5) \Rightarrow r = \sqrt{200} = 10\sqrt{2}$.

49 With $a = 125$ and $p = 50$, $S = \dfrac{8\pi p^2}{3}\left[\left(1 + \dfrac{a^2}{4p^2}\right)^{3/2} - 1\right] \approx 64{,}968 \text{ ft}^2$.

51 Depending on the type of calculator or software used, we may need to solve for y in

terms of x. $x = -y^2 + 2y + 5 \;\Rightarrow\; y^2 - 2y + (x - 5) = 0$.   This is a quadratic

equation in y. Using the quadratic formula to solve for y yields

$$y = \frac{-(-2) \pm \sqrt{(-2)^2 - 4(1)(x-5)}}{2(1)} = \frac{2 \pm \sqrt{24 - 4x}}{2} = \frac{2 \pm \sqrt{4}\,\sqrt{6-x}}{2} = 1 \pm \sqrt{6 - x}.$$

$[-11,\, 10,\, 2]$ by $[-7,\, 7]$ $[-2,\, 4]$ by $[-3,\, 3]$

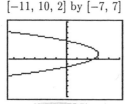   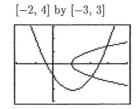

Figure 51   Figure 53

53 $x = y^2 + 1 \;\Rightarrow\; y = \pm\sqrt{x - 1}$. From the graph, we can see that there are 2 points

of intersection. Their coordinates are approximately $(2.08,\, -1.04)$ and $(2.92,\, 1.38)$.

10.2 Exercises

Note: Let C, V, F, and M denote the center, the vertices, the foci, and the endpoints of

the minor axis, respectively. Let c denote the distance from the center of the

ellipse to a focus.

1 $\dfrac{x^2}{9} + \dfrac{y^2}{4} = 1$ • The x-intercepts are $\pm\sqrt{9} = \pm 3$.   The y-intercepts are

$\pm\sqrt{4} = \pm 2$. The major axis { the longer of the two axes } is the horizontal axis and

has length $2(3) = 6$. The minor axis { the shorter } is the vertical axis and has length

$2(2) = 4$. To find the foci, it is helpful to remember the relationship

$$\left[\tfrac{1}{2}(\text{minor axis})\right]^2 + \left[\,c\,\right]^2 = \left[\tfrac{1}{2}(\text{major axis})\right]^2.$$

Using the values from above we have

$\left[\tfrac{1}{2}(4)\right]^2 + c^2 = \left[\tfrac{1}{2}(6)\right]^2 \;\Rightarrow\; 4 + c^2 = 9 \;\Rightarrow$

$c^2 = 5 \;\Rightarrow\; c = \pm\sqrt{5}$. Simply put, c^2 is equal to

difference of the denominators.

$V(\pm 3,\, 0);\; F(\pm\sqrt{5},\, 0);\; M(0,\, \pm 2)$

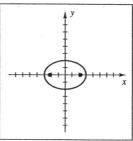

Figure 1

3 $\frac{x^2}{15} + \frac{y^2}{16} = 1$ • Since the 16 in the denominator of the term with the variable y is

larger than the 15 in the denominator of the term with the variable x, the vertices

and foci are on the y-axis and the major axis is the vertical axis. $c^2 = 16 - 15 \Rightarrow$

$c = \pm 1.$ $V(0, \pm 4); F(0, \pm 1); M(\pm \sqrt{15}, 0)$

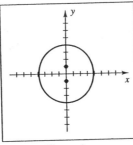

Figure 3

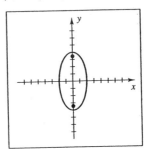

Figure 5

5 We first divide by 16 to obtain the "1" on the right side of the equation.

$4x^2 + y^2 = 16 \Rightarrow \frac{x^2}{4} + \frac{y^2}{16} = 1.$ Since the 16 in the denominator of the term with

the variable y is larger than the 4 in the denominator of the term with the variable x,

the vertices and foci are on the y-axis and the major axis is the vertical axis.

$4 + c^2 = 16 \Rightarrow c^2 = 16 - 4 \Rightarrow c = \pm \sqrt{12} = \pm 2\sqrt{3}.$

$V(0, \pm 4); F(0, \pm 2\sqrt{3}); M(\pm 2, 0)$

7 First write $4x^2 + 25y^2 = 1$ in standard form: $\frac{x^2}{\frac{1}{4}} + \frac{y^2}{\frac{1}{25}} = 1.$

$\frac{1}{25} + c^2 = \frac{1}{4} \Rightarrow c^2 = \frac{1}{4} - \frac{1}{25} = \frac{21}{100} \Rightarrow c = \pm \frac{1}{10}\sqrt{21}.$

$V(\pm \frac{1}{2}, 0); F(\pm \frac{1}{10}\sqrt{21}, 0); M(0, \pm \frac{1}{5})$

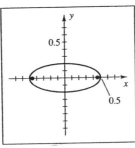

Figure 7

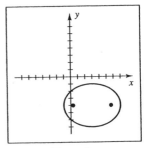

Figure 9

$\boxed{9}$ $\dfrac{(x-3)^2}{16} + \dfrac{(y+4)^2}{9} = 1$ • The effect of the factors $(x-3)$ and $(y+4)$ is to shift the center of the ellipse from $(0, 0)$ to $(3, -4)$. Since the larger denominator, 16, is in the term with x, the major axis will be horizontal. The endpoints are 4 units in either direction of the center. Their coordinates are the points $(3 \pm 4, -4)$, or, equivalently, $(7, -4)$ and $(-1, -4)$. The minor axis will be vertical with endpoints $(3, -4 \pm 3)$, or, equivalently, $(3, -1)$ and $(3, -7)$. $9 + c^2 = 16 \Rightarrow c^2 = 16 - 9 \Rightarrow c = \pm\sqrt{7}$. Remember that c is the distance *from the center* to a focus. Hence, the coordinates of the foci are $(3 \pm \sqrt{7}, -4)$. See *Figure 9* on the previous page.

$$C(3, -4); \; V(3 \pm 4, -4); \; F(3 \pm \sqrt{7}, -4); \; M(3, -4 \pm 3)$$

$\boxed{11}$ $4x^2 + 9y^2 - 32x - 36y + 64 = 0 \Rightarrow (4x^2 - 32x) + (9y^2 - 36y) = -64 \Rightarrow$

{ first group the x terms and the y terms } $\quad 4(x^2 - 8x) + 9(y^2 - 4y) = -64 \Rightarrow$

{ factor out the coefficients of x^2 and y^2 }

$$4(x^2 - 8x + \underline{\;16\;}) + 9(y^2 - 4y + \underline{\;4\;}) = -64 + \underline{\;64\;} + \underline{\;36\;} \Rightarrow$$

{ Complete the squares—remember that we have added $\underline{\;4\;}$(16) and $\underline{\;9\;}$(4), not 16 and 4. Thus, we add 64 and 36 to the right side of the equation. }

$4(x-4)^2 + 9(y-2)^2 = 36 \Rightarrow \dfrac{(x-4)^2}{9} + \dfrac{(y-2)^2}{4} = 1$. $\; c^2 = 9 - 4 \Rightarrow c = \pm\sqrt{5}$.

$$C(4, 2); \; V(4 \pm 3, 2); \; F(4 \pm \sqrt{5}, 2); \; M(4, 2 \pm 2)$$

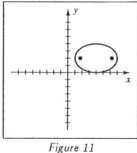

Figure 11

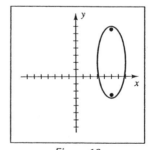

Figure 13

$\boxed{13}$ $25x^2 + 4y^2 - 250x - 16y + 541 = 0 \Rightarrow$

$25(x^2 - 10x + \underline{\;25\;}) + 4(y^2 - 4y + \underline{\;4\;}) = -541 + \underline{\;625\;} + \underline{\;16\;} \Rightarrow$

$25(x-5)^2 + 4(y-2)^2 = 100 \Rightarrow \dfrac{(x-5)^2}{4} + \dfrac{(y-2)^2}{25} = 1$.

$c^2 = 25 - 4 \Rightarrow c = \pm\sqrt{21}$. $\qquad C(5, 2); \; V(5, 2 \pm 5); \; F(5, 2 \pm \sqrt{21}); \; M(5 \pm 2, 2)$

Note: Let b denote the distance equal to $\frac{1}{2}$(minor axis length) and

let a denote the distance equal to $\frac{1}{2}$(major axis length).

15 $a = 2$ and $b = 6$ $\Rightarrow$ $\frac{x^2}{a^2} + \frac{y^2}{b^2} = 1$ is $\frac{x^2}{4} + \frac{y^2}{36} = 1$.

17 The center of the ellipse is $(-2, 1)$. $a = 5$ and $b = 2$ give us $\frac{(x+2)^2}{25} + \frac{(y-1)^2}{4} = 1$.

19 Since the vertices are at $(\pm 8, 0)$, $\frac{1}{2}$(major axis) $= 8$. Since the foci are at $(\pm 5, 0)$,

$c = 5$. Using the relationship $\left[\frac{1}{2}(\text{minor axis})\right]^2 + \left[c\right]^2 = \left[\frac{1}{2}(\text{major axis})\right]^2$, we have

$\left[\frac{1}{2}(\text{minor axis})\right]^2 + 5^2 = 8^2$ $\Rightarrow$ $\left[\frac{1}{2}(\text{minor axis})\right]^2 = 64 - 25 = 39$.

An equation is $\frac{x^2}{64} + \frac{y^2}{39} = 1$.

21 If the length of the minor axis is 3, then $b = \frac{1}{2}(3) = \frac{3}{2}$. The vertices are $V(0, \pm 5)$,

so an equation is $\frac{x^2}{(\frac{3}{2})^2} + \frac{y^2}{5^2} = 1$, or, equivalently, $\frac{4x^2}{9} + \frac{y^2}{25} = 1$.

23 With the vertices at $(0, \pm 6)$, an equation of the ellipse is $\frac{x^2}{b^2} + \frac{y^2}{6^2} = 1$. Substituting

$x = 3$ and $y = 2$ and solving for b^2 yields $\frac{9}{b^2} + \frac{4}{36} = 1$ $\Rightarrow$ $\frac{9}{b^2} = \frac{8}{9}$ $\Rightarrow$ $b^2 = \frac{81}{8}$.

An equation is $\frac{x^2}{\frac{81}{8}} + \frac{y^2}{36} = 1$, or, equivalently, $\frac{8x^2}{81} + \frac{y^2}{36} = 1$.

25 With vertices $V(0, \pm 4)$, an equation of the ellipse is $\frac{x^2}{b^2} + \frac{y^2}{16} = 1$.

Remember the formula for the eccentricity:

$$e = \frac{\text{distance from center to focus}}{\text{distance from center to vertex}} = \frac{c}{a} = \frac{\sqrt{a^2 - b^2}}{a}$$

Hence, $e = \frac{c}{a} = \frac{3}{4}$ and $a = 4$ $\Rightarrow$ $c = 3$. Thus, $b^2 + c^2 = a^2$ $\Rightarrow$

$b^2 = a^2 - c^2 = 4^2 - 3^2 = 16 - 9 = 7$. An equation is $\frac{x^2}{7} + \frac{y^2}{16} = 1$.

27 x-intercepts $= \pm 2$ and y-intercepts $= \pm \frac{1}{3}$ $\Rightarrow$

$$\frac{x^2}{2^2} + \frac{y^2}{(\frac{1}{3})^2} = 1 \Rightarrow \frac{x^2}{4} + \frac{y^2}{\frac{1}{9}} = 1 \Rightarrow \frac{x^2}{4} + 9y^2 = 1.$$

29 Remember to divide the lengths of the major and minor axes by 2.

$$\frac{x^2}{(\frac{1}{2} \cdot 8)^2} + \frac{y^2}{(\frac{1}{2} \cdot 5)^2} = 1 \Rightarrow \frac{x^2}{16} + \frac{y^2}{\frac{25}{4}} = 1 \Rightarrow \frac{x^2}{16} + \frac{4y^2}{25} = 1.$$

31 The graph of $x^2 + 4y^2 = 20$, or, equivalently, $\dfrac{x^2}{20} + \dfrac{y^2}{5} = 1$,

is that of an ellipse with x-intercepts at $\pm\sqrt{20}$ and y-

intercepts $\pm\sqrt{5}$. The graph of $x + 2y = 6$, or,

equivalently, $y = -\frac{1}{2}x + 3$, is that of a line with y-

intercept 3 and slope $-\frac{1}{2}$. Substituting $x = 6 - 2y$ into

$x^2 + 4y^2 = 20$ yields $(6 - 2y)^2 + 4y^2 = 20 \Rightarrow$

$8y^2 - 24y + 16 = 0 \Rightarrow 8(y^2 - 3y + 2) = 0 \Rightarrow$

$8(y - 1)(y - 2) = 0 \Rightarrow y = 1, 2; x = 4, 2.$

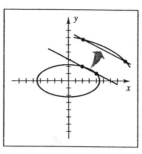

Figure 31

The two points of intersection are $(2, 2)$ and $(4, 1)$.

33 Refer to the definition of an ellipse. As in the discussion on page 774, we will let the

positive constant, k, equal $2a$. $k = 10 \Rightarrow 2a = 10 \Rightarrow a = 5$.

$F(3, 0)$ and $F'(-3, 0) \Rightarrow c = 3.$

$$b^2 = a^2 - c^2 = 25 - 9 = 16. \text{ An equation is } \frac{x^2}{25} + \frac{y^2}{16} = 1.$$

35 $k = 34 \Rightarrow 2a = 34 \Rightarrow a = 17.$ $F(0, 15)$ and $F'(0, -15) \Rightarrow c = 15.$

$$b^2 = a^2 - c^2 = 289 - 225 = 64. \text{ An equation is } \frac{x^2}{64} + \frac{y^2}{289} = 1.$$

37 $y = 11\sqrt{1 - \dfrac{x^2}{49}} \Rightarrow \dfrac{y}{11} = \sqrt{1 - \dfrac{x^2}{49}} \Rightarrow \dfrac{x^2}{49} + \dfrac{y^2}{121} = 1.$

Since $y \geq 0$ in the original equation, its graph is the upper half of the ellipse.

39 $x = -\frac{1}{3}\sqrt{9 - y^2} \Rightarrow -3x = \sqrt{9 - y^2} \Rightarrow 9x^2 = 9 - y^2 \Rightarrow x^2 + \dfrac{y^2}{9} = 1.$

Since $x \leq 0$ in the original equation, its graph is the left half of the ellipse.

41 $x = 1 + 2\sqrt{1 - \dfrac{(y + 2)^2}{9}} \Rightarrow \dfrac{x - 1}{2} = \sqrt{1 - \dfrac{(y + 2)^2}{9}} \Rightarrow \dfrac{(x - 1)^2}{4} + \dfrac{(y + 2)^2}{9} = 1.$

Since $x \geq 1$ in the original equation, its graph is the right half of the ellipse.

43 $y = 2 - 7\sqrt{1 - \dfrac{(x + 1)^2}{9}} \Rightarrow \dfrac{y - 2}{-7} = \sqrt{1 - \dfrac{(x + 1)^2}{9}} \Rightarrow \dfrac{(x + 1)^2}{9} + \dfrac{(y - 2)^2}{49} = 1.$

Since $y \leq 2$ in the original equation, its graph is the lower half of the ellipse.

45 Model this problem as an ellipse with $V(\pm 15, 0)$ and $M(0, \pm 10)$.

Substituting $x = 6$ into $\dfrac{x^2}{15^2} + \dfrac{y^2}{10^2} = 1$ yields $\dfrac{y^2}{100} = \dfrac{189}{225} \Rightarrow y^2 = 84.$

The desired height is $\sqrt{84} = 2\sqrt{21} \approx 9.165$ ft.

47 $e = \dfrac{c}{a} = 0.017 \Rightarrow c = 0.017a = 0.017(93{,}000{,}000) = 1{,}581{,}000.$ The maximum and

minimum distances are $a + c = 94{,}581{,}000$ miles and $a - c = 91{,}419{,}000$ miles.

49 (a) Let c denote the distance from the center of the hemi-ellipsoid to F.

Hence, $(\frac{1}{2}k)^2 + c^2 = h^2 \Rightarrow c^2 = h^2 - \frac{1}{4}k^2 \Rightarrow c = \sqrt{h^2 - \frac{1}{4}k^2}$.

$d = d(V, F) = h - c \Rightarrow d = h - \sqrt{h^2 - \frac{1}{4}k^2}$ and

$d' = d(V, F') = h + c \Rightarrow d' = h + \sqrt{h^2 - \frac{1}{4}k^2}$.

(b) From part (a), $d' = h + c \Rightarrow c = d' - h = 32 - 17 = 15$. $c = \sqrt{h^2 - \frac{1}{4}k^2} \Rightarrow$

$15 = \sqrt{17^2 - \frac{1}{4}k^2} \Rightarrow 225 = 289 - \frac{1}{4}k^2 \Rightarrow \frac{1}{4}k^2 = 64 \Rightarrow k^2 = 256 \Rightarrow$

$k = 16$ cm. $d = h - c = 17 - 15 = 2 \Rightarrow F$ should be located 2 cm from V.

51 $c^2 = (\frac{1}{2} \cdot 50)^2 - 15^2 = 625 - 225 = 400 \Rightarrow c = 20$.

Their feet should be $25 - 20 = 5$ ft from the vertices.

53 First determine an equation of the ellipse for the orbit of Earth.

$e = \frac{c}{a} \Rightarrow c = ae = 0.093 \times 149.6 = 13.9128$.

$b^2 = a^2 - c^2 = 149.6^2 - 13.9128^2 \Rightarrow b \approx 148.95 \approx 149.0$.

An approximate equation for the orbit of Earth is $\dfrac{x^2}{149.6^2} + \dfrac{y^2}{149.0^2} = 1$.

Solving for y gives us $\dfrac{y^2}{149^2} = 1 - \dfrac{x^2}{149.6^2} \Rightarrow y^2 = 149^2\left(1 - \dfrac{x^2}{149.6^2}\right) \Rightarrow$

$y = \pm 149\sqrt{1 - \dfrac{x^2}{149.6^2}}$. Graph $Y_1 = 149\sqrt{1 - (x^2/149.6^2)}$ and $Y_2 = -Y_1$.

The sun is at $(\pm 13.9128, 0)$. Plot the point $(13.9128, 0)$ for the sun.

[−300, 300, 100] by [−200, 200, 100]   [−6, 6] by [−2, 6]

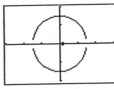

Figure 53

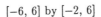
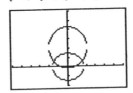

Figure 55

55 $\dfrac{x^2}{2.9} + \dfrac{y^2}{2.1} = 1 \Rightarrow \dfrac{y^2}{2.1} = 1 - \dfrac{x^2}{2.9} \Rightarrow y^2 = 2.1\left(1 - \dfrac{x^2}{2.9}\right) \Rightarrow y = \pm\sqrt{2.1(1 - x^2/2.9)}$.

$\dfrac{x^2}{4.3} + \dfrac{(y - 2.1)^2}{4.9} = 1 \Rightarrow \dfrac{(y - 2.1)^2}{4.9} = 1 - \dfrac{x^2}{4.3} \Rightarrow (y - 2.1)^2 = 4.9\left(1 - \dfrac{x^2}{4.3}\right) \Rightarrow$

$y - 2.1 = \pm\sqrt{4.9(1 - x^2/4.3)} \Rightarrow y = 2.1 \pm \sqrt{4.9(1 - x^2/4.3)}$.

From the graph, the points of intersection are approximately $(\pm 1.540, 0.618)$.

57 $\dfrac{(x+0.1)^2}{1.7} + \dfrac{y^2}{0.9} = 1 \;\Rightarrow\; \dfrac{y^2}{0.9} = 1 - \dfrac{(x+0.1)^2}{1.7} \;\Rightarrow\; y^2 = 0.9[1 - (x+0.1)^2/1.7] \;\Rightarrow$

$$y = \pm\sqrt{0.9[1 - (x+0.1)^2/1.7]}.$$

$\dfrac{x^2}{0.9} + \dfrac{(y-0.25)^2}{1.8} = 1 \;\Rightarrow\; \dfrac{(y-0.25)^2}{1.8} = 1 - \dfrac{x^2}{0.9} \;\Rightarrow\; (y-0.25)^2 = 1.8\left(1 - \dfrac{x^2}{0.9}\right) \;\Rightarrow$

$$y - 0.25 = \pm\sqrt{1.8(1 - x^2/0.9)} \;\Rightarrow\; y = 0.25 \pm \sqrt{1.8(1 - x^2/0.9)}.$$

$[-3, 3]$ by $[-2, 2]$

From the graph, the points of intersection are
approximately $(-0.88, 0.76)$, $(-0.48, -0.91)$,
$(0.58, -0.81)$, and $(0.92, 0.59)$.

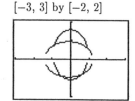

Figure 57

10.3 Exercises

Note: Let C, V, F, and W denote the center, the vertices, the foci, and the endpoints of
the conjugate axis, respectively. Let c denote the distance from the center of the
hyperbola to a focus.

1 $\dfrac{x^2}{9} - \dfrac{y^2}{4} = 1$ • The hyperbola will have a right branch and a left branch since the

term containing x is positive. The vertices will be on the horizontal transverse axis,

$\pm\sqrt{9} = \pm 3$ units from the center. The endpoints of the vertical conjugate axis are

$(0, \pm\sqrt{4}) = (0, \pm 2)$. The asymptotes have equations

$$y = \pm\left[\frac{\sqrt{\text{constant under } y^2}}{\sqrt{\text{constant under } x^2}}\right](x) \quad \text{or} \quad y = \pm\left[\frac{\frac{1}{2}(\text{vertical axis length})}{\frac{1}{2}(\text{horizontal axis length})}\right](x).$$

Note that the terms are "vertical" and "horizontal" and not transverse and conjugate
since the latter can be either vertical or horizontal. In this case, we have

$y = \pm\dfrac{\sqrt{4}}{\sqrt{9}}(x) = \pm\dfrac{2}{3}x.$ The positive sign corresponds to the asymptote with positive

slope and the negative sign corresponds to the asymptote with negative slope.
To find the foci, it is helpful to remember the relationship

$$\left[\tfrac{1}{2}(\text{transverse axis})\right]^2 + \left[\tfrac{1}{2}(\text{conjugate axis})\right]^2 = \left[c\right]^2,$$

which is the same as $c^2 = (\text{constant under } x^2) + (\text{constant under } y^2)$. Thus,
$c^2 = 9 + 4 \;\Rightarrow\; c = \pm\sqrt{13}.$ $\qquad$ (continued)

$V(\pm 3,\, 0);\ F(\pm \sqrt{13},\, 0);\ W(0,\ \pm 2);\ y = \pm \tfrac{2}{3}x$

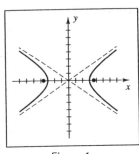

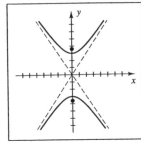

Figure 1   *Figure 3*

$\boxed{3}\ \dfrac{y^2}{9} - \dfrac{x^2}{4} = 1$ • The hyperbola will have an upper branch and a lower branch since the term containing y is positive. The vertices will be on the vertical transverse axis, $\pm \sqrt{9} = \pm 3$ units from the center. The endpoints of the horizontal conjugate axis are $(\pm \sqrt{4},\, 0) = (\pm 2,\, 0)$. The asymptotes have equations $y = \pm \dfrac{\sqrt{9}}{\sqrt{4}}(x) = \pm \tfrac{3}{2}x$.

Using the foci relationship given in Exercise 1, we have $c^2 = 9 + 4 \ \Rightarrow\ c = \pm \sqrt{13}$.

$$V(0,\ \pm 3);\ F(0,\ \pm \sqrt{13});\ W(\pm 2,\, 0);\ y = \pm \tfrac{3}{2}x$$

$\boxed{5}\ x^2 - \dfrac{y^2}{24} = 1\ \Leftrightarrow\ \dfrac{x^2}{1} - \dfrac{y^2}{24} = 1$ • The asymptotes have equations $y = \pm \dfrac{\sqrt{24}}{\sqrt{1}}(x) =$

$\pm \sqrt{24}x$. For the foci: $c^2 = 1 + 24 \ \Rightarrow\ c = \pm 5$.

$$V(\pm 1,\, 0);\ F(\pm 5,\, 0);\ W(0,\ \pm \sqrt{24});\ y = \pm \sqrt{24}x$$

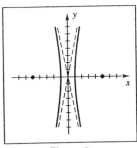

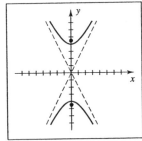

Figure 5   *Figure 7*

$\boxed{7}\ y^2 - 4x^2 = 16$ {divide by 16} $\Rightarrow\ \dfrac{y^2}{16} - \dfrac{x^2}{4} = 1$ • The vertices will be $\pm \sqrt{16} = \pm 4$ units from the center on the vertical transverse axis. The endpoints of the horizontal conjugate axis are $(\pm \sqrt{4},\, 0) = (\pm 2,\, 0)$. The asymptotes have equations $y = \pm \dfrac{\sqrt{16}}{\sqrt{4}}(x) = \pm \tfrac{4}{2}x = \pm 2x$. For the foci: $c^2 = 4 + 16 \ \Rightarrow$

$c^2 = 20 \ \Rightarrow\ c = \pm \sqrt{20} \ \Rightarrow\ c = \pm 2\sqrt{5}$.

$$V(0,\ \pm 4);\ F(0,\ \pm 2\sqrt{5});\ W(\pm 2,\, 0);\ y = \pm 2x$$

⑨ $16x^2 - 36y^2 = 1$ {rewrite in standard form} $\Rightarrow \frac{x^2}{\frac{1}{16}} - \frac{y^2}{\frac{1}{36}} = 1$.

The asymptotes have equations $y = \pm \frac{\sqrt{\frac{1}{36}}}{\sqrt{\frac{1}{16}}}(x) = \pm \frac{\frac{1}{6}}{\frac{1}{4}}(x) = \pm \frac{4}{6}(x) = \pm \frac{2}{3}x$.

For the foci: $c^2 = \frac{1}{16} + \frac{1}{36} \Rightarrow c^2 = \frac{9}{144} + \frac{4}{144} \Rightarrow c^2 = \frac{13}{144} \Rightarrow c = \pm \frac{1}{12}\sqrt{13}$.

$$V(\pm \tfrac{1}{4}, 0); \; F(\pm \tfrac{1}{12}\sqrt{13}, 0); \; W(0, \pm \tfrac{1}{6}); \; y = \pm \tfrac{2}{3}x$$

Note that the branches of the hyperbola almost coincide with the asymptotes.

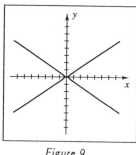

Figure 9

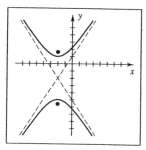

Figure 11

⑪ $\frac{(y+2)^2}{9} - \frac{(x+2)^2}{4} = 1$ •   The effect of the factors $(x+2)$ and $(y+2)$ is to shift

the center of the hyperbola from $(0, 0)$ to $(-2, -2)$. Since the term involving y is

positive, the transverse axis will be vertical. The vertices are 3 units in either

direction of the center. Their coordinates are $(-2, -2 \pm 3)$ or equivalently, $(-2, 1)$

and $(-2, -5)$. The conjugate axis will be horizontal with endpoints $(-2 \pm 2, -2)$, or,

equivalently, $(0, -2)$ and $(-4, -2)$. Remember that c is the distance *from the center*

to a focus. $c^2 = 9 + 4 \Rightarrow c = \pm\sqrt{13}$.   Hence, the coordinates of the foci are

$(-2, -2 \pm \sqrt{13})$. If the center of the hyperbola was at the origin, we would have

asymptote equations $y = \pm \frac{3}{2}x$. Since the center of the hyperbola has been shifted by

the factors $(y+2)$ and $(x+2)$, we can shift the asymptote equations using the same

factors. Hence, these equations are $(y+2) = \pm \frac{3}{2}(x+2)$.

$$C(-2, -2); \; V(-2, -2 \pm 3); \; F(-2, -2 \pm \sqrt{13}); \; W(-2 \pm 2, -2)$$

13 $144x^2 - 25y^2 + 864x - 100y - 2404 = 0 \Rightarrow$

$144(x^2 + 6x + \underline{9\ }) - 25(y^2 + 4y + \underline{4\ }) = 2404 + \underline{1296\ } - \underline{100\ } \Rightarrow$

$144(x+3)^2 - 25(y+2)^2 = 3600 \Rightarrow \dfrac{(x+3)^2}{25} - \dfrac{(y+2)^2}{144} = 1.$

$c^2 = 25 + 144 \Rightarrow c = \pm 13.$

$C(-3, -2); V(-3 \pm 5, -2); F(-3 \pm 13, -2); W(-3, -2 \pm 12); (y+2) = \pm\frac{12}{5}(x+3)$

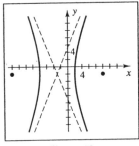

Figure 13

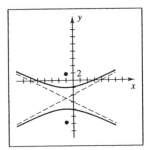

Figure 15

15 $4y^2 - x^2 + 40y - 4x + 60 = 0 \Rightarrow$

$4(y^2 + 10y + \underline{25\ }) - 1(x^2 + 4x + \underline{4\ }) = -60 + \underline{100\ } - \underline{4\ } \Rightarrow$

$4(y+5)^2 - (x+2)^2 = 36 \Rightarrow \dfrac{(y+5)^2}{9} - \dfrac{(x+2)^2}{36} = 1.$

$c^2 = 9 + 36 \Rightarrow c = \pm 3\sqrt{5}.$

$C(-2, -5); V(-2, -5 \pm 3); F(-2, -5 \pm 3\sqrt{5}); W(-2 \pm 6, -5); (y+5) = \pm\frac{1}{2}(x+2)$

Note: Let b denote the distance equal to $\frac{1}{2}$(conjugate axis length) and

let a denote the distance equal to $\frac{1}{2}$(transverse axis length).

17 $a = 3$ and $c = 5 \Rightarrow b^2 = c^2 - a^2 = 16.$ $\dfrac{x^2}{a^2} - \dfrac{y^2}{b^2} = 1$ is then $\dfrac{x^2}{9} - \dfrac{y^2}{16} = 1.$

19 The center of the hyperbola is $(-2, -3)$. $a = 1$ and $c = 2 \Rightarrow b^2 = 2^2 - 1^2 = 3$ and

an equation is $\dfrac{(y+3)^2}{1^2} - \dfrac{(x+2)^2}{3} = 1$, or, equivalently, $(y+3)^2 - \dfrac{(x+2)^2}{3} = 1.$

21 $F(0, \pm 4) \Rightarrow c = 4.$ $V(0, \pm 1) \Rightarrow a = 1.$ $a^2 + b^2 = c^2 \Rightarrow b^2 = 4^2 - 1^2 = 15.$

Since the vertices are on the y-axis, the "1^2" is associated with the y^2 term.

An equation is $\dfrac{y^2}{1} - \dfrac{x^2}{15} = 1.$

23 $F(\pm 5, 0)$ and $V(\pm 3, 0) \Rightarrow b^2 = c^2 - a^2 = 5^2 - 3^2 = 16 \Rightarrow W(0, \pm 4).$

An equation is $\dfrac{x^2}{9} - \dfrac{y^2}{16} = 1.$

25 Conjugate axis of length 4 implies that $b = 2.$ $F(0, \pm 5) \Rightarrow c = 5.$

$a^2 + b^2 = c^2 \Rightarrow a^2 = 5^2 - 2^2 = 21.$ An equation is $\dfrac{y^2}{21} - \dfrac{x^2}{4} = 1.$

$\boxed{27}$ Since the asymptote equations are $y = \pm 2x$ and we know that the point $(3, 0)$ is on the hyperbola, we conclude that the upper right corner of the rectangle formed by the transverse and conjugate axes has coordinates $(3, 6)$ {substitute $x = 3$ in $y = 2x$ to obtain the 6}. Thus we have endpoints of the conjugate axis at $(0, \pm 6)$.

$$\text{An equation is } \frac{x^2}{3^2} - \frac{y^2}{6^2} = 1, \text{ or } \frac{x^2}{9} - \frac{y^2}{36} = 1.$$

$\boxed{29}$ $a = 5$, $b = 2(5) = 10$. $\frac{x^2}{5^2} - \frac{y^2}{10^2} = 1 \Rightarrow \frac{x^2}{25} - \frac{y^2}{100} = 1$.

$\boxed{31}$ Since the transverse axis is vertical, the y^2 term will be positive.

$$\text{An equation is } \frac{y^2}{(\frac{1}{2} \cdot 10)^2} - \frac{x^2}{(\frac{1}{2} \cdot 14)^2} = 1, \text{ or } \frac{y^2}{25} - \frac{x^2}{49} = 1.$$

$\boxed{33}$ $\frac{1}{3}(x + 2) = y^2$ • We have a first degree x-term and a second degree y-term, so this is a parabola with a horizontal axis.

$\boxed{35}$ $x^2 + 6x - y^2 = 7$ • The coefficient of x^2 is positive and the coefficient of y^2 is negative, so this is a hyperbola.

$\boxed{37}$ $-x^2 = y^2 - 25 \Leftrightarrow x^2 + y^2 = 25$, a circle with center at the origin and radius 5.

$\boxed{39}$ $4x^2 - 16x + 9y^2 + 36y = -16$ • The coefficients of x^2 and y^2 are positive, so this is an ellipse.

$\boxed{41}$ $x^2 + 3x = 3y - 6$ • We have a first degree y-term and a second degree x-term, so this is a parabola with a vertical axis.

$\boxed{43}$ The graph of $y^2 - 4x^2 = 16$, or, equivalently, $\frac{y^2}{16} - \frac{x^2}{4} = 1$,

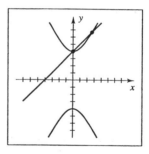

is that of a hyperbola with y-intercepts ± 4. The graph of $y - x = 4$, or, equivalently, $y = x + 4$, is that of a line with y-intercept 4 and slope 1. Substituting $y = x + 4$ into $y^2 - 4x^2 = 16$ yields $(x + 4)^2 - 4x^2 = 16 \Rightarrow$ $3x^2 - 8x = 0 \Rightarrow x(3x - 8) = 0 \Rightarrow x = 0, \frac{8}{3}; y = 4, \frac{20}{3}$.

The two points of intersection are $(0, 4)$ and $(\frac{8}{3}, \frac{20}{3})$.

Figure 43

$\boxed{45}$ Refer to the definition of a hyperbola. As in the discussion on page 788, we will let the positive constant, k, equal $2a$. $k = 24 \Rightarrow 2a = 24 \Rightarrow a = 12$.

$F(13, 0)$ and $F'(-13, 0) \Rightarrow c = 13$. $b^2 = c^2 - a^2 = 13^2 - 12^2 = 169 - 144 = 25$.

$$\text{An equation is } \frac{x^2}{144} - \frac{y^2}{25} = 1.$$

$\boxed{47}$ $k = 16 \Rightarrow 2a = 16 \Rightarrow a = 8$. $F(0, 10)$ and $F'(0, -10) \Rightarrow c = 10$.

$b^2 = c^2 - a^2 = 10^2 - 8^2 = 100 - 64 = 36$. $\qquad$ $\text{An equation is } \frac{y^2}{64} - \frac{x^2}{36} = 1$.

49 $x = \frac{5}{4}\sqrt{y^2 + 16} \Rightarrow \frac{4}{5}x = \sqrt{y^2 + 16} \Rightarrow \frac{16}{25}x^2 = y^2 + 16 \Rightarrow \frac{16}{25}x^2 - y^2 = 16 \Rightarrow$

$\frac{x^2}{25} - \frac{y^2}{16} = 1$, which is an equation of a hyperbola with right and left branches.

Since $x > 0$ in the original equation, its graph is the right branch of the hyperbola.

51 $y = \frac{3}{7}\sqrt{x^2 + 49} \Rightarrow \frac{7}{3}y = \sqrt{x^2 + 49} \Rightarrow \frac{49}{9}y^2 = x^2 + 49 \Rightarrow \frac{y^2}{9} - \frac{x^2}{49} = 1.$

Since $y > 0$ in the original equation, its graph is the upper branch of the hyperbola.

53 $y = -\frac{9}{4}\sqrt{x^2 - 16} \Rightarrow -\frac{4}{9}y = \sqrt{x^2 - 16} \Rightarrow \frac{16}{81}y^2 = x^2 - 16 \Rightarrow \frac{x^2}{16} - \frac{y^2}{81} = 1.$

Since $y \leq 0$ in the original equation,

its graph is the lower halves of the branches of the hyperbola.

55 $x = -\frac{2}{3}\sqrt{y^2 - 36} \Rightarrow -\frac{3}{2}x = \sqrt{y^2 - 36} \Rightarrow \frac{9}{4}x^2 = y^2 - 36 \Rightarrow \frac{y^2}{36} - \frac{x^2}{16} = 1.$

Since $x \leq 0$ in the original equation,

its graph is the left halves of the branches of the hyperbola.

57 Their equations are $\frac{x^2}{25} - \frac{y^2}{9} = 1$ and $\frac{x^2}{25} - \frac{y^2}{9} = -1$,

or, equivalently, $\frac{y^2}{9} - \frac{x^2}{25} = 1.$

Conjugate hyperbolas have the same asymptotes

and exchange transverse and conjugate axes.

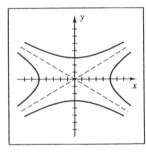

Figure 57

59 The path is a hyperbola with $V(\pm 3, 0)$ and $W(0, \pm\frac{3}{2})$.

An equation is $\frac{x^2}{(3)^2} - \frac{y^2}{(\frac{3}{2})^2} = 1$ or equivalently, $x^2 - 4y^2 = 9.$

If only the right branch is considered, then $x = \sqrt{9 + 4y^2}$ is an equation of the path.

61 Set up a coordinate system like the one in Example 6. Let the origin be located on

the shoreline halfway between A and B and let $P(x, y)$ denote the coordinates of the

ship. The coordinates of A and B (which can be thought of as the foci of the

hyperbola) are $(-100, 0)$ and $(100, 0)$, respectively. Hence, $c = 100$. As in Exercise

45, the difference in distances is a constant—that is, $d(P, A) - d(P, B) = 2a = 160$

$\Rightarrow a = 80.$ $b^2 = c^2 - a^2 = 100^2 - 80^2 \Rightarrow b = 60.$ An equation of the hyperbola is

$\frac{x^2}{80^2} - \frac{y^2}{60^2} = 1.$ Now, $y = 100 \Rightarrow \frac{x^2}{80^2} = 1 + \frac{100^2}{60^2} \Rightarrow x^2 = 80^2 \cdot \frac{13{,}600}{60^2} \Rightarrow$

$x = 80 \cdot \frac{10}{60}\sqrt{136} = \frac{80}{3}\sqrt{34}.$ The ship's coordinates are $(\frac{80}{3}\sqrt{34}, 100) \approx (155.5, 100).$

63 $\dfrac{(y-0.1)^2}{1.6} - \dfrac{(x+0.2)^2}{0.5} = 1 \;\Rightarrow\; \dfrac{(y-0.1)^2}{1.6} = 1 + \dfrac{(x+0.2)^2}{0.5} \;\Rightarrow$

$(y-0.1)^2 = 1.6[1+(x+0.2)^2/0.5] \;\Rightarrow\; y - 0.1 = \pm\sqrt{1.6[1+(x+0.2)^2/0.5]} \;\Rightarrow$

$$y = 0.1 \pm \sqrt{1.6[1+(x+0.2)^2/0.5]}.$$

$\dfrac{(y-0.5)^2}{2.7} - \dfrac{(x-0.1)^2}{5.3} = 1 \;\Rightarrow\; \dfrac{(y-0.5)^2}{2.7} = 1 + \dfrac{(x-0.1)^2}{5.3} \;\Rightarrow$

$(y-0.5)^2 = 2.7[1+(x-0.1)^2/5.3] \;\Rightarrow\; y - 0.5 = \pm\sqrt{2.7[1+(x-0.1)^2/5.3]} \;\Rightarrow$

$$y = 0.5 \pm \sqrt{2.7[1+(x-0.1)^2/5.3]}.$$

From the graph,

the point of intersection in the first quadrant is approximately $(0.741, 2.206)$.

[−15, 15] by [−10, 10]   [−15, 15, 2] by [−10, 10, 2]

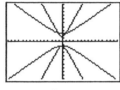   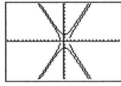

Figure 63   *Figure 65*

65 $\dfrac{(x-0.3)^2}{1.3} - \dfrac{y^2}{2.7} = 1 \;\Rightarrow\; \dfrac{(x-0.3)^2}{1.3} - 1 = \dfrac{y^2}{2.7} \;\Rightarrow\; y^2 = 2.7[-1+(x-0.3)^2/1.3] \;\Rightarrow$

$$y = \pm\sqrt{2.7[-1+(x-0.3)^2/1.3]}.$$

$\dfrac{y^2}{2.8} - \dfrac{(x-0.2)^2}{1.2} = 1 \;\Rightarrow\; \dfrac{y^2}{2.8} = 1 + \dfrac{(x-0.2)^2}{1.2} \;\Rightarrow\; y^2 = 2.8[1+(x-0.2)^2/1.2] \;\Rightarrow$

$$y = \pm\sqrt{2.8[1+(x-0.2)^2/1.2]}.$$

The two graphs nearly intersect in the second and fourth quadrants,

but there are no points of intersection.

67 (a) The comet's path is hyperbolic with $a^2 = 26 \times 10^{14}$ and $b^2 = 18 \times 10^{14}$.

$c^2 = a^2 + b^2 = 26 \times 10^{14} + 18 \times 10^{14} = 44 \times 10^{14} \;\Rightarrow\; c \approx 6.63 \times 10^7.$

The coordinates of the sun are approximately $(6.63 \times 10^7, 0)$.

(b) The minimum distance between the comet and the sun will be

$$c - a = \sqrt{44 \times 10^{14}} - \sqrt{26 \times 10^{14}} = 1.53 \times 10^7 \text{ mi.}$$

Since r must be in meters, 1.53×10^7 mi $\times$ 1610 m/mi $\approx 2.47 \times 10^{10}$ m. At this

distance, v must be greater than $\sqrt{\dfrac{2k}{r}} \approx \sqrt{\dfrac{2(1.325 \times 10^{20})}{2.47 \times 10^{10}}} \approx 103{,}600$ m/sec.

$\boxed{1}$ For this exercise (and others), we solve for t in terms of x, and then substitute that

expression for t in the equation that relates y and t.

$x = t - 2 \;\Rightarrow\; t = x + 2.$ $y = 2t + 3 = 2(x + 2) + 3 = 2x + 7.$

As t varies from 0 to 5, $(x,\, y)$ varies from $(-2,\, 3)$ to $(3,\, 13)$.

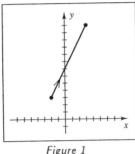

Figure 1

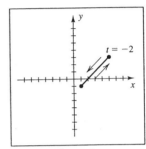

Figure 3

$\boxed{3}$ $x = t^2 + 1 \;\Rightarrow\; t^2 = x - 1.$ $y = t^2 - 1 = x - 2.$ As t varies from -2 to 2,

$(x,\, y)$ varies from $(5,\, 3)$ to $(1,\, -1)$ { when $t = 0$ } and back to $(5,\, 3)$.

$\boxed{5}$ Since y is linear in t, it is easier to solve the second equation for t than it is to solve

the first equation for t. Hence, we solve for t in terms of y, and then substitute that

expression for t in the equation that relates x and t. $y = 2t + 3 \;\Rightarrow\; t = \frac{1}{2}(y - 3).$

$x = 4\left[\frac{1}{2}(y - 3)\right]^2 - 5 \;\Rightarrow\; (y - 3)^2 = x + 5.$ This is a parabola with vertex at $(-5,\, 3)$.

Since t takes on all real values, so does y, and the curve C is the entire parabola.

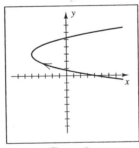

Figure 5

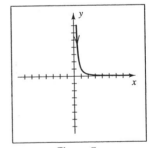

Figure 7

$\boxed{7}$ $y = e^{-2t} = (e^t)^{-2} = x^{-2} = 1/x^2.$ This is a rational function. Only the first quadrant

portion is used and as t varies from $-\infty$ to ∞, x varies from 0 to ∞, excluding 0.

9 $x = 2\sin t$ and $y = 3\cos t \Rightarrow \frac{x}{2} = \sin t$ and $\frac{y}{3} = \cos t \Rightarrow$

$\frac{x^2}{4} + \frac{y^2}{9} = \sin^2 t + \cos^2 t = 1$. As t varies from 0 to 2π,

(x, y) traces the ellipse from $(0, 3)$ in a clockwise direction back to $(0, 3)$.

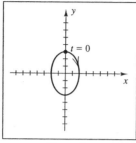

Figure 9

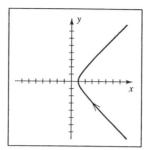

Figure 11

11 $x = \sec t$ and $y = \tan t \Rightarrow x^2 - y^2 = \sec^2 t - \tan^2 t = 1$.

As t varies from $-\frac{\pi}{2}$ to $\frac{\pi}{2}$, (x, y) traces the right branch of the hyperbola along the

asymptote $y = -x$ to $(1, 0)$ and then along the asymptote $y = x$.

13 $y = 2\ln t = \ln t^2 \{\text{since } t > 0\} = \ln x$.

As t varies from 0 to ∞, so does x, and y varies from $-\infty$ to ∞.

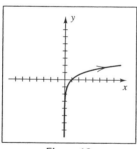

Figure 13

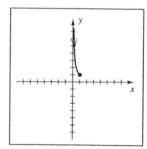

Figure 15

15 $y = \csc t = \frac{1}{\sin t} = \frac{1}{x}$.

As t varies from 0 to $\frac{\pi}{2}$, (x, y) varies asymptotically from the positive y-axis to $(1, 1)$.

17 $x = t$ and $y = \sqrt{t^2 - 1}$ $\Rightarrow$ $y = \sqrt{x^2 - 1}$ $\Rightarrow$ $x^2 - y^2 = 1$.

    Since y is nonnegative, the graph is the top half of both branches of the hyperbola.

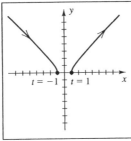

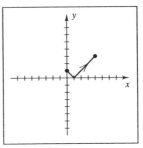

Figure 17   *Figure 19*

19 $x = t$ and $y = \sqrt{t^2 - 2t + 1}$ $\Rightarrow$ $y = \sqrt{x^2 - 2x + 1} = \sqrt{(x-1)^2} = |x - 1|$.

    As t varies from 0 to 4, (x, y) traces $y = |x - 1|$ from $(0, 1)$ to $(4, 3)$.

21 $x = (t + 1)^3$ $\Rightarrow$ $t = x^{1/3} - 1$. $y = (t + 2)^2 = (x^{1/3} + 1)^2$.

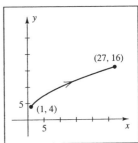

    This is probably an unfamiliar graph. The graph of $y = x^{1/3}$ is similar to the graph of $y = x^{1/2}$ $\{y = \sqrt{x}\}$ but is symmetric with respect to the origin. The "+1" shifts $y = x^{1/3}$ up 1 unit and the "squaring" makes all y values nonnegative. Since we have restrictions on the variable t, we only have a portion of this graph. As t varies from 0 to 2, (x, y) varies from $(1, 4)$ to $(27, 16)$.

Figure 21

23 All of the curves are a portion of the parabola $x = y^2$.

    C_1: $x = t^2 = y^2$. y takes on all real values and we have the entire parabola.

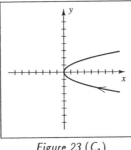

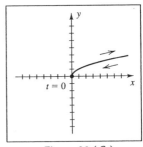

Figure 23 (C_1)   *Figure 23 (C_2)*

    C_2: $x = t^4 = (t^2)^2 = y^2$. C_2 is only the top half since $y = t^2$ is nonnegative.

    As t varies from $-\infty$ to ∞, the top portion is traced twice.

C_3: $x = \sin^2 t = (\sin t)^2 = y^2$. C_3 is the portion of the curve from $(1, -1)$ to $(1, 1)$.

The point $(1, 1)$ is reached at $t = \frac{\pi}{2} + 2\pi n$ and the point $(1, -1)$ when

$$t = \frac{3\pi}{2} + 2\pi n.$$

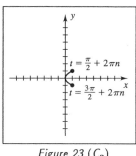

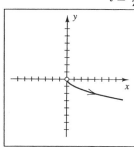

Figure 23 (C_3)   *Figure 23 (C_4)*

C_4: $x = e^{2t} = (e^t)^2 = (-e^t)^2 = y^2$. C_4 is the bottom half of the parabola since y

is negative. As t approaches $-\infty$, the parabola approaches the origin.

$\boxed{25}$ In each part, the motion is on the unit circle since $x^2 + y^2 = 1$.

(a) For $0 \le t \le \pi$, x { $\cos t$ } varies from 1 to -1 and y { $\sin t$ } is nonnegative.

$P(x, y)$ moves from $(1, 0)$ counterclockwise to $(-1, 0)$.

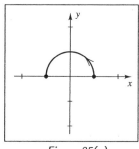

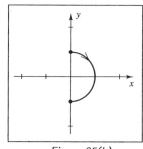

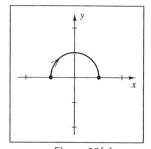

Figure 25(a)   *Figure 25(b)*   *Figure 25(c)*

(b) For $0 \le t \le \pi$, y { $\cos t$ } varies from 1 to -1 and x { $\sin t$ } is nonnegative.

$P(x, y)$ moves from $(0, 1)$ clockwise to $(0, -1)$.

(c) For $-1 \le t \le 1$, x { t } varies from -1 to 1 and y { $\sqrt{1 - t^2}$ } is nonnegative.

$P(x, y)$ moves from $(-1, 0)$ clockwise to $(1, 0)$.

$\boxed{27}$ $x = a \cos t + h$ and $y = b \sin t + k$ $\Rightarrow$ $\frac{x - h}{a} = \cos t$ and $\frac{y - k}{b} = \sin t$ $\Rightarrow$

$\frac{(x - h)^2}{a^2} + \frac{(y - k)^2}{b^2} = \cos^2 t + \sin^2 t = 1$. This is the equation of an ellipse with

center (h, k) and semiaxes of lengths a and b (axes of lengths $2a$ and $2b$).

29 Some choices for parts (a) and (b) are given—there are an infinite number of choices.

   (a) (1) $x = t$, $y = t^2$; $t \in \mathbb{R}$

   Letting x equal t is usually the simplest choice we can make.

   (2) $x = \tan t$, $y = \tan^2 t$; $-\frac{\pi}{2} < t < \frac{\pi}{2}$

   (3) $x = t^3$, $y = t^6$; $t \in \mathbb{R}$

   (b) (1) $x = e^t$, $y = e^{2t}$; $t \in \mathbb{R}$

   This choice only gives $x > 0$ since e^t is always positive for $t \in \mathbb{R}$.

   (2) $x = \sin t$, $y = \sin^2 t$; $t \in \mathbb{R}$

   This choice only gives $-1 \le x \le 1$ since $-1 \le \sin t \le 1$ for $t \in \mathbb{R}$.

   (3) $x = \tan^{-1} t$, $y = (\tan^{-1} t)^2$; $t \in \mathbb{R}$

   This choice only gives $-\frac{\pi}{2} < x < \frac{\pi}{2}$ since $-\frac{\pi}{2} < \tan^{-1} t < \frac{\pi}{2}$ for $t \in \mathbb{R}$.

31 We use $x(t) = (s\cos\alpha)t$ and $y(t) = -\frac{1}{2}gt^2 + (s\sin\alpha)t + h$ with $s = 256\sqrt{3}$, $\alpha = 60°$, and $h = 400$ to give us

$$x = 256\sqrt{3}(\tfrac{1}{2})t \quad \text{and} \quad y = -\tfrac{1}{2}(32)t^2 + 256\sqrt{3}(\sqrt{3}/2)t + 400 \quad \Rightarrow$$

$$x = 128\sqrt{3}t \quad \text{and} \quad y = -16t^2 + 384t + 400.$$

To find the range, we let $y = 0 \Rightarrow -16t^2 + 384t + 400 = 0 \Rightarrow$
$t^2 - 24t - 25 = 0$ { divide by -16 } $\Rightarrow (t-25)(t+1) = 0 \Rightarrow t = 25$ seconds, and
then substitute 25 for t in the equation for x. $x = 128\sqrt{3}(25) = 3200\sqrt{3} \approx 5542.56$
feet. To find the maximum altitude, we'll determine the values of t for which the
height is 400, and then use the symmetry property of a parabola to find the vertex.
$y = 400 \Rightarrow -16t^2 + 384t + 400 = 400 \Rightarrow -16t^2 + 384t = 0 \Rightarrow$
$-16t(t - 24) = 0 \Rightarrow t = 0, \; 24$.   Thus, the maximum altitude occurs when
$t = \frac{1}{2}(24) = 12$. This value of h is $y = -16(12)^2 + 384(12) + 400 = 2704$ feet.

33 $x = 704(\sqrt{2}/2)t$ and $y = -\frac{1}{2}(32)t^2 + 704(\sqrt{2}/2)t + 0 \Rightarrow$

$$x = 352\sqrt{2}\,t \quad \text{and} \quad y = -16t^2 + 352\sqrt{2}\,t.$$

$y = 0 \Rightarrow -16t^2 + 352\sqrt{2}\,t = 0 \Rightarrow -16t(t - 22\sqrt{2}) = 0 \Rightarrow t = 22\sqrt{2}$ seconds.

$$x = 352\sqrt{2}(22\sqrt{2}) = 15{,}488 \text{ feet.}$$

The maximum altitude occurs when $t = \frac{1}{2}(22\sqrt{2}) = 11\sqrt{2}$ and has value

$$y = -16(11\sqrt{2})^2 + 352\sqrt{2}(11\sqrt{2}) = 3872 \text{ feet.}$$

35 (a) $x = a \sin \omega t$ and $y = b \cos \omega t$ $\Rightarrow$ $\frac{x}{a} = \sin \omega t$ and $\frac{y}{b} = \cos \omega t$ $\Rightarrow$ $\frac{x^2}{a^2} + \frac{y^2}{b^2} = 1$.

The figure is an ellipse with center $(0, 0)$ and axes of lengths $2a$ and $2b$.

(b) $f(t + p) = a \sin[\omega_1(t + p)] = a \sin[\omega_1 t + \omega_1 p] = a \sin[\omega_1 t + 2\pi n] =$

$$a \sin \omega_1 t = f(t).$$

$g(t + p) = b \cos[\omega_2(t + p)] = b \cos[\omega_2 t + \frac{\omega_2}{\omega_1} 2\pi n] = b \cos[\omega_2 t + \frac{m}{n} 2\pi n] =$

$$b \cos[\omega_2 t + 2\pi m] = b \cos \omega_2 t = g(t).$$

Since f and g are periodic with period p,

the curve retraces itself every p units of time.

37 (a) Let $x = 3 \sin(240\pi t)$ and $y = 4 \sin(240\pi t)$ for $0 \le t \le 0.01$.

(b) From the graph, $y_{\text{int}} = 0$ and $y_{\max} = 4$.

Thus, the phase difference is $\phi = \sin^{-1} \frac{y_{\text{int}}}{y_{\max}} = \sin^{-1} \frac{0}{4} = 0°$.

$[-9, 9]$ by $[-6, 6]$ $[-120, 120, 10]$ by $[-80, 80, 10]$

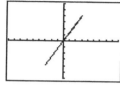

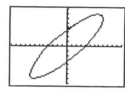

Figure 37   Figure 39

39 (a) Let $x = 80 \sin(60\pi t)$ and $y = 70 \cos(60\pi t - \pi/3)$ for $0 \le t \le 0.035$.

(b) See *Figure 39*. From the graph, $y_{\text{int}} = 35$ and $y_{\max} = 70$.

Thus, the phase difference is $\phi = \sin^{-1} \frac{y_{\text{int}}}{y_{\max}} = \sin^{-1} \frac{35}{70} = 30°$.

Note: Tstep must be sufficiently small to obtain the maximum of 70 on the

graph. $70 \cos(60\pi t - \pi/3) = 70 \sin(60\pi t + \pi/6)$

41 $x(t) = \sin(6\pi t)$, $y(t) = \cos(5\pi t)$ for $0 \le t \le 2$

$[-1, 1, 0.5]$ by $[-1, 1, 0.5]$

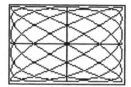

Figure 41

43 Let $\theta = \angle FDP$ and $\alpha = \angle GDP = \angle EDP$. Then $\angle ODG = \left(\frac{\pi}{2} - t\right)$ and

$\alpha = \theta - \left(\frac{\pi}{2} - t\right) = \theta + t - \frac{\pi}{2}$. Arcs AF and PF are equal in length since each is

the distance rolled. Thus, $at = b\theta$, or $\theta = \left(\frac{a}{b}\right)t$ and $\alpha = \frac{a+b}{b}t - \frac{\pi}{2}$.

Note that $\cos \alpha = \sin\left(\frac{a+b}{b}t\right)$ and $\sin \alpha = -\cos\left(\frac{a+b}{b}t\right)$.

For the location of the points as illustrated, the coordinates of P are:

$x = d(O,\,G) + d(G,\,B) = d(O,\,G) + d(E,\,P) = (a+b)\cos t + b\,\sin\alpha$

$$= (a+b)\cos t - b\,\cos\left(\frac{a+b}{b}t\right).$$

$y = d(B,\,P) = d(G,\,D) - d(D,\,E) = (a+b)\sin t - b\,\cos\alpha$

$$= (a+b)\sin t - b\,\sin\left(\frac{a+b}{b}t\right).$$

It can be verified that these equations are valid for all locations of the points.

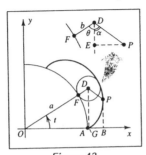

Figure 43

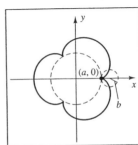

Figure 45

45 $b = \frac{1}{3}a \;\Rightarrow\; a = 3b$. Substituting into the equations from Exercise 43 yields:

$$x = (3b + b)\cos t - b\,\cos\left(\frac{3b+b}{b}t\right) = 4b\,\cos t - b\,\cos 4t$$

$$y = (3b + b)\sin t - b\,\sin\left(\frac{3b+b}{b}t\right) = 4b\,\sin t - b\,\sin 4t$$

As an aid in graphing, to determine where the path of the smaller circle will intersect the path of the larger circle (for the original starting point of intersection at $A(a,\,0)$), we can solve $x^2 + y^2 = a^2$ for t.

$x^2 + y^2 = 16b^2 \cos^2 t - 8b^2 \cos t \cos 4t + b^2 \cos^2 4t +$

$$16b^2 \sin^2 t - 8b^2 \sin t \sin 4t + b^2 \sin^2 4t$$

$$= 17b^2 - 8b^2 (\cos t \cos 4t + \sin t \sin 4t)$$

$$= 17b^2 - 8b^2 [\cos(t - 4t)] = 17b^2 - 8b^2 \cos 3t.$$

Thus, $x^2 + y^2 = a^2 \;\Rightarrow\; 17b^2 - 8b^2 \cos 3t = a^2 = 9b^2 \;\Rightarrow\; 8b^2 = 8b^2 \cos 3t \;\Rightarrow$

$1 = \cos 3t \;\Rightarrow\; 3t = 2\pi n \;\Rightarrow\; t = \frac{2\pi}{3}n$.

It follows that the intersection points are at $t = \frac{2\pi}{3}, \frac{4\pi}{3}$, and 2π.

47 Change "Param" from "Function" under ⎡MODE⎤. Make the assignments $3(\sin T)^5$ to X_{1T}, $3(\cos T)^5$ to Y_{1T}, 0 to Tmin, 6.28 to Tmax, and 0.105 to Tstep. Algebraically, we have $x = 3\sin^5 t$ and $y = 3\cos^5 t \Rightarrow \frac{x}{3} = \sin^5 t$ and $\frac{y}{3} = \cos^5 t \Rightarrow \left(\frac{x}{3}\right)^{2/5} + \left(\frac{y}{3}\right)^{2/5} = \sin^2 t + \cos^2 t = 1$. The graph traces an astroid.

$[-6, 6]$ by $[-4, 4]$ 　　　　　　$[-30, 30, 5]$ by $[-20, 20, 5]$

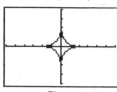　　　　　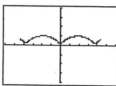

Figure 47 　　　　　　　　*Figure 49*

49 The graph traces a curtate cycloid.

51 The figure is a mask with a mouth, nose, and eyes. This graph may be obtained with a graphing utility that has the capability to graph 5 sets of parametric equations.

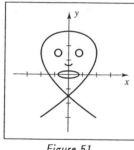　　　　　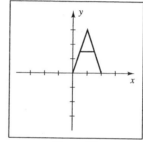

Figure 51 　　　　　　　　*Figure 53*

53 C_1 is the line $y = 3x$ from $(0, 0)$ to $(1, 3)$. For C_2, $x - 1 = \tan t$ and $1 - \frac{1}{3}y = \tan t \Rightarrow x - 1 = 1 - \frac{1}{3}y \Rightarrow y = -3x + 6$. This line is sketched from $(1, 3)$ to $(2, 0)$. C_3 is the horizontal line $y = \frac{3}{2}$ from $\left(\frac{1}{2}, \frac{3}{2}\right)$ to $\left(\frac{3}{2}, \frac{3}{2}\right)$. The figure is the letter A.

10.5 Exercises

Note: For the following exercises, the substitutions $y = r\sin\theta$, $x = r\cos\theta$, $r^2 = x^2 + y^2$, and $\tan\theta = \frac{y}{x}$ are used without mention. We have found it helpful to find the "pole" values {when the graph intersects the pole} to determine which values of θ should be used in the construction of an r-θ chart. The numbers listed on each line of the r-θ chart correspond to the numbers labeled on the figures.

$\boxed{1}$ (a) Since $\frac{7\pi}{3}$ is coterminal with $\frac{\pi}{3}$, $(3, \frac{7\pi}{3})$ represents the same point as $(3, \frac{\pi}{3})$.

(b) The point $(3, -\frac{\pi}{3})$ is in QIV, not QI, as is $(3, \frac{\pi}{3})$.

(c) The angle $\frac{4\pi}{3}$ is π radians larger than $\frac{\pi}{3}$, so its terminal side is on the same line as the terminal side of the angle $\frac{\pi}{3}$. $r = -3$ in the $\frac{4\pi}{3}$ direction is equivalent to $r = 3$ in the $\frac{\pi}{3}$ direction, so $(-3, \frac{4\pi}{3})$ represents the same point as $(3, \frac{\pi}{3})$.

(d) The point $(3, -\frac{2\pi}{3})$ is diametrically opposite the point $(3, \frac{\pi}{3})$.

(e) From part (d), we deduce that the point $(-3, -\frac{2\pi}{3})$ would represent the same point as $(3, \frac{\pi}{3})$.

(f) The point $(-3, -\frac{\pi}{3})$ is in QII, not QI.

Thus, choices (a), (c), and (e) represent the same point as $(3, \pi/3)$.

$\boxed{3}$ Use (1) in the "relationship between rectangular and polar coordinates."

(a) $x = r\cos\theta = 3\cos\frac{\pi}{4} = 3\left(\frac{\sqrt{2}}{2}\right) = \frac{3}{2}\sqrt{2}$. $y = r\sin\theta = 3\sin\frac{\pi}{4} = 3\left(\frac{\sqrt{2}}{2}\right) = \frac{3}{2}\sqrt{2}$.

Hence, rectangular coordinates for $(3, \frac{\pi}{4})$ are $(\frac{3}{2}\sqrt{2}, \frac{3}{2}\sqrt{2})$.

(b) $x = -1\cos\frac{2\pi}{3} = -1(-\frac{1}{2}) = \frac{1}{2}$. $y = -1\sin\frac{2\pi}{3} = -1\left(\frac{\sqrt{3}}{2}\right) = -\frac{1}{2}\sqrt{3}$.

$\boxed{5}$ (a) $x = 8\cos\left(-\frac{2\pi}{3}\right) = 8(-\frac{1}{2}) = -4$. $y = 8\sin\left(-\frac{2\pi}{3}\right) = 8\left(-\frac{\sqrt{3}}{2}\right) = -4\sqrt{3}$.

(b) $x = -3\cos\frac{5\pi}{3} = -3(\frac{1}{2}) = -\frac{3}{2}$. $y = -3\sin\frac{5\pi}{3} = -3\left(-\frac{\sqrt{3}}{2}\right) = \frac{3}{2}\sqrt{3}$.

$\boxed{7}$ Let $\theta = \arctan\frac{3}{4}$. Then we have $\cos\theta = \frac{4}{5}$ and $\sin\theta = \frac{3}{5}$.

$$x = 6\cos\theta = 6(\tfrac{4}{5}) = \tfrac{24}{5}.\quad y = 6\sin\theta = 6(\tfrac{3}{5}) = \tfrac{18}{5}.$$

$\boxed{9}$ Use (2) in the "relationship between rectangular and polar coordinates."

(a) $r^2 = x^2 + y^2 = (-1)^2 + (1)^2 = 2 \;\Rightarrow\; r = \sqrt{2}$ { since we want $r > 0$ }.

$$\tan\theta = \frac{y}{x} = \frac{1}{-1} = -1 \;\Rightarrow\; \theta = \tfrac{3\pi}{4}\ \{\theta \text{ in QII}\}.$$

(b) $r^2 = (-2\sqrt{3})^2 + (-2)^2 = 16 \;\Rightarrow\; r = \sqrt{16} = 4$.

$$\tan\theta = \frac{-2}{-2\sqrt{3}} = \frac{1}{\sqrt{3}} \;\Rightarrow\; \theta = \tfrac{7\pi}{6}\ \{\theta \text{ in QIII}\}.$$

$\boxed{11}$ (a) $r^2 = 7^2 + (-7\sqrt{3})^2 = 196 \;\Rightarrow\; r = \sqrt{196} = 14$.

$$\tan\theta = \frac{-7\sqrt{3}}{7} = -\sqrt{3} \;\Rightarrow\; \theta = \tfrac{5\pi}{3}\ \{\theta \text{ in QIV}\}.$$

(b) $r^2 = 5^2 + 5^2 = 50 \;\Rightarrow\; r = \sqrt{50} = 5\sqrt{2}$. $\tan\theta = \frac{5}{5} = 1 \;\Rightarrow\; \theta = \frac{\pi}{4}\ \{\theta \text{ in QI}\}$.

$\boxed{13}$ $x = -3 \;\Rightarrow\; r\cos\theta = -3 \;\Rightarrow\; r = \frac{-3}{\cos\theta} \;\Rightarrow\; r = -3\sec\theta$

$\boxed{15}$ $x^2 + y^2 = 16 \;\Rightarrow\; r^2 = 16 \;\Rightarrow\; r = \pm 4$ { both are circles with radius 4 }.

$\boxed{17}$ $y^2 = 6x$ $\Rightarrow$ $r^2 \sin^2\theta = 6r \cos\theta$ $\Rightarrow$ $r^2 \sin^2\theta - 6r \cos\theta = 0$ $\Rightarrow$

$r(r \sin^2\theta - 6 \cos\theta) = 0$ $\Rightarrow$ $r \sin^2\theta - 6 \cos\theta = 0$ $\Rightarrow$ $r \sin^2\theta = 6 \cos\theta$ $\Rightarrow$

$r = \dfrac{6 \cos\theta}{\sin^2\theta} = 6 \cdot \dfrac{\cos\theta}{\sin\theta} \cdot \dfrac{1}{\sin\theta} = 6 \cot\theta \csc\theta.$

Note that $r = 0$ { the pole } is included in the graph of $r = 6 \cot\theta \csc\theta.$

$\boxed{19}$ $x + y = 3$ $\Rightarrow$ $r\cos\theta + r\sin\theta = 3$ $\Rightarrow$ $r(\cos\theta + \sin\theta) = 3$ $\Rightarrow$ $r = \dfrac{3}{\cos\theta + \sin\theta}$

$\boxed{21}$ $2y = -x$ $\Rightarrow$ $\dfrac{y}{x} = -\dfrac{1}{2}$ $\Rightarrow$ $\tan\theta = -\dfrac{1}{2}$ $\Rightarrow$ $\theta = \tan^{-1}\left(-\dfrac{1}{2}\right)$

$\boxed{23}$ $y^2 - x^2 = 4$ $\Rightarrow$ $r^2 \sin^2\theta - r^2 \cos^2\theta = 4$ $\Rightarrow$ $-r^2(\cos^2\theta - \sin^2\theta) = 4$ $\Rightarrow$

$$-r^2 \cos 2\theta = 4 \ \Rightarrow \ r^2 = \dfrac{-4}{\cos 2\theta} \ \Rightarrow \ r^2 = -4\sec 2\theta$$

$\boxed{25}$ $(x - 1)^2 + y^2 = 1$ $\Rightarrow$ $x^2 - 2x + 1 + y^2 = 1$ $\Rightarrow$ $x^2 + y^2 = 2x$ $\Rightarrow$

$$r^2 = 2r \cos\theta \ \Rightarrow \ r = 2\cos\theta$$

$\boxed{27}$ $r \cos\theta = 5$ $\Rightarrow$ $x = 5.$ This is a vertical line with x-intercept $(5, 0).$

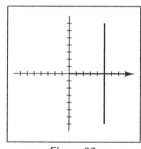

Figure 27

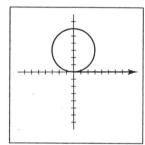

Figure 29

$\boxed{29}$ $r - 6\sin\theta = 0$ $\Rightarrow$ $r^2 = 6r \sin\theta$ $\Rightarrow$ $x^2 + y^2 = 6y$ $\Rightarrow$ $x^2 + y^2 - 6y + \underline{9} = \underline{9}$ $\Rightarrow$

$$x^2 + (y - 3)^2 = 9.$$

$\boxed{31}$ $\theta = \dfrac{\pi}{4}$ $\Rightarrow$ $\tan\theta = \tan\dfrac{\pi}{4}$ $\Rightarrow$ $\dfrac{y}{x} = 1$ $\Rightarrow$ $y = x.$

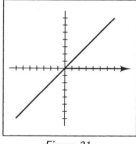

Figure 31

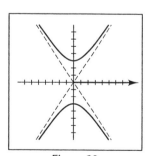

Figure 33

$\boxed{33}$ $r^2(4\sin^2\theta - 9\cos^2\theta) = 36$ $\Rightarrow$ $4r^2 \sin^2\theta - 9r^2 \cos^2\theta = 36$ $\Rightarrow$

$$4y^2 - 9x^2 = 36 \ \Rightarrow \ \dfrac{y^2}{9} - \dfrac{x^2}{4} = 1.$$

35 $r^2\cos 2\theta = 1 \ \Rightarrow \ r^2(\cos^2\theta - \sin^2\theta) = 1 \ \Rightarrow \ r^2\cos^2\theta - r^2\sin^2\theta = 1 \ \Rightarrow \ x^2 - y^2 = 1.$

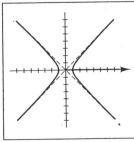

Figure 35

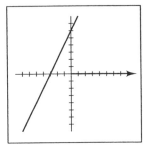

Figure 37

37 $r(\sin\theta - 2\cos\theta) = 6 \ \Rightarrow \ r\sin\theta - 2r\cos\theta = 6 \ \Rightarrow \ y - 2x = 6.$

39 $r(\sin\theta + r\cos^2\theta) = 1 \ \Rightarrow \ r\sin\theta + r^2\cos^2\theta = 1 \ \Rightarrow \ y + x^2 = 1, \text{ or } y = -x^2 + 1.$

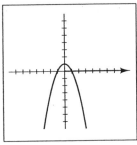

Figure 39

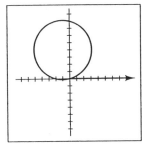

Figure 41

41 $r = 8\sin\theta - 2\cos\theta \ \Rightarrow \ r^2 = 8r\sin\theta - 2r\cos\theta \ \Rightarrow \ x^2 + y^2 = 8y - 2x \ \Rightarrow$

$$x^2 + 2x + \underline{1} + y^2 - 8y + \underline{16} = \underline{1} + \underline{16} \ \Rightarrow \ (x+1)^2 + (y-4)^2 = 17.$$

43 $r = \tan\theta \ \Rightarrow \ r^2 = \tan^2\theta \ \Rightarrow \ x^2 + y^2 = \dfrac{y^2}{x^2} \ \Rightarrow$

$x^4 + x^2 y^2 = y^2 \ \Rightarrow \ y^2 - x^2 y^2 = x^4 \ \Rightarrow$

$y^2(1 - x^2) = x^4 \ \Rightarrow \ y^2 = \dfrac{x^4}{1 - x^2}.$

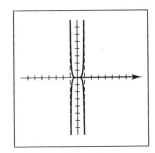

Figure 43

This is a tough one even after getting the equation in x and y. Since we have solved for y^2, the right side must be positive. Since x^4 is always nonnegative, we must have $1 - x^2 > 0$, or equivalently, $|x| < 1$.

We have vertical asymptotes at $x = \pm 1$, that is, when the denominator is 0.

45 $r = 5 \Rightarrow r^2 = 25 \Rightarrow x^2 + y^2 = 25$, a circle centered at the origin with radius 5.

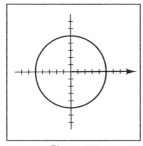

Figure 45

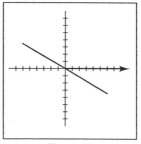

Figure 47

47 $\theta = -\frac{\pi}{6} \Rightarrow \tan\theta = \tan\left(-\frac{\pi}{6}\right) \Rightarrow \frac{y}{x} = -\frac{1}{\sqrt{3}} \Rightarrow y = -\frac{1}{3}\sqrt{3}\,x.$

This is a line through the origin with slope $-\frac{1}{3}\sqrt{3}$.

49 $r = 3\cos\theta \Rightarrow \{\text{multiply by } r \text{ to obtain } r^2\}$

$r^2 = 3r\cos\theta \Rightarrow x^2 + y^2 = 3x \Rightarrow$

$\{\text{recognize this as an equation of a circle and complete the square}\}$

$(x^2 - 3x + \frac{9}{4}) + y^2 = \frac{9}{4} \Rightarrow (x - \frac{3}{2})^2 + y^2 = \frac{9}{4}.$

This is a circle with center $(\frac{3}{2}, 0)$ and radius $\sqrt{\frac{9}{4}} = \frac{3}{2}$. From the table, we see that as θ varies from 0 to $\frac{\pi}{2}$, r will vary from 3 to 0. This corresponds to the portion of the circle in the first quadrant. As θ varies from $\frac{\pi}{2}$ to π, r varies from 0 to -3. Remember that -3 in the π direction is the same as 3 in the 0 direction. This corresponds to the portion of the circle in the fourth quadrant.

Variation of θ		Variation of r	
1) 0	$\to \frac{\pi}{2}$	$3 \to$	0
2) $\frac{\pi}{2}$	$\to \pi$	$0 \to$	-3

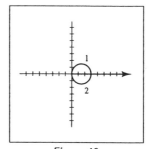

Figure 49

51 $r = 4\cos\theta + 2\sin\theta \Rightarrow r^2 = 4r\cos\theta + 2r\sin\theta \Rightarrow$

$x^2 + y^2 = 4x + 2y \Rightarrow$

$x^2 - 4x + \underline{4} + y^2 - 2y + \underline{1} = \underline{4} + \underline{1} \Rightarrow$

$(x - 2)^2 + (y - 1)^2 = 5.$

Variation of θ		Variation of r	
1) 0	$\to \frac{\pi}{2}$	$4 \to$	2
2) $\frac{\pi}{2}$	$\to \pi$	$2 \to$	-4

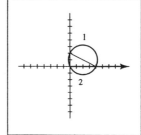

Figure 51

53 $r = 4(1 - \sin\theta)$ is a cardioid since the coefficient of $\sin\theta$

has the same magnitude as the constant term.

$0 = 4(1 - \sin\theta) \Rightarrow \sin\theta = 1 \Rightarrow \theta = \frac{\pi}{2} + 2\pi n$. The "v"

in the heart-shaped curve corresponds to the pole value $\frac{\pi}{2}$.

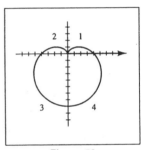

Figure 53

Variation of θ			Variation of r	
1)	0	$\to$ $\frac{\pi}{2}$	$4 \to$	0
2)	$\frac{\pi}{2}$	$\to$ π	$0 \to$	4
3)	π	$\to$ $\frac{3\pi}{2}$	$4 \to$	8
4)	$\frac{3\pi}{2}$	$\to$ 2π	$8 \to$	4

55 $r = -6(1 + \cos\theta)$ is a cardioid. $0 = -6(1 + \cos\theta) \Rightarrow$

$\cos\theta = -1 \Rightarrow \theta = \pi + 2\pi n$. The "v" in the

heart-shaped curve corresponds to the pole value π.

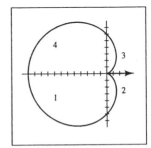

Figure 55

Variation of θ			Variation of r	
1)	0	$\to$ $\frac{\pi}{2}$	$-12 \to$	-6
2)	$\frac{\pi}{2}$	$\to$ π	$-6 \to$	0
3)	π	$\to$ $\frac{3\pi}{2}$	$0 \to$	-6
4)	$\frac{3\pi}{2}$	$\to$ 2π	$-6 \to$	-12

57 $r = 2 + 4\sin\theta$ is a limaçon with a loop since the constant term has a smaller

magnitude than the coefficient of $\sin\theta$. $0 = 2 + 4\sin\theta \Rightarrow \sin\theta = -\frac{1}{2} \Rightarrow$

$\theta = \frac{7\pi}{6} + 2\pi n, \frac{11\pi}{6} + 2\pi n$. Trace through the table and the figure to make sure you

understand what values of θ form the loop.

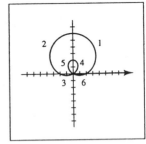

Figure 57

Variation of θ			Variation of r	
1)	0	$\to$ $\frac{\pi}{2}$	$2 \to$	6
2)	$\frac{\pi}{2}$	$\to$ π	$6 \to$	2
3)	π	$\to$ $\frac{7\pi}{6}$	$2 \to$	0
4)	$\frac{7\pi}{6}$	$\to$ $\frac{3\pi}{2}$	$0 \to$	-2
5)	$\frac{3\pi}{2}$	$\to$ $\frac{11\pi}{6}$	$-2 \to$	0
6)	$\frac{11\pi}{6}$	$\to$ 2π	$0 \to$	2

59 $r = \sqrt{3} - 2\sin\theta$ is a limaçon with a loop. $0 = \sqrt{3} - 2\sin\theta \Rightarrow \sin\theta = \sqrt{3}/2 \Rightarrow$

$\theta = \frac{\pi}{3} + 2\pi n, \frac{2\pi}{3} + 2\pi n.$ Let $a = \sqrt{3} - 2 \approx -0.27$ and $b = \sqrt{3} + 2 \approx 3.73.$

Variation of θ			Variation of r		
1)	0	$\rightarrow \frac{\pi}{3}$	$\sqrt{3}$	$\rightarrow$	0
2)	$\frac{\pi}{3}$	$\rightarrow \frac{\pi}{2}$	0	$\rightarrow$	a
3)	$\frac{\pi}{2}$	$\rightarrow \frac{2\pi}{3}$	a	$\rightarrow$	0
4)	$\frac{2\pi}{3}$	$\rightarrow \pi$	0	$\rightarrow$	$\sqrt{3}$
5)	π	$\rightarrow \frac{3\pi}{2}$	$\sqrt{3}$	$\rightarrow$	b
6)	$\frac{3\pi}{2}$	$\rightarrow 2\pi$	b	$\rightarrow$	$\sqrt{3}$

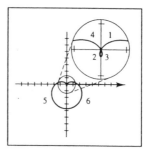

Figure 59

61 $r = 2 - \cos\theta$ •

$0 = 2 - \cos\theta \Rightarrow \cos\theta = 2 \Rightarrow$ no pole values.

Variation of θ			Variation of r		
1)	0	$\rightarrow \frac{\pi}{2}$	1	$\rightarrow$	2
2)	$\frac{\pi}{2}$	$\rightarrow \pi$	2	$\rightarrow$	3
3)	π	$\rightarrow \frac{3\pi}{2}$	3	$\rightarrow$	2
4)	$\frac{3\pi}{2}$	$\rightarrow 2\pi$	2	$\rightarrow$	1

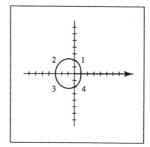

Figure 61

63 $r = 4\csc\theta \Rightarrow r\sin\theta = 4 \Rightarrow y = 4.$

r is undefined at $\theta = \pi n.$

This is a horizontal line with y-intercept $(0, 4)$.

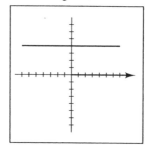

Figure 63

65 $r = 8\cos 3\theta$ is a 3-leafed rose since 3 is odd. $0 = 8\cos 3\theta \Rightarrow$

$\cos 3\theta = 0 \Rightarrow 3\theta = \frac{\pi}{2} + \pi n \Rightarrow \theta = \frac{\pi}{6} + \frac{\pi}{3}n.$

Variation of θ			Variation of r		
1)	0	$\rightarrow \frac{\pi}{6}$	8	$\rightarrow$	0
2)	$\frac{\pi}{6}$	$\rightarrow \frac{\pi}{3}$	0	$\rightarrow$	-8
3)	$\frac{\pi}{3}$	$\rightarrow \frac{\pi}{2}$	-8	$\rightarrow$	0
4)	$\frac{\pi}{2}$	$\rightarrow \frac{2\pi}{3}$	0	$\rightarrow$	8
5)	$\frac{2\pi}{3}$	$\rightarrow \frac{5\pi}{6}$	8	$\rightarrow$	0
6)	$\frac{5\pi}{6}$	$\rightarrow \pi$	0	$\rightarrow$	-8

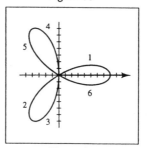

Figure 65

67 $r = 3\sin 2\theta$ is a 4-leafed rose. $0 = 3\sin 2\theta \Rightarrow \sin 2\theta = 0 \Rightarrow 2\theta = \pi n \Rightarrow \theta = \frac{\pi}{2}n.$

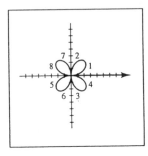

Figure 67

Variation of θ	Variation of r
1) $0 \to \frac{\pi}{4}$	$0 \to 3$
2) $\frac{\pi}{4} \to \frac{\pi}{2}$	$3 \to 0$
3) $\frac{\pi}{2} \to \frac{3\pi}{4}$	$0 \to -3$
4) $\frac{3\pi}{4} \to \pi$	$-3 \to 0$
5) $\pi \to \frac{5\pi}{4}$	$0 \to 3$
6) $\frac{5\pi}{4} \to \frac{3\pi}{2}$	$3 \to 0$
7) $\frac{3\pi}{2} \to \frac{7\pi}{4}$	$0 \to -3$
8) $\frac{7\pi}{4} \to 2\pi$	$-3 \to 0$

69 $r^2 = 4\cos 2\theta$ (lemniscate) •

$0 = 4\cos 2\theta \Rightarrow \cos 2\theta = 0 \Rightarrow 2\theta = \frac{\pi}{2} + \pi n \Rightarrow \theta = \frac{\pi}{4} + \frac{\pi}{2}n.$

Note that as θ varies from 0 to $\frac{\pi}{4}$, we have $r^2 = 4$ or $r = \pm 2$ to $r^2 = 0$—both parts labeled "1" are traced with this range of θ. When θ varies from $\frac{\pi}{4}$ to $\frac{3\pi}{4}$, 2θ varies from $\frac{\pi}{2}$ to $\frac{3\pi}{2}$, and $\cos 2\theta$ is negative. Since r^2 can't equal a negative value, no portion of the graph is traced for these values of θ.

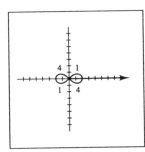

Figure 69

Variation of θ	Variation of r
1) $0 \to \frac{\pi}{4}$	$\pm 2 \to 0$
2) $\frac{\pi}{4} \to \frac{\pi}{2}$	undefined
3) $\frac{\pi}{2} \to \frac{3\pi}{4}$	undefined
4) $\frac{3\pi}{4} \to \pi$	$0 \to \pm 2$

71 $r = 2^\theta$, $\theta \geq 0$ (spiral) •

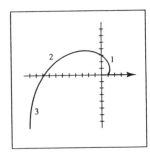

Figure 71

Variation of θ	Variation of r
1) $0 \to \frac{\pi}{2}$	$1 \to 2.97$
2) $\frac{\pi}{2} \to \pi$	$2.97 \to 8.82$
3) $\pi \to \frac{3\pi}{2}$	$8.82 \to 26.22$
4) $\frac{3\pi}{2} \to 2\pi$	$22.62 \to 77.88$

$\boxed{73}$ $r = 2\theta$, $\theta \geq 0$ •

Variation of θ			Variation of r		
1)	0	$\rightarrow$ $\frac{\pi}{2}$	0	$\rightarrow$	π
2)	$\frac{\pi}{2}$	$\rightarrow$ π	π	$\rightarrow$	2π
3)	π	$\rightarrow$ $\frac{3\pi}{2}$	2π	$\rightarrow$	3π

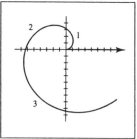

Figure 73

$\boxed{75}$ To simplify $\sin^2(\frac{\theta}{2})$, recall the half-angle identity for the

sine. $r = 6\sin^2\left(\frac{\theta}{2}\right) = 6\left(\frac{1 - \cos\theta}{2}\right) = 3(1 - \cos\theta)$ is a

cardioid. $0 = 3(1 - \cos\theta) \Rightarrow \cos\theta = 1 \Rightarrow \theta = 2\pi n$.

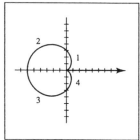

Variation of θ			Variation of r	
1)	0	$\rightarrow$ $\frac{\pi}{2}$	0 $\rightarrow$	3
2)	$\frac{\pi}{2}$	$\rightarrow$ π	3 $\rightarrow$	6
3)	π	$\rightarrow$ $\frac{3\pi}{2}$	6 $\rightarrow$	3
4)	$\frac{3\pi}{2}$	$\rightarrow$ 2π	3 $\rightarrow$	0

Figure 75

$\boxed{77}$ Note that $r = 2\sec\theta$ is equivalent to $x = 2$. If $0 < \theta < \frac{\pi}{2}$ or $\frac{3\pi}{2} < \theta < 2\pi$, then

$\sec\theta > 0$ and the graph of $r = 2 + 2\sec\theta$ is to the right of $x = 2$. If $\frac{\pi}{2} < \theta < \frac{3\pi}{2}$,

$\sec\theta < 0$ and $r = 2 + 2\sec\theta$ is to the left of $x = 2$. r is undefined at $\theta = \frac{\pi}{2} + \pi n$.

$0 = 2 + 2\sec\theta \Rightarrow \sec\theta = -1 \Rightarrow \theta = \pi + 2\pi n$.

Variation of θ			Variation of r	
1)	0	$\rightarrow$ $\frac{\pi}{2}$	4 $\rightarrow$	∞
2)	$\frac{\pi}{2}$	$\rightarrow$ π	$-\infty \rightarrow$	0
3)	π	$\rightarrow$ $\frac{3\pi}{2}$	0 $\rightarrow$	$-\infty$
4)	$\frac{3\pi}{2}$	$\rightarrow$ 2π	$\infty \rightarrow$	4

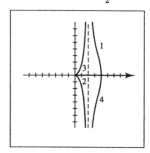

Figure 77

$\boxed{79}$ Let $P_1(r_1, \theta_1)$ and $P_2(r_2, \theta_2)$ be points in an $r\theta$-plane.

Let $a = r_1$, $b = r_2$, $c = d(P_1, P_2)$, and $\gamma = \theta_2 - \theta_1$.

Substituting into the law of cosines,

$c^2 = a^2 + b^2 - 2ab\cos\gamma$, gives us the formula.

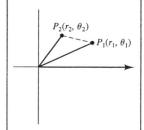

Figure 79

81 (a) $I = \frac{1}{2}I_0[1 + \cos(\pi \sin\theta)] \Rightarrow$

$r = 2.5[1 + \cos(\pi \sin\theta)]$ for $\theta \in [0, 2\pi]$.

(b) The signal is maximum in an east–west direction

and minimum in a north–south direction.

[−9, 9] by [−6, 6]

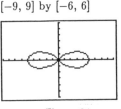

Figure 81

83 Change to "Pol" mode under $\boxed{\text{MODE}}$, assign $2(\sin\theta)^2(\tan\theta)^2$ to r1 under $\boxed{Y =}$,

and $-\pi/3$ to θmin, $\pi/3$ to θmax, and 0.04 to θstep under $\boxed{\text{WINDOW}}$. The graph is

symmetric with respect to the polar axis.

[−9, 9] by [−6, 6]

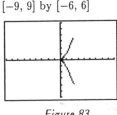

Figure 83

[−12, 12] by [−9, 9]

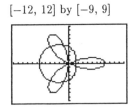

Figure 85

85 Assign $8\cos(3\theta)$ to r1, $4 - 2.5\cos\theta$ to r2, 0 to θmin, 2π to θmax, and $\pi/30$ to θstep.

From the graph, there are six points of intersection. The approximate polar

coordinates are $(1.75, \pm 0.45)$, $(4.49, \pm 1.77)$, and $(5.76, \pm 2.35)$.

10.6 Exercises

Note: For the ellipse, the major axis is vertical if the denominator contains $\sin\theta$,

horizontal if the denominator contains $\cos\theta$. For the hyperbola, the transverse

axis is vertical if the denominator contains $\sin\theta$, horizontal if the denominator

contains $\cos\theta$. The focus at the pole is called F and V is the vertex associated

with (or closest to) F. $d(V, F)$ denotes the distance from the vertex to the focus.

The foci are not asked for in the directions, but are listed. For the parabola, the

directrix is on the right, on the left, above, or below the focus depending on the

term "$+\cos$", "$-\cos$", "$+\sin$", or "$-\sin$", respectively, appearing in the

denominator.

$\boxed{1}$ Divide the numerator and denominator by the constant term in the denominator, i.e.,

6. $r = \dfrac{12}{6 + 2\sin\theta} = \dfrac{2}{1 + \frac{1}{3}\sin\theta} \;\Rightarrow\; e = \frac{1}{3} < 1$, ellipse. From the previous note, we see

that the denominator has "$+\sin\theta$" and we have vertices when $\theta = \frac{\pi}{2}$ and $\frac{3\pi}{2}$. They

are $V(\frac{3}{2}, \frac{\pi}{2})$ and $V'(3, \frac{3\pi}{2})$. The distance from the focus at the pole to the vertex V is

$\frac{3}{2}$. The distance from V' to F' must also be $\frac{3}{2}$ and we see that $F' = (\frac{3}{2}, \frac{3\pi}{2})$. We will

use the following notation to summarize this in future problems:

$$d(V, F) = \tfrac{3}{2} \;\Rightarrow\; F' = (\tfrac{3}{2}, \tfrac{3\pi}{2}).$$

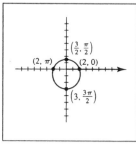

Figure 1

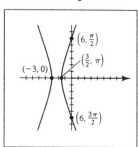

Figure 3

$\boxed{3}$ Divide both the numerator and the denominator by 2 to obtain the "1" in the

standard form. $r = \dfrac{12}{2 - 6\cos\theta} = \dfrac{6}{1 - 3\cos\theta} \;\Rightarrow\; e = 3 > 1$, hyperbola.

$$V(\tfrac{3}{2}, \pi) \text{ and } V'(-3, 0). \;\; d(V, F) = \tfrac{3}{2} \;\Rightarrow\; F' = (-\tfrac{9}{2}, 0).$$

$\boxed{5}$ $r = \dfrac{3}{2 + 2\cos\theta} = \dfrac{\frac{3}{2}}{1 + 1\cos\theta} \;\Rightarrow\; e = 1$, parabola. Note that the expression is

undefined in the $\theta = \pi$ direction. The vertex is in the $\theta = 0$ direction, $V(\frac{3}{4}, 0)$.

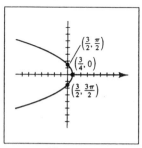

Figure 5

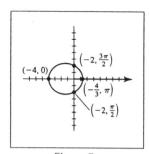

Figure 7

$\boxed{7}$ $r = \dfrac{4}{\cos\theta - 2} = \dfrac{-2}{1 - \frac{1}{2}\cos\theta} \;\Rightarrow\; e = \frac{1}{2} < 1$, ellipse.

$$V(-\tfrac{4}{3}, \pi) \text{ and } V'(-4, 0). \;\; d(V, F) = \tfrac{4}{3} \;\Rightarrow\; F' = (-\tfrac{8}{3}, 0)$$

$\boxed{9}$ We multiply by $\frac{\sin\theta}{\sin\theta}$ to obtain the standard form of a conic.

$$r = \frac{6\csc\theta}{2\csc\theta + 3} \cdot \frac{\sin\theta}{\sin\theta} = \frac{6}{2 + 3\sin\theta} = \frac{3}{1 + \frac{3}{2}\sin\theta} \;\Rightarrow\; e = \frac{3}{2} > 1, \text{ hyperbola.}$$

$V(\frac{6}{5}, \frac{\pi}{2})$ and $V'(-6, \frac{3\pi}{2})$. $d(V, F) = \frac{6}{5} \;\Rightarrow\; F' = (-\frac{36}{5}, \frac{3\pi}{2})$.

Since the original equation is undefined when $\csc\theta$ is undefined { which is

when $\theta = \pi n$ }, the points $(3, 0)$ and $(3, \pi)$ are excluded from the graph.

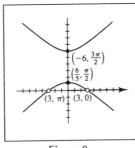

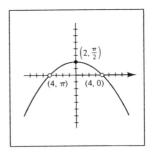

Figure 9   Figure 11

$\boxed{11}$ $r = \frac{4\csc\theta}{1 + \csc\theta} \cdot \frac{\sin\theta}{\sin\theta} = \frac{4}{1 + 1\sin\theta} \;\Rightarrow\; e = 1$, parabola. The vertex is in the $\theta = \frac{\pi}{2}$

direction, $V(2, \frac{\pi}{2})$. Since the original equation is undefined when $\csc\theta$ is undefined,

the points $(4, 0)$ and $(4, \pi)$ are excluded from the graph.

$\boxed{13}$ $r = \frac{12}{6 + 2\sin\theta} \;\Rightarrow\; 6r + 2r\sin\theta = 12 \;\Rightarrow\; 6r + 2y = 12 \;\Rightarrow$

{ Isolate the term with the "r" so that by squaring both sides of the equation,

we can make the substitution $r^2 = x^2 + y^2$. }

$3r = 6 - y \;\Rightarrow\; 9r^2 = 36 - 12y + y^2 \;\Rightarrow\; 9(x^2 + y^2) = 36 - 12y + y^2 \;\Rightarrow$
$$9x^2 + 8y^2 + 12y - 36 = 0$$

Note: For the following exercises, the substitutions

$x = r\cos\theta$, $y = r\sin\theta$, and $r^2 = x^2 + y^2$ are made without mention.

$\boxed{15}$ $r = \frac{12}{2 - 6\cos\theta} \;\Rightarrow\; 2r - 6x = 12 \;\Rightarrow\; r = 3x + 6 \;\Rightarrow\; r^2 = 9x^2 + 36x + 36 \;\Rightarrow$
$$8x^2 - y^2 + 36x + 36 = 0$$

$\boxed{17}$ $r = \frac{3}{2 + 2\cos\theta} \;\Rightarrow\; 2r + 2x = 3 \;\Rightarrow\; 2r = 3 - 2x \;\Rightarrow\; 4r^2 = 4x^2 - 12x + 9 \;\Rightarrow$
$$4y^2 + 12x - 9 = 0$$

$\boxed{19}$ $r = \frac{4}{\cos\theta - 2} \;\Rightarrow\; x - 2r = 4 \;\Rightarrow\; x - 4 = 2r \;\Rightarrow\; x^2 - 8x + 16 = 4r^2 \;\Rightarrow$
$$3x^2 + 4y^2 + 8x - 16 = 0$$

$\boxed{21}$ $r = \dfrac{6\csc\theta}{2\csc\theta + 3} \cdot \dfrac{\sin\theta}{\sin\theta} = \dfrac{6}{2 + 3\sin\theta}$ $\Rightarrow$ $2r + 3y = 6$ $\Rightarrow$ $2r = 6 - 3y$ $\Rightarrow$

$4r^2 = 36 - 36y + 9y^2$ $\Rightarrow$ $4x^2 - 5y^2 + 36y - 36 = 0$.

r is undefined when $\theta = 0$ or π. For the rectangular equation, these points

correspond to $y = 0$ (or $r\sin\theta = 0$). Substituting $y = 0$ into the above rectangular

equation yields $4x^2 = 36$, or $x = \pm 3$. $\therefore$ exclude $(\pm 3, 0)$

$\boxed{23}$ $r = \dfrac{4\csc\theta}{1 + \csc\theta} \cdot \dfrac{\sin\theta}{\sin\theta} = \dfrac{4}{1 + 1\sin\theta}$ $\Rightarrow$ $r + y = 4$ $\Rightarrow$ $r = 4 - y$ $\Rightarrow$

$r^2 = y^2 - 8y + 16$ $\Rightarrow$ $x^2 + 8y - 16 = 0$. r is undefined when $\theta = 0$ or π.

For the rectangular equation, these points correspond to $y = 0$ (or $r\sin\theta = 0$).

Substituting $y = 0$ into the above rectangular equation yields $x^2 = 16$, or $x = \pm 4$.

$\therefore$ exclude $(\pm 4, 0)$

$\boxed{25}$ $r = 2\sec\theta$ $\Rightarrow$ $r\cos\theta = 2$ $\Rightarrow$ $x = 2$. Remember that d is the distance from the

focus at the pole to the directrix. Thus, $d = 2$ and since the directrix is on the

right of the focus at the pole, we use "$+\cos\theta$". $r = \dfrac{2\left(\frac{1}{3}\right)}{1 + \frac{1}{3}\cos\theta} \cdot \dfrac{3}{3} = \dfrac{2}{3 + \cos\theta}$.

$\boxed{27}$ $r\cos\theta = -3$ $\Rightarrow$ $x = -3$. Thus, $d = 3$ and since the directrix is on the left of the

focus at the pole, we use "$-\cos\theta$". $r = \dfrac{3\left(\frac{4}{3}\right)}{1 - \frac{4}{3}\cos\theta} \cdot \dfrac{3}{3} = \dfrac{12}{3 - 4\cos\theta}$.

$\boxed{29}$ $r\sin\theta = -2$ $\Rightarrow$ $y = -2$. Thus, $d = 2$ and since the directrix is below the focus at

the pole, we use "$-\sin\theta$". $r = \dfrac{2(1)}{1 - 1\sin\theta} = \dfrac{2}{1 - \sin\theta}$.

$\boxed{31}$ $r = 4\csc\theta$ $\Rightarrow$ $r\sin\theta = 4$ $\Rightarrow$ $y = 4$. Thus, $d = 4$ and since the directrix is above

the focus at the pole, we use "$+\sin\theta$". $r = \dfrac{4\left(\frac{2}{5}\right)}{1 + \frac{2}{5}\sin\theta} \cdot \dfrac{5}{5} = \dfrac{8}{5 + 2\sin\theta}$.

$\boxed{33}$ For a parabola, $e = 1$. The vertex is 4 units above the focus at the pole,

so $d = 2(4)$ and we should use "$+\sin\theta$" in the denominator. $r = \dfrac{8}{1 + \sin\theta}$

35 (a) See *Figure 35*. $e = \frac{c}{a} = \frac{d(C, F)}{d(C, V)} = \frac{3}{4}$.

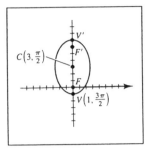

(b) Since the vertex is below the focus at the pole,

use "$-\sin\theta$". $r = \dfrac{d(\frac{3}{4})}{1 - \frac{3}{4}\sin\theta}$ and $r = 1$ when

$\theta = \frac{3\pi}{2} \Rightarrow 1 = \dfrac{d(\frac{3}{4})}{1 - \frac{3}{4}(-1)} \Rightarrow 1 = \dfrac{\frac{3}{4}d}{\frac{7}{4}} \Rightarrow$

$d = \frac{7}{3}$. Thus, $r = \dfrac{(\frac{7}{3})(\frac{3}{4})}{1 - \frac{3}{4}\sin\theta} \cdot \frac{4}{4} = \dfrac{7}{4 - 3\sin\theta}$.

Figure 35

An equivalent rectangular equation is $\dfrac{x^2}{7} + \dfrac{(y-3)^2}{16} = 1$.

37 (a) Let V and C denote the vertex closest to the sun and the center of the ellipse, respectively. Let s denote the distance from V to the directrix to the left of V.

$d(O, V) = d(C, V) - d(C, O) = a - c = a - ea = a(1 - e)$.

Also, by the first theorem in §10.6, $\dfrac{d(O, V)}{s} = e \Rightarrow s = \dfrac{d(O, V)}{e} = \dfrac{a(1-e)}{e}$.

Now, $d = s + d(O, V) = \dfrac{a(1-e)}{e} + a(1-e) = \dfrac{a(1-e^2)}{e}$ and $de = a(1-e^2)$.

Thus, the equation of the orbit is $r = \dfrac{(1-e^2)a}{1 - e\cos\theta}$.

(b) The minimum distance occurs when $\theta = \pi$. $r_{\text{per}} = \dfrac{(1-e^2)a}{1 - e(-1)} = a(1-e)$.

The maximum distance occurs when $\theta = 0$. $r_{\text{aph}} = \dfrac{(1-e^2)a}{1 - e(1)} = a(1+e)$.

39 (a) Since $e = 0.9673 < 1$, the orbit of Halley's Comet is elliptical.

(b) The polar equation for the orbit of Saturn is $r = \dfrac{9.006(1 + 0.056)}{1 - 0.056\cos\theta}$.

The polar equation for Halley's comet is $r = \dfrac{0.5871(1 + 0.9673)}{1 - 0.9673\cos\theta}$.

$[-36, 36, 3]$ by $[-24, 24, 3]$ $[-18, 18, 3]$ by $[-12, 12, 3]$

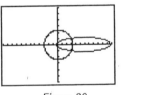

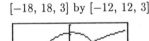
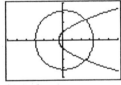

Figure 39   *Figure 41*

41 (a) Since $e = 1.003 > 1$, the orbit of Comet 1959 III is hyperbolic.

(b) The polar equation for Comet 1959 III is $r = \dfrac{1.251(1 + 1.003)}{1 - 1.003\cos\theta}$.

Chapter 10 Review Exercises

Note: Let the notation be the same as in §10.1–10.3.

$\boxed{1}$ $y^2 = 64x \;\Rightarrow\; x = \frac{1}{64}y^2 \;\Rightarrow\; a = \frac{1}{64}.\; p = \dfrac{1}{4(\frac{1}{64})} = 16.\;\; V(0,\,0);\; F(16,\,0);\; l{:}\,x = -16$

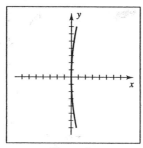

Figure 1

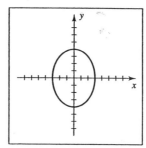

Figure 3

$\boxed{3}$ $9y^2 = 144 - 16x^2 \;\Rightarrow\; 16x^2 + 9y^2 = 144 \;\Rightarrow\; \dfrac{16x^2}{144} + \dfrac{9y^2}{144} = \dfrac{144}{144} \;\Rightarrow\; \dfrac{x^2}{9} + \dfrac{y^2}{16} = 1.$

$c^2 = 16 - 9 \;\Rightarrow\; c = \pm\sqrt{7}.$ $\qquad\qquad V(0,\,\pm 4);\; F(0,\,\pm\sqrt{7});\; M(\pm 3,\,0)$

$\boxed{5}$ $x^2 - y^2 - 4 = 0 \;\Rightarrow\; \dfrac{x^2}{4} - \dfrac{y^2}{4} = 1.\; c^2 = 4 + 4 \;\Rightarrow\; c = \pm\sqrt{8} \;\Rightarrow\; c = \pm 2\sqrt{2}.$

$V(\pm 2,\,0);\; F(\pm 2\sqrt{2},\,0);\; W(0,\,\pm 2);\; y = \pm x$

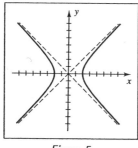

Figure 5

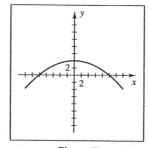

Figure 7

$\boxed{7}$ $25y = 100 - x^2 \;\Rightarrow\; y = 4 - \frac{1}{25}x^2 \;\Rightarrow\; a = -\frac{1}{25}.\; p = \dfrac{1}{4(-\frac{1}{25})} = -\dfrac{25}{4}.$

The vertex is at $(0,\,4)$ and the focus is $\frac{25}{4}$ units below the vertex. $4 - \frac{25}{4} = -\frac{9}{4}.$

The directrix is $\frac{25}{4}$ units above the vertex. $4 + \frac{25}{4} = \frac{41}{4}.$

$V(0,\,4);\; F(0,\,-\frac{9}{4});\; l{:}\,y = \frac{41}{4}$

$\boxed{9}$ $x^2 - 9y^2 + 8x + 90y - 210 = 0 \Rightarrow$

$(x^2 + 8x + \underline{16}) - 9(y^2 - 10y + \underline{25}) = 210 + \underline{16} - \underline{225} \Rightarrow$

$(x+4)^2 - 9(y-5)^2 = 1 \Rightarrow \dfrac{(x+4)^2}{1} - \dfrac{(y-5)^2}{\frac{1}{9}} = 1. \ c^2 = 1 + \frac{1}{9} \Rightarrow c = \pm\frac{1}{3}\sqrt{10}.$

$C(-4, 5); \ V(-4 \pm 1, 5); \ F(-4 \pm \frac{1}{3}\sqrt{10}, 5); \ W(-4, 5 \pm \frac{1}{3}); \ (y-5) = \pm\frac{1}{3}(x+4)$

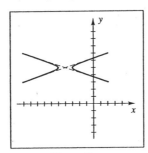

Figure 9

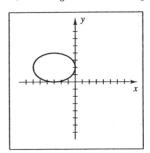

Figure 11

$\boxed{11}$ $4x^2 + 9y^2 + 24x - 36y + 36 = 0 \Rightarrow$

$4(x^2 + 6x + \underline{9}) + 9(y^2 - 4y + \underline{4}) = -36 + \underline{36} + \underline{36} \Rightarrow$

$4(x+3)^2 + 9(y-2)^2 = 36 \Rightarrow \dfrac{(x+3)^2}{9} + \dfrac{(y-2)^2}{4} = 1. \ c^2 = 9 - 4 \Rightarrow c = \pm\sqrt{5}.$

$C(-3, 2); \ V(-3 \pm 3, 2); \ F(-3 \pm \sqrt{5}, 2); \ M(-3, 2 \pm 2)$

$\boxed{13}$ $y^2 - 8x + 8y + 32 = 0 \Rightarrow x = \frac{1}{8}y^2 + y + 4 \Rightarrow a = \frac{1}{8}. \ p = \dfrac{1}{4(\frac{1}{8})} = 2.$

$V(2, -4); \ F(4, -4); \ l: x = 0$

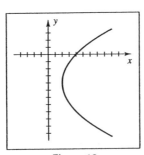

Figure 13

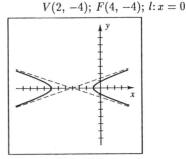

Figure 15

$\boxed{15}$ $x^2 - 9y^2 + 8x + 7 = 0 \Rightarrow$

$(x^2 + 8x + \underline{16}) - 9(y^2) = -7 + \underline{16} \Rightarrow (x+4)^2 - 9(y^2) = 9 \Rightarrow$

$\dfrac{(x+4)^2}{9} - \dfrac{y^2}{1} = 1. \ c^2 = 9 + 1 \Rightarrow c = \pm\sqrt{10}.$

$C(-4, 0); \ V(-4 \pm 3, 0); \ F(-4 \pm \sqrt{10}, 0); \ W(-4, 0 \pm 1); \ y = \pm\frac{1}{3}(x+4)$

$\boxed{17}$ The vertex is halfway between the x-intercepts, so it is of the form $V(-7, k)$.

$y = a(x+10)(x+4)$ and $x = 0, \ y = 80 \Rightarrow 80 = a(10)(4) \Rightarrow a = 2.$

$x = -7 \Rightarrow y = 2(3)(-3) = -18.$ Hence, $y = 2(x+7)^2 - 18.$

$\boxed{19}$ An equation of a hyperbola with vertices $V(0, \pm 7)$ and endpoints of the conjugate

axis $(\pm 3, 0)$ is $\dfrac{y^2}{7^2} - \dfrac{x^2}{3^2} = 1$ or $\dfrac{y^2}{49} - \dfrac{x^2}{9} = 1$.

$\boxed{21}$ $F(0, -10)$ and $l: y = 10 \;\Rightarrow\; p = -10$ and $V(0, 0)$.

$\qquad\qquad$ An equation is $(x - 0)^2 = [4(-10)](y - 0)$, or $x^2 = -40y$.

$\boxed{23}$ $V(0, \pm 10)$ and $F(0, \pm 5) \;\Rightarrow\; b^2 = 10^2 - 5^2 = 100 - 25 = 75$.

$\qquad\qquad$ An equation is $\dfrac{x^2}{75} + \dfrac{y^2}{10^2} = 1$ or $\dfrac{x^2}{75} + \dfrac{y^2}{100} = 1$.

$\boxed{25}$ Hyperbola, with vertices $V(0, \pm 6)$ and asymptotes $y = \pm 9x$ $\quad\bullet\quad$ The equations of

the asymptotes are $y = \pm \dfrac{a}{b}x$ and $a = 6$, so $9 = \dfrac{6}{b} \;\Rightarrow\; b = \dfrac{6}{9} = \dfrac{2}{3}$.

$\qquad\qquad$ An equation is $\dfrac{y^2}{6^2} - \dfrac{x^2}{(\frac{2}{3})^2} = 1$ or $\dfrac{y^2}{36} - \dfrac{x^2}{\frac{4}{9}} = 1$.

$\boxed{26}$ $F(\pm 2, 0) \;\Rightarrow\; c^2 = 4$. Now $\dfrac{x^2}{a^2} + \dfrac{y^2}{b^2} = 1$ can be written as $\dfrac{x^2}{a^2} + \dfrac{y^2}{a^2 - 4} = 1$ since

$b^2 = a^2 - c^2$. Substituting $x = 2$ and $y = \sqrt{2}$ into that equation yields

$\dfrac{4}{a^2} + \dfrac{2}{a^2 - 4} = 1 \;\Rightarrow\; 4a^2 - 16 + 2a^2 = a^4 - 4a^2 \;\Rightarrow\; a^4 - 10a^2 + 16 = 0 \;\Rightarrow$

$(a^2 - 2)(a^2 - 8) = 0 \;\Rightarrow\; a^2 = 2, 8$. Since $a > c$, a^2 must be 8 and b^2 is equal to 4.

$\qquad\qquad$ An equation is $\dfrac{x^2}{8} + \dfrac{y^2}{4} = 1$.

$\boxed{27}$ $M(\pm 5, 0) \;\Rightarrow\; b = 5$. $e = \dfrac{c}{a} = \dfrac{\sqrt{a^2 - b^2}}{a} = \dfrac{\sqrt{a^2 - 25}}{a} = \dfrac{2}{3} \;\Rightarrow\; \dfrac{2}{3}a = \sqrt{a^2 - 25} \;\Rightarrow$

$\dfrac{4}{9}a^2 = a^2 - 25 \;\Rightarrow\; \dfrac{5}{9}a^2 = 25 \;\Rightarrow\; a^2 = 45$. An equation is $\dfrac{x^2}{25} + \dfrac{y^2}{45} = 1$.

$\boxed{29}$ (a) Substituting $x = 2$ and $y = -3$ in $Ax^2 + 2y^2 = 4 \;\Rightarrow\; A(2)^2 + 2(-3)^2 = 4 \;\Rightarrow$

$\qquad\qquad 4A + 18 = 4 \;\Rightarrow\; 4A = -14 \;\Rightarrow\; A = -\dfrac{7}{2}$.

(b) The equation is $-\dfrac{7}{2}x^2 + 2y^2 = 4 \;\Rightarrow\; -\dfrac{7}{8}x^2 + \dfrac{2}{4}y^2 = 1 \;\Rightarrow\; \dfrac{y^2}{2} - \dfrac{7x^2}{8} = 1$,

$\qquad\qquad\qquad\qquad\qquad\qquad\qquad\qquad\qquad\qquad$ a hyperbola.

$\boxed{31}$ The focus is a distance of $p = \dfrac{1}{4a} = \dfrac{1}{4 \cdot \frac{1}{8}} = \dfrac{1}{\frac{1}{2}} = 2$ units from the origin. So the

center of the circle is $(0, 2)$ and since the circle passes through the origin, the radius

of the circle must be 2. An equation of the circle is $x^2 + (y - 2)^2 = 2^2 = 4$.

$\boxed{33}$ $y = t - 1$ $\Rightarrow$ $t = y + 1$. $x = 3 + 4t = 3 + 4(y + 1) = 4y + 7$.

As t varies from -2 to 2, (x, y) varies from $(-5, -3)$ to $(11, 1)$. See *Figure 33* below.

$\boxed{35}$ $x = \cos^2 t - 2$ $\Rightarrow$ $x + 2 = \cos^2 t$; $y = \sin t + 1$ $\Rightarrow$ $(y - 1)^2 = \sin^2 t$.

$\sin^2 t + \cos^2 t = 1 = x + 2 + (y - 1)^2$ $\Rightarrow$ $(y - 1)^2 = -(x + 1)$.   This is a parabola

with vertex at $(-1, 1)$ and opening to the left. $t = 0$ corresponds to the vertex and as

t varies from 0 to 2π, the point (x, y) moves to $(-2, 2)$ at $t = \frac{\pi}{2}$, back to the vertex

at $t = \pi$, down to $(-2, 0)$ at $t = \frac{3\pi}{2}$, and finishes at the vertex at $t = 2\pi$.

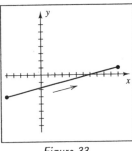
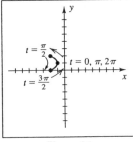
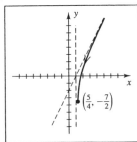

Figure 33	*Figure 35*	*Figure 37*

$\boxed{37}$ $x = \frac{1}{t} + 1$ $\Rightarrow$ $x - 1 = \frac{1}{t}$ $\Rightarrow$ $t = \frac{1}{x - 1}$ and

$$y = \frac{2}{t} - t = 2(x - 1) - \left(\frac{1}{x - 1} \right) = \frac{2(x^2 - 2x + 1) - 1}{x - 1} = \frac{2x^2 - 4x + 1}{x - 1}.$$

This is a rational function with a vertical asymptote at $x = 1$ and an oblique

asymptote of $y = 2x - 2$. The graph has a minimum point at $\left(\frac{5}{4}, -\frac{7}{2} \right)$ when $t = 4$ and

then approaches the oblique asymptote as t approaches 0.

$\boxed{39}$ $x = 1024(\sqrt{3}/2)t$ and $y = -\frac{1}{2}(32)t^2 + 1024(\frac{1}{2})t + 5120$ $\Rightarrow$

$$x = 512\sqrt{3}\,t \quad \text{and} \quad y = -16t^2 + 512t + 5120.$$

$y = 0$ $\Rightarrow$ $-16t^2 + 512t + 5120 = 0$ $\Rightarrow$ $t^2 - 32t - 320 = 0$ $\Rightarrow$

$(t - 40)(t + 8) = 0$ $\Rightarrow$ $t = 40$ seconds. $x = 512\sqrt{3}(40) = 20{,}480\sqrt{3} \approx 35{,}472.40$ feet.

$y = 5120$ $\Rightarrow$ $-16t^2 + 512t + 5120 = 5120$ $\Rightarrow$ $-16t^2 + 512t = 0$ $\Rightarrow$

$-16t(t - 32) = 0$ $\Rightarrow$ $t = 0, 32$. The maximum altitude occurs when

$$t = \tfrac{1}{2}(32) = 16 \text{ and has value } y = -16(16)^2 + 512(16) + 5120 = 9216 \text{ feet.}$$

$\boxed{41}$ $x = r \cos \theta = 5 \cos \frac{7\pi}{4} = 5\left(\frac{\sqrt{2}}{2} \right) = \frac{5}{2}\sqrt{2}$. $y = r \sin \theta = 5 \sin \frac{7\pi}{4} = 5\left(-\frac{\sqrt{2}}{2} \right) = -\frac{5}{2}\sqrt{2}$.

$\boxed{43}$ $y^2 = 4x$ $\Rightarrow$ $r^2 \sin^2\theta = 4r \cos\theta$ $\Rightarrow$

$$r = \frac{4r \cos\theta}{r \sin^2\theta} = 4 \cdot \frac{\cos\theta}{\sin\theta} \cdot \frac{1}{\sin\theta} \Rightarrow r = 4 \cot\theta \csc\theta.$$

$\boxed{45}$ $2x - 3y = 8$ $\Rightarrow$ $2r \cos\theta - 3r \sin\theta = 8$ $\Rightarrow$ $r(2\cos\theta - 3\sin\theta) = 8$.

$\boxed{46}$ $x^2 + y^2 = 2xy$ $\Rightarrow$ $r^2 = 2r^2 \cos\theta \sin\theta$ $\Rightarrow$ $1 = 2\sin\theta \cos\theta$ $\Rightarrow$ $\sin 2\theta = 1$ $\Rightarrow$

$\qquad$ $2\theta = \frac{\pi}{2} + 2\pi n$ $\Rightarrow$ $\theta = \frac{\pi}{4}, \frac{5\pi}{4}$ on $[0, 2\pi)$, which are the same lines.

In rectangular coordinates: $x^2 + y^2 = 2xy$ $\Rightarrow$ $x^2 - 2xy + y^2 = 0$ $\Rightarrow$

$\qquad\qquad\qquad\qquad\qquad (x - y)^2 = 0$ $\Rightarrow$ $x - y = 0$, or $y = x$.

$\boxed{47}$ $r^2 = \tan\theta$ $\Rightarrow$ $x^2 + y^2 = \frac{y}{x}$ $\Rightarrow$ $x^3 + xy^2 = y$.

$\boxed{49}$ $r^2 = 4\sin 2\theta$ $\Rightarrow$ $r^2 = 4(2\sin\theta \cos\theta)$ $\Rightarrow$ $r^2 = 8\sin\theta \cos\theta$ $\Rightarrow$

$\qquad\qquad r^2 \cdot r^2 = 8(r\sin\theta)(r\cos\theta)$ $\Rightarrow$ $(x^2 + y^2)^2 = 8xy$.

$\boxed{50}$ $\theta = \sqrt{3}$ $\Rightarrow$ $\tan^{-1}\left(\frac{y}{x}\right) = \sqrt{3}$ $\Rightarrow$ $\frac{y}{x} = \tan\sqrt{3}$ $\Rightarrow$ $y = (\tan\sqrt{3})x$.

Note that $\tan\sqrt{3} \approx -6.15$. This is a line through the origin making an angle of

$\qquad$ approximately $99.24°$ with the positive x-axis. The line is *not* $y = \frac{\pi}{3}x$.

$\boxed{51}$ $r = 5\sec\theta + 3r\sec\theta$ $\Rightarrow$ $r\cos\theta = 5 + 3r$ $\Rightarrow$ $x - 5 = 3r$ $\Rightarrow$

$\qquad x^2 - 10x + 25 = 9r^2$ $\Rightarrow$ $x^2 - 10x + 25 = 9x^2 + 9y^2$ $\Rightarrow$ $8x^2 + 9y^2 + 10x - 25 = 0$

$\boxed{53}$ $r = -4\sin\theta$ $\Rightarrow$ $r^2 = -4r\sin\theta$ $\Rightarrow$ $x^2 + y^2 = -4y$ $\Rightarrow$

$\qquad\qquad\qquad x^2 + y^2 + 4y + \underline{4} = \underline{4}$ $\Rightarrow$ $x^2 + (y + 2)^2 = 4$.

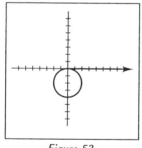

Figure 53

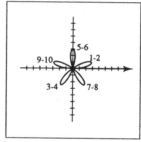

Figure 55

$\boxed{55}$ $r = 3\sin 5\theta$ is a 5-leafed rose. $0 = 3\sin 5\theta$ $\Rightarrow$ $\sin 5\theta = 0$ $\Rightarrow$ $5\theta = \pi n$ $\Rightarrow$ $\theta = \frac{\pi}{5}n$.

The numbers 1–10 correspond to θ ranging from 0 to π in $\frac{\pi}{10}$ increments. One leaf is

centered on the line $\theta = \frac{\pi}{2}$ and the others are equally spaced apart $\left(\frac{360°}{5} = 72°\right)$.

$\boxed{57}$ $r = 3 - 3\sin\theta$ is a cardioid since the coefficient of $\sin\theta$

has the same magnitude as the constant term.

$0 = 3 - 3\sin\theta$ $\Rightarrow$ $\sin\theta = 1$ $\Rightarrow$ $\theta = \frac{\pi}{2} + 2\pi n$.

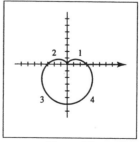

Figure 57

Variation of θ			Variation of r	
1)	0	$\rightarrow$ $\frac{\pi}{2}$	$3 \rightarrow$	0
2)	$\frac{\pi}{2}$	$\rightarrow$ π	$0 \rightarrow$	3
3)	π	$\rightarrow$ $\frac{3\pi}{2}$	$3 \rightarrow$	6
4)	$\frac{3\pi}{2}$	$\rightarrow$ 2π	$6 \rightarrow$	3

59 $r^2 = 9\sin 2\theta$ •

$0 = 9\sin 2\theta \;\Rightarrow\; \sin 2\theta = 0 \;\Rightarrow\; 2\theta = \pi n \;\Rightarrow\; \theta = \frac{\pi}{2}n.$

Variation of θ	Variation of r
1) $\quad 0 \;\to\; \frac{\pi}{4}$	$0 \to \pm 3$
2) $\quad \frac{\pi}{4} \;\to\; \frac{\pi}{2}$	$\pm 3 \to \quad 0$
3) $\quad \frac{\pi}{2} \;\to\; \frac{3\pi}{4}$	undefined
4) $\quad \frac{3\pi}{4} \;\to\; \pi$	undefined

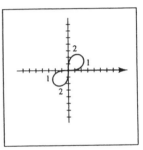

Figure 59

61 $r = \dfrac{8}{1 - 3\sin\theta} \;\Rightarrow\; e = 3 > 1$, hyperbola. See §10.6 for more details on this problem.

$$V\!\left(2, \tfrac{3\pi}{2}\right) \text{ and } V'\!\left(-4, \tfrac{\pi}{2}\right).\quad d(V, F) = 2 \;\Rightarrow\; F'\!\left(-6, \tfrac{\pi}{2}\right).$$

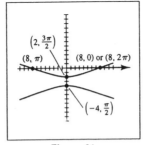

Figure 61

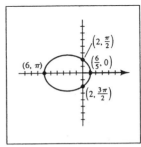

Figure 63

63 $r = \dfrac{6}{3 + 2\cos\theta} = \dfrac{2}{1 + \frac{2}{3}\cos\theta} \;\Rightarrow\; e = \frac{2}{3} < 1$, ellipse.

$$V\!\left(\tfrac{6}{5}, 0\right) \text{ and } V'(6, \pi).\quad d(V, F) = \tfrac{6}{5} \;\Rightarrow\; F' = \left(\tfrac{24}{5}, \pi\right).$$

Chapter 10 Discussion Exercises

$\boxed{1}$ For $y = ax^2$, the horizontal line through the focus is $y = p$. Since $a = 1/(4p)$, we have $p = (1/(4p))x^2 \Rightarrow x^2 = 4p^2 \Rightarrow x = 2|p|$. Doubling this value for the width gives us $w = 4|p|$.

$\boxed{3}$ Refer to Figure 2 and the derivation on text page 775.

$\boxed{5}$ $P(x, y)$ is a distance of $(2+d)$ from $(0, 0)$ and a distance of d from $(4, 0)$. The difference of these distances is $(2+d) - d = 2$, a <u>positive constant</u>. By the definition of a hyperbola, $P(x, y)$ lies on the right branch of the hyperbola with foci $(0, 0)$ and $(4, 0)$. The center of the hyperbola is halfway between the foci, i.e., $(2, 0)$. The vertex is halfway from $(2, 0)$ to $(4, 0)$ since the distance from the circle to P equals the distance from P to $(4, 0)$.

Thus, the vertex is $(3, 0)$ and $a = 1$.

$b^2 = c^2 - a^2 = 2^2 - 1^2 = 3$ and

an equation of the right branch of the hyperbola is

$\dfrac{(x-2)^2}{1} - \dfrac{y^2}{3} = 1,\ x \ge 3 \quad \text{or} \quad x = 2 + \sqrt{1 + \dfrac{y^2}{3}}.$

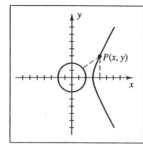

Figure 5

$\boxed{7}$ Solve $\quad y = -16t^2 + 1024\,(\sin \alpha)\,t + 2304 \quad$ for $\quad t \quad$ when $\quad y = 0$. This gives $t_1 = 32 \sin \alpha + 4\sqrt{64 \sin^2 \alpha + 9}$. The range r is given by $r = 1024\,(\cos \alpha)\,t_1$. Table r as a function of α to find that the maximum value of r occurs when $\alpha \approx 0.752528$ radians, or about $43.12°$.

$\boxed{9}$

$x = \sin 2t$	given
$x = 2 \sin t \cos t$	double-angle formula
$x^2 = 4 \sin^2 t \, \cos^2 t$	square both sides
$x^2 = 4(1 - \cos^2 t)\cos^2 t$	Pythagorean identity
$x^2 = 4(1 - y^2)y^2$	$y = \cos t$
$4y^4 - 4y^2 + x^2 = 0$	equivalent equation
$y^2 = \dfrac{4 \pm \sqrt{16 - 16x^2}}{8}$	use the quadratic formula to solve for y^2
$\quad = \dfrac{1 \pm \sqrt{1 - x^2}}{2}$	simplify
$y = \pm\sqrt{\dfrac{1 \pm \sqrt{1 - x^2}}{2}}$	take the square root

These complicated equations should indicate the advantage of expressing the curve in parametric form.

[11] **n even:** There are $2n$ leaves, each having a leaf angle of $(180/n)°$. There is no open space between the leaves.

n odd: There are n leaves, each having a leaf angle of $(180/n)°$. There is $180°$ of open space—each space is $(180/n)°$, equispaced between the leaves. If $n = 4k - 1$, where k is a natural number, there is a leaf centered on the $\theta = 3\pi/2$ axis; and if $n = 4k + 1$, there is a leaf centered on the $\theta = \pi/2$ axis.

For $r = \sin n\theta$, the pole values start at $0°$ and occur every $(180/n)°$. For $r = \cos n\theta$, the pole values start at $(90/n)°$ and occur every $(180/n)°$.